ALLE ZEIT WACH
SJ
1842

A. Hartmann F. Yatsu W. Kuschinsky
(Eds.)

Cerebral Ischemia and Basic Mechanisms

With 130 Figures and 30 Tables

Springer-Verlag
Berlin Heidelberg New York
London Paris Tokyo
Hong Kong Barcelona
Budapest

Professor Dr. med. ALEXANDER HARTMANN
Neurologische Universitätsklinik
Sigmund-Freud-Straße 25
53127 Bonn, Germany

Professor Dr. med. FRANK YATSU
Department of Neurology, Texas Medical Center
The University of Texas
6431 Fannin, Suite 7.004 MSB
Houston, TX 77030, USA

Professor Dr. med. WOLFGANG KUSCHINSKY
Institut für Physiologie
Universität Heidelberg
Im Neuenheimer Feld 326
69120 Heidelberg, Germany

ISBN-13: 978-3-642-78153-7 e-ISBN-13: 978-3-642-78151-3
DOI: 10.1007/978-3-642-78151-3

Library of Congress Cataloging-in-Publication Data. Cerebral ischemia and basic mechanisms / A. Hartmann, F. Yatsu, W. Kuschinsky (eds.). p. cm. Includes bibliographical references and index. ISBN-13: 978-3-642-78153-7 1. Cerebral ischemia—Pathophysiology. I. Hartmann, A. (Alexander), 1943– . II. Yatsu, Frank M. III. Kuschinsky, Wolfgang, 1944– [DNLM: 1. Cerebral Ischemia. WL 355 C411375 1993] RC388.5.C65 1994 616.8'1—dc20 DNLM/DLC for Library of Congress 93-24247

Softcover reprint of the hardcover 1st edition 1994

Typesetting: Best-set Typesetter Ltd., Hong Kong
25/3130/SPS-5 4 3 2 1 0 – Printed on acid-free paper

List of Contributors

You will find the addresses at the beginning of the respective contribution

Preface

In this volume, one of a series of monographs devoted to the problems of cerebral ischemia and related topics, we present the proceedings of an international conference on Cerebral Ischemia and Basic Mechanisms held in Bad Schachen/Lake Constance, Germany in June 1992.

The enormous progress in research recently on the basic mechanisms associated with cerebral ischemia has provided greater insight into the pathophysiological mechanisms of reduced brain perfusion and decreased cerebral metabolism. The high technology instrumentation used to unravel the intricacies of cerebral blood flow and metabolism includes positron emission tomography and magnetic resonance imaging. A description of sophisticated neurophysiological techniques will give the reader insight into new models of reversible and irreversible tissue damage and changes at the molecular level have been described. The therapeutic approaches which have developed from this research have been or will be used in clinical trials and will open new avenues in the treatment of stroke.

The organizers of the meeting would like to thank the advisory board for its helpful suggestions and the Deutsche Forschungsgemeinschaft and other sponsors for their important support.

Bonn, Houston, Heidelberg
December 1993

Alexander Hartmann
Frank Yatsu
Wolfgang Kuschinsky

Contents

Intracellular Messengers

Excitatory Amino Acids and Excitotoxicity

Free Radicals, Protein Synthesis

Gene Expression

The Window of Opportunity for the Treatment of Cerebral Ischaemia

M.D. O'Brien

For many years the window of opportunity for effective treatment of cerebral ischaemia was thought to be too short to be of any practical significance but this depends to some extent on the definition of the term; it is useful and less nihilistic to take a broader view. In the past two decades much evidence has accumulated on the many parameters which define this window, or rather these windows.

Failure of cerebral perfusion is associated with a continuous series of events leading to complete dissolution of tissue, and, although this is a continuum, there are many thresholds for different functions and for survival, not only for the individual cell but also for neuronal networks. Intervention in the early stages might result in complete recovery with no residual deficit, but later intervention might only result in arresting progression, and at some stage any intervention will have no effect at all. So it is important to define the parameters which determine these stages.

The window of opportunity for the treatment of cerebrovascular disease really starts before the brain becomes ischaemic, and the opening of the window should be extended to include attention to risk factors, since prevention is much better than any treatment. The next opportunity begins during the phase of physiological compensation to a fall in perfusion pressure. Cerebral blood flow (CBF) is maintained by progressive vasodilatation as measured by a rise in blood volume. When vasodilatation is maximal, the CBF starts to fall, but metabolism is maintained by increasing oxygen extraction. When this compensation fails, the brain becomes ischaemic and metabolism fails (Powers and Raichle 1985). The pace of this process is usually very rapid, a matter of minutes, and the clinical circumstances in which it is possible to intervene therapeutically before ischaemia starts are few, though of course they are important to recognise since effective treatment during these early stages would prevent any loss of neurones. Thus apart from prophylaxis, the window of opportunity really starts with the first failure of perfusion pressure and not at the onset of ischaemia.

Once the brain becomes ischaemic, a sequence of events occur, the pace, extent and degree of which depend on a large number of factors, most

[1] Department of Neurology, Guy's Hospital, London, SE1 9RT, UK

Cerebral Ischemia and Basic Mechanisms
Ed. by A. Hartmann, F. Yatsu, and W. Kuschinsky

notably the depth of ischaemia, which is dependent on the degree of residual blood flow and the duration of ischaemia. Since most of the data quantifying these factors come from animal experiments, it is very important to realise that species variation in the response to ischaemia can be very considerable, and the results of experiments in small animals cannot necessarily be applied to man. Assuming that all other variables are optimal, it is the duration and depth of ischaemia which are the critical factors. However, other factors that may play a major part in altering the outcome include the perfusion pressure, the collateral potential, temperature, blood viscosity, blood glucose content, oxygen saturation, local pH, the effect of excitotoxins and free radicals. Clenenden et al. (1971) reviewed the work published during the 1950s and 1960s on brain slices. Complete permanent ischaemia in small animals results in loss of spontaneous cortical electrical activity in 10–20 s, and immediately the lactate starts to rise. By 30 s the glucose levels are down to about 15%, and there is some shift of sodium. By 3 min the neuronal membrane permeability is affected, and there is a more marked shift of electrolytes. At 10 min the lactate level has risen fivefold and the glucose and glycogen levels are zero. Between 10 and 20 min, if brain slices are incubated in optimal media, there can be some restoration of electrical activity, resynthesis of ATP and restoration of glycogen levels. This is the fastest pace of the ischaemic process and might reflect the rate of change in the ischaemic core of an infarct. These observations support the long-held view that once ischaemia starts the process is so rapid that intervention is unlikely to have much effect.

However, some blood flow persists in most of an infarct, and more physiological experiments have shown the levels of blood flow at which different processes occur. In 1977 Astrup et al. reported that in progressive ischaemia the evoked responses fail at higher levels of residual flow than are required for the maintenance of cellular integrity. In baboons this level was around 18 ml/100 g/min, and this was potentially reversible with recovery of the evoked responses if the blood flow was restored. The residual blood flow had to fall below 5–8 before there was failure of cell membranes. If these experiments are put together, a blood flow window between 8 and 18 becomes evident. These acute animal experiments led to the concept of an ischaemic core (Astrup et al. 1981) surrounded by a penumbra in which cells are not functioning but are still viable, surrounded in turn by normal brain. This concept led to the hope that even if the window of opportunity of the ischaemic core is a matter of minutes, the window for the less ischaemic peripheral areas of an infarct might be very much longer.

Further observations in baboons using the middle cerebral artery occlusion model (Harris et al. 1987) showed no abnormality with flows at 58 ml/100 g/min. At 30 the tissue PO_2 is down to 30%; at 20 there is a fall in pH, evoked responses are inhibited, and there is an increase in intracellular water. Cellular disruption with extracellular potassium rise occurs at 10 ml/100 g/min.

Progressive hypotension in gerbils (Naritomi et al. 1988) shows no change with CBF down to 51 ml/100 g/min. At 21–27 there is a fall in pH. The level 18–23 is associated with an isoelectric electrocorticogram, but ATP stores remain normal. Between 12 and 14 the ATP levels start to fall, and by 4–5 the ATP levels have vanished and sodium pump fails. These experiments show that cortical electrical activity fails while the ATP stores remain normal, and that the sodium pump fails when ATP depletion is complete.

There appears to be some confusion over what the term ischaemic penumbra actually means. This may have arisen because the concept of an ischaemic core surrounded by a penumbra is often described in geographical terms, with the implication that cells in this area are viable but not functioning, and this is not correct. There is a wide range in vulnerability between cells, so that some cells die at flow rates normally associated with uninterrupted activity, and others continue to function normally at flow rates usually associated with cell death. The CBF window between function and survival therefore applies to an individual cell and not to a volume of brain, so that within the so-called penumbra there are cells which are already dead, some surviving but unable to function, and others that are functioning normally. The level of residual flow determines the proportion of cells in each of these three categories: the greater the ischaemia the more cell death (Mies et al. 1983).

As might be expected, the cerebral metabolic rate for oxygen is a more reliable indicator than the local CBF as measured in 50 subjects by positron emission tomography (Powers et al. 1985). The critical level was found to be a cerebral metabolic rate for oxygen of 1.3, but the thresholds for blood flow were much more variable. Thus the penumbra is not a volume of brain in which all the cells are in a state of suspended animation, but rather one in which some cells, perhaps rather few, are in this state. A cell at the periphery of this area might be dead, and a cell close to the ischaemic core might be alive and well, but if it does not have any other cells around it to talk to, it might as well be dead because it cannot be part of a functional network, and it is of course the restoration of functional neuronal networks which determines the degree of recovery, rather than the function of an individual cell.

The concept of selective vulnerability has been well established and is usually applied to the fact that different cells are affected at different levels of ischaemia. Rosner et al. (1986), using the middle cerebral artery occlusion model in the cat, showed that spontaneous activity ceased immediately in neurones in which the local CBF was less than 10. With reperfusion after 15 min of ischaemia, recovery occurred in anything from 10 min to 3 h, indicating that different neurones respond quite differently to similar levels of ischaemia. Purkinje's cells appear to be the most sensitive to ischaemia, followed by large and then small neurones, oligodendrocytes and finally astrocytes, with the vascular endothelium even more resistant. But in the clinical situation, selective vulnerability should also include those areas of the brain which are particularly at risk because of the vascular architecture, such as watershed infarcts with global ischaemia and in focal ischaemia where the collateral potential is poor, as in the deep white matter infarcts. Cells may thus be selectively vulnerable because of their location as well their intolerance to ischaemia.

The next essential development of this concept was the addition of the time factor; eventual survival depends on a product of the depth of ischaemia and time, so that the residual blood flow window must be integrated with time (Heiss and Rosner 1983). The observation that cells do not die in short-term animal experiments if the blood flow is much above 20 ml 100 g^{-1} min^{-1} led to the suggestion that cells would survive indefinitely as long as the flow remained above this level, and this concept carries with it the idea that neurones with a blood supply of around 10–20 ml 100 g^{-1} min^{-1} could also survive indefinitely, even if they were not functioning. This was a matter of some correspondence under the heading "Is the sleeping beauty dead?" (Stroke 1984 15:755–758).

Unfortunately, this does not equate with the observation that mature infarcts have clearly demarcated boundaries, and that positron emission tomography has failed to identify a penumbral zone around mature infarcts. This may be partly a matter of resolution, since the penumbra is a state for the individual neurone and not really the state of a volume of brain. The question is therefore what factors cause the loss of the penumbra with time, and what is the time scale over which these changes occur? Cavanaugh et al. (1992), using the middle cerebral artery occlusion model in the rat and sacrificing the animals at 1, 3, 6, 12, 15 and 24 h after occlusion showed no detectable damage at 1 h, but a progressive increase in infarct size to a maximum by 12 h. The reasons for this are likely to be multiple. Several possible mechanisms include acidosis, largely as a result of lactate production.

Nedergaard et al. (1986) showed that the cerebral metabolic rate for glucose was increased in the border zone of an infarct at 15 min, but by 6 h only 50% of the border zone area showed an increase in the cerebral metabolic rate for glucose, and by 20 h this had gone completely. Blood flow declines during this time, but this is more likely to be due to a fall in metabolic demand than to be the cause of further ischaemia. However, a major factor is likely to be the inhibition of protein synthesis. Mies et al. (1991), using the rat middle cerebral artery occlusion model, showed that protein synthesis is inhibited at flow rates very much higher than is necessary to inhibit function. Protein synthesis is a cellular activity which does not necessarily have an instant effect on cell function but would certainly be associated with delayed cell death; it therefore seems likely that the inhibition of protein synthesis at residual flow levels far higher than are thought necessary for survival in acute experiments is one of the parameters which define the closing stages of the therapeutic window for the treatment of acute ischaemia. Effective treatment during this time might restore the penumbra to normality rather than allow it to become part of the completed stroke.

It is therefore possible to identify four panes of glass in the therapeutic window in stroke: (a) the treatment of risk factors; (b) the treatment of pre-ischaemic low flow states or treatment in anticipation of a neurosurgical procedure; (c) the treatment of acute ischaemia, which is different in the first few hours than in the first few days (for example, ischaemic cerebral oedema does not reach its peak for about 2 days); and (d) the treatment of a patient with a mature infarct by rehabilitation. It should not be assumed that no therapeutic endeavour is worthwhile after the first few days, the management of the patient continues as long as neurological deficit persists.

References

Astrup J, Symon L, Branston NM, Lassen MA (1977) Cortical evoked potential and extreacellular potasssium and hydrogen at critical levels of brain ischaemia. Stroke 8:51–57

Astrup J, Siesjö BK, Symon L (1981) Thresholds in cerebral ischaemia in the ischaemic in the ischaemic penumbra. Stroke 12:723–725

Cavanaugh K, Dean R, Eveleth D, Bartus R (1992) Time related neurodegenerative changes following MCA occlusions: evidence for a therapeutic window of opportunity and a role of calpain. Stroke 23:18

Clenenden NR, Allen N, Komatsu T, Liss L, Gordon WA, Heimberger K (1971) Biochemical alterations in anoxic ischemic lesions of rat brain. Arch Neurol 25:432–448

Harris RJ, Richards PG, Symon L, Habib AHA, Rosenstein J (1987) pH, K and PO_2 of the extracellular space during ischaemia of primate. J Cereb Blood Flow Metab 7:599–604

Heiss WD, Rosner G (1983) Functional recovery of cortical neurones as related to degree and duration of ischaemia. Ann Neurol 14:294–301

Mies G, Auer LM, Ebhardt G, Traupe H, Heiss WD (1983) Flow and neuronal density in tissues surrounding chronic infarction. Stroke 14:22–27

Mies G, Ishimaru S, Xie Y, Seo K, Hossmann KA (1991) Ischemic thresholds of cerebral protein synthesis and energy state following middle cerebral artery occlusion in rat. J Cereb Blood Flow Metab 11:753–761

Naritomi H, Sasaki M, Kanashiro M, Kitani M, Sawada T (1988) Flow thresholds for cerebral energy disturbance and Na^+ pump failure as studied by in vivo P-31 and Na-23 nuclear magnetic resonance spectroscopy. J Cereb Blood Flow Metab 8:16–23

Nedergaard M, Gjedde A, Diemer NH (1986) Focal ischemia of the rat brain – autoradiographic determination of cerebral glucose utilization, glucose content, and blood flow. J Cereb Blood Flow Metab 6:405–413

Powers WJ, Raichle ME (1985) Positron emission tomography and its application to the study of cerebrovascular disease in man. Stroke 16:361–376

Powers WJ, Grubb RL, Darriet D, Raichle ME (1985) Cerebral blood flow and cerebral metabolic rate of oxygen requirements for cerebral function and viability in humans. J Cereb Blood Flow Metab 5:600–608

Rosner G, Graf R, Kataoka K, Heiss WD (1986) Selective funcitonal vulnerability of cortical neurons following transient MCA-occlusion in the cat. Stroke 17:76–82

Energy Metabolism

Uncoupling of Flow and Metabolism in Early Ischemic Stroke

W.-D. Heiss, G.R. Fink, M. Huber, K. Herholz, and U. Pietrzyk

Introduction

The therapeutic window in ischemic stroke, as may be deducted from experimental results of functional recovery (reviews in Heiss 1992), could be effective for different therapeutic approaches: one short period during which improvement of perfusion is successful in reversing functional deficits and preventing morphological damage, and a probably longer period during which the progressive infarction could be ameliorated by intervention with ongoing deleterious biochemical alterations. Clinical experience indicates that similar conditions do exist in ischemic stroke; their assessment in a clinical setting, however, is extremely difficult since most routine examinations do not yield evidence for viability of tissue, and regional studies of physiologic variables, for example, by positron emission tomography (PET) are prevented in most instances by their logistic complexity.

The perfusion failure during initial stages after an ischemic attack manifests itself as a decrease in regional cerebral blood flow (rCBF) with regional cerebral metabolic rate for oxygen ($rCMRO_2$) and that for glucose (rCMRGl) remaining fully or relatively preserved. This condition, described as early mismatch (Kuhl et al. 1980) and usually called misery perfusion (Baron et al. 1981), implies that blood flow is inadequate relative to the energy metabolic demand for oxygen and substrate of still viable tissue. It is indicative of the precarious condition of the tissue, but it also holds promise for a full recovery. Since the initial observations misery perfusion, as defined by increased oxygen extraction fraction (OEF), has been observed regularly (Wise et al. 1983; Powers et al. 1985; Hakim et al. 1989). Tissue affected by misery perfusion occurs in 45%–57% of cases studied within 4 days after acute stroke (Baron 1987). However, the further fate of this tissue cannot be predicted; it may ultimately become necrotic or may finally surround the infarct as a viable rim. From these studies it was concluded that definitely necrotic tissue is defined by rCBF below $12\,ml\,100\,g^{-1}\,min^{-1}$ and/or $rCMRO_2$ below $65\,\mu mol\,100\,g^{-1}\,min^{-1}$ (= $1.5\,ml\,100\,g^{-1}\,min^{-1}$; Powers et al. 1985;

Max-Planck-Institut für Neurologische Forschung, and Neurologische Klinik der Universität zu Köln, Gleueler Str. 50, 50931 Köln, Germany

Cerebral Ischemia and Basic Mechanisms
Ed. by A. Hartmann, F. Yatsu, and W. Kuschinsky

Baron et al. 1983). That the lowest flow values determined after transient ischemic attacks (TIAs), i.e., by definition after a reversible flow disturbance within penumbra values, was $22\,ml\,100\,g^{-1}\,min^{-1}$ (Powers et al. 1987), a brain region with rCBF between 12 and $22\,ml\,100\,g^{-1}\,min^{-1}$ has unstable metabolic function, and infarction is probable in such a region if these low flow values persist.

Metabolic Changes in Ischemic Tissue

The observations in early stroke suggest that tissue affected by misery perfusion but with rCBF and $rCMRO_2$ still above the critical thresholds ($12\,ml\,100\,g^{-1}\,min^{-1}$ and $65\,\mu mol\,100\,g^{-1}\,min^{-1}$, or 45% of contralateral CBF and 65% of contralateral $CMRO_2$; Baron 1987) are viable. However, the evidence that acutely misery-perfused tissue recovers is rather scarce and is based on a few single observations, while most studies demonstrate that such tissue is metabolically damaged in the further course and turns into necrosis. In 24 studies to 17 patients comparing early PET and late computer tomography (CT) ultimately viable areas with preserved $CMRO_2$ (>85%) and maximal OEF indicating a precarious but still compensated situation were observed in five cases between days 2 and 4. Only one patient recovered completely without damage shown on late CT in an area when $CMRO_2$ was depressed to 75% and CBF to 50%; maximal OEF was at 0.85 on day 2. From the results in 131 areas Baron (1987) plotted a curve of local CBF versus local $CMRO_2$ in which the morphologically intact regions on late CT were separated from the necrotic areas by the thresholds of 45% and 65% $CMRO_2$. Regions with increased OEF located above these thresholds might ultimately survive.

In this, as in most studies of ischemic stroke, only data obtained at one time at highly variable intervals after the attack were analyzed and related to the clinical course or to final lesions on CT. Additionally, most studies selected only one slice of the brain to demonstrate the infarct and surrounding tissue. Therefore, observations were to some extent incidental and lacked the three-dimensional representation of the infarct and peri-infarct tissue as well as the assessment of changes in critically perfused tissue with time. In a series of nine stroke patients, Wise et al. (1983) demonstrated a high OEF in critically perfused tissue early after the attacks that was characterized by low $CMRO_2$ and CBF at later studies. These changes in OEF observed in the course of stroke were interpreted as evidence for the transition from viable ischemic tissue to infarction. However, due to technical limitations of this early study only large tissue volumes could be evaluated, preventing the identification of changes in the center and border zones of ischemia. To avoid erroneous conclusions from the comparison of single slices that might not be exactly aligned and to include potentially useful information from neighboring brain slices, Hakim et al. (1989) generated superimposable surface maps of metabolism and perfusion for two sets of PET studies 7 days apart. By a surface contour technique with sampling from up to six acceptable transverse images and a linear interpolation of the 6-mm inter-plane separations, the cortical representation of the area of disturbed perfusion could be outlined and matched in repeat measurements with acceptable accuracy. By categorizing the cortical areas according to flow values below 12 ml 100 g^{-1} min^{-1}, between 12 and 18 ml 100 g^{-1} min^{-1}, and above 18 ml 100 g^{-1} min^{-1}, these authors demonstrated that the $CMRO_2$ in the "penumbra zone" (flow of 12–18 ml 100 g^{-1} min^{-1}) deteriorated with a decrease in CBF between the first (<48 h after ictus) and the second (7 days) PET study, bringing this tissue into the category of infarction.

Concomitant changes in perfusion and metabolism early after ischemia were studied in 16 patients by multitracer PET within 6–48 h (mean 23) of onset of a hemispheric stroke and again 13–25 days (mean 15.6) later (Heiss et al. 1992). CBF, CBV, $CMRO_2$, (OER) and CMRGl were measured each time by standard methods following the guidelines given by Baron et al. (1989).

The set of brain slices obtained in the two studies (Fig. 1) were matched using a three-dimensional alignment procedure which also permitted the comparison to CT or magnetic resonance imaging (MRI; Pietrzyk et al. 1990; Fig. 2). On the matched brain slices regions of interest for infarct and peri-infarct tissue, for contralateral mirror regions, and for major structures were outlined. In the core of infarction – defined in the first study as tissue at or below the threshold value of $CMRO_2$ (Powers et al. 1985) – blood flow and metabolism were significantly reduced compared to the corresponding contralateral regions ($CMRO_2$ by 55% ± 24.8%, OEF 26% ± 34.1%, CBF 42% ± 26.5%, and CMRGl 29% ± 20.2%) in the first study, and did not change during the observation period. In the peri-infarct tissue, $CMRO_2$ was moderately decreased at the first measurement (17% ± 15.1%); over time the $CMRO_2$ deteriorated progressively (to 26% ± 12.8% side difference) while flow did not change. When peri-infarct regions were selected on the basis of increased OEF (25% ± 29.8% above corresponding contralateral regions) on the early scans as an indicator of viable tissue, the CBF was significantly decreased (23% ± 6.6%) while the $CMRO_2$ showed only a slight difference from the mirror region.

Within the observation period, the CBF improved, but the $CMRO_2$, OEF, and CMRGl deteriorated. Only in a few regions with increased OEF and slightly impaired $CMRO_2$ was metabolism preserved close to normal value (Fig. 1, anterior rim). Regions with relatively increased CMRGl and reduced $CMRO_2$ indicative of anaerobic glycolysis were found only in two instances and deteriorated over time (Fig. 1, posterior rim). These data from repeat PET studies in reproducibly defined tissue compartments furnish evidence of viable tissue in the border zone of ischemia up to 48 h after stroke. With a few exceptions (Fig. 3) these tissue compartments suffer progressive metabolic derangement and turn into necrosis during the following

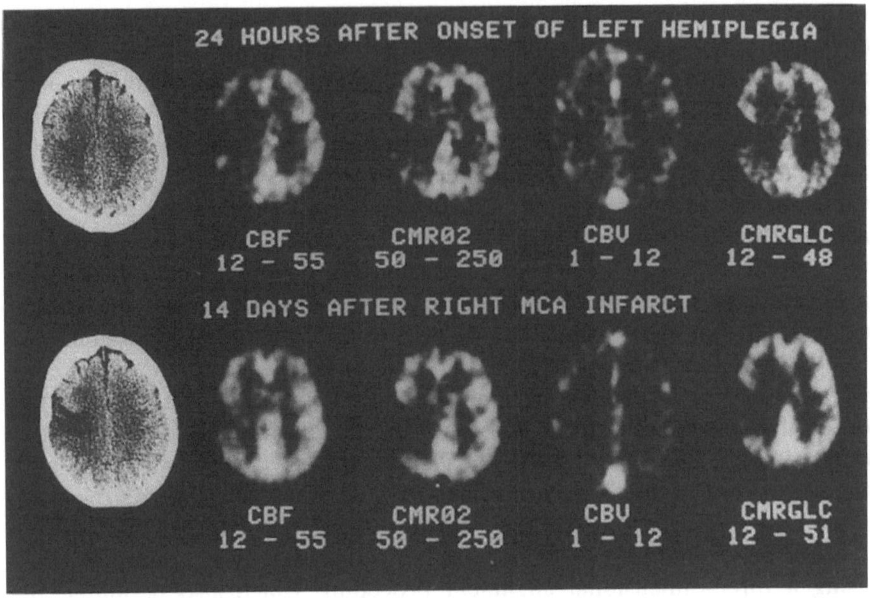

Fig. 1. Set of data from two CT studies (days 1 and 4) and two PET studies (days 1 and 15) of CBF, $CMRO_2$, CBV, and CMRGl in one brain slice just above the basal ganglia. Image matching between the two studies and among various modalities was achieved by three-dimensional alignment procedure. Values give range of individual variable as $ml/100\,g^{-1}\,min^{-1}$ (CBF), $\mu mol\,100\,g^{-1}\,min^{-1}$ ($CMRO_2$, CMRGl) and percentage (CBV). Infarcted tissue with low values extend posteriorly (low $CMRO_2$ at first study), while anterior rim of peri-infarct tissue survives (low CBF, preserved $CMRO_2$ and CMRGl)

2 weeks. While this viable peri-infarct tissue exhibits some potential for effective treatment of ischemic stroke, therapeutic routines usually applied today cannot prevent subsequent metabolic derangement and progression to necrosis. However, studies identifying viable tissue could be of value in the development of effective therapeutic strategies.

Metabolic Changes in Hyperperfused Tissue

The fate of tissue with "misery perfusion" was studied repeatedly, and, as described, this condition indicates a potential for recovery which usually cannot be utilized. Increased perfusion beyond the metabolic demand, on the other hand, is a well-documented condition ("luxury perfusion"; Lassen 1966) and can be observed within or in the surrounding of infarcted tissue in up to 33% of patients examined during the first 48 h after an ischemic stroke (Ackerman et al. 1985; Herholz and Heiss 1991). Since reestablishment of blood supply or even forced reperfusion are strategies applied in early

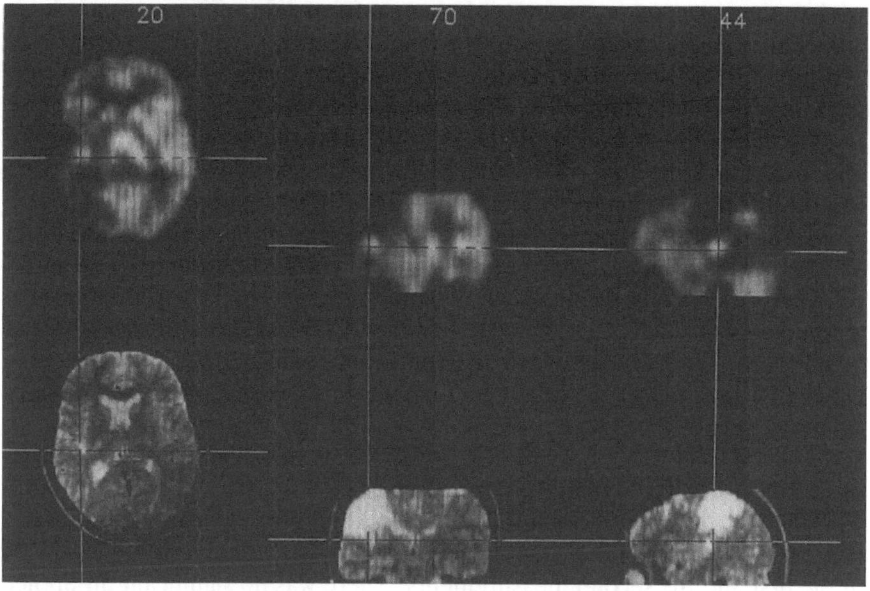

Fig. 2. Typical display of the three-dimensional alignment procedure with cerebral blood flow PET (*upper row*: transaxial, coronar, and sagittal planes, 24 h after the onset of symptoms) and corresponding T2-weighted MRI images (*lower row*: transaxial, coronar, and sagittal planes, 6 months after the ictus) after reslicing of MRI images as to correspond with PET images

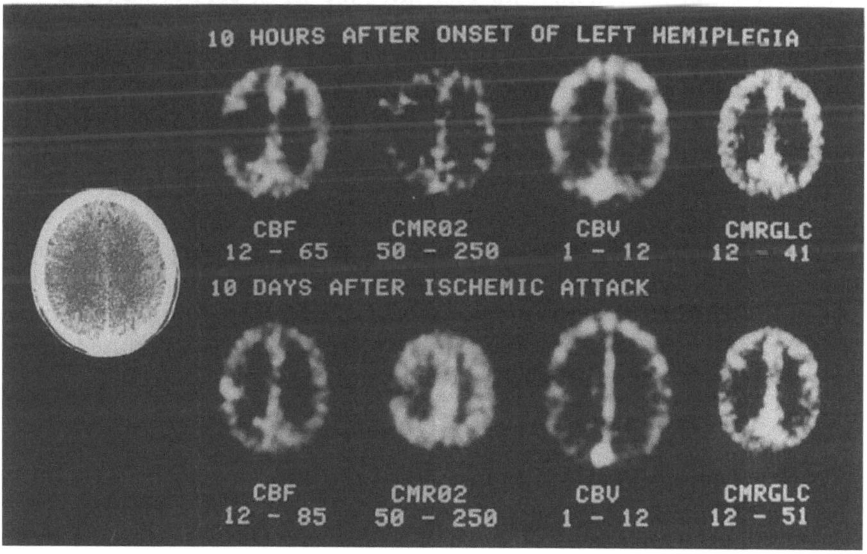

Fig. 3. Set of data from two PET studies (days 1 and 17) in a patient who recovered completely from hemiparalysis after an ischemic stroke. Details as in Fig. 1. Severe CBF decrease at first study turns into hyperperfusion at second measurement. $CMRO_2$, and CMRGl remain above a critical level

stroke, the spontaneous course of tissue afflicted by different pathologic perfusional states should be known. The influence of differently perfused peri-infarct tissue on metabolism, morphological damage, and clinical outcome was therefore studied in 22 patients (Fink et al. 1993). CBF of these cases was examined by $[^{15}O]H_2O$ PET 6–48 h after onset of neurological symptoms, and the fate of the tissue was judged in follow-up by repeated measurements of glucose metabolism (FDG-PET at initial flow study and again 2 weeks later) and by estimation of the size of the permanent infarct on CT or MRI which were three-dimensionally aligned to the PET images.

Early after the ischemic attack, 9 patients showed hyperperfused areas in the peri-infarct tissue (i.e., flow greater than CBF + 2 SD of the unaffected hemisphere), and 18 cases had hypoperfused regions (i.e., flow less than mean CBF − 2 SD of the unaffected hemisphere). Metabolism of the hypoperfused peri-infarct tissue (CMRGl 1 = $18.5 \pm 3.98\,\mu mol\,100\,g^{-1}\,min^{-1}$, CMRGl 2 = $18.1 \pm 3.65\,\mu mol\,100\,g^{-1}\,min^{-1}$) was significantly lower (analysis of variance, $p < 0.01$) than that of hyperperfused (CMRGl 1 = 26.0 ± 5.93, CMRGl 2 = $24.0 \pm 4.52\,\mu mol\,100\,g^{-1}\,min^{-1}$) and normoperfused peri-infarct tissue (CMRGl 1 = 24.4 ± 3.64, CMRGl 2 = $23.4 \pm 2.81\,\mu mol\,100\,g^{-1}\,min^{-1}$) in the first and the second measurements. There was no significant difference between hyper- and normoperfused regions. Hyperperfused peri-infarct regions did not become necrotic, as confirmed by late CT or MRI. In some cases, however, MRI signal intensity changes indicated partial ischemic damage (Fig. 4). Some of the hypoperfused regions were finally included into the area of infarction. One year after the ictus, clinical outcome was evaluated using the Barthel index. Patients with hyperperfusion in peri-infarct tissue scored 98.7 ± 3.53, patients without hyperperfused peri-infarct regions reached a Barthel index of 93.2 ± 10.8. This difference was not significant, and the statistical analyses was further complicated by the fact that all patients with hyperperfused regions also showed hypoperfusion in other peri-infarct areas.

It might be concluded from these results that as early as 6–48 h after an ischemic attack spontaneous hyper- and normoperfusion of peri-infarct tissue indicates at least partial tissue survival, while hypoperfusion is an early sign of irreversible tissue damage. However, necrosis of single vulnerable cells and the simultaneous occurrence of hyper- and hypoperfused tissue around an infarct obscure the predictability of the clinical outcome.

PET Evaluation of Therapeutic Intervention

The spontaneous course of TIA, where by definition flow is increased above the functional threshold within the effective time window, and a number of observations of the effectiveness of vascular surgery (Baron et al. 1981; Powers et al. 1984; Gibbs et al. 1987) suggest that improvement of blood supply can reverse misery perfusion and prevent progressive damage of

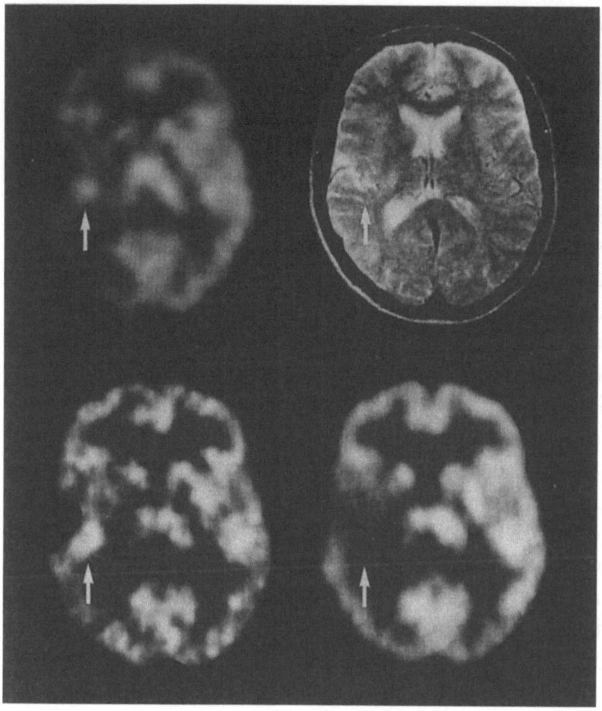

Fig. 4. PET images across the basal ganglia plane in a 42-year-old woman with right middle cerebral artery infarction presenting with hemipareses. *Upper row*: CBF (*left*) and MRI (*right*, 6 months after the ictus); *lower row*: CMRGl2 (*right*). Note large posterior peri-infarct hyperperfusion. *Arrows* point to early hyperperfusion which is in low metabolic region in late CMRGC image (*right*)

initially viable tissue. However, the uncoupled restoration of perfusion with further metabolic derangement of tissue and the lack of permanent effects of surgical revascularization on oxygen utilization and clinical course (Powers et al. 1989) indicate that other mechanisms than low blood supply determine the further fate of these tissue compartments. One could even argue that critical components of this "viable" tissue, for example, large neurons substantial in the functional network, are already irreversibly damaged, or that most still active cells are in the process of delayed neuronal death (Kirino 1982).

Therefore, only strategies intervening with pathologic biochemical mechanisms in the process of ischemic cell damage might be effective in the treatment of stroke in this stage, and since these secondary mechanisms take time, the therapeutic window for intervention with biochemical changes might be longer and extend to 48 h – as long as viable tissue is indicated in PET studies. Despite the promising features of PET application to monitoring of treatment effects, experience on this topic is still rather scarce, but a few examples show that predictions made from PET studies coincided with the results of clinical trials (Hakim 1989).

The increase in intracellular Ca^{2+} concentration is one mechanism contributing to ischemic cell damage, and Ca^{2+} channel blockers were able to reduce histological damage and neurological deficits in animal experiments and to improve neurological outcome in ischemic stroke (Gelmers et al. 1988). Hakim et al. (1989) measured CBF, CBV, $CMRO_2$, and OEF on the 1st and 7th days after ischemic stroke and found that nimodipine reversed the decline in $CMRO_2$ and increased CBF in the densely ischemic zone. In the penumbra (CBF between 12 and 18 ml 100 g^{-1} min^{-1}) $CMRO_2$ was also improved, but the difference between the two groups did not reach statistical significance, and the patients were not followed clinically or by late CTs to establish the clinical reliability of drug-induced changes.

Heiss et al. (1990) compared changes in rCMRGl from the first 48 h to 3 weeks after the stroke with the clinical outcome after 6 months. In the nimodipine-treated group metabolism outside the infarct improved significantly compared to the placebo-treated control group, and the treated patients reached significantly higher scores on the Barthel index for late assessment of daily life activities. These studies support findings from multicenter trials of the efficacy of nimodipine in ischemic stroke (American Nimodipine Study Group 1992). They also indicate that metabolic changes as observed by PET in the early course around the infarct predict the final outcome reached after 6 months.

In another double-blind controlled clinical trial, propentofylline which by enhancing extracellular adenosine and gamma-aminobutyric acid and lowering glutamate (Andiné et al. 1990) interferes with disturbed amino acid activity during ischemia was given for 2 weeks and the clinical course was followed for 3 months. rCMRGl increased significantly in the infarct and in the ipsilateral gray matter in the treatment group compared to the placebo group, but due to the high variability in small groups (14 versus 13 patients) the differences in the clinical recovery up to 3 months (change in Barthel index: propentofylline 45, placebo 25 points) did not reach statistical significance.

The ganglioside GM_1, injected intravenously over 3 weeks, increased CMRGl values, which were 40%–10% below normal, but had no effect on higher values (Heiss et al. 1989). GM_1-treated patients improved significantly better than controls up to 42 days after the stroke (stroke index difference $p = 0.04$, Wilcoxon/Mann-Whitney test). However, in a multicenter study (Argentino et al. 1989) this difference in improvement between GM_1 and placebo group vanished in the further course and was not statistically significant at evaluation 3 and 6 months after the stroke.

With the increasing need to select drugs of potential efficacy for expensive large-scale clinical trials the role of PET in evaluation of therapy might become more established. Further studies should use PET criteria for selection of patients by which the viability of the tissue at initiation of treatment is shown. As demonstrated, these criteria based only on a few follow-up observations are vague but still the best to indicate the existence of a window for therapy in an individual case. Additionally, drugs might be selected according to specific action in an individual case; activated [^{11}C]nimodipine binding sites might identify vulnerable ischemic tissue and receptor binding kinetics might be related to outcome (Hakim and Hogan 1991). Such PET studies could demonstrate the effect of a proposed therapy on fundamental membrane properties and biochemical mechanisms involved in ischemic cell damage.

References

Ackerman RH, Correia JA, Alpert NM, Haley EC Jr, Buxton RB, Elmaleh DR, Taveras JM (1985) PET studies of stroke. In: Reivich M, Alavi A (eds) Positron emission tomography. Liss, New York, pp 249–262

American Nimodipine Study Group (1992) Clinical trial of nimodipine in acute ischemic stroke. Stroke 23:3–8

Andiné P, Rudolphi KA, Fredholm BB, Hagberg H (1990) Effect of propentofylline (HWA 285) on extracellular purines and excitatory amino acids in CA1 of rat hippocampus during transient ischaemia. Br J Pharmacol 100:814–818

Argentino C, Toni D, Sacchetti ML, Girmenia F, Savoini G, Erminio F et al. (1989) Hemodilution and monosialoganglioside GM_1 in the treatment of acute ischemic stroke. In: Krieglstein J (ed) Pharmacology of cerebral ischemia 1988. Wissenschaftliche Verlagsgesellschaft, Stuttgart, pp 449–453

Baron JC (1987) Ischemic stroke studied by ^{15}O-labeled compounds: misery perfusion and luxury perfusion. In: Heiss W-D, Pawlik G, Herholz K, Wienhard K (eds) Clinical efficacy of positron emission tomography. Nijhoff, Dordrecht, pp 15–23

Baron JC, Bousser MG, Rey A, Guillard A, Comar D, Castaigne P (1981) Reversal of focal "misery-perfusion syndrome" by extra-intracranial arterial bypass in hemodynamic cerebral ischemia. Stroke 12:454–459

Baron JC, Rougemont D, Bousser MG, Lebrun-Grandié P, Iba-Zizen MT, Chiras J (1983) Local CBF, oxygen extraction fraction (OEF), and $CMRO_2$: prognostic value in recent supratentorial infarction in humans. J Cereb Blood Flow Metab 3 Suppl 1:S1–S2

Baron JC, Frackowiak RSJ, Herholz K, Jones T, Lammertsma AA, Mazoyer B, Wienhard K (1989) Use of PET methods for measurement of cerebral energy metabolism and hemodynamics in cerebrovascular disease. J Cereb Blood Flow Metab 9:723–724

Fink GR, Herholz K, Pietrzyk U, Huber M, Heiss W-D (1993) Peri-infarct perfusion in human ischemia: its relation to tissue metabolism, morphology and clinical outcome. J Stroke Cerebrovasc Dis 3:123–131

Gelmers HG, Gorter K, De Weerdt CJ, Wiezer HJA (1988) A controlled trial of nimodipine in acute ischemic stroke. N Engl J Med 318:203–207

Gibbs JM, Wise RJS, Thomas DJ, Mansfield AO, Ross Russel RW (1987) Cerebral hemodynamic changes after extracranial-intracranial bypass surgery. J Neurol Neurosurg Psychiatry 50:140–150

Hakim AM (1989) Hemodynamic and metabolic studies in stroke. Semin Neurol 9:286–292

Hakim AM, Evans AC, Berger L, Kuwabara H, Worsley K, Marchal G, Biel C, Pokrupa R, Diksic M, Meyer E, Gjedde A, Marrett S (1989) The effect of nimodipine on the evolution of human cerebral infarction studied by PET. J Cereb Blood Flow Metab 9:523–534

Hakim AM, Hogan MJ (1991) In vivo binding of nimodipine in the brain. I. The effect of focal cerebral ischemia. J Cereb Blood Flow Metab 11:762–770

Heiss W-D (1992) Experimental evidence for ischemic thresholds and functional recovery. Stroke 23:1668–1672

Heiss W-D, Pawlik G, Hebold I, Beil C, Herholz K, Szelies B, von Einsiedel R, Wienhard K (1989) Can PET estimate functional recovery and indicate therapeutic strategy in stroke? In: Krieglstein J (ed) Pharmacology of cerebral ischemia 1988. Wissenschaftliche Verlagsgesellschaft, pp 433–438

Heiss W-D, Holthoff V, Pawlik G, Neveling M (1990) Effect of nimodipine on regional cerebral glucose metabolism in patients with acute ischemic stroke as measured by positron emission tomography. J Cereb Blood Flow Metab 10:127–132

Heiss W-D, Huber M, Fink G, Herholz K, Pietrzyk U, Wagner R, Wienhard K (1992) Progressive derangement of peri-infarct viable tissue in ischemic stroke. J Cereb Blood Flow Metab 12:193–203

Herholz K, Heiss W-D (1991) Use of PET to evaluate acute stroke and other cerebrovascular disorders. In: Diksic M, Reba RC (eds) Radiopharmaceuticals and brain pathology studied with PET and SPECT. CRC, Boca Raton, pp 217–239

Kirino T (1982) Delayed neuronal death in the gerbil hippocampus following ischemia. Brain Res 239:57–59

Kuhl DE, Phelps ME, Kowell AP, Metter EJ, Selin C, Winter J (1980) Effects of stroke on local cerebral metabolism and perfusion: mapping by emission computed tomography of ^{18}FDG and ^{13}NH3. Ann Neurol 8:47–60

Lassen NA (1966) The luxury-perfusion syndrome and its possible relation to acute metabolic acidosis localized within the brain. Lancet 2:1113–1115

Pietrzyk U, Herholz K, Heiss W-D (1990) Three-dimensional alignment of functional and morphological tomograms. J Comput Assist Tomogr 14:51–59

Powers WJ, Martin WRW, Herscovitch P, Raichle ME, Grubb RL Jr (1984) Extracranial - intracranial bypass surgery: hemodynamic and metabolic effects. Neurology (Cleveland) 34:1168–1174

Powers WJ, Grubb RL Jr, Darriet D, Raichle ME (1985) Cerebral blood flow and cerebral metabolic rate of oxygen requirements for cerebral function and viability in humans. J Cereb Blood Flow Metab 5:600–608

Powers WJ, Press GW, Grubb RL Jr et al. (1987) The effect of hemodynamically significant carotid artery disease on the hemodynamic status of the cerebral circulation. Ann Intern Med 106:27–35

Powers WJ, Grubb RL Jr, Raichle ME (1989) Clinical results of extracranial-intracranial bypass surgery in patients with hemodynamic cerebrovascular disease. J Neurosurg 70:61–67

Wise RJS, Bernardi S, Frackowiak RSJ, Legg NJ, Jones T (1983) Serial observations on the pathopysiology of acute stroke. Brain 106:197–222

Recovery from Motor Stroke: Different Patterns of Functional Reorganisation in Individual Patients

C. Weiller[1] and R.S.J. Frackowiak[2]

Introduction

Recovery of function after stroke is often observed although the morphological lesion that originally caused the deficit persists [14]. Two principle mechanisms of gross functional reorganisation may account for the recovery of function in the case of an underbroken major efferent pathway. In the undamaged hemisphere uncrossed ipsilateral projecting descending motor pathways may be accessed, possibly via subcortical loops, to channel signals coding for movement down to the spinal cord. This hypothesis is most dramatically supported by the motor recovery observed in children or young adults after surgical hemispherectomy [12]. Secondly, alternative cortical areas in the damaged hemisphere may contain representations of movements, and these may access spinal cord neurones via pathways that bypass the pyramidal tract lesion (e.g., the polysynaptic corticulo-reticulo-spinal tract). If the pyramidal tract is not completely disrupted, preserved corticospinal tract fibres with a different destination under normal conditions may bypass the interruption of efferent traffic from the motor cortex and may subserve the transport of information for fractionated finger movements to the hand and forearm motor neurones. All these mechanisms imply a functional reorganisation of the adult human brain.

Opportunities to investigate the functional reorganisation of the *human* brain after injury directly in vivo have arisen with the development of increasingly sophisticated, non-invasive, functional mapping techniques. We have used functional activation studies with regional cerebral blood flow changes (rCBF) detected by positron emission tomography (PET) as an index of synaptic function [3, 22].

Systematic and idiosyncratic components of a functional reorganisation may be distinguished. Observations in studies of groups of patients reflect the common patterns of reorganisation. Averaging across subjects with similar infarcts results in a greater sensitivity for detecting differences in rCBF elicited by the task. However, infarcts vary in size and shape, and

[1] Neurologische Klinik, Universität GHS, Essen, Germany
[2] MRC Cyclotron Unit, Hammersmith Hospital, London, UK

Cerebral Ischemia and Basic Mechanisms
Ed. by A. Hartmann, F. Yatsu, and W. Kuschinsky

there is therefore interest in attempting to discover individual patterns of reorganisation and relating these to the neural structures damaged.

This chapter describes work on the functional anatomy of recovery in the motor system. Two studies are reported. The findings in a group of patients in recovery from striatocapsular infarcts deals with the systematic components of recovery. In the second study we developed a technique to study individual patients.

Clinical Material

Patients were selected with first-ever ischemic events, clearly characterised by a single, appropriately sited lesion on structural imaging, with no other antecedent neurological or significant medical history. The primary criterion for inclusion was the presence of a hemiparesis of acute onset affecting one upper limb (at least), followed by substantial recovery of power and dexterity, sufficient to allow the performance of a rapid (three movements per 2s), sequential finger-to-thumb opposition task. Only patients with subcortical infarcts were selected. In the group study, ten patients (seven men, three women; age range 21–62 years; mean 41 years) had suffered from a mostly embolic striatocapsular infarction with normal extra- and intracranial large vessels. In the single-patient study (five men, three women; age range 21–67 years; mean 43 years), five of the ten patients with striatocapsular infarcts (the ones with complete recovery) were investigated in addition to three patients with lacunar infarcts within the internal capsule. There are three arguments that the deficits had indeed been caused by disruption of pyramidal tract fibres. The paradigm studied was sequential finger opposition, a task that the patients initially were unable to perform. The pyramidal tract is thought of as the major efferent pathway for these fractionated finger movements. The lesions were located in the internal capsule at a site through which the pyramidal tract passes [19]. Thus, we conclude on clinical and anatomical grounds that the lesions did indeed disrupt the pyramidal tract. Indeed, anterograde and retrograde degeneration of the pyramidal tract was demonstrated directly on magnetic resonance imaging in patient number 2 (a case previously reported by Danek et al. [5]). Therefore, the lesions disrupted the output of Betz' cells from the hand area in the primary sensorimotor cortex (SMC), but despite this, recovery of fine finger function occurred.

Ten healthy, right-handed, age-matched volunteers served as controls in both studies.

Scanning and Image Analysis

Each subject underwent six consecutive measurements of rCBF, using a $C^{15}O_2$ dynamic/integral inhalation technique [13] and the ECAT 931-08/12 PET scanner (CTI, Knoxville, TN, USA). Two scans were performed at rest, two during finger opposition of the right hand and two during finger opposition of the left hand. The rhythm was driven by a metronome at a rate of three movements every 2 s. To control for auditory input the metronome was also on during the rest scans. Scans were performed with the eyes shut in a balanced order – A (rest), B (left hand), C (right hand), C, B, A – to avoid habituation and learning effects.

Each subject's scans were reorientated parallel to the intercommissural line and then normalised using a plastic transformation into the stereotactic space of the anatomical atlas of Talairach and Tournoux [21] (see also [9, 10]). The normalised images were then smoothed with a gaussian filter to compensate for individual variations in gyral and functional anatomy. The blood flow estimate in each voxel corresponded to a weighted mean rCBF centred on a spherical domain 20 mm in diameter.

As the study was designed to examine regional changes of cerebral activity under different behavioural conditions, global flow differences between the scans were normalised to a mean of $50\,\mathrm{ml\,dl^{-1}\,min^{-1}}$ by analysis of covariance with measured global flow as covariate. The scans were then averaged across the ten normal subjects to give normal average blood flow maps for each condition, with an adjusted mean rCBF and the associated error variance for each voxel [8]. The same procedure was used for the ten patients of the group study. In this study we compared the changes in rCBF between baseline and task between the groups to identify pixels at which these changes were significantly different in normal subjects and patients. The error variance for these comparisons was pooled from both groups.

In the second study individual patients were compared with the group of normals in terms of the rCBF increase that was significantly greater than that predicted by the group data. For all pixels in parallel a Z score was calculated using the difference between group and individual rCBF increases, divided by the standard deviation that would be expected by chance (derived from the analysis of covariance error variance). The Z scores were thresholded according to the normal distribution. For both studies, any pixel which showed a significantly greater increase in rCBF during the motor task as compared to the ten normal controls at an omnibus significance level of $p < 0.001$ was recorded, and the location in Tailarach space was used to correlate the functional change with anatomy.

Results

Group Study

In this study the patterns of activation elicited by finger opposition were compared in ten patients in recovery from striatocapsular motor stroke with ten normal controls. The contralateral sensorimotor cortex, the premotor cortex and the ipsilateral cerebellum were activated to the same extent in the patients as in the controls. However, there was a significantly stronger

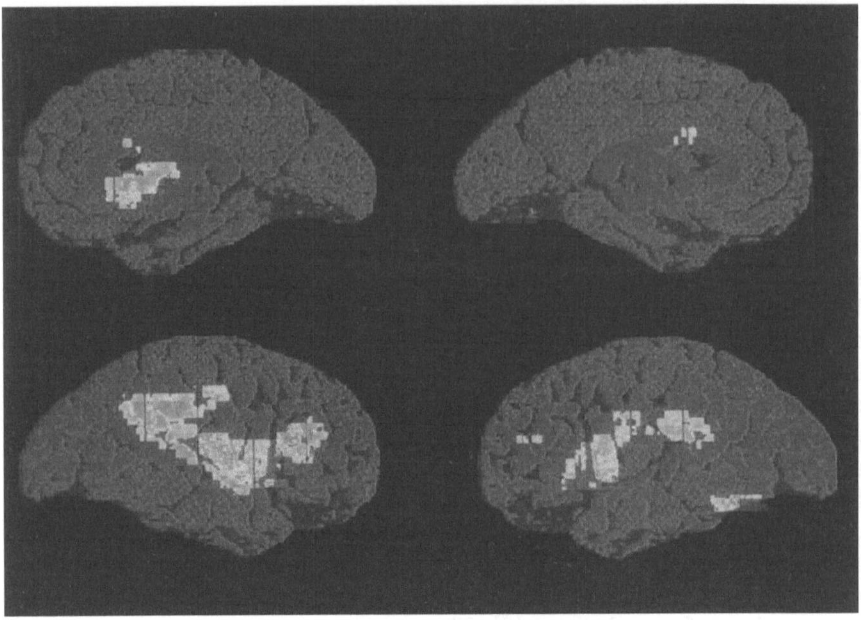

Fig. 1. Differences in activation during sequential finger opposition between ten patients in recovery from motor stroke due to striatocapsular infarction and ten normal controls. The recovered hand was the right hand in all cases. Voxels with significant greater rCBF increases in the patients during the task with the recovered hand relative to the group of normals constitute the statistical parametric maps (SPMs) in the standard, proportional, stereotactic space of Tailarach and Tournoux [21]. Significant voxels within a rim of 2 cm from the lateral and medial surfaces of the SPM have been projected onto lateral and medial cortical renditions of the brain. There was significantly greater rCBF increase in bilateral anterior insula and low premotor cortices, bilateral area 49, lateral prefrontal cortices and anterior cingulate cortices – and in ipsilateral premotor cortex, basal ganglia and contralateral cerebellum

increase in the patients during the task in the ipsilateral premotor cortex (PMC), basal ganglia and contralateral cerebellum – and bilateral in an area comprising the anterior aspects of the insula and the most ventral part of the PMC. In the patients there was also significantly greater activation in bilateral parietal cortices (area 40), lateral prefrontal cortices and anterior cingulate cortices (Fig. 1).

Area 40 in the parietal cortex and the insula are functionally connected, and both project to the PMC (area 6) [2, 15]. In turn, there are projections from inferior area 6 to area 4 [20]. Area 40 and the insula both showed an increase in rCBF during finger opposition in normals as well, but to a lesser degree than in the patients. Thus we concluded that area 40 and the insula together with the most ventral aspect of PMC constitute an additional motor system. This additional motor system is recruited by patients after stroke above its normal contribution to the execution of a sequential motor task.

Whether the bilateral activation of the motor system really indicates recruitment of ipsilateral descending corticospinal fibres remains unclear, as the ipsilateral primary motor cortex itself was not activated.

The lateral prefrontal cortex and the cingulate cortices are related less obviously and directly to the executive motor system. The anterior and posterior cingulate cortices appear to form an extensive, reciprocally interconnected system dealing with selective attention or vigilance [17]. The lateral prefrontal cortex belongs to a brain system concerned with internally generated as opposed to stimulus driven responses [11]. Activation of these areas may indicate that for our patients, finger opposition to the rhythm of a metronome is no longer a simple, automated, highly overlearned, stimulus-driven task, but requires additional selective attention and intention.

Single Patients

Individual patients showed different patterns of cortical activation compared to normal subjects. We found a significantly greater increase of rCBF in the ipsilateral PMC in all patients and in the contralateral cerebellum in five, which indicated activation of ipsilateral motor pathways. Most patients showed increased and bilateral activations in various combinations; in middle to anterior insula together with the most ventral aspects of the PMC in five out of eight; in parietal cortex (area 40) in six out of eight; in prefrontal cortex in six out of eight; and in anterior cingulate cortex in three out of eight.

In addition, we found increased activation in the mesial premotor cortex (SMA) in six of the eight patients; unfortunately, no data were available from this brain region in one patient.

The ipsilateral primary SMC was activated more than in normal subjects in only four patients, all of whom exhibited associated movements of the unaffected hand during finger opposition by the recovered hand (Fig. 2).

In four patients we found an additional activation, above that in the normal group, in the contralateral SMC along its ventral extent (Fig. 2). In normal subjects the contralateral SMC was activated during finger opposition from 44 to 60 mm above the inter-commissural (ACPC) line, with the maximum at 52 mm – which is in agreement with previous results [4]. In the SMC of the patients, significantly greater increases in rCBF were found from as low as 20 mm above the ACPC line extending up to 52 mm. Thus, in these four patients the SMC extended more than 1 cm beyond its normal inferior boundary. In all these four patients the lesions were located in the posterior limb of the internal capsule, sparing the anterior limb, the genu and the anterior part of the posterior limb. No extension of activation in the contralateral SMC was found in three patients with anterior lesions of the posterior limb of the internal capsule. No relationship between the pattern of changed activation and time from stroke was found.

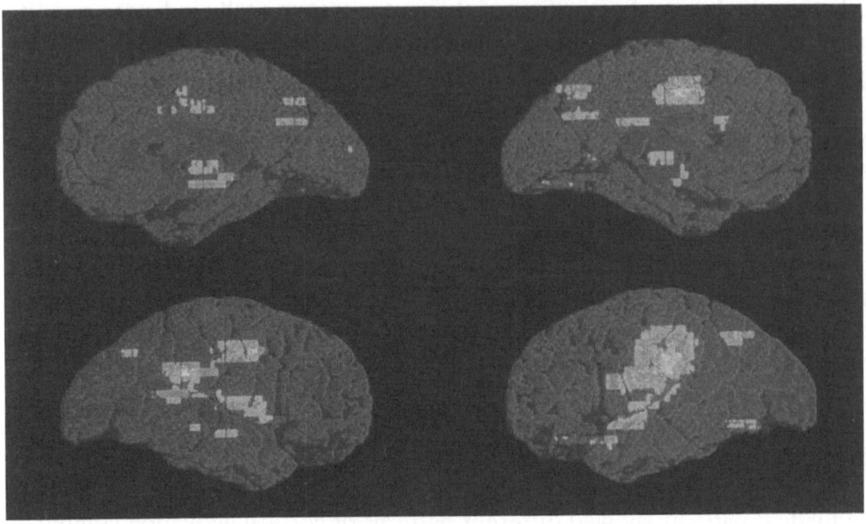

Fig. 2. Example of functional reorganisation in a patient with a posterior capsular infarction. The grid and display are as in Fig. 1. Extension of the hand area of the contralateral primary sensorimotor cortex is shown. Activation of ipsilateral sensorimotor cortex is associated with strong associated movements of the unaffected hand when this patient attempted to move the previously plegic hand. There is also bilateral activation of insula, parietal cortex and supplementary motor area and ipsilateral premotor cortex and contralateral cerebellum

The areas found activated in most patients (insula, area 40) correspond with the findings in the group studies and constitute the systematic components of functional reorganisation after capsular stroke. In addition, there was an abnormally strong activation in the supplementary motor area (mesial BA 6) in most patients. Activation of the SMA has been reported in primates after recovery from lesions of the primary motor cortex [1]. In normals, several functions within the motor system have been attributed to the SMA; activation of the SMA in association with simple repetitive finger movements has been interpreted as indicating a "readiness to move". Activation of the SMA has been associated with the initiation, programming or learning of movements, or with the selection of the appropriate neuronal populations necessary to execute a movement [4, 7, 18]. Any or all of these features could explain the findings in SMA in our patients.

The role of ipsilateral descending corticospinal fibres in the proces of recovery remains unclear. Activation of the ipsilateral primary SMC was in all cases associated with movements of the unaffected hand.

We found an extension of the contralateral primary SMC during finger oppositon in four of the patients. There is growing evidence that cortical maps may undergo plastic changes even in adult *animals*, after upper limb deafferentation or motor nerve lesions [6, 16]. We were now able to demonstrate similar changes within the adult *human* brain. There was no difference between these four and the other four patients in terms of time or quality of recovery of function. The reason why only these four patients showed an extension of the cortical fields may be related to the somatotopical organisation of the corticospinal tract fibres within the internal capsule. All these four patients had lesions of the posterior limb of the internal capsule, sparing the anterior limb, the genu and the anterior part of the posterior limb. Although the exact organisation of the fibres within the posterior limb of the internal capsule is disputed, there is agreement about the somatotopic organisation in that fibres destined for arm and leg are positioned posteriorly to those for tongue and face [19]. Thus a ventral extension of the hand fields of the SMC into the field originally destined for the face may be possible only in more posteriorly located lesions of the internal capsule.

Summary and Conclusions

There is considerable scope for functional plasticity within the adult human motor cortex. Common patterns of functional reorganisation that follow striatocapsular infarctions in man include the activation of an additional motor system, consisting of insula and area 40 in the parietal cortex, possibly recruiting non-pyramidal descending pathways within the damaged hemisphere that bypass the lesion. Whether recruitment of ipsilateral descending fibres of the corticospinal tract within the undamaged hemisphere really subserves recovery or is an epiphenomenon, possibly responsible for the occurrence of associated movements of the primarily unaffected hand remains unclear. Because of the somatotopic organisation of the pyramidal tract, the site of a subcortical lesion determines the pattern of reorganisation in the cerebral cortex. If fibres of the corticospinal tract are spared by the lesion, plastic changes within the human cortex may allow recruitment of fibres originally destined for face movements for the execution of fractionated finger movements. Further studies should emphasize on the temporal evolution of reorganisational changes in relation to the progress of recovery of function.

Acknowledgements. Dr. Weiller was a Feodor Lynen Research Fellow of the Alexander von Humboldt Foundation, Bonn-Bad Godesberg, FRG. The authors thank Dr. A. Danek, Neurologische Klinik, Klinikum Großhadern, Munich, FRG, who referred patient 2 for study. We thank colleagues from the Radiochemistry and Methods Sections at the Medical Research Council Cyclotron Unit, without whose assistance this study would not have been possible. We thank especially the radiography staff, Ms. Claire Taylor, Mrs. Andreanna Williams and Mr. Graham Lewington.

References

1. Aizawa H, Inase M, Mushiake H, Shima K, Tanji J (1991) Reorganization of activity in the supplementary motor area associated with motor learning and functional recovery. Exp Brain Res 84:668–671
2. Cavada C, Goldman-Rakic PS (1989) Posterior parietal cortex in rhesus monkeys: evidence for segregated corticocortical networks linking sensory and limbic areas with the frontal lobe. J Comp Neurol 287:422–445
3. Chollet F, DiPiero V, Wise RJS, Brooks DJ, Dolan RJ, Frackowiak RSJ (1991) The functional anatomy of motor recovery after stroke in humans: a study with positron emission tomography. Ann Neurol 29:63–71
4. Colebatch JG, Deiber MP, Passingham RE, Friston K, Frackowiak RSJ (1991) Regional cerebral blood flow during voluntary arm and hand movements in human subjects. J Neurophysiol 65:1392–1401
5. Danek A, Bauer M, Fries W (1990) Tracing of neuronal connections in the human brain by magnetic resonance imaging in vivo. Eur J Neurosci 2:112–115
6. Donoghue JP, Suner S, Sanes JN (1990) Dynamic organization of primary motor cortex output to traget muscles in adult rats. II. Rapid reorganization following motor nerve lesions. Exp Brain Res 79:492–503
7. Fox PT, Fox JM, Raichle ME, Burde RM (1985) The role of cerebral cortex in the generation of voluntary saccades: a positron emission tomography study. J Neurophysiol 54:348–369
8. Friston KJ, Frith CD, Liddle PF, Dolan RJ, Lammertsma AA, Frackowiak RSJ (1990) The relationship between global and local changes in PET scans. J Cereb Blood Flow Metab 10:458–466
9. Friston KJ, Frith CD, Liddle PF, Frackowiak RSJ (1991) Plastic transformation of PET images. J Comput Assist Tomogr 15:634–639
10. Friston KJ, Passingham RE, Nutt JG, Heather JD, Sawle GV, Frackowiak RS (1989) Localisation in PET images: direct fitting of the intercommisural (AC-PC) line. J Cereb Blood Flow Metab 9:690–695
11. Frith CD, Friston K, Liddle PF, Frackowiak RSJ (1991) Willed action and the prefrontal cortex in man: a study with PET. Proc R Soc Lond [Biol] 244:241–246
12. Gardner WJ (1933) Removal of the right hemisphere for infiltrating glioma. JAMA 101:823–826
13. Lammertsma AA, Cunnigham VJ, Deiber MP, Heather JD, Bloomfield P, Nutt J, Frackowiak RSJ, Jones T (1989) Combination of dynamic and integral methods for generating reproducible functional CBF images. J Cereb Blood Flow Metab 9:461–470
14. Merill EG, Wall PD (1987) Plasticity of connections in the adult nervous system. In: Cotman CW (ed) Neuronal plasticity. Raven, New York, pp 97–111
15. Mesulam MM, Mufson EJ (1985) The insula of Reil in man and monkey: architectonics, connectivity, and function. In: Peters A, Jones EG (eds) Cerebral cortex: association and auditory cortices. Plenum, New York, pp 179–226
16. Pons T, Garraghty PE, Ommaya AK, Kaas JH, Taun E, Mishkin M (1991) Massive cortical reorganization after sensory deafferentation in adult macaques. Science 252:1857–1860
17. Posner MI, Peterson SE (1990) The attention system of the human brain. Annu Rev Neurosci 13:25–42
18. Roland PE, Larsen B, Lassen NA, Skinhoj E (1980) Supplementary motor cortex and other cortical areas in organization of voluntary movements in man. J Neurophysiol 43:118–136
19. Ross ED (1980) Localization of the pyramidal tract in the internal capsule by whole brain dissection. Neurology (NY) 30:59–64
20. Strick P (1985) How do the basal ganglia and cerebellum gain access to the cortical motor areas? Behav Brain Res 18:107–124

21. Talairach J, Tournoux P (1988) Coplanar stereotaxic atlas of the human brain. Thieme, New York
22. Weiller C, Chollet F, Friston KJ, Wise RJS, Frackowiak RSJ (1992) Functional reorganization of the brain in recovery from striatocapsular infarction in man. Ann Neurol 31:463–472

Head Injuries: A Target for Neuroprotection

G.M. Teasdale,[1] R. Bullock,[1] and J. McCulloch[2]

Introduction

The importance of head injuries as a target for clinical neuroprotection derives from several factors. These include the high incidence of head injury and the increasing frequency with which younger patients are injured. There is evidence from several sources that ischaemic brain damage is the most important component of secondary brain damage either due to cerebral compression, or from poor oxygenation and perfusion, due to post-traumatic lung or circulatory dysfunction.

Incidence and Importance of Head Injury

Head injuries are very common; studies in Scotland have shown that approximately 2% of the population attend a hospital with a head injury each year (Jennett and MacMillan 1981). Most of the victims are young, with the peak incidence of severe injury in the 15–20 years age group. As a result, head injury is now the leading cause of death under the age of 45 in westernised countries and is becoming so in many developing countries. More years of life are lost in males under 65 years of age from physical injury than from cancer, in the USA, UK or Europe and Japan. Head injury accounts for many of these deaths, in particular for the majority of cases who survive to reach hospital yet die afterwards. Estimates of the incidence of head injuries indicate that some 2 million patients attend hospital each year in the USA, with 75 000 deaths and with a similar number left permanently disabled. The total cost is estimated to be 25 billion dollars per year. In the UK, 1 in 300 families has a member permanently disabled by head injury.

[1] University Department of Neurosurgery, Institute of Neurological Sciences, Southern General Hospital, Glasgow G51 4TF, UK
[2] Wellcome Surgical Institute, University of Glasgow.

Cerebral Ischemia and Basic Mechanisms
Ed. by A. Hartmann, F. Yatsu, and W. Kuschinsky

The Frequency and Importance of Ischaemic Damage in Head-Injured Patients

The importance of ischaemic damage after a head injury was highlighted by neuropathological studies in the 1970s and 1980s (Graham et al. 1978; Graham 1989). These showed that in more than 85% of fatal cases ischaemic brain damage was found at post mortem. About half of these patients had spoken at some time after injury, a clear indication that the damage had developed secondarily, as a result of complications of the injury (Reilly et al. 1975). A range of complications of injury, either systemic or intracranial can lead to secondary ischaemic damage (Table 1). Clinical studies show that such events are common in severe injuries (Andrews et al. 1990) and are associated with a worse outcome (Kohi et al. 1983; Gentleman and Jennett 1990; Table 2). Experimental models which replicate components of a head injury show clear evidence of early ischaemic damage, for example, in association with an intracranial haematoma (Mendelow et al. 1984; Miller 1990; Inglis 1990).

Table 1. Secondary insults leading to hypoxic–ischaemic brain damage after head injuries

Systemic	Intracranial
Hypotension	Haematoma
Hypoxia	Raised ICP
Anaemia	Brain oedema
Hyperpyrexia	Vasospasm
Hypocarbia	Epilepsy

Table 2. Effect of systemic insults (hypoxia–ischaemia) on the outcome of severe head injuries (Gentleman and Jennett 1990)

	No insult		Insult	
Severity and type to injury	Total (*n*)	DVS (%)	Total (*n*)	DVS (%)
GCS 3–5	68	60	15	100
GCS 6–8	90	28	18	56
Diffuse injury	101	35	26	73
Haematoma	57	61	7	86
All cases	158	44	33	76

GCS, Glasgow Coma Score; DVS, death or vegetative state

Initial clinical studies, using overall measures of cerebral blood flow, indicated that ischaemic levels of flow were uncommon in patients who had been admitted to neurosurgery and stabilised (Enevoldsen and Jensen 1977; Mendelow et al. 1985; Obrist et al. 1979). It is now clear that such studies were misleading on three counts: data needed to be obtained sooner, continuously or at least frequently and more focal methods of measurement were required. Recent studies of cerebral blood flow soon after admission to Neurosurgery show low levels of flow in a high proportion of patients within the first six to 12h (Bouma et al. 1991; Yoshinoe 1985; Muizelaar et al. 1989; Robertson et al. 1992). Patients with ischaemia at this early stage are more likely to have ischaemic events subsequently. Moreover, they have a worse outcome, even if their cerebral blood flow rises to levels similar to those of other head injuries, over the next 3 days.

Intermittent measurements of cerebral blood flow could overlook transient episodes of ischaemia. The monitoring of cerebral arteriovenous–oxygen difference is now possible on a continuous basis, using fibreoptic methods and a catheter sited in the internal jugular vein. Such continuous methods show that transient episodes of widened $AVDO_2$, indicative of cerebral oligemia, occur in many severely head-injured patients (Robertson et al. 1987, 1992). The appearance of such episodes soon after injury predicts occurrence of cerebral oligemia in the subsequent days and is associated with a worse outcome.

Advances in technology have also made it practicable to perform tomographic studies of blood flow after acute head injury (Darby et al. 1988; Marion and Bouma 1991; Bullock et al. 1989). Methods used include xenon-enhanced CT scanning and mapping of perfusion by radioisotope tracer and single photon emission computed tomography. These show that under a subdural haematoma, or surrounding a contusion or intracerebral haematoma, blood flow is severely depressed (Bullock et al. 1989; Fig. 1). Evacuation of the haematoma leads to improved blood flow but specimens taken at operation from the margins of a resected contusion have shown that ischaemic brain damage is present beyond the haemorrhagic area and is associated with oedema and raised intracranial pressure after operation (Bullock et al. 1991; Galbraith 1984).

Several studies show a risk of ischaemia when head-injured patients are being transported, either between hospitals or within a hospital (Gentleman and Jennett 1990; Andrews et al. 1990). Gentleman and Jennett (1990) have shown that ischaemic insults are present in a quarter of head-injured patients transferred from general hospitals to a neurosurgical unit and Andrews and colleagues (1990) report the occurrence of such episodes in a quarter of patients during transfer from the emergency room to CT scanning.

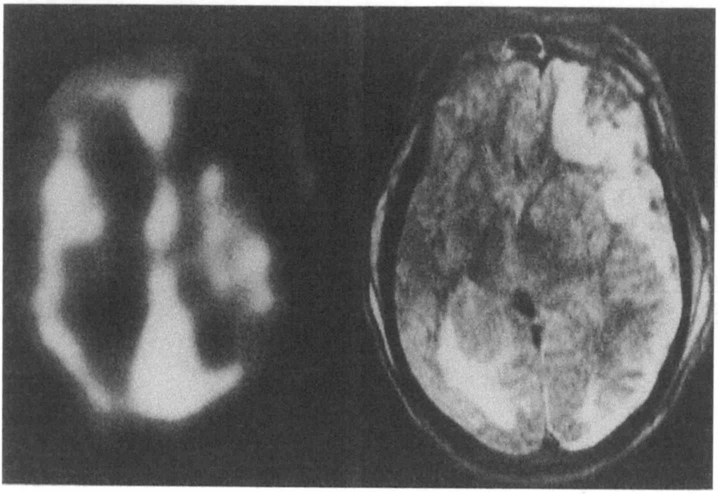

Fig. 1. Depression of local cerebral perfusion by intracranial haematoma. (From Bullock and Fujisawa 1992)

Increased Susceptibility

Experimental and clinical studies indicate that the brain is susceptible to ischaemic damage following a head injury. Studies of the regulation of cerebral blood flow in response to changes in arterial pressure provide experimental evidence for disturbance of autoregulation and vascular reactivity after experimental injury (Lewelt et al. 1980, 1982) and in patients (Enevoldsen and Jensen 1978). Synergy between the effects of injury and ischaemia were shown by Jenkins et al. (1989) who demonstrated that a mild ischaemic insult, tolerated without damage in the normal brain, resulted in permanent ischaemic damage when it was preceded by diffuse traumatic injury.

Microdialysis studies in experimental models of head injury (Katayama et al. 1989; Bullock et al. 1991b) show that concussive head injury and subdural haematoma in rat models are each followed by a surge in the extracellular concentration of excitatory amino acids, which remain persistently raised in the hippocampus when there is a subdural haematoma (Fig. 2).

Current Management of Head Injuries

Management aimed at rapidly optimising the intracranial milieu is better established in severely head-injured patients than, for example, in victims of a stroke. Rapid transport from the scene of the accident is usual with

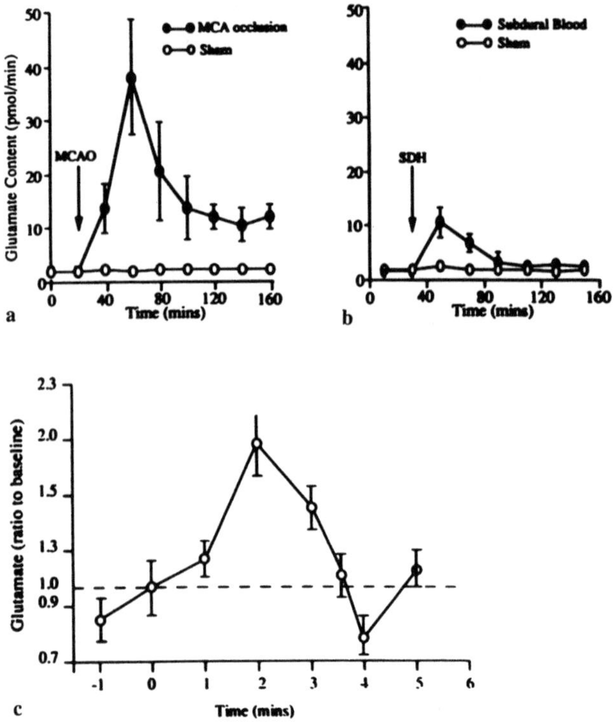

Fig. 2a–c. Glutamate release in focal cerebral ischaemia (**a**), subdural haematoma (**b**) and fluid percussion injuries (**c**) (From Bullock and Fujisawa 1992)

either direct admission or prompt referral to a neurosurgical centre. Brain imaging, either with computed tomography or magnetic resonance imaging is performed urgently to identify the patterns of traumatic damage. Intensive care routinely includes monitoring of the adequacy of cerebral oxygenation by continuous measurement of intracranial pressure, systemic arterial pressure, cerebral perfusion pressure and jugular venous oxygen saturation. Noninvasive assessment of the cerebral circulation is being extended by methods such as transcranial Doppler and cerebral $AVDO_2$ differences.

Many severely injured patients are managed by sedation and ventilation in order to minimise hypoxia from respiratory dysfunction. This strategy facilitates the administration of neuroactive agents and minimises the possibility of psychomimetic side effects, which may be important when effective "neuroprotective" doses are given (Graham et al. 1988; Park et al. 1988)

Prognosis of Head Injury

The establishment of the efficacy of a new treatment depends upon comparison with the outcome of cases of similar severity, treated with conventional measures. The prognostic features that relate initial severity to outcome are well established in head injury and include age, clinical signs indicating the depth and duration of impaired consciousness (Teasdale and Jennett 1974), disturbance of brain stem function and the findings of investigations such as CT scan (Eisenberg et al. 1988) and measures of intracranial haemodynamics and metabolism such as CBF and $CMRO_2$ (Robertson et al. 1992). Knowledge of these facilitates the choice of the most appropriate patient population to study; stratification or statistical modelling can ensure strict comparability of patient populations and may also allow patient by patient prediction of outcome for assessment of the effect of the new treatment.

Conclusion

The foregoing considerations indicate that head injuries are at least as important as stroke as a target for neuroprotection (McCulloch et al. 1991). Indeed, the development of effective neuroprotective methods for head-injured patients may have even greater benefits, both in the broad social context or, in particular, from the point of view of the victims and their families.

References

Andrews PJD, Piper IR, Dearden NM, Miller JD (1990) Secondary insults during intrahospital transport of head injured patients. Lancet 335:327–334

Bouma GJ, Muizelaar JP, Choi SC et al. (1991) Cerebral circulation and metabolism after severe traumatic brain injury: the elusive role of ischaemia. J Neurosurg 75:685–693

Bullock R, Fujisawa H (1992) The role of glutamate antagonists for the treatment of CNS injury. J Neurotrauma 9(2):S443–S462

Bullock R, Teasdale GM, Wyper D, Patterson J, Teasdale E (1989) Tomographic mapping of CBF and blood brain barrier changes after focal head injury using SPECT: mechanisms for late deterioration. In: Hoff J (ed) Proceedings: VIIth International Symposium on Intracranial Pressure and Brain Injury, pp 637–639

Bullock R, Maxwell WL, Graham DI, Teasdale GM, Adams JH (1991a) Glial swelling following human cerebral contusion: an ultrastructural study. J Neurol Neurosurg Psychiatry 54:427–434

Bullock R, Butcher SP, Chen M-H, Kendall L, McCulloch J (1991b) Correlation of the extracellular glutamate concentration with extent of blood flow reduction after subdural haematoma in the rat. J Neurosurg 74:794–802

Darby JM, Yonas H, Marion DW et al. (1988) Local "inverse steal" induced by hyperventilation in head injury. Neurosurgery 23:84–88

Eisenberg HM, Marshall L, Marmarou A (1988) Initial CT findings in 753 patients with severe head injury. A report from the NIH Traumatic Coma Data Bank. J Neurosurg 33:688–598

Enevoldsen EM, Jensen FT (1977) Compartmental analysis of regional cerebral blood flow in patients with severe head injuries. J Neurosurg 47:699–712

Enevoldsen EM, Jensen FT (1978) Autoregulation and CO_2 responses of cerebral blood flow in patients with acute severe head injuries. J Neurosurg 48:689–703

Galbraith S, Cardoso E, Patterson J, Marmarou T (1984) The water content of white matter after head injury in man. In: Go KG, Baethmann A (eds) Recent progress in the study and therapy of brain oedema. Plenum, New York, pp 323–329

Gentleman D, Jennett B (1990) Audit of transfer of unconscious head-injured patients to a neurosurgical unit. Lancet 335:330–334

Graham DI, Adams JH, Doyle D (1978) Ischaemic brain damage in fatal non-missile head injuries. J Neurol Sci 39:213–234

Graham DI, Ozyurt E, Park CK, Nehls DG, Teasdale GM, McCulloch J (1988) Protective effects of the NMDA antagonist, MK-801, on focal ischaemic brain damage. Third European Meeting of Neuropathology. Clin Neuropathol 7:168

Graham DI, Ford I, Adams HJ, Doyle D, Teasdale GM, Lawrence AE, McLellan DR (1989) Ischaemic brain damage is still common in fatal non-missile head injury. J Neurol Neurosurg Psychiat 52:346–350

Inglis FM, Bullock R, Chen MH, Graham DI, Miller JD, McCulloch J, Teasdale GM (1990) Ischaemic brain damage associated with tissue hypermetabolism in acute subdural haematoma. Reduction by a glutamate antagonist. In: Reulen H-J, Baethmann A, Fenstermacher J, Marmarou A, Spatz M (eds) "Brain Oedema 1990" VIII. Acta Neurochir [Suppl] 51:277–279

Jenkins LW, Lewelt W, Young HF, Clifton GL, Hayes AL (1989) Muscarinic and NMDA receptor blockade attenuates increased post traumatic vulnerability to cerebral ischaemia. J CBF Metabol 9:S750

Jennett B, MacMillan R (1981) Epidemiology of head injury. Br Med J 282:101–107

Katayama Y, Cheung MK, Alves A, Becker DP (1989) Ion fluxes and cell swelling in experimental traumatic brain injury: the role of excitatory amino acids. In: Hoff JT, Betz AL (eds) Intracranial pressure 7. Springer, Berlin Heidelberg New York, pp 584–586

Kohi YM, Teasdale GM, Mendelow AD (1983) Assessment of patients with traumatic brain damage. Afr J Neurol Sci 2:27–35

Lewelt W, Jenkins LW, Miller JD (1980) Autoregulation of cerebral blood flow after experimental fluid percussion injury of the brain. J Neurosurg 53:500–511

Lewelt W, Jenkins LW, Miller JD (1982) Effect of experimental fluid percussion injury to the brain on cerebrovascular reactivity to hypoxia and to hypocapnia. J Neurosurg 56:332–337

Marion DW, Bouma GJ (1991) The use of stable Xenon CT CBF studies to define changes in cerebral CO_2 vasoresponsivity caused by severe head injury. Neurosurgery 29:869–873

McCulloch J, Bullock R, Teasdale GM (1991) Excitatory amino acid antagonists: opportunities for the treatment of ischaemic brain damage. In: Meldrum B (ed) Excitatory amino acid antagonists. Blackwell, Oxford, pp 287–326

Mendelow AD, Bullock R, Teasdale GM, Graham DI, McCulloch J (1984) Intracranial haemorrhage induced at arterial pressure in the rat. Part 2. Short-term changes in local cerebral blood flow measured by autoradiography. Neurol Res 6:189–194

Mendelow AD, Teasdale GM, Russell T et al. (1985) Effect of mannitol on cerebral blood flow and cerebral perfusion pressure in human head injury. J Neurosurg 63:43–48

Miller JD, Bullock R, Graham DI, Chen M-H, Teasdale G (1990) Ischaemic brain damage in a model of acute subdural haematoma. Neurosurgery 27:433–439

Muizelaar JP, Marmarou J, Wachi A (1989) ICP and PVI with blood pressure alterations and relation with CBF autoregulation. In: Hoff JT, Betz AL (eds) Intracranial pressure 7. Springer, Berlin Heidelberg New York, pp 825–828

Muizelaar JP, Marmarou J, Ward JD et al. (1991) Adverse effects of prolonged hyperventilation in patients with severe head injury: a randomised clinical trial. J Neurosurg 75: 731–739

Murray GD, Murray LS, Barlow P et al. (1986) Assessing the performance and clinical impact of a computerised prognostic system in severe head injury. Stat Med 5:403–410

Obrist WD, Gennarelli TA, Segawa H et al. (1979) Relations of cerebral blood flow to neurological status and outcome in head injured patients. J Neurosurg 61:292–300

Park CK, Nehls DG, Graham DI, Teasdale GM, McCulloch J (1988) Focal cerebral ischaemia in the cat: treatment with the glutamate antagonist MK-801 after induction of ischaemia. Short Communication. J CBF Metabol 8:757–762

Reilly PL, Adams JH, Graham DI, Jennett B (1975) Patients with head injury who talk and die. Lancet 2:375–377

Robertson CS, Grossman RG, Goodman J, Narayan RK (1987) The predictive value of cerebral anaerobic metabolism with cerebral infarction after head injury. J Neurosurg 60:361–368

Robertson CS, Contant CF, Gokaslan ZL et al. (1992) Cerebral blood flow, $AVDO_2$, and neurological outcome in head injured patients. J Neurol Neurosurg Psychiat (In press)

Teasdale GM, Bullock, R, McCulloch J (1993) Pharmacological therapy of human head injury – can laboratory efficacy by translated to improved clinical outcome? Neurotrauma Meeting, Fukishima, Japan (In press)

Teasdale G, Jennett B (1974) Assessment of coma and impaired consciousness. A practical scale. Lancet 2:81–84

Yoshino E, Yamaki T, Higuchi T et al. (1985) Acute brain edema in fatal head injury: analysis by dynamic CT-scanning. J Neurosurg 63:830–839

Changes in the Biophysical Environment of Water Following Focal Brain Ischemia in the Rat

J.A. Helpern[1,2], R.J. Ordidge[1,2], and R.A. Knight[1]

Introduction

The biophysical environment of water in living systems is of great interest to the study of cell function. Magnetic resonance imaging (MRI) offers the ability non-invasively to assess specific parameters which are sensitive to changes in the biophysical environment of water in tissue. The degree by which water is "influenced" by cellular constituents is defined as that level of influence which is sufficient to result in a measurable difference in MRI parameters. The MRI parameters of interest include the measurement of ^{1}H spin-spin (T2) and spin-lattice (T1) relaxation times, water density (ρ), the apparent diffusion coefficient of water (ADC_w), and exchange rates (K_w) between "bound" and "bulk" water.

Early results of the application of MRI to acute stroke have demonstrated that ^{1}H T2 relaxation times in ischemically affected brain are markedly increased [1–6]. This is the main reason why MRI has proven to be more sensitive than computed tomography in identifying acute stroke [5]. This increase in T2 has been associated with the development of edema [3, 4, 7]; however, an explicit relationship has not been demonstrated. The rationale for this relationship is based on the fact that pure water has a relatively long T2, and an increase in the fraction of pure water should therefore increase the observed T2. Other evidence, however, indicates that the formation of edema cannot be entirely responsible for the observed increase in T2 seen in acute stroke [8–10].

Protons exist in a multiplicity of environments in biological tissue. However, in terms of the measurable MRI resonance linewidth, protons exist in at least two distinct environments. Bulk protons are relatively mobile and typically have a linewidth of less than 100 Hz. Bound protons are incorporated into membranes and proteins and have limited mobility and relatively long correlation times. Their MRI resonance linewidths are therefore broad (10 kHz or greater). Water with limited mobility may also exist as

[1] NIH Center for Stroke Research, Department of Neurology, Henry Ford Hospital and Health Sciences Center, Detroit, Michigan, USA
[2] Department of Physics, Oakland University, Rochester, Michigan, USA

Cerebral Ischemia and Basic Mechanisms
Ed. by A. Hartmann, F. Yatsu, and W. Kuschinsky

hydration layers, and these are also included in the broad proton category since the lack of motional freedom causes a broad MRI resonance.

Unfortunately, the broad MRI resonance cannot be easily detected in a standard high-resolution MRI spectrometer. Since the bulk and bound protons are constantly exchanging magnetization, however, a perturbation of the magnetization state of the bound protons can be transferred to the observable bulk proton magnetization. This technique, called magnetization transfer contrast (MTC) imaging [11], saturates the broad resonance of the bound proton component using a prolonged off-resonance radiofrequency (RF) irradiation within the broad resonance linewidth. Since this irradiation is applied outside the narrow resonance linewidth of the bulk proton signal, it has little direct effect on the bulk proton resonance. However, because exchange of magnetization between the bound and bulk protons occurs during the prolonged saturation pulse, there is a reduction in the net magnetization of bulk protons. The magnetization exchange mechanism, which could either be cross-relaxation between bulk and bound protons and/or chemical exchange of the protons [11], can influence the observed MRI relaxation times T1 and T2, and the investigation of this process may lead to a better understanding of the biophysical environment of water in normal and diseased tissue.

Finally, integral to the investigation of the biophysical environment of water in tissue is assessment of the ADC_w. Diffusion is a process whereby H_2O molecules move in a random fashion due to thermal energy (Brownian motion). The diffusion coefficient of brain water reflects mainly the degree of translational motion of intracellular water since this forms the bulk of the MRI visible proton signal in brain tissue.

Assessment of the biophysical environment of H_2O by MRI has the potential to yield valuable information related to the pathophysiological evolution and/or viability of brain tissue in stroke and may someday be used to predict the severity of stroke-induced brain damage in humans. Toward this end, we report here the results of our investigations into changes in the biophysical environment of water following experimental cerebral ischemia.

Methods

Experimental Cerebral Ischemia

Ischemic injury in the rat brain was produced by permanent surgical occlusion of the middle cerebral artery (MCA-O) similar to that performed by Brint et al. [12]. The model utilizes a tandem occlusion of the left common carotid artery and the left middle cerebral artery. Adult male Fisher rats (n = 5, 200–250 g) were anesthetized with 69% nitrous oxide, 30% oxygen, and 1%–1.5% halothane during the 30 min of MCA-O surgery. Body temperature was maintained at 37° ± 0.5°C during surgery using a rectal thermistor feedback-controlled thermal pad. The animal's head was rigidly fixed in a holder. Initially, the left common carotid artery was isolated, tied, and cut between two sutures. In order to isolate the MCA, an incision was made in the left temporal-parietal region and the temporalis muscle and lower jaw bone partially removed. Under direct visualization with a dissecting microscope (Zeiss) the left MCA was exposed through a 2-mm burr hole drilled about 1 mm rostral to the fusion of the zygomatic arch with the squamosal bone. Drilling was done under a continuous drip of 0.9% saline to prevent heat injury of the underlying cerebral cortex. The dura was retracted to expose the MCA. The tip of an 80-μm diameter stainless steel wire, bent at 90° to form a hook, was inserted under the MCA just superior to the inferior cortical vein using a micromanipulator. The MCA was raised 0.5–1 mm above the cortical surface with the micromanipulator, and, using bipolar coagulator forceps, the MCA was coagulated and cut at the position of the rhinal fissure.

The MCA was cauterized with minimal thermal trauma to the underlying cortex. Gelfoam (Upjohn, Kalamazoo, MI) was placed in the area of the craniotomy, and the incision was sutured shut. Following surgery the animals were placed in a plexiglass holder which included a 20-cm-diameter RF shield and a water-heated support platform. The holder is also equipped with an RF transmit/receive coil tuned to 80.3 MHz into which the animal's head was inserted. The coil was constructed as a modified Helmholtz design and consisted of two circular coils of 5 cm diameter, spaced 2.5 cm apart, connected in a parallel configuration. Stereotaxic ear bars were used to hold the animal's head in a well-defined and motionless position during the imaging procedure. Halothane anesthesia was decreased to 0.75% and rectal temperature maintained at 36°–37° C using a feedback-controlled water bath during the imaging experiments.

Acquisition of MRI Data

All MRI data were obtained using a 1.9-T 60-cm bore magnet (Oxford Magnet Technology, Eynsham, UK) in combination with a Biospec I MRI spectrometer (Bruker Instruments, Billerica, MA). A 20-cm-bore gradient coil insert was used to produce the necessary magnetic field gradients (maximum capability of 2 Gs/cm). Images were obtained with a 3-mm coronal section through the rat brain with a 128 × 128 image matrix and a field of view of 6 cm, corresponding to an in-plane resolution of approximately 470 μm.

Correct reproducible positioning of the animal was essential to studying the same imaging slice in all animals. A modified FLASH [13] interleaved imaging sequence was used to align the animal correctly in the imaging system. The sequence effectively saturates the selected slice in both a coronal and sagittal section of the brain which are simultaneously imaged in

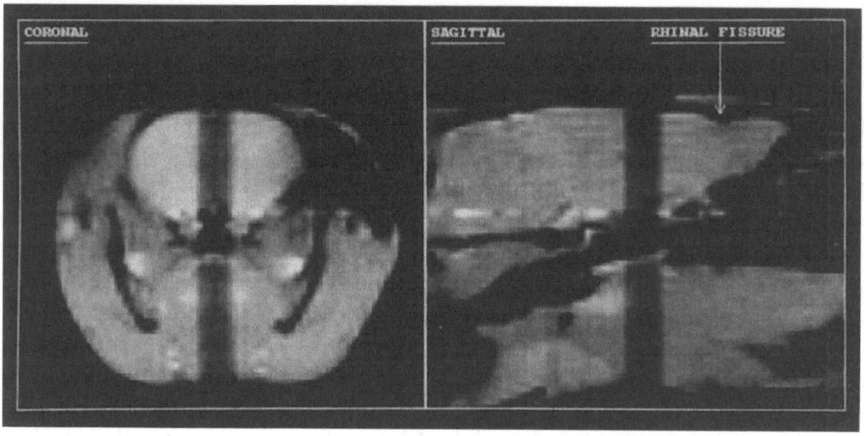

Fig. 1. Coronal and sagittal MRI images of the rat brain obtained by the interleaved FLASH imaging technique. The dark band in the coronal image indicates the position from which the sagital image was obtained and vice versa. These dark bands serve as markers for positioning the animal relative to the rhinal tissue (indicated on sagittal image)

an experiment taking less than 1 min (Fig. 1). The dark bands on the images produced by this method serve as markers for the axial position and angle of the selected slice through the rat brain, thus allowing rapid repositioning of the animal in an iterative manner. The animal's position was adjusted until the imaging section was 5 mm posterior to the rhinal fissure with the head of the rat held in a flat skull position.

T1-weighted images were obtained by saturation recovery with time-to-repeat (TR) delays of 0.75, 1, 1.5, and 3 s and a time-to-echo (TE) of 10 ms. T2-weighted images were obtained with TE values of 10, 29, 59, 109, and 209 ms (TR = 3 s).

The ADC_w was measured by the use of seven diffusion-weighted MRI sequences using the intra-voxel incoherent motion technique [14]. The diffusion-weighting gradients were applied for two intervals of 60 ms either side of the 180° refocusing RF pulse and were nonlinearly varied from 0 to approximately 1 Gs/cm during the sequence. Taking account of the timing of imaging gradients the equivalent b values for this sequence [14] were 0, 18, 73, 292, 659, 1179, and 1825 s/mm^2.

Measurements of ρ were performed with a TE of 10 ms and a TR of 4 s. The measurement of ρ should ideally be performed with a TE greater than five times the expected T1 value of the MRI spins to be observed. Our use of TR = 4 s therefore caused a slight distortion of relative ρ values. In the worst case, this corresponded to an underestimate of ρ in ischemic brain by approximately 3%. Measurements of T2 values are largely insensitive to this source of error provided TR >> TE.

MTC imaging was performed using the approach of Wolff and Balaban [11]. Off-resonance irradiation was applied for 3 s immediately prior to the

MRI experiment (TR = 4 s, TE = 10 ms) at a resonance frequency offset of 10 kHz and an RF power level of approximately 1-W. This corresponds to an angular frequency of nutation of 2240 radians/s. Phantom studies showed that the proton signal from bulk water was unaffected by saturation with resonance frequencies of greater than 1 kHz. The frequency dependence of the off-resonance saturation procedure was determined experimentally in a separate sequence of MTC images.

Measurements of the spin-lattice relaxation time in the absence and presence of off-resonance saturation, T1 and $T1_{sat}$, respectively, were performed using an inversion-recovery (IR) sequence with TR = 4 s, TE = 10 ms. The IR delay was stepped in consecutive images in the sequence 0, 100, 200, 400, 600, 800, 1200, and 1800 ms. Off-resonance saturation was achieved by use of a 1-W RF pulse of 3-s duration and 10-kHz frequency offset applied immediately prior to the IR imaging experiment and replacing part of the TR delay. Off-resonance saturation at the same frequency and power level was also applied during the IR delay as in the IR_{sat} sequence described by Grad et al. [15].

Image pixels were combined (2 × 2) to produce an average value from a region of interest within an isointense region of the brain MRI images. Analysis was performed using a least-squares fit to a straight-line plot of the natural logarithm of normalized image intensity versus TR, TE, and gradient *b* value for T1, T2, and ADC_w calculations, respectively. Images were obtained at 2, 4, 6, and 8 h and 1, 2, 4, and 7 days following MCA-O. Because the MCA-O was performed outside the magnet, we were not able to confirm changes in MRI parameters within the 1st h.

Results

Figure 2 shows a sequence of T2-weighted images (TE = 59 ms) obtained from a single animal at 2, 4, 8, 24 h, 2 days, and 1-week following MCA-O. The images clearly depict the evolution of a hyperintense region on the right hand side of the image (corresponding to left-side of animal) which maximizes in size and intensity at 24 h. Average T2 values obtained from measurements in five animals are shown as a function of time in Fig. 3. Control values for T2 in normal brain were obtained from the contralateral side of the brain. The T2 value for normal brain does not show a significant variation over the 1-week experiment. T2 values for ischemic brain are significantly elevated, even at the earliest time points. The T2 value maximized at approximately 150% of control values at 24 h after MCA-O and had not returned to normal by 1 week after MCA-O. T1 values demonstrated a similar time course to changes in T2 (data not shown).

Figure 4 shows diffusion-weighted images (b = 266) at the same time points and in the same animal as in Fig. 2. The ischemic tissue is immediately apparent as a region of hyperintensity on the right side of these brain

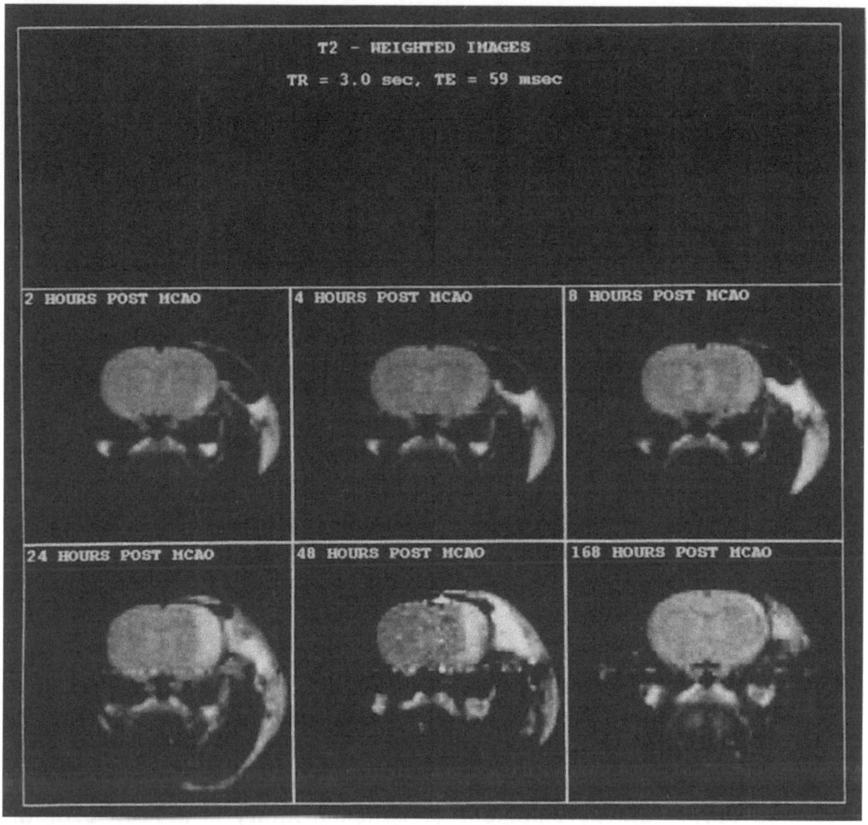

Fig. 2. T2-weighted images (TE = 59 ms, TR = 3 s) of coronal sections through the rat brain showing the development of hyperintensity in the ischemic region (*right side of images*) at 2, 4, 8, 24, and 48 h and 1 week after MCA-O

images. The hyperintense region expands in size during the first 24 h followed by a reduction in image contrast at subsequent time points. Average diffusion values calculated from five animals are presented as a function of time in Fig. 5. At early time points there is a marked reduction in ADC_w to 50% of control values obtained from the contralateral side. ADC_w values show an increase toward control values in the time period 18–48 h followed by either a stabilization of ADC_w or possibly a gradual increase above control values.

The observed proton density in ischemic tissue in the absence and presence of saturation (ρ_i and MTC-ρ_i, respectively) was normalized to control values (ρ_c and MTC-ρ_c, respectively) and is plotted as a function of time in Fig. 6 ($n = 4$). Normalized proton density for ischemic tissue (ρ_i/ρ_c) increased to a maximum value of approximately 115% around day 2 after MCAO. The observed ratio is an underestimate of the true change in proton

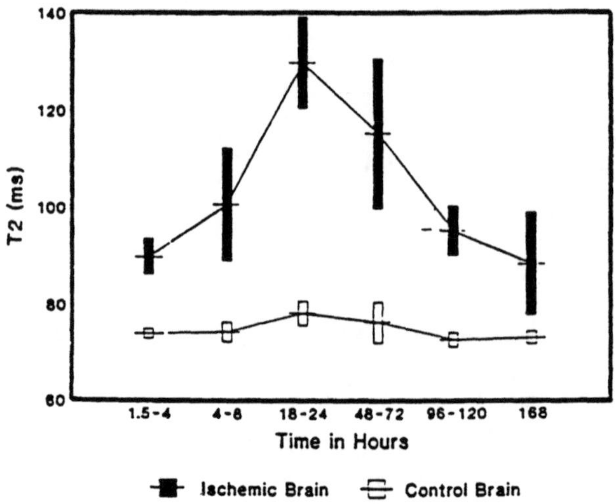

Fig. 3. Mean T2 values (±SD) for five animals plotted as a function of time for ischemic and control brain (contralateral area)

density based on measured T1 values (maximum error of 3%) as previously described. Normalized MTC-ρ values show a maximum increase of 135% at day 1 after MCA-O, with a maximum underestimate of less than 0.5% based on measured $T1_{sat}$ values. The variation of normalized MTC-ρ values measured during the development of ischemic injury is indicative of the temporal evolution of changes in magnetization exchange rate.

Analysis of magnetization transfer kinetics has been described by Forsen and Hoffman [16]. This analysis allows an estimate of the exchange rate of water from the bulk to the bound state. The data required for this analysis include the measurement of M_s/M_o and $T1_{sat}$, where M_s is the magnetization of the bulk water during complete saturation of the bound water, M_o is the magnetization of the bulk water in the absence of the saturation of the bound water, and $T1_{sat}$ is the spin-lattice relaxation time (T1) of the bulk water in the presence of complete saturation of the bound water. With this data, the exchange rate of bluk to bound water can be calculated from the equation:

$$K_w = \frac{1}{T1_{sat}}\left(1 - \frac{M_s}{M_o}\right) \quad (1)$$

Using values of M_s/M_o and $T1_{sat}$, measured at a saturation frequency offset of 10 kHz, Eq. 1 was used to estimate the magnetization exchange rate constant from bulk to bound protons (K_f). For equivalent regions in the rat brain K_f was $1.25 \pm 0.25\,s^{-1}$ in control brain and $0.54 \pm 0.09\,s^{-1}$ in ischemic brain at 24 h after MCAO and $1.0 \pm 0.2\,s^{-1}$ in ischemic brain 1

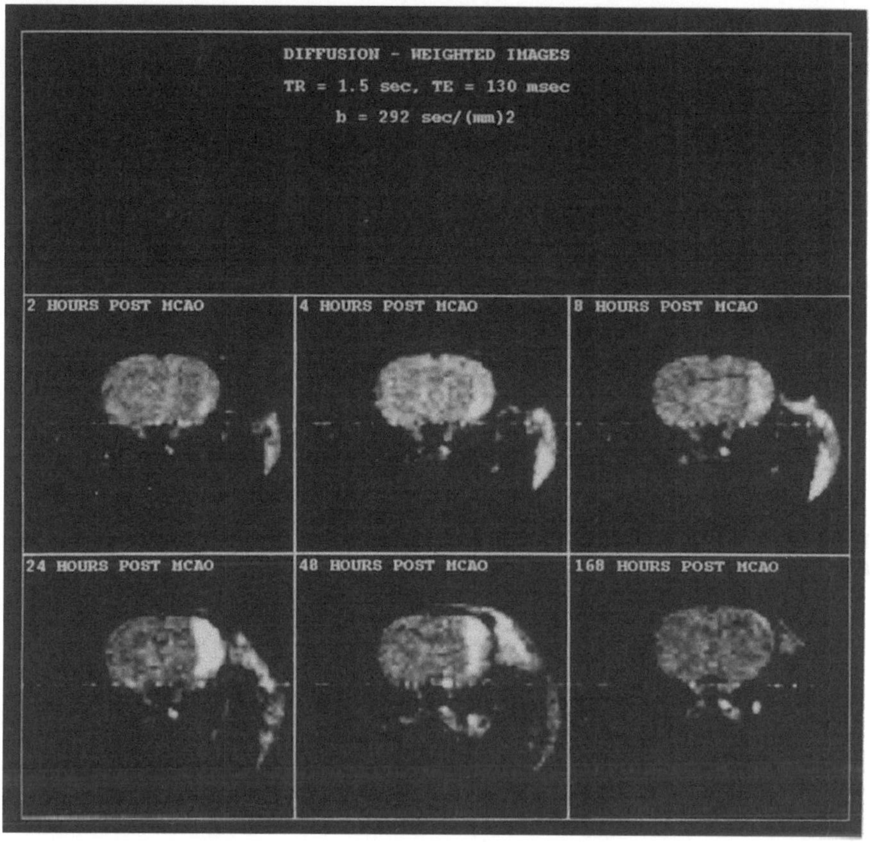

Fig. 4. Diffusion-weighted coronal images (b = 226, TE = 120 ms, TR = 1.5 s) of rat brain at progressive intervals following MCA-O. Ischemic injury appears as hyperintensity on the right side of the brain images

week after MCAO. These data suggest a significant decrease in the amount of bound water at 24 h after MCA-O.

Discussion

The exchange of nuclear spin magnetization between bulk and bound protons is known to influence both T1 and T2 values in biological systems [17]. A contributory mechanism to the observed increase in T1 and T2 following ischemia can thus be formulated by considering a change in the ratio of bulk-to-bound water, without a net change in the total amount of water [18]. This decrease in the amount of bound water demonstrated in the present study is also consistent with the commonly held belief that bound water

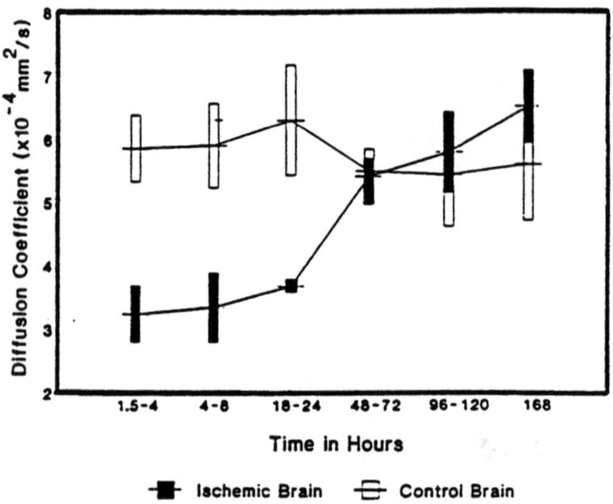

Fig. 5. Mean ADC_w values ($\pm$SD) for five animals plotted as a function of time for ischemic and control brain (contralateral area)

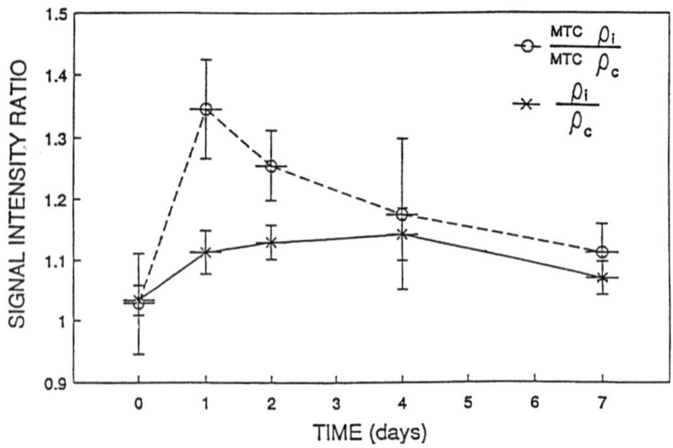

Fig. 6. Average normalized ischemic spin density (ρ_i/ρ_c) and average normalized ischemic MTC spin density (MTC-ρ_i/MTC-ρ_c) for four animals at progressive time periods following MCA-O

fractions are reduced in most disease states [17]. The key point here is that, although the net amount of H_2O has not increased, the mechanisms by which T2 relaxation occurs have been altered.

An implicit assumption in the calculation of exchange rates is that bulk and bound tissue protons are in equilibrium so that:

$$\rho_f K_f = \rho_b K_b$$

where ρ_f and ρ_b are the relative molar concentrations of bulk and bound protons, respectively, and K_f and K_b are the magnetization exchange rate constants from bulk to bound and bound to bulk protons, respectively. In this study, measurements of ρ_f and K_f show that in ischemic tissue 24 h after MCA-O the product ρ_b K_b is reduced to approximately 50% of control values. Since the bound proton component cannot be directly imaged, it is not possible to establish categorically whether ρ_b, K_b, or a combination of these parameters is responsible for this change. However, an increase in bulk proton density to approximately 116% of control values and a reduction in bound proton density to approximately 54% of control values have been reported for infarcted brain tissue following autopsy [19]. This suggests that a decrease in ρ_b is mainly responsible for the observed changes in exchange rate during ischemia. Other studies have shown that an increase in bulk water density during ischemia (edema) is accompanied by a reduction in the tissue density of various proteins as assessed by histopathology in the first few days following MCA-O [20]. Therefore, an associated decrease in ρ_b bound by these proteins and other cell structures may be partly responsible for the observed MRI changes. A reduction in ρ_b may also contribute to the observed elevation in ρ_f since bound water fraction may be liberated by structural breakdown to become part of the bulk water pool.

The potential influence of magnetization exchange on T1 relaxation is much greater than the observed influence on T2 relaxation, since longitudinal relaxation is a slower process, during which more exchange may occur. The data presented here support the idea that magnetization exchange is important in tissue and is consistent with previous reports that cross-relaxation is the dominant T1 shortening mechanism for tissue matter [21].

The Apparent Diffusion Coefficient of H_2O

The ADC_w as measured by MRI decreases by up to a factor of 2 following permanent focal cerebral ischemia, and remains depressed at this level for a period of 18–24 h, at which time it begins to increase toward more normal values. Recently, Benveniste and Johnson [22] reported a 45% decline in ADC_w in rat brain immediately following direct microdialysis treatment of the brain with ouabain, a known inhibitor of [Na,K]-ATPase. Additionally, Busza et al. [23] observed that the decline in ADC_w following cerebral ischemia in the gerbil did not begin until cerebral blood flow had reached a level of approximately 15 ml 100 g^{-1} min^{-1}, a threshold of cerebral blood flow generally associated with energy failure and loss of membrane potential.

Benveniste et al. [24] have suggested that the decline in ADC_w following ischemia may be due to the movement of water from the extracellular space to the more "diffusion-restricted" intracellular space (i.e., cytotoxic edema). Alternatively, changes in ADC_w may be related to changes in cell membrane permeability to water. Alterations in membrane permeability to water have been associated with several disease states. For example, Benga and Morariu [25] demonstrated an abnormally low membrane permeability to water in erythrocytes from epileptics and suggested that this may be indicative of a generalized membrane defect. More importantly, Fritz and Swift [26] have demonstrated a decline in membrane permeability to water with loss of nerve cell membrane polarization.

The results of Benveniste et al. [24] and Busza et al. [23] may point toward a relationship between the loss of cell membrane potential and a decline in ADC_w. These results, taken together with those of Fritz and Swift [26], further suggest a relationship between the decline in ADC_w following ischemia and a decrease in membrane water permeability.

The increase in ADC_w seen at later times following MCA-O may be interpreted as resulting from a loss in cell membrane integrity. As the membranes begin to break apart, the influence of the membrane on the translational motion of the water would be reduced. If this is the case, the assessment of ADC_w values may be valuable in the assessment of therapeutic interventions in human stroke.

Conclusions

MTC-ρ measurements provide a relative indication of magnetization exchange rate. The temporal evolution of T1 and T2 values during ischemia are similar to the observed changes in normalized MTC-ρ values even though the magnetization transfer contribution to T2 relaxation is less. Breakdown of cellular structure in ischemic tissue may result in an increase in bulk proton mobility. This may be the main reason for the observed elevation in T2 values. Cellular breakdown of structure also causes an increase in bulk proton density. A reduction in magnetization exchange rate, caused by reductions in K_b and/or the bound water density, are probably further consequences of this cellular breakdown process and provide additional contributions to relaxation rate changes during ischemia.

ADC_w in ischemic brain tissue shows an immediate reduction to approximately 50% of control values at the earliest time points after MCA-O, followed by a gradual increase toward a value which is either normal or possibly slightly elevated 1 week after MCA-O. The reasons for this decline in ADC_w are unknown at this time. These changes may be associated with changes in cell membrane water permeability or, as proposed by Benveniste et al., this decline may be related to the development of cytotoxic edema [24]. Finally, increases in ADC_w, measured at later times following stroke, may be related to loss of cell membrane integrity indicating irreversible damage. Further investigations which correlate the histopathological status of tissue with changes in MRI parameters may eventually allow the assessment of stroke-induced tissue damage noninvasively in humans.

References

1. Buonanno FS, Pykett IL, Brady TJ, Vielma J, Burt CT, Goldman MR, Hinshaw WS, Pohost GM, Kistler JP (1983) Proton NMR imaging in experimental ischemic infarction. Stroke 14:178–184
2. Mano I, Levy RM, Crooks LE, Hosobuchi Y (1983) Proton nuclear magnetic resonance imaging of acute experimental cerebral ischemia. Invest Radiol 17:345–351
3. Naruse S, Horikawa Y, Tanaka C, Hirakawa K, Hiroyasu N, Yoshizaki K (1982) Proton nuclear magnetic resonance studies on brain edema. J Neurosurg 56:747–752
4. Bradley WG (1984) Magnetic resonance imaging of the central nervous system. Neurol Res 6:91–106
5. Black SE, Helpern JA, Kertesz A, Smith MB, Chopp M, Welch KMA (1987) Nuclear magnetic resonance imaging and spectroscopy in stroke. In: Moore WS (ed) Surgery and cerebrovascular disease. Churchill Livingston, New York, pp 217–253
6. Knight RA, Ordidge RJ, Helpern JA, Chopp M, Rodolosi LC, Peck D (1991) Temporal evolution of ischemic damage in rat brain measured by proton nuclear magnetic resonance imaging. Stroke 22:802–808
7. Bederson JB, Bartkowski HM, Moon K, Halks-Miller M, Nishimura MC, Brant-Zawadiski M, Pitts LH (1986) Nuclear magnetic resonance imaging and spectroscopy in experimental brain edema in a rat model. J Neurosurg 64:795–802
8. Go GK, Edzes HT (1975) Water in brain edema. Arch Neurol 32:462–465
9. Kato H, Kogure K, Ohtomo H, Izumiyama M, Tobita M, Matsui S, Yamamoto E, Kohno H, Ikebe Y, Watanabe T (1986) Characterization of experimental ischemic brain edema utilizing proton nuclear magnetic resonance inaging. J Cereb Blood Flow Metab 6:212–221
10. Horikawa Y, Naruse S, Tanaka C, Hirakawa K, Hiroyasu N (1982) Proton NMR relaxation times in ischemic brain edema. Stroke 56:747–752
11. Wolff S, Balaban R (1989) Magnetization transfer contrast (MTC) and tissue water proton relaxation in vivo. Magn Reson Med 10:135–144
12. Brint S, Jacewicz M, Kiessling M, Tanabe T, Pulsinelli W (1988) Focal brain ischemia in the rat: methods for reproducible neocortical infarction using tandem occlusion of the distal middle cerebral and ipsilateral common carotid arteries. J Cereb Blood Flow Metab 8:474–485
13. Frahm J, Haase A, Matthaei D (1986) Rapid NMR imaging of dynamic processes using the FLASH technique. Magn Reson Med 3:321–327
14. Le Bihan D, Breton E, Lallemard D, Grenier P, Cabanis E, Laval-Jeantet M (1986) MR imaging of intravoxel incoherent motions: applications to diffusion and perfusion in neurologic disorders. Radiology 161:401–407
15. Grad T, Mendelson D, Hyder F, Bryant RG (1990) Direct measurements of longitudinal relaxation and magnetization transfer in heterogeneous systems. J Magn Reson 86:416–419
16. Forsen S, Hoffman RA (1963) Study of moderately rapid chemical exchange reactions by means of nuclear magnetic double resonance. J Chem Phys 39:2892–2901
17. Lynch LJ (1983) Water relaxation in heterogeneous and biological systems. In: Cohen JS (ed) Magnetic resonance in biology, vol 2. Wiley, New York, pp 280–296
18. Ordidge RJ, Helpern JA, Knight RA, Qing Z, Welch KMA (1991) Investigation of cerebral ischemia using magnetization transfer contrast (MTC) MR imaging. Magn Reson Imaging 9:895–902
19. Kaneoke Y, Masahiro F, Inao S, Saso K, Yoshida K, Motegi Y, Mizuno M, Izawa A (1987) Spin-lattice relaxation times of bound water – its determination and implications for tissue discrimination. Magn Reson Imaging 5:415–420
20. Inuzuka T, Tamura A, Sato S, Kirino T, Yanagisawa K, Toyoshima I, Miyatake T (1990) Changes in the concentrations of cerebral proteins following occlusion of the middle cerebral artery in rats. Stroke 21:917–922

21. Edzes HT, Samulski E (1978) The measurement of cross-relaxation effects in the proton NMR spin-lattice relaxation of water in biological systems: hydrated collagen and muscle. J Magn Reson 31:207–229
22. Benveniste H, Johnson GA (1991) Mechanisms of ischemia-induced changes in the brain water diffusion coefficient studied by magnetic resonance imaging and brain microdialysis. J Cereb Blood Flow Metab 11(2):S862
23. Busza AL, Allen KL, Gadian DG, Crockard HA (1991) Early changes demonstrated by diffusion-weighted MR imaging in experimental cerebral ischemia. In: Proceedings of the tenth annual meeting of the Society for Magnetic Resonance in Medicine, vol 1, 328
24. Benveniste H, Hedlund LW, Johnson GA (1992) Mechanism of detection of acute cerebral ischemia in rats by diffusion-weighted magnetic resonance microscopy. Stroke 23:746–754
25. Benga G, Morariu VV (1977) Membrane defect affecting water permeability in human epilepsy. Nature 265:636–638
26. Fritz OG Jr, Swift TJ (1967) The state of water in polarized and depolarized frog nerves. Biophys J 7:675–687

Acidosis

Aspects of Acid-Base Homeostasis in Ischemia

L. Symon, D.L. Taylor, and T.P. Obrenovitch

Introduction

Experimental and clinical data strongly support the concept that brain tissue acidification is a common and major factor responsible for neural tissue damage resulting from inadequate oxygen supply to the brain or head trauma (see Rosner and Becker 1984 for review). Intracellular acidosis has been extensively investigated and nuclear magnetic resonance spectroscopy has provided new insights (Siesjö 1985; Allen et al. 1988), but it is also important to assess the events occurring in the extracellular fluid as it is the neuronal microenvironment where neurotransmitters and drugs are available to receptors. This chapter concentrates on new data related to changes detected in the extracellular fluid.

Changes Associated with Early Ischemia

Direct measurements of H^+ activity in the extracellular space of initial changes in acid-base homeostasis. ($[H^+]_o$) during ischemia have shown that progressive acidification of brain tissue starts immediately after the onset of ischemia (Harris and Symon 1984; Kraig et al. 1985; Mutch and Hansen 1984; Obrenovitch et al. 1990), despite the fact that acidosis is initiated intracellularly. The increase in $[H^+]_o$ and disappearance of neuronal activity are among the earliest changes in cerebral ischemia (Branston et al. 1974; Harris and Symon 1984), occurring much earlier than the massive transmembrane ionic shifts which result from dramatic changes in cellular membrane permeability to ions and are associated with brain cell depolarization. The initial acidification is associated with a small and progressive rise in the extracellular activity of K^+ (Harris and Symon 1984; Hansen 1985) and Ca^{2+} (Siemkowicz and Hansen 1981). The rise in $[K^+]_o$ may result from a progressive failure of the ATP-dependent Na^+/K^+ pump and increase in the cell membrane permeability to K^+. The increase in Ca^{2+}

Department of Neurosurgery, The National Hospital, Queens Square, London WC1N 3BG, UK

Cerebral Ischemia and Basic Mechanisms
Ed. by A. Hartmann, F. Yatsu, and W. Kuschinsky

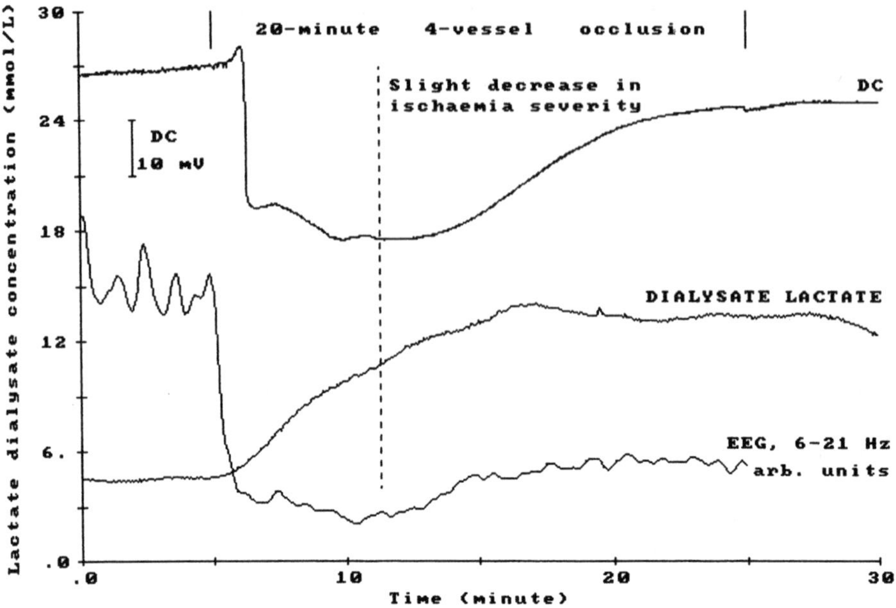

Fig. 1. Changes in extracellular lactic acid monitored using intracerebral microdialysis with on-line detection in rat model of four-vessel occlussion. The gradual positive shift in DC potential which occurred during ischemia probably resulted from a decrease in the severity of ischemia subsequent to improved blood supply

may be due to an increasing extrusion of Ca^{2+} from an intracellular store (Obrenovitch et al. 1990).

The classical concept is that acidification of the extracellular space is considered to occur as a result of a rise in tissue CO_2, whereas the predominant cause of intracellular acidosis is lactic acid accumulation subsequent to stimulation of anaerobic metabolism (Siesjö 1985). However, in vivo monitoring of lactic acid with intracerebral microdialysis has demonstrated that an efflux of lactate into the extracellular space occurs from the onset of ischemia (Fig. 1), contributing directly to the acidification of this compartment. Lactic acid accumulated in the intracellular space during ischemia dissociates into lactate and H^+ which are transported out of the cell via lactate and Na^+/H^+ transporters, respectively. Eventually, when ATP is depleted, the sodium gradient is lost, and therefore H^+ would be trapped in the cell, but lactate would be extruded into the extracellular space (Assaf et al. 1990; Fig. 1).

Other data suggest that translocation of H^+ and HCO_3^- across the cell membrane due to the activation of Na^+/H^+ and HCO_3^-/Cl^- exchangers which regulate intracellular pH under normal conditions may influence this early acidification (Mutch and Hansen 1984; Chesler and Kraig 1989). Direct

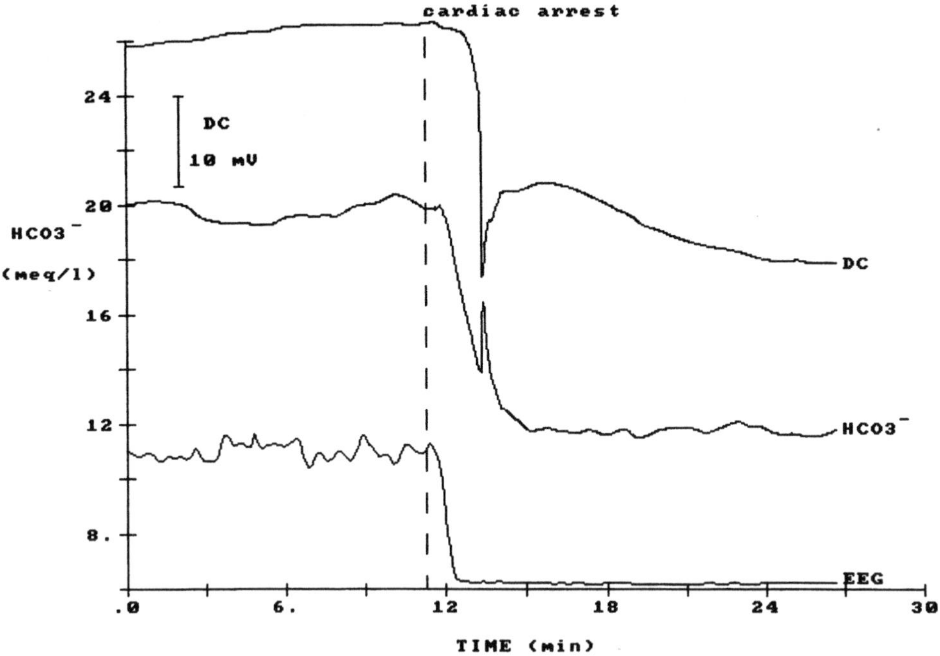

Fig. 2. Reperesentative recording of the changes in $[HCO_3^-]_o$ (0–25 mmol/l scale), DC potential (−30 to 20 mV scale) and EEG (mean amplitude of linear spectrum in the frequency window 6–21 Hz; arbitrary units) in the rat cortex during complete ischemia

measurement of $[CO_3^{2-}]_o$ and subsequently $[HCO_3^-]_o$ has shown that both ions decrease with the onset of ischemia (Fig. 2; Taylor et al. 1991). The data relating to the amplitude of the alkalotic shift associated with anoxic depolarization (see below) also support this hypothesis.

Acid-Base Changes Associated with Anoxic Depolarization

The occurrence of anoxic depolarization is associated with abrupt changes in the extracellular ionic composition, which approaches that of the intracellular compartment with efflux of K^+ and influxes of Na^+ Cl, and Ca^{2+} (Harris and Symon 1984; Hansen 1985; Obrenovitch et al. 1990). Direct measurements of $[H^+]_o$ during anoxia/ischemia consistently reveal a sudden alkalotic shift, superimposed on the steady $[H^+]_o$ increase and intimately related to cell membrane depolarization (Fig. 3; Kraig et al. 1983; Harris and Symon 1984; Mutch and Hansen 1984). Kraig et al. (1983) proposed that the fall in $[H^+]_o$ may be due to an increase in $[HCO_3^-]_o$ as a consequence of a rapid reduction in the extracellular compartment, with

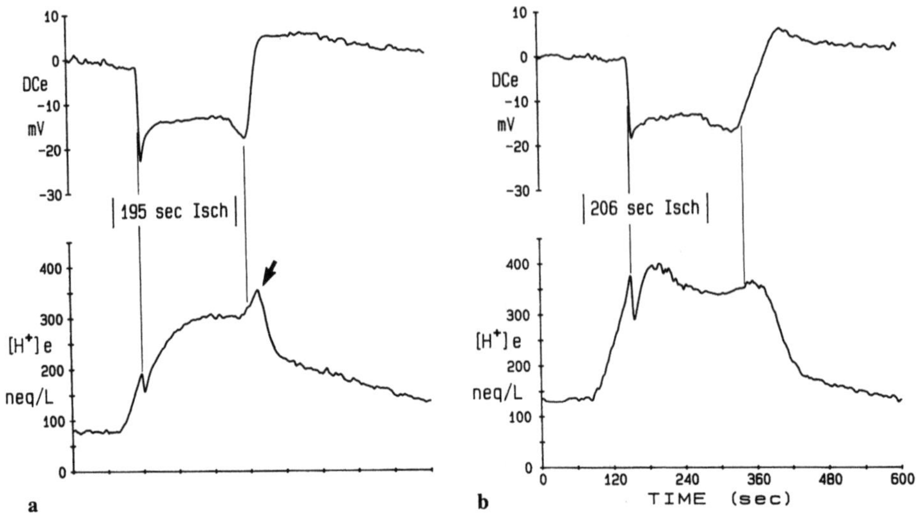

Fig. 3a,b. Typical changes in $[H^+]_o$ and DC potential recorded in the cerebral cortex of rats subjected to 3-min four-vessel occlusion followed by reperfusion associated with ischemia and reperfusion. The relationship between the magnitude of the alkalotic shift and the level of acidosis can be seen. Early during recirculation, a further increase in $[H^+]_o$ associated with membrane repolarization was clearly visible (**a** *arrow*)

selective permeability of the cellular membrane to HCO_3^-. The size of the extracellular space changes during ischemia (Hansen and Zeuthen 1981; Nicholson and Kraig 1981), and it rapidly shrinks to half its initial volume. However, this hypothesis is contradicted by Phillips and Nicholson (1979) who demonstrated that changes in ionic distribution, during spreading depression, are due to the opening of channels with a pore diameter of 6.6 A, which is larger than the diameter of the HCO_3^- ion (4.2 A; Bormann et al. 1987). Furthermore, γ-aminobutyric acid (GABA) and glycine-gated anion channels have been shown to be permeable to HCO_3^- when activated (Bormann et al. 1987; Kaila et al. 1989).

In accordance with our data (Fig. 3), the amplitude of this drop increases with the level of extracellular acidosis at the time it occurs; this may reflect activation of the processes regulating $[H^+]_i$ by tissue acidosis. This theory is supported by the finding that the cell membrane permeability to H^+ markedddly increases in depolarized voltage-clamped snail neurones (Thomas and Meech 1982), to such extent that $[H^+]_i$ becomes determined simply by $[H^+]_o$ and the potential across the cell membrane. However, the above speculative mechanism conflicts with data obtained with drugs which inhibit Na^+/H^+ or HCO_3^-/Cl^- antiports. Mutch and Hansen (1984) observed that pretreatment with amiloride or 4-4′-diisothiocyanostilbene-2,2′-disulfonic acid (DIDS) increases the magnitude of the alkaline shifts of $[H^+]_o$ evoked by spreading depression or anoxic depolarization. According to the mechanism proposed above , inhibition of Na^+/H^+ or HCO_3^-/Cl^-exchanges should, rather, descrease these transients.

Another possibility is that the correlation between the magnitude of the transient fall in $[H^+]_o$ and the level of extracellular acidosis may reflect alteration of the extracellular buffering capacity during ischemia, which is directly related to $[HCO_3^-]_o$ (Siesjö 1985). Since $[HCO_3^-]_o$ decreases as $[H^+]_o$ increases during ischemia (Kraig et al. 1986; Siesjö 1985), the extracellular buffering power becomes diminished, and a given acid-base perturbation produces a larger change in $[H^+]_o$. One acid-base alteration could well be an efflux of HCO_3^- mediated via GABA-gated channels (Bormann et al. 1987), although recent results (Chesler and Chan 1988) show that stimulus-evoked alkalotic shifts are not bicarbonate dependent. The latter mechanism is compatible with the data of Mutch and Hansen (1984; see above), for example, increased alkalotic shift following DIDS pretreatment. These authors propose that, during the alkaline transient, increased $[HCO_3^-]_o$ and decreased $[C^-]_o$ stimulate the Cl^-/HCO_3^- exchanger, which attempts to maintain a set $[HCO_3^-]_o/[HCO_3^-]_i:[Cl^-]_o/[Cl^-]_i$ ratio, which attenuates the magnitude of the alkaline shift. Abolition of this reduction by DIDS results in an enhanced alkaline shift.

It is therefore possible that the transient fall in $[H^+]_o$ may result from a rapid redistribution of H^+ and/or HCO_3^- between the intra- and extracellular space subsequent to the opening of ionic channels. In support of this, measurements of $PtCO_2$ have shown a discontinuity in the ischemia-induced $PtCO_2$ increase, and a steeper rise synchronous with membrane depolarization (Obrenovitch et al. 1990).

The sudden negative shift in DC, indicating anoxic depolarization, is also associated with a transient shift in $[CO_3^{2-}]_o$ superimposed onto its steady decrease. From this it can be seen that there is also a transient increase in with $[HCO_3^-]_o$ with depolarization, which may indicate that a sudden transmembrane redistribution in H^+ is accompanied by a shift in HCO_3^-.

Early in recirculation an acidotic shift was associated with membrane repolarisation (arrow in Fig. 3) whenever the ionic gradients recovered rapidly. This event reflects active recovery of $[H^+]_i$ regulation with normalization of cellular membrane permeability and vigourous extrusion of H^+ and/or HCO_3^- influx.

These mechanisms imply that an acid shift fo $[H^+]_i$ may occur simultaneously with transient falls in $[H^+]_o$, suddenly exposing intracellular biochemical processes to a more severe acidosis, which is sustained until energy again becomes available to restore ionic gradients. If such a step increase in $[H^+]_i$ associated with ischemic depolarization is confirmed in nerve cells of the CNS, it may be a potential threat to the survival of depolarized neurones, along with other well-established damaging events known to be associated with depolarization, in particular: net movement of Na^+ and Ca^{2+} into the cells leading to a toxic intracellular accumulation of these ions

(Goldberg et al. 1986), dramatic releases of neurotransmitters resulting in deleteriously high extracellular levels of these substances (Hagberg et al. 1985; Richards et al. 1992), and osmotic swelling of the intracellular compartment at the expense of the extracellular space (Hansen and Zeuthen 1981; Korf and Postema 1988; Matsuoka and Hossmann 1982). Recent data also suggest that $[H^+]_i$ recovery resulting from the sudden resumption of Na^+ uptake and extrusion of H^+ eraly on during reperfusion may participate in postischemic cell swelling (Jakubovicz et al. 1987; Kraig and Petito 1989). Cl^- uptake associated with extrusion of HCO_3^- may also play a part in this phenomenon.

In summary, the sudden and transient increase in $[H^+]_o$ associated with anoxic depolarization suggests that a rapid intra-/extracellular redistribution of H^+ and HCO_3^- may occur as the cell membrane depolarizes and ionic gradients break down. These data support the concept that some degree of cellular acid-base homeostasis persists as long as anoxic depolarization does not occur. This strengthens the theory that cellular depolarization may be a critical event in the development of injury to brain cells during ischemia and other related insults.

References

Allen K, Busza AL, Crockard HA, Frackowiak RSJ, Gadian DG, Proctor E, Russell RWR, Williams SR (1988) Acute cerebral ischaemia: concurrent changes in cerebral blood flow, energy metabolites, pH and lactate measured with hydrogen clearance and ^{31}P and ^{1}H unclear magnetic resonance spectroscopy. III. Changes following ischaemia. J Cereb Blood Flow Metab 8:816–821

Assaf HM, Ricci AJ, Whittingham TS, LaManna JC, Ratcheson RA, Lust WD (1990) Lactate compartmenation in hippocampal slices: evidence for a transporter. Metab Brain Dis 5:143–154

Bormann J, Hamill OP Sakmann, B (1987) Mechanisms of anion permeation through channels gated by glycine and gamma-aminobutyric acid in mouse cultured spinal neurones. J Physiol (Lond) 385:243–286

Branston NM, Symon L, Crockard, HA, Pasztor E (1974) Relationship between the cortical evoked potential and local cortical blood flow following acute middle cerebral artery occlusion in the baboon. Exper Neurol 45:195–208

Branston NM, Harris RJ (1988) Measurement of cerebral ions. In: Boulton AA, Baker GB, Boisvert DPJ (eds) Imaging and correlative physicochemical techniques. Humana, Clifton, pp 299–342 (Neuromethods 8)

Chesler M, Chan CY (1988) Stimulus-induced extracellular pH transients in the in vivo turtle cerebellum. J Neurosci 27:941–948

Chesler M, Kraig RP (1989) Intracellular pH transients of mammalian astrocytes. J Neurosci 9:2011–2019

Goldberg WJ, Kadingo RM, Barrett JN (1986) Effects of ischemia-like conditions on cultured neurons: protection by low Na^+, low Ca^{2+} solutions. J Neurosci 6:3144–3151

Hagberg H, Lehmann A, Sandberg M, Nyström B, Jacobson I, Hamberger A (1985) Ischemia-induced shift of inhibitory and excitatory aminoacids from intra- and extracellular compartments. J Cereb Blood Flow Metab 5:413–419

Hansen AJ (1985) Effects of anoxia on ion distribution in brain. Physiol Rev 65:101–148

Hansen AJ, Zeuthen T (1981) Extracellular ion concentrations during spreading depression and ischemia in the rat brain cortex, Acta Physiol Scand 113:437–445
Harris RJ, Symon L (1984) Extracellular pH, potassium and calcium actvities in progressive ischemia of rat cortex. J Cereb Blood Flow Metab 4:178–186
Jakubovicz DE, Grinstein S, Klip A (1987) Cell swelling following recovery from acidification in C6 glioma cells: an in vitro model of postischemic brain edema. Brain Res 435:138–146
Kaila K, Pasternack M, Saarikoski I, Voipio V (1989) Influence of GABA-gated bicarbonate conductance on potential, current and intracellular chloride in crayfish muscle fibres. J Physiol (Lond) 416:161–181
Korf J, Postema F (1988) Rapid shrinkage of rat striatal extracellular space after local kainate application and ischemia as recorded by impedance. J Neurosci Res 19:504–510
Kraig RP, Petito CK (1989) Astroglial acidosis and swelling occur after (normoglycemic) global ischemia. J Cereb Blood Flow Metab 9 Suppl 1:S58
Kraig RP, Ferreira-Filho CR, Nicholson C (1983) Alkaline and acid transients in cerebellar microenvironment. J Neurophysiol 49:831–850
Kraig RP, Pulsinelli WA, Plum F (1985) Hydrogen ion buffering during complete ischemia. Brain Res 342:281–290
Kraig RP, Pulsinelli WA, Plum F (1986) Carbonic acid buffer changes during complete brain ischemia. Am J Physiol 250 19:R348–R357
Matsuoka Y, Hossmann K-A (1982) Cortical impedance and extracellular volume changes following middle cerebral artery occlusion in cats. J Cereb Blood Flow Metab 2:466–474
Mutch WAC, Hansen AJ (1984) Extracellular pH changes during spreading depression and cerebral ischemia: mechanisms of brain pH regulation. Cereb Blood Flow Metab 4:17–27
Nicholson C, Kraig RP (1981) The behavior of extracellular ions during spreading depression. In: Zeuthen T (ed) The applicationof ion-selective microelectrodes. Elsevier, Amsterdam, pp 217–238
Obrenovitch TP, Scheller D, Matsumoto T, Tegtmeier F, Höller M, Symon L (1990) A rapid redistribution of hydrogen ions is associated with depolarisation and repolarisation subsequent to cerebral ischemia reperfusion. J Neurophysiol 64:1125–1133
Phillips JM, Nicholson C (1979) Anion permeability in spreading depression investigated with ion-sensitive microelectrode. Brain Res 173:567–571
Richards DA, Obrenovitch TP, Symon L, Curzon G (1992) Extracellular dopamine and seratonin in the rat striatum during transient ischaemia of different severities: a microdialysis study. J Neurochem (in press)
Rosner MJ, Becker DP (1984) Experimental brain injury: successful therapy with the weak base tromethamine. With an overiew of CNS acidosis. J Neurosurg 60:961–971
Siemkowicz E, Hansen AJ (1981) Brain extracellular ion composition and EEG activity following 10 minutes ischemia in normo-hyperglycemic rats. Stroke 12:236–240
Siesjö BK (1985) Acid-base homeostasis in the brain: physiology, chemistry, and neurochemical pathology. Prog Brain Res 63:121–154
Taylor DL, Obrenovitch TP, Ueda Y, Symon L (1991) Direct evidence for a rapid redistribution of bicarbonate ions during anoxic depolarisation. J Cereb Blood Flow Metab 11(2): S519
Thomas RC, Meech RW (1982) Hydrogen ion currents and intracellular pH in depolarised voltage-clamped snail neurones. Nature 299:826–828

Acidosis: A Marker for Penumbra or a Target for Treatment?

A.J. Strong and E. Monteiro

Introduction

The recognition that lactate and proton accumulation in the ischaemic brain develop at relatively superficial levels of ischaemia has prompted several lines of investigation in recent years. Two examples are discussed in this chapter, one in the area of basic pathophysiology of focal ischaemia and the second in that of therapy.

Pathophysiology: Metabolic Mapping of Ischaemic Boundary Zones

Numerous regional imaging methods have been applied to coronal sections of freeze-trapped brain in order to examine both normal and ischaemic physiology. In addition to single-variable methods such as flow or deoxyglucose autoradiography, double-label methods have been used to compare, for example, flow and metabolism on a region-by-region or pixel-by-pixel basis. In 1985 Hossmann et al. elaborated the concept of multiparametric imaging and applied it to the study of focal cerebral ischaemia. Regional cerebral blood flow, pH, potassium, glucose and ATP contents, and rates of glucose utilisation and of protein synthesis were examined, with the conclusion that in the acute phase imaging of blood flow and glucose utilisation do not always given an accurate measure of the extent of functional disturbance.

In recent work we have been examining a particular application of multiparametric imaging in which the topographical distribution of acidosis and potassium depletion is compared in freeze-trapped coronal brain sections following middle cerebral artery occlusion (Strong et al. 1987), in order to delineate a region which might have some properties closely resembling those of penumbra. In this paper the basic approach, methods and results are re-assessed and discussed critically in the light of current interest in the potential contribution of excitotoxicity and related phenomena to ischaemic brain damage.

Institute of Psychiatry, London SE5 8AZ, UK

Cerebral Ischemia and Basic Mechanisms
Ed. by A. Hartmann, F. Yatsu, and W. Kuschinsky

Differential Markers to Delineate Penumbra in Coronal Sections. We have previously described (Strong et al. 1987) the combined use of the sodium cobaltinitrite staining method for potassium content (Mies et al. 1984) as a potential marker of core ischaemia together with, in the adjacent frozen section, imaging for regional pH using umbelliferone-impregnated paper (Csiba et al. 1983). Since acidosis is an early change characteristic of both core and penumbral areas (Astrup et al. 1977; Gibson et al. 1983; Harris et al. 1987), the extent of abnormality seen on the sections embraces both core *and* boundary zones, whereas the depletion of potassium revealed by the sodium cobaltinitrate technique is believed to indicate the extent of *core* ischaemia. This reasoning is based on the well-established finding that sustained increases in extracellular potassium occur only at core levels of ischaemia (Branston et al. 1977), and that such an extracellular increase is necessary for potassium depletion, by elution through residual perfusion.

Methods

Preparation. Acute experiments were carried out in cats anaesthetised with halothane and nitrous oxide and paralysed and ventilated to physiological $PaCO_2$. The brain was exposed, protected with warm mineral oil maintained at 37°C and the middle cerebral artery occluded by the transorbital route for 2 h. The paraffin was removed and the brain frozen with chilled isopentane followed by liquid nitrogen using the funnel method. After storage of the brain at −70°C, 20-μm coronal sections were cut in a cryostat at −20°C and pairs of adjacent sections examined by one of two methods.

Potassium Content Imaging. The method of Mies et al. (1984) was used without modification. In brief, a solution of sodium cobaltinitrite reacts with brain potassium to form potassium cobaltinitrite (yellow). This is counterstained with ammonium sulphide, converting the coloration to dark rown/black, with density proportional to potassium content.

Regional pH. The method of Csiba et al. (1983) was used: this relies on the pH sensitivity of umbelliferone (pK $\approx$ 6.9). Sections were cut onto umbelliferone-impregnated filter paper, transferred to a Peltier stack, and allowed to thaw to 0°C before photography of 450-nm fluorescence with 370-nm excitation.

Analysis. For the purpose of the present studies neither technique was calibrated, since attention was focused entirely on the relative distribution of abnormality in pH and potassium content in adjacent sections. Photographs of K^+ content and of pH were digitised and examined on an image analyser. In the case of pH lesion boundaries, the interface between acidotic and normal brain was relatively distinct in the majority of cases, and observers had no difficulty in assigning lesion boundaries. This proved not to be so for potassium lesion delineation, and the data have now been analysed on two separate occasions. In the first case, three independent observers measured the potassium lesions and found positive, significant linear correlations of K^+ content with pH lesion sizes. Each observer appeared internally consistent, but there was significant inter-observer variability. More recently, the same data was re-analysed by a new observer.

Results

pH lesion boundaries were abrupt in some 70% of experiments, but a more gradual transition from normal pH to acidosis was seen in the remaining experiments. Although a significant linear relationship between pH lesion area and that of potassium could be demonstrated, based on the potassium data generated by each observer in turn, the inter-observer variability did not encourage confidence in the method, and the material was therefore rigorously re-analysed by the new observer. After exclusion of experiments in which (a) the imaging techniques showed no distinction between grey and white matter, (b) lesion boundaries could not be identified in both pH and K^+ images, (c) anatomical integrity was not reasonably maintained, or (d) significant potassium depletion was not seen in adjacent core samples examined for potassium content by flame photometry, ten 2-h experiments involving occlusion of the middle cerebral artery were available for analysis. In five animals lesion boundaries appeared coincident but in the other five pH lesions were larger than K^+ lesions, with a mean difference(pH $-K^+$) of 1.72% whole coronal section area (95% CI 0.26%–3.16%).

Discussion

Relevance of Differential Imaging Techniques to Analysis of the Penumbra. The concept of "penumbra" was based on serial physiological observations conducted over a substantial period of time and was defined explicitly in terms of physiological suppression, with preservation of extracellular potassium homeostasis, coupled with the potential for recovery if reperfusion was established within a period of time (Astrup et al. 1977). Both flow and duration thresholds, or an integral of the two, were defined. These early studies, and numerous subsequent ones, have emphasised the physiological instability of ischaemic boundary zones, especially when these are examined in subprimate mammals, principally the cat, rabbit and rat. Measurement of DC potential or extracellular potassium has been a revealing index of this instability, and spontaneous, transient changes in DC potential or extracellular potassium are now widely regarded as representing episodes of spreading depression (SD). The long-established association of spreading depression with increases in extracellular glutamate (Van Harreveld and Fifkova 1970) suggests that spreading depression may serve as an index of excitotoxic activity in boundary zones. The purpose of this discussion is to draw attention to the implications of these ideas for the interpretation of coronal section images. A coronal section of brain must necessarily present the topography of a lesion at a single point in time during the evolution of an ischaemic lesion. Transient changes such as spreading depression are usually viewed as being propagated across the cortex. This propagation is likely to be in progress at one or more sites *within* the penumbra at the time when freeze trapping is carried out, but is not necessarily present at the *outer* border of the territory to which it is potentially capable of spreading.

A Solution? The model of penumbra described above, a “kaleidoscope” of transients propagating from time to time across the cortical surface but not necessarily lying at the outer border of the penumbra, is based on inference from isolated electrode data. A method is required which maps the propagation of these phenomena over the territory as a whole both spatially and with respect to time. Surface fluorescence methods coupled with image intensifier technology and digital subtraction software offer the prospect of investigating this concept more directly.

Insulin as a Protective Agent in Cerebral Ischaemia

Rehncrona and colleagues (1981) showed that measures which served to restrict glucose availability to the ischaemic brain in incomplete ischaemia enhanced metabolic recovery upon reperfusion. A prospective study of insulin hypoglycaemia in unilateral carotid occlusion with hypotension in the rat demonstrated partial protection of mitochondrial function (Strong et al. 1985), and neurophysiological evidence of protection with insulin came with the work of Robertson and Grossman (1987), who demonstrated preservation of somato-sensory evoked potentials in spinal cord ischaemia in the rabbit. Recognising the need for evidence of insulin protection using end-points based on cognitive or neurological function, a number of groups have confirmed this using global ischaemia models (LeMay et al. 1988; Fukuoka et al. 1989; Voll et al. 1989; Strong et al. 1990).

Mechanisms of Insulin Protection. There seems little doubt that the pretreatment insulin protection regimens are capable of operating at least by restriction of glucose availability and hence of brain acidosis. However, Auer and his group have been able to show that treatment with insulin *following* transient forebrain ischaemia in the rat reduces both spatial learning deficit and histological damage (cerebral cortex; Voll et al. 1989). Moreover, the cytoprotective effect of insulin post-treatment was maintained when plasma glucose levels were sustained with glucose infusion at levels equivalent to those in the untreated group (Voll and Auer 1991). They suggest that this finding points to the possibility that at least two other protection mechanisms might operate, namely neuromodulation and CNS growth stimulation. A probable route of access of insulin to the brain exists in the form of insulin binding sites on cerebral endothelial cells (Duffy and Pardridge 1987), with subsequent endocytosis and intravesicular transport. The extensive body of evidence which now exists on the distribution of insulin binding sites in the brain, particularly in areas rich in synapses, and its inhibitory effect on neuronal firing have been thoroughly reviewed (Voll and Auer 1991). In addition, the brain possesses a number of binding sites for insulin-like growth factors (Baskin et al. 1988), and there is considerable evidence for the production of these factors in the neonatal brain. It is tempting to speculate that increased expression of these factors might be a component of the brain's response to injury during the recovery phase. As a further possible mechanism of action insulin is known to influence ion transport mechanisms, its systemic hypokalaemic affect being the most obvious example. Insulin stimulates the sodium-proton antiport mechanism in muscle, leading to an increase in intracellular pH (Klip et al. 1986).

Clinical Application. The close association of hyperglycaemia with enhancement of ischaemic damage, particularly where glucose delivery is preserved in focal ischaemia – the clinical situation most familiar to neurologists and neurosurgeons – suggests that the value of insulin regimens in the management of cerebral ischaemia as it presents in man should now be examined in formal studies. The demonstration that protection can be achieved with post-lesion treatment in an experimental model without resort to hypoglycaemia is of great practical value, and the potential importance of the concept is increasingly recognised by anaesthetists and intensivists.

Acknowledgments. We thank the Medical Research Council (G8313957N), The Bethlem and Maudsley Hospitals Research Committee, and CAPES (Brazil) for support for this study.

References

Astrup J, Symon L, Branston NM, Lassen NA (1977) Cortical evoked potential and extracellular K^+ and H^+ at critical levels of brain ischaemia. Stroke 8:51–57

Baskin DG, Wilcox BJ, Figlewicz DP, Dorsa DM (1988) Insulin and insulin-like growth factors in the CNS. Trends Neurosci 11:107–111

Branston NM, Strong AJ, Symon L (1977) Extracellular potassium activity, evoked potential and tissue blood flow: relationships during progressive ischaemia in baboon cerebral cortex. J Neurol Sci 32:305–321

Csiba L, Paschen W, Hossmann KA (1983) A topographic quantitative method for measuring brain tissue pH under physiological and pathophysiological conditions. Brain Res 289: 34–337

Duffy KR, Pardridge WM (1987) Blood-brain barrier transcytosis of insulin in developing rabbits. Brain Res 420:32–38

Fukuoka S, Yeh H, Mandybur TI, Tew JM Jr (1989) Effect of insulin on acute experimental cerebral ischemia in gerbils. Stroke 20:396–399

Gibson G, Miller SA, Venables GS, Strong AJ (1983) Evidence of acidosis in the ischaemic penumbra. J Cereb Blood Flow Metab 3 Suppl 1:s401–s402

Harris RJ, Richards PG, Symon L, Habib AH, Rosenstein J (1987) pH, K^+, and PO_2 of the extracellular space during ischaemia of primate cerebral cortex. J Cereb Blood Flow Metab 7:599–604

Hossmann KA, Mies G, Paschen W, Csiba L, Bodsch W, Rapin JR, Le Poncin-Lafitte M, Takahashi K (1985) Multiparametric imaging of blood flow and metabolism after middle cerebral artery occlusion in cats. J Cereb Blood Flow Metab 5:97–107

Klip A, Ramlal T, Cragoe EJ Jr (1986) Insulin-induced cytoplasmic alkalinization and glucose transport in muscle cells. Am J Physiol 250:C720–C728

LeMay DR, Gehua L, Zelenock GB, D'Alecy LG (1988) Insulin administration protects neurologic function in cerebral ischemia in rats. Stroke 19:1411–1419

Mies G, Kloiber O, Drewes LR, Hossmann KA (1984) Cerebral blood flow and regional potassium distribution during focal ischemia of gerbil brain. Ann Neurol 16:232–237

Rehncrona S, Rosen I, Siesjö BK (1981) Brain lactic acidosis and ischemic cell damage. I. Biochemistry and neurophysiology. J Cereb Blood Flow Metab 1:297–311

Robertson CS, Grossman RG (1987) Protection against spinal cord ischemia with insulin-induced hypoglycemia. J Neurosurg 67:739–744

Strong AJ, Miller SA, West IC (1985) Protection of respiration of a crude mitochondrial preparation in cerebral ischaemia by control of blood glucose. J Neurol Neurosurg Psychiatry 48:450–454

Strong AJ, Wild AR, Obrenovitch TP (1987) Visual demonstration of the penumbra by differential imaging of acidosis and K^+ depletion. J Cereb Blood Flow Metab 7 Suppl 1:S98

Strong AJ, Fairfield JE, Monteiro E, Kirby M, Hogg AR, Snape M, Rossfield L (1990) Insulin protects cognitive function in experimental stroke. J Neurol Neurosurg Psychiatry 53: 847–853

Van Harreveld A, Fifkova E (1970) Glutamate release from the retina during spreading depression. J Neurobiol 2:13–29

Voll CL, Auer RN (1991) Insulin attenuates ischemic brain damage independent of its hypoglycemic effect. J Cereb Blood Flow Metab 11:1006–1014

Voll CL, Whishaw IQ, Auer RN (1989) Postischemic insulin reduces spatial learning deficit following transient forebrain ischemia in rats. Stroke 20:646–651

Astrocytic H^+ and Ca^{2+} Changes: Implications for Ischemic Brain Injury

R.P. Kraig and C.D. Lascola

Introduction

Astrocytes closely interact with all other brain cell elements. Therefore the consequences of brain injury from ischemia may be reflected specifically in the behavior of these cells. Astrocytes are either transformed into so-called "reactive species" after a reduction in blood flow [1, 33], or they are destroyed by the process, along with other brain cell elements [1]. Reactive astrocytosis is a process that may influence mechanisms by which the brain develops [28], functions under normal circumstances, and responds to global ischemia with selective loss of vulnerable neurons [18]. Alternatively, destruction of astrocytes from ischemia may not only contribute to selective loss of vulnerable neurons but also play a key role in the development of brain infarction [20, 23]. Two well-conserved and essential physiologic processes of eukaryotic cells that sensitively reflect their functional status are the regulation of intracellular pH (pH_i) and calcium ($Ca^{2+}{}_i$) [6]. For example, a rise in pH_i heralds the onset of increased cellular activity while a fall in pH_i is associated with a reduction in vital processes in numerous cell types [5]. Furthermore, a rise in $Ca^{2+}{}_i$ is a ubiquitous trigger for both constructive [41] and destructive cellular events [10, 11, 14, 36]. We are studying the role that astroglial changes in pH_i and $Ca^{2+}{}_i$ may have in causing the functional changes from ischemia that are seen in astrocytes.

Reactive Astrocytosis

Introduction

Reactive astrocytosis has been defined as a hypertrophy and hyperplasia of these cells which is associated with increased glial fibrillary acidic protein (GFAP) immunohistochemical staining [1]. This classical definition has been extended by Petito et al. [33] who have shown that two forms of reactive astrocytosis are seen after global ischemia. One is associated with

Departments of Neurology and Pharmacological and Physiological Sciences, University of Chicago, Chicago, IL 60637, USA

Cerebral Ischemia and Basic Mechanisms
Ed. by A. Hartmann, F. Yatsu, and W. Kuschinsky

hypertrophy, increased GFAP staining, and hyperplasia. It is seen in areas of brain where neurons are destroyed after global ischemia, such as the lateral striatum and hippocampus. The second form is also associated with astrocytic hypertrophy and increased GFAP staining but not hyperplasia. This latter type of reactive astrocytosis is seen in the paramedian neocortex, an area that is not associated with neuronal death from global ischemia, although infrequent loss of neurons does occur more laterally in layers 3, 5, and 6 [1].

Mechanisms by which astrocytes are transformed into reactive species are incompletely defined. Primary cultures of astrocytes are commonly used to examine potential cellular or molecular mechanisms that may induce hypertrophy and hyperplasia of these cells. However, results from such studies, when supported by data derived only from in vitro systems, should be interpreted with some care. For example, astrocytes in culture do not fully mature. They continue to express the cytoskeletal element vimentin [13]. This protein is seen in astrocytes in vivo during ontogenesis and in specific geographic and temporal zones associated with destructive lesions [13, 33]. Thus, cultured astrocytes may already act in many ways as reactive species and exhibit unique responses to potential structural and functional stimulants as compared to their more quiescent counterparts under normal in vivo conditions. On the other hand, one can not deny the experimental advantages that isolated systems provide. Since we are interested in the mechanisms that transform astrocytes into reactive species, the caveat here is not to avoid in vitro systems but to study them in parallel with astrocytes in vivo. In this way analogies can potentially be drawn between in vitro and in vivo systems. Furthermore, observations generated in one system can be tested for their applicability in the other. This philosophy has driven our initial work in vivo.

Methods

Spreading depression (SD) is defined as a propagating, transient loss of spontaneous and evoked neuronal activity, coupled to a large, negative DC potential in the interstitial space. SD propagates at a rate of 3 mm/min through susceptible gray matter areas and is now known to be associated with a host of metabolic and physiologic changes [4]. Single episodes of SD result in large changes in interstitial potassium, sodium, hydrogen, calcium, chloride [32], bicarbonate and carbon dioxide [21] as well as in glutamate and lactate [25, 30]. SD causes many electrophysiologic, ionic, and metabolic changes that are similar to those seen during ischemia [4]. Yet, there is no evidence (from hematoxylin and eosin staining) that recurrent SD causes neuronal death [24, 31]. Thus, SD may be an opportune, physiologic, and stereotypic perturbation with which to examine mechanisms by which astrocytes are transformed into reactive species.

SD was elicited by topical application of 1 *M* KCl to exposed parietal cortex of halothane anesthetized, spontaneously breathing rats ($n = 24$). Contralateral exposed parietal cortex was exposed to 1 *M* NaCl and served as a sham control. Microelectrodes were inserted in frontal cortices to determine whether SD was induced. Within minutes of the administration of salt solutions to the parietal cortices, SDs began to appear in those cortices exposed to KCl while no SD was seen in the cortices exposed to NaCl. After 3 h of recurrent SD, animals were allowed to recover for 48 h, and then their brains were processed for semiquantitative and computer-based image analyses of GFAP staining intensity [24]. Other animals were allowed to survive for 1–4 weeks after 3 h of SD before being processed similarly.

In other experiments [8, 37] we utilized double-barrel and pH-sensitive microelectrodes to measure astroglial pH_i during SD in halothane-anesthetized, artificially ventilated rats ($n = 39$). Tridodecylamine served as the ionophore for the pH-sensitive microelectrodes [2]. SD was elicited by bipolar, surface electrical stimulation.

Finally, we have successfully begun to use the Ca^{2+}-sensitive ionophore ETH-129 [35] in the fabrication of double-barrel and Ca^{2+}-sensitive intracellular microelectrodes [19]. For these studies, rats ($n = 8$) were anesthetized with pentobarbital but otherwise prepared as in the experiments in which astrocytic pH_i changes were measured [8, 9, 37].

Results

Recurrent SD produces a significant ($n = 8$; $p < 10^{-4}$) 43% increase in neocortical GFAP staining intensity compared to contralateral control values (Fig. 1). Furthermore, thin (1-μm) sections cut by a simple cooling procedure from paraffin-embedded brains and stained with toluidine blue and for GFAP show that recurrent SD also results in a hypertrophy of astrocytes (Fig. 1). The increase in GFAP staining from SD gradually returns toward normal, contralateral levels but is still significantly greater than normal for at least 2 weeks (Fig. 2). Finally, when SD is inhibited by

Fig. 1. Increased GFAP staining and astrocytic hypertrophy after recurrent SD. *Above*, bright-field photomicrograph of GFAP staining in a frontal coronal section showing experimental (*left*) and control (*right*) neocortices. KCl was administered to the left parietal cortex for 3 h to elicit SD while NaCl was administered to the right parietal cortex as a sham, control procedure for a similar period of time. Animals were allowed to recover for 48 hours and then were processed for immunohistochemical visualization of GFAP. In this example 37 SDs occurred in the left neocortex, and no SDs occurred in the right neocortex. *Below*, GFAP- and toluidine-stained neocortical sections from an animal that was processed 48 h after undergoing 25 SDs on the left, experimental neocortex (*left*) and no SDs on the contralateral, control neocortex (*right*). SD produced more extensive GFAP staining and hypertrophy of astrocytes (*left*) compared to control conditions (*right*). *Calibration bar* (*lower left*), 25 μm. (Adapted from [24])

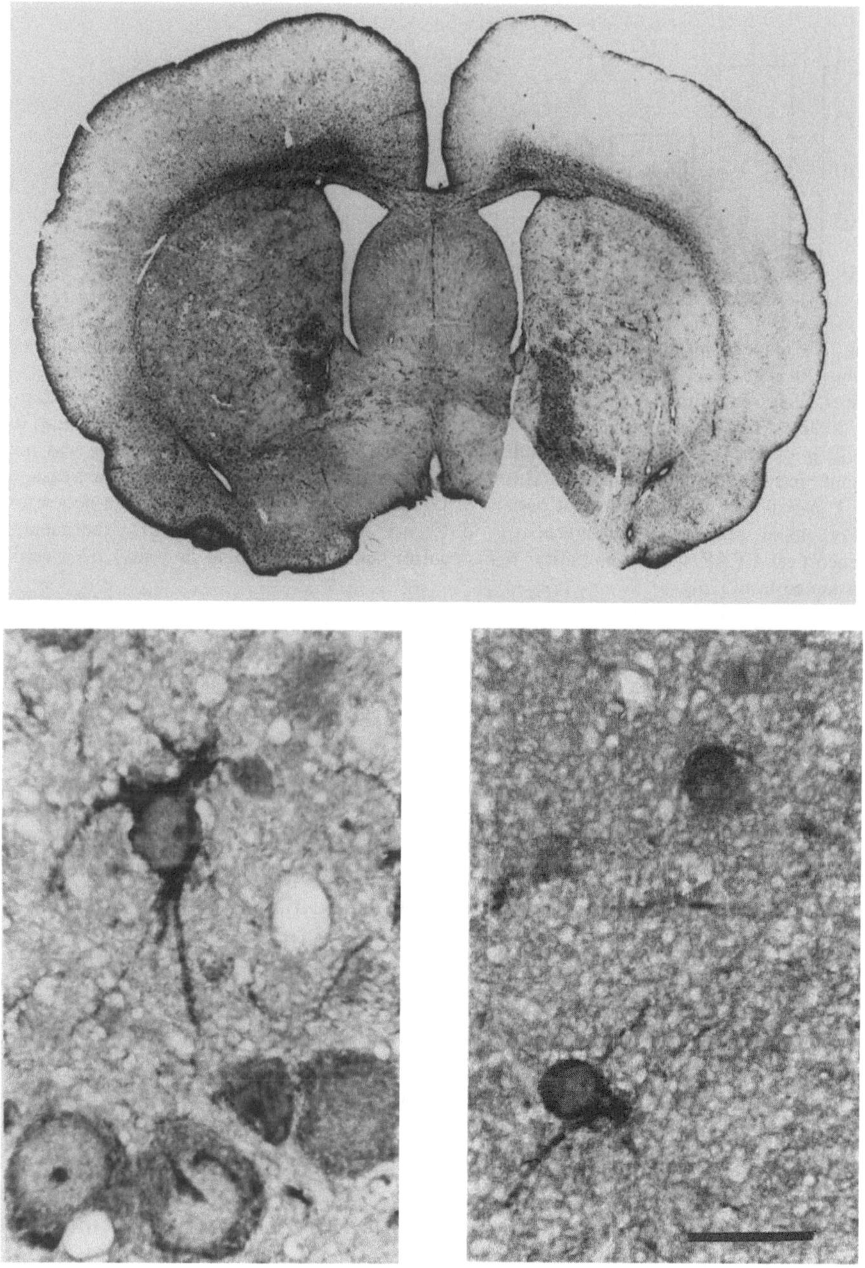

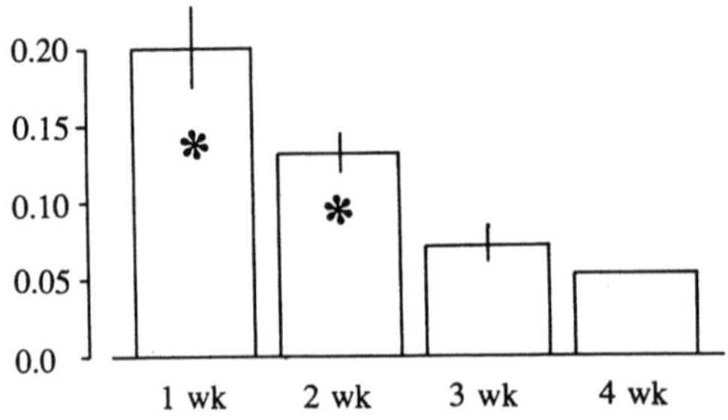

Fig. 2. Temporal profile of increased GFAP staining after recurrent SD. Recurrent SD causes a transient increase in neocortical GFAP staining intensity as compared to contralateral neocortices. One week after 3h of SD, GFAP staining intensity was significantly elevated ($n = 2$, $pn < 0.002$). At 2 weeks, the increase of 14% was still significantly greater than control values ($n = 2$; $p < 0.005$). However, by 3 and 4 weeks after SD, increased GFAP staining was not significantly greater than contralateral controls (7% increase, $n = 2$, $p = 0.089$; 5% increase, $n = 1$, $p = 0.259$). Such a transient pattern of GFAP staining change is also seen in neocortex after global ischemia [33]. *Vertical axis*, difference of the log ratio of (*left*, experimental neocortical GFAP staining intensity; *right*, control neocortical staining intensity) from zero. (Adapted from [24])

combined hyperoxia and hypercarbia, GFAP staining is not significantly ($n = 6$; $p > 0.200$) increased (7%), illustrating that little of the increased GFAP staining measured is due to local necrosis at the site of KCl application [24].

Astrocytic pH_i rises during SD (Fig. 3, left) [8, 9, 37]. Indeed, astrocytic pH_i rises when these cells are depolarized by evoked neuronal activity [8, 9], SD [8, 9, 37], and even nearly complete ischemia (under normoglycemic conditions) [22]. Under normal conditions astrocytic pH_i is 7.04 ± 0.04 ($n = 65$), and their membrane potential is 75 ± 6 mV ($n = 65$) [8, 9, 37]. During SD astrocytic pH_i rises to 7.6–7.8 when the cells are maximally depolarized before returning to baseline as the cells repolarize.

During SD astrocytic intracellular calcium activity ($aCa^{2+}{}_i$) rises from a resting level of a 100 nM by 1–2 orders of magnitude ($n = 5$) before returning to normal when the cells repolarize (Fig. 3, right). The resting $aCa^{2+}{}_i$ of 100 nM was determined with electrodes based on ETH-129. This is considerably lower than the micromolar range for resting astrocytes that was previously reported with the use of Ca^{2+}-sensitive microelectrodes based on a different ionophore (i.e., ETH-1001) [29]. Recent measurements of astrocytic $aCa^{2+}{}_i$ using fluorescent dyes suggest that normal levels of astrocytic $aCa^{2+}{}_i$ in vitro are in the 100-nM range [15, 27]. Thus, our recent measurements using ETH-129 (Fig. 3, right) [19] are perhaps more accurate than those previously reported from the use of ETH-1001 [29].

Electrodes based on ETH-1001 measure more interference from co-ions than microelectrodes based on ETH-129 [35]. Conceivably, such interference contribute to what appears to be the artificially high levels of $aCa^{2+}{}_i$

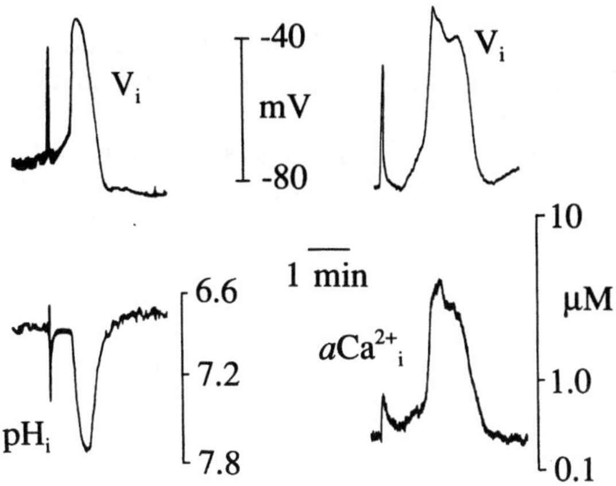

Fig. 3. Astrocytic pH_i and $Ca^{2+}{}_i$ changes during SD. *Record at left.* astrocytic membrane potential (V_i; *top*) and pH_i (*bottom*) change during spreading depression that was elicited by 100-Hz (100-μs pulse; 100-V) bipolar electrical stimulation for 1 s (initial, vertical artifact in both records). Records shown are from second of two SDs elicited in close succession. Therefore, baseline pH_i is more acidic (6.85) than usual (i.e., 7.04) before SD. Nonetheless, the typical profound rise in astrocytic pH_i to 7.4–7.8 during maximal cellular depolarization is evident. In this example, membrane potential declined from −75 mV to −34 mV before repolarization occurred (adapted from [9]). *Records at right*, similar intracellular recordings during SD using a calcium sensitive microelectrodes based on ETH-129 [35]. In this example, a 100-Hz (300-μs pulse; 100-V) bipolar electrical stimulation for 3 s was needed to elicit SD. Therefore, the typical electrical artifact (*vertical line before SD*) was longer than in the pH_i recording to the left. Perhaps this larger stimulus caused SD itself to be somewhat more prolonged than in the example to the left. Astrocytic membrane potential declined from −82 mV to −30 mV during SD while astrocytic intracellular calcium activity ($Ca^{2+}{}_i$) rose from about 0.2 μ*M* up to more than 3 μ*M* before returning to baseline. In other cells astrocytic $aCa^{2+}{}_i$ rose well past 10 μ*M* during SD. (Adapted from [19])

originally reported in vivo. In addition, we have noticed that only our best cellular penetrations (i.e., highest membrane potential) are associated with a low baseline level of $aCa^{2+}{}_i$ in the 100-n*M* range. Perhaps, less than optimal cellular penetration of cells induces a leak of Ca^{2+} from the interstitial space to the intracellular compartment, resulting in an artificially elevated baseline level. At this juncture, we need to increase the number of successful penetrations of astrocytes in vivo with electrodes based on ETH-129 to confirm these possibilities.

Discussion

The above studies illustrate that SD can induce reactive astrocytosis that consists of hypertrophy and increased GFAP staining. Studies are underway to determine whether actual GFAP content rises with immunohistochemical staining for GFAP, and whether astrocytes undergo increased mitotic activity from SD. At present we have no clear evidence that SD induces increased protein synthesis or mitotic activity in astrocytes. Nonetheless, the above studies suggest that SD may be an extremely useful model for examination of the mechanisms of reactive astrocytosis.

One potential stimulus for the induction of reactive astrocytosis must be the rise in pH_i that occurs during SD. A rise in pH_i is often correlated with increased metabolic activity while a decrease in pH_i accompanies a reduction in vital processes [6]. This correlation has prompted Busa and Nuccitelli to suggest that pH_i may be a permissive cellular signal that "sets the tone for subsequent processes" [6]. This notion is supported by considerable in vitro data that indicate a rise in pH_i accompanies fertilization of sea urchin and frog eggs, initiation of protein synthesis in a cell-free system derived from sea urchin eggs, insulin stimulation of glycolysis in frog muscle, and exposure of various cultured cells to mitogens [6]. However, the applicability of these results to in vivo conditions has been questioned in light of more recent studies that indicate the rise in pH_i in the above instances to have occurred under experimental conditions that are associated with reduced extracellular levels of bicarbonate or after replacement of bicarbonate with a more convenient extracellular buffer [7, 16, 26, 38, 41]. When cells normally accustomed to living in an extracellular milieu that includes a bicarbonate buffer in the 10 m*M* range are bathed in vitro by a solution containing a similar level of bicarbonate buffer, their pH_i does not rise upon exposure to mitogens [16]. Instead, the efficiency with which plasma membrane-based regulatory systems modulate pH_i is enhanced [16]. This fact led Ganz and coworkers to suggest that a rise in pH_i is not essential for mitosis but, instead, reflects induction of increased pH_i regulatory capacities, triggered in preparation for greater acid loads associated with growth [16].

In light of the fact that pH_i of cells is higher when they are in a bicarbonate-based buffer as opposed to when they are not, Whitfield has added the suggestion that cells need not elevate their pH_i in response to mitogens if their pH_i is already sufficiently high [41]. Although clearly consistent with all in vitro observations, the applicability of these suggestions to in vivo conditions and astrocyte behavior in particular is uncertain. Astrocytes are surrounded by interstitial fluid in vivo that is high in bicarbonate concentration, and yet these cells undergo a profound rise in their pH_i during SD, a perturbation resulting in the hypertrophy of these cells that continues for at least 2 weeks. How a rise in astrocytic pH_i influences the structure and function of these cells is undefined. Obviously a rise in pH_i itself could influence the activities of various vital enzymes. However, the fact that astrocytic $aCa^{2+}{}_i$ also rises when pH_i is elevated during SD raises the exciting possibility that simultaneous elevation of these two physiologic parameters may act as synergists to promote anabolic activities. To our knowledge, this was first suggested by Busa and Nuccitelli [6]. These authors suggested that a rise in pH_i might affect the degree to which a rise in $aCa^{2+}{}_i$ influences calmodulin-sensitive enzymatic activities. They drew on the observation that calmodulin is highly pH sensitive [39]. Accordingly, a rise in pH_i of 0.5 would have the same effect as a fivefold rise in $aCa^{2+}{}_i$ [6]. If both pH_i and $aCa^{2+}{}_i$ rise, their combined effect might induce processes that would otherwise remain quiescent when either change occurs individually.

Astrocytes and Selective Neuronal Vulnerability

Introduction

If a rise in pH_i and $aCa^{2+}{}_i$ can act as synergistic signals to enhance cellular activity, what might happen to cell function if the rise in $aCa^{2+}{}_i$ occurs when cells are more acidotic than normal? This question stems from a fascinating observation that was recently made in regard to astrocytic $aCa^{2+}{}_i$ homeostasis and the viability of these cells after exposure to reduced levels of extracellular Ca^{2+} [18].

Similar transient exposure of isolated hearts to reduced Ca^{2+} via perfusion results in a large influx of Ca^{2+} to myocardial intracellular space when the perfusate is returned to a normal Ca^{2+} concentration. This phenomenon, termed the "Ca^{2+} paradox," was initially described in myocardium and subsequently extended to traumatic and ischemic injury of nervous tissue by Young (for review see [42, 43]). Kim-Lee et al. [18] examined the potential role of a Ca^{2+} paradox in vitro using cultured neural cells. These workers exposed cultures of astrocytoma cells, spinal glia, and neuroblastoma cells to a transient period of low extracellular Ca^{2+}. They found that $aCa^{2+}{}_i$ rose only in the astrocytoma and glia cells and not in the neuroblastoma cells when the extracellular Ca^{2+} concentration was returned to normal. Furthermore, phase contrast microscopy was used to suggest that the glial cells were irreversibly injured by this manipulation of extracellular Ca^{2+}. They went on to suggest that if such a Ca^{2+} paradox occurs in vivo, perhaps perineuronal astrocytes are primarily killed. Then, their death precipitates the subsequent loss of selectively vulnerable neurons. This scenario suggests that astroglial cell death, and not neuronal death, may be the initial cellular focus of irreversible injury associated with selective neuronal vulnerability.

Methods

We have completed pilot experiments that may be relevant to this hypothesis [22]. Rats (n = 15) were fasted over night but given free access to water. Animals were then anesthetized with halothane, the vertebral arteries cauterized, and reversible clasps placed around their common carotids arteries and cervical neck musculature as described by Pulsinelli and coworkers [34]. Animals were warmed only by a body-water jacket. Rectal temperature was monitored and kept at 37° ± 0.5°C. Astrocytic pH_i was measured with double-barrel and pH-sensitive microelectrodes [8, 9, 37] based on tridodecylamine [2]. Astrocytes were identified by electrophysiologic criteria (high membrane potential, low membrane impedance, and absence of any spontaneous or evoked synaptic activity) before and sufficiently after the onset of reperfusion when the cells had repolarized. During ischemia and shortly after reperfusion, when brain cells are nearly completely depolarized, we can only presume that we penetrated astrocytes. Some confidence in this presumption is provided by the realization that we chose electrode tip configurations that predominantly penetrated only astrocytes [8, 9, 37].

Results

Our measurements show that resting astrocytic pH_i was slightly greater than 7.00, and that the pH_i of these cells rose modestly during ischemia. However, astrocytic pH_i became markedly more acidic than normal 30–80 min after reperfusion (Fig. 4).

Discussion

If astrocytes in vivo have similar characteristics to those studied in vitro, it is likely that $aCa^{2+}{}_i$ rises in these cells during global ischemia. Increased extracellular glutamate elevates astrocytic $aCa^{2+}{}_i$ in vitro [12], and interstitial glutamate rises during global ischemia [3]. Under resting conditions astrocytes in vitro have a $aCa^{2+}{}_i$ level in the 100 n*M* range. Upon exposure to glutamate, astrocytic $aCa^{2+}{}_i$ can rise significantly. Whether it also rises in vivo during reperfusion is unclear. Many workers have shown an increased uptake of Ca^{2+} into brain zones involved in selective neuronal destruction [36]. However, it is still uncertain in most of these instances whether the Ca^{2+} increase occurs before or after cell death. Greenberg and coworkers, using a Ca^{2+}-sensitive dye in vivo do provide some suggestion that tissue $aCa^{2+}{}_i$ may further rise after reperfusion from focal ischemia [17]. Interestingly, this secondary rise is on a similar time frame as the appearance of astrocytic maximal acidosis after nearly complete ischemia (under normoglycemic conditions; Fig. 4) [22]. This raises the question of whether astrocytic $aCa^{2+}{}_i$ might rise at a time when they are most acidic. If true, perhaps astrocytes can be irreversibly injured as can mitochondria [40] by conditions associated with excessive levels of Ca^{2+} and protons.

Mitochondria are known to be able to accumulate large quantities of Ca^{2+}. Ca^{2+} uptake by mitochondria, if excessive, can irreversibly injure these organelles. The amount of Ca^{2+} that can be bound by mitochondria is pH dependent. For example, Ca^{2+} binding by mitochondria can be in-

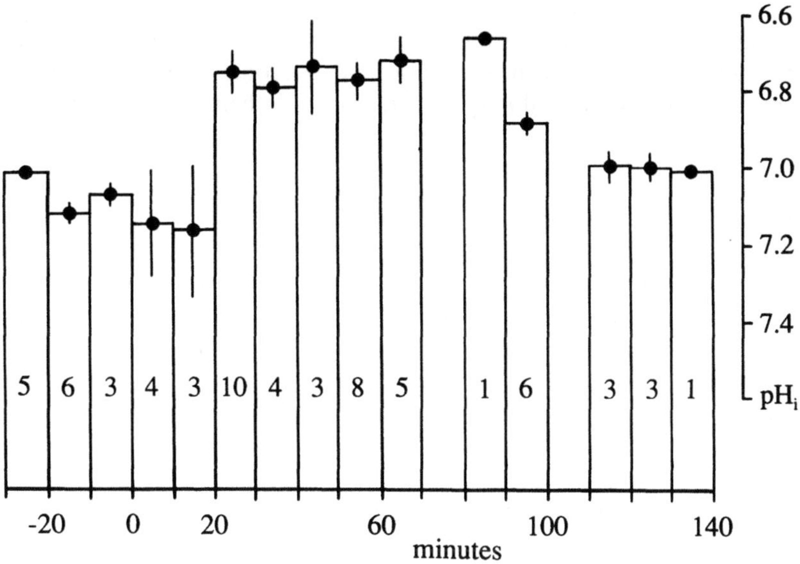

Fig. 4. Effect of global ischemia on pH_i brain cells presumed to be astrocytes. Nearly complete forebrain ischemia was induced by the method of four-vessel occlusion [34]. Briefly, common carotid arteries were reversibly occluded (−20 to 0 min) in anesthetized rats which, 1 day previously, had their vertebral arteries permanently occluded. In addition, a reversible tie was placed around the cervical musculature to enhance forebrain ischemia during carotid occlusion. Brains were then reperfused for approximately 2 h (0–140 min). During this entire procedure animals were placed in a standard stereotaxic setup so that intracellular recordings could be completed. Cells were identified as glia before and approximately 30 min after ischemia by their high membrane potential (i.e., greater than 65 mV), low membrane impedance (i.e., less than 10 MΩ), and absence of any synaptic or injury discharges. The taper of the microelectrode tips was such that astrocytes were the predominant cell type that was penetrated. Therefore, during and shortly after ischemia when brain cells are maximally depolarized and neuronal electrical activity was absent, we presume our electrodes continued to penetrate only astrocytes. Recordings were grouped into 10-min intervals and averaged. In each histogram bar: *vertical line*, standard error of the mean; *numeral*, number of cells penetrated for that 10-min interval. During ischemia (−20 to 0 min), when cells first depolarized, astrocytic pH_i rose slightly. Astrocytic pH_i became most acidic during reperfusion (i.e., at 30–70 min after recirculation began) before returning to normal by 2 h of recirculation. (Adapted from [22])

creased by increasing the pH of their bathing medium. Furthermore, Ca^{2+} uptake by mitochondria is accompanied by ejection of proton equivalents to the bathing medium (for review see [40]). Therefore, it is conceivable that if astrocytic $aCa^{2+}{}_i$ rises during reperfusion from global ischemia when these cells are most acidic, irreversible mitochondrial injury may contribute to the destruction of these cells. This in turn, as suggested by Kim-Lee et al. [18], may then enhance the loss of adjacent neurons. Loss of only perineuronal astrocytes would be difficult to detect because more remote astrocytes are likely to survive, be transformed into reactive species, and come to occupy the space of their lost comrades. Experimental verification of these suggestions will be exceedingly difficult. Nonetheless, documentation of such a close interdependency between astrocytic and neuronal vitality would be important to disclosing the cellular pathogenesis of ischemic brain injury.

Conclusion

Techniques are now available to examine pH_i and $aCa^{2+}{}_i$ using either fluorescent dyes or ion-selective microelectrodes. The particular restrictions of individual biological preparations, the technical limitations of these measuring techniques, and the specific expectations of the investigators need to be kept in mind when designing experiments. It is likely that examination of the manner and mechanisms by which astrocytes concomitantly alter their pH_i and $aCa^{2+}{}_i$ will shed important new insights into how these cells function under normal and pathological conditions.

Acknowledgements. This work was supported by National Institute of Neurological Diseases and Stroke (NINDS) grant NS-19108 (R.P.K.) as well as by an Established Investigator Award (R.P.K.) from the American Heart Association. C.L. was supported by the training grant HD-07009 from the USPHS.

References

1. Adams JH, Corellis JAAN, Duchen LW (1984) Greenfield's neuropathology. Wiley, New York
2. Ammann DF, Lanter RA, Steiner P, Schulthess P, Shijo Y, Simon W (1981) Neutral carrier based hydrogen ion selective microelectrodes for extra- and intracellular studies. Anal Chem 53:2267–2269
3. Benveniste H, Drejer J, Schousboue A, Diemer NH (1984) Elevation of the extracellular concentrations of glutamate and aspartate in rat hippocampus during transient cerebral ischemia monitored by intracerebral microdialysis. J Neurochem 43:1369–1374
4. Bures J, Buresova O, Krivanek J (1974) The mechanism and applications of Leao's spreading depression. Wiley, New York
5. Busa WB (1986) The proton as an integrating effector in metabolic activation. Curr Top Memb Transport 26:291
6. Busa WB, Nuccitelli R (1984) Metabolic regulation via intracellular pH. Am J Physiol 246:R409–R439
7. Cassel D, Whitely B, Zhuang YX, Glaser L (1984) Mitogen-independent activation of Na^+/H^+ exchange in human epidermoid carcinoma A431 cells: regulation by medium osmolarity. J Cell Physiol 122:178–190
8. Chesler M, Kraig RP (1987) Intracellular pH of astrocytes increases rapidly with cortical stimulation. Am J Physiol 253:R666–R670
9. Chesler M, Kraig RP (1989) Intracellular pH transients of mammalian astrocytes. J Neurosci 9:2011–2019
10. Cheung JY, Bonvente JV, Malis Leaf A (1986) Calcium and ischemic injury. N Engl J Med 314:1670–1676

11. Choi DW (1988) Calcium-mediated neurotoxicity: relationship to specific channel types and role in ischemic damage. Trends Neurosci 11:465–469
12. Cornell-Bell AH, Finkbeiner SM, Cooper MS, Smith SJ (1990) Glutamate induces calcium waves in cultured astrocytes: long-range glial signaling. Science 247:470–473
13. De Vellis J, Wu DK, Kumar S (1986) Enzyme induction and regulation of protein synthesis. In: Federoff S, Vernadakis A (eds) Astrocytes, vol 2. Academic, New York, pp 209–237
14. Farber JL (1981) The role of calcium in cell death. Life Sci 29:1289–1295
15. Fatatis A, Russell JT (1992) Spontaneous changes in intracellular calcium concentration in type I astrocytes from rat cerebral cortex in primary culture. Glia 5:95–104
16. Ganz MB, Boyarsky G, Sterzel RB, Boron WF (1989) Arginine vasopressin enhances pH_i regulation in the presence of HCO_3 by stimulating three acid-base transport systems. Nature 246:648–650
17. Greenberg JH, Uematsu D, Araki N, Reivich M (1991) Intracellular calcium and pathophysiological changes in cerebral ischemia. Arzneimittel forschung 41:324–331
18. Kim-Lee MH, Stokes BT, Yates AJ (1992) Reperfusion paradox: a novel mode of glial cell injury. Glia 5:56–64
19. Kraig RP, Lascola CD (1992) Electrodes based on ETH-129 show astrocytic Ca^{2+} rises during spreading depression. Soc Neurosci Abstr 18:1589
20. Kraig RP, Chesler M (1990) Astrocytic acidosis in hyperglycemic and complete ischemia. J Cereb Blood Flow Metab 10:104–114
21. Kraig RP, Cooper AJL (1987) Bicarbonate and ammonia changes in brain during spreading depression. Can J Physiol Pharmacol 65:1099–1104
22. Kraig RP, Petito CK (1989) Interrelation of proton and volume regulation in astrocytes. In: Ginsberg MD, Dietrich WD (eds) Cerebrovascular diseases. Raven, New York, pp 239–246
23. Kraig RP, Pulsinelli WA, Plum F (1986) Carbonic acid buffer changes during complete ischemia. Am J Physiol 250:R348–R357
24. Kraig RP, Dong L, Thisted R, Jaeger CB (1991) Spreading depression increases immunohistochemical staining of glial fibrillary acidic protein. J Neurosci 11:2187–2198
25. Krivanek J (1961) Some metabolic changes accompanying Leao's spreading cortical depression in the rat. J Neurochem 6:183–189
26. L'Allemain G, Pairs S, Pouyssegaur J (1984) Growth factor action and intracellular pH regulation in fibroblasts. J Biol Chem 259:5809–5815
27. MacVicar BA, Hochman D, DeLay MJ, Weiss S (1991) Modulation of intracellular Ca^{++} in cultured astrocytes by influx through voltage-activated Ca^{++} channels. Glia 4:448–455
28. Manthrope M, Rudge JS, Varon S (1986) Astroglial cell contributions to neuronal survival and neurite growth. In: Federoff S, Vernadakis A (eds) Astrocytes, vol 2. Academic, New York, pp 315–376
29. Morris ME, Krnjevic K, MacDonald JF (1985) Changes in intracellular free Ca ion concentration evoked by electrical activity in cat spinal neurons in situ. Neuroscience 14:563–580
30. Mutch WAC, Hansen AJ (1984) Extracellular pH changes during spreading depression and cerebral ischemia: mechanisms of brain pH regulation. J Cereb Blood Flow Metab 4:17–27
31. Nedergaard M, Hansen AJ (1987) Spreading depression is not associated with neuronal injury in rat brain. Brain Res 449:395–398
32. Nicholson C, Kraig RP (1981) The behavior of extracellular ions during spreading depression. In: Zeuthen T (ed) The application of ion-selective microelectrodes. Elsevier/-orth-Holland, Amsterdam, pp 217–238
33. Petito CK, Morgello S, Felix JC, Lesser MI (1990) The two patterns of reactive astrocytosis in postischemic brain. J Cereb Blood Flow Metab 10:850–859
34. Pulsinelli WA, Duffy T (1983) Regional energy balance in rat brain after transient forebrain ischemia. J Neurochem 40:1500–1503

35. Schefer U, Ammann D, Pretsch E, Oesch U, Simon W (1986) Neutral carrier based Ca^{2+}-selective electrode with detection limit in the sub-nanomolar range. Anal Chem 58:2282–2285
36. Siesjö BK, Bengtsson F (1989) Calcium fluxes, calcium antagonists, and calcium-related pathology in brain ischemia, hypoglycemia, and spreading depression: a unifying hypothesis. J Cereb Blood Flow Metab 9:127–140
37. Swain MS, Blei AT, Butterworth RF, Kraig RP (1991) Intracellular pH rises and astrocytes swell after portacaval anastomosis in rats. Am J Physiol 261:R1491–R1496
38. Szergold BS, Brown TR, Freed JJ (1989) Bicarbonate abolishes intracellular alkalinization in mitogen-stimulated 3T3 cells. J Cell Physiol 138:227–235
39. Tkachuk VA, Men'shikov MY (1981) Effect of pH on Ca-binding properties of calmodulin and its interaction with the Ca^{2+}-dependent form of cyclic nucleotide phosphodiesterase. Biokhimiia 46:779–788
40. Wainio WW (1970) The mammalian mitochondrial respiratory chain. Academic, New York
41. Whitfield JF (1990) Calcium, cell cycles, and cancer. CRC, Boca Raton
42. Young W (1986) Ca paradox in neural injury. CNS Trauma 3:235–251
43. Young W (1987) The post-injury responses in trauma and ischemia: secondary injury or protective mechanisms? CNS Trauma 4:27–51

Anoxic Injury of Central Myelinated Axons: Nonsynaptic Ionic Mechanisms

B.R. RANSOM, S.G. WAXMAN, and P.K. STYS

Introduction

The pathophysiology of stroke and central nervous system (CNS) trauma can now be effectively studied at a molecular level. This research is concerned with understanding how cells in the brain, devoid of oxygen and/or metabolic substrates, are injured and ultimately destroyed. The reasonable presumption is that knowledge about the fundamental mechanisms of cell injury will yield clinically applicable insights relevant to how the brain may be protected during periods of disrupted perfusion or metabolism. This work can be subdivided into the study of how each of the major cellular compartments in the brain, i.e., neuronal cell bodies and dendrites, axons and glial cells, are injured by anoxia/ischemia. While great progress has been made in analyzing the mechanisms of neuronal injury in gray matter (GM) areas such as cortex, much less is known about how anoxia/ischemia damages glial cells and axons. We have been interested in the pathophysiology of CNS axonal injury and have developed a reliable model system for studying the basic mechanisms of injury to CNS-myelinated axons caused by anoxia (Stys et al. 1990a; Ransom et al. 1993). The nonsynaptic ionic mechanisms which are critical in the development of irreversible anoxic injury in white matter (WM) are the focus of this review.

Ischemia of the mammalian CNS, including the secondary vascular embarrassment that frequently accompanies traumatic brain and spinal cord insults, damages both GM and WM. In fact, about 20% of ischemic strokes involve predominantly WM, as a result of occlusion of small penetrating arteries that supply the deep parenchymal areas of the cerebral hemispheres (Fisher 1979). Clinical deficits in spinal cord injury are in large part due to damage to WM. Ischemia of WM (delayed WM hypoperfusion) also plays a prominent role in the response of the spinal cord to trauma (Young 1985). The pathophysiology of anoxic injury is likely to be different in WM than in GM, since WM contains no neuronal cell bodies or synapses, but does contain myelinated axons which have a unique, highly specialized structure (Waxman and Ritchie 1985). It follows that the strategies needed to confer

[1] Department of Neurology, Yale University School of Medicine, New Haven, CT 06510 and Neuroscience Research Center, VA Hospital, West Haven, CT 06516, USA

Cerebral Ischemia and Basic Mechanisms
Ed. by A. Hartmann, F. Yatsu, and W. Kuschinsky

protection against anoxic injury may also vary in these vastly different areas of the CNS.

WM Injury Can Be Quantitatively Studied Using the Isolated Rat Optic Nerve

The rat optic nerve has proven to be a highly reproducible, quantitative model system for studying WM injury (Davis and Ransom 1987). Nerves from Long-Evans rats (50–70 days old) are dissected free and placed in a tissue slice chamber of interface design (Stys et al. 1990a). Optic nerves from animals of this age have mature physiological properties (Foster et al. 1982; Connors et al. 1982; Ransom et al. 1985). Experiments are conducted at 37°C and the nerves are perfused continuously with a physiological saline containing 2 m*M* Ca^{2+} (Stys et al. 1990a). The tissue is oxygenated in a 95% O_2–5% CO_2 atmosphere; anoxia is achieved by switching to a 95% N_2–5% CO_2 gas mixture, which results in a fall in chamber O_2 tension from 95% to zero in approximately 2.5 min (Ransom et al. 1992). After switching to anoxia, perfusion of the nerve with glucose-containing solution continues so the anoxic tissue still has the same access to glucose as before. Thus, the circumstances of our experiments are such that the effects of oxygen deprivation are studied in isolation.

The effects of anoxia on optic nerve function are assessed by measuring the compound action potential (CAP), using quantitative electrophysiological techniques (Stys et al. 1991a). Orthodromic stimulation and recording from optic nerves are accomplished using suction electrodes. The stimulus strength was set to 25% above the strength that elicited a maximum CAP. In most experiments, area under the CAP is examined before and after experimental manipulations using a method that corrects for changes in recording electrode impedance (Stys et al. 1991a); CAP amplitude is shown as percentage of the control CAP integral.

The CAP rapidly diminishes after the start of anoxia and virtually disappears after 8–10 min (Fig. 1). The standard anoxic insult used in most experiments is 60 min. During reoxygenation, the CAP area shows partial recovery and attains a plateau value within 1 h (Fig. 1; Stys et al. 1990a; Ransom et al. 1993). The degree of functional recovery depends on duration of the anoxic period and is, in general, much greater in this WM structure than in GM (Kass and Lipton 1982; Taylor et al. 1985; Ransom et al. 1990a). For example, 15 min of anoxia in the hippocampal brain slice preparation results in complete, irreversible loss of evoked synaptic activity (Kass and Lipton 1982; Taylor et al. 1985), whereas the optic nerve exposed to a similar period of anoxia shows, following reoxygenation, only a 10%–15% loss of CAP area (B.R. Ransom and P. Davis, unpublished observations). The average CAP recovery from 60 min of anoxia is about 30% (Fig. 1; Stys et al. 1990a).

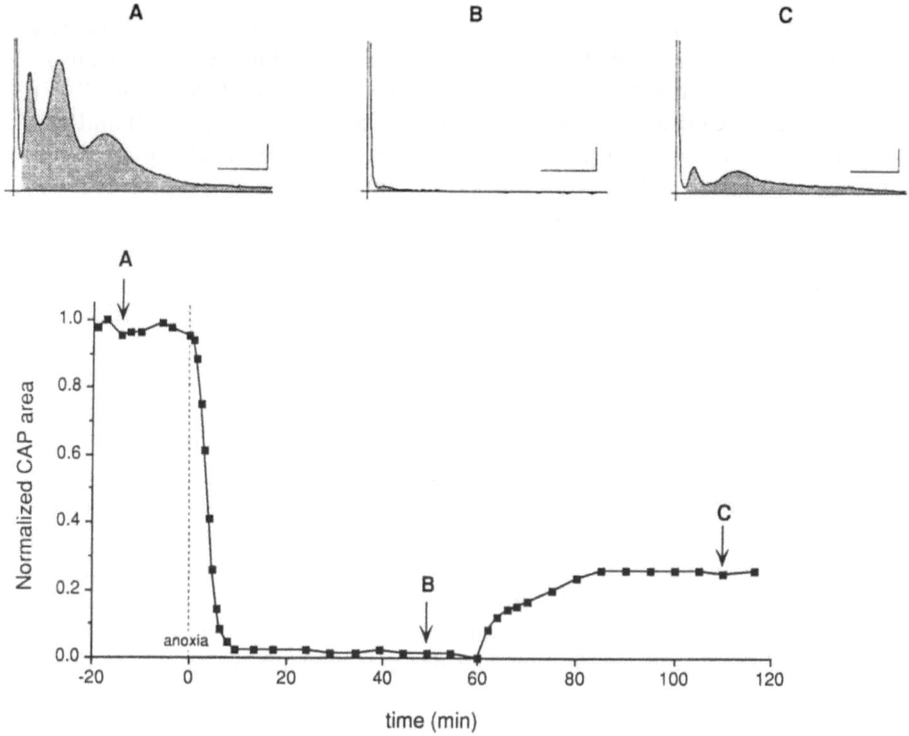

Fig. 1. Effects of anoxia on WM function. The function of the optic nerve was monitored as the area under the CAP (*A*, *B*, *C*); this is shown graphically as the percentage of the control CAP integral. Anoxia was begun at time zero by switching from a 95% O_2/5% CO_2 to a 95% N_2/5% CO_2 atmosphere. CAP area rapidly declined, becoming virtually zero after 10 min of anoxia (residual area was mostly stimulus artifact). A standard 60-min period of anoxia was used for all experiments. After O_2 was reintroduced, CAP area gradually recovered to a mean of 28.5% ± 10.6% (±1% S.D.) of control in perfusate containing 2 m*M* $[Ca^{2+}]_o$. For quantification, post-anoxic CAP measurements were routinely made 60 min after the end of anoxia since recovery always reached a plateau by this time. Specimen records of the CAP under control (*A*), anoxic (*B*) and post-anoxic (*C*) conditions are shown. *Calibration marks* are 1 ms and 1 mV. (Modified from Stys et al. 1990a)

Anoxia Causes Rapid Increases in Extracellular K^+ and H^+ in WM

The rapid changes in brain extracellular ion concentrations that occur with anoxia are important in understanding the pathophysiology of anoxic/-ischemic brain injury (Hansen 1985). These changes convey information about the metabolic state of local brain tissue (Siesjö 1981) and can have direct effects on neural behavior. Elevated $[K^+]_o$ depolarizes neuronal membranes, reducing and then blocking action potentials, causes uncontrolled transmitter release (Benveniste et al. 1984), reduces electrogenic glial uptake of neurotransmitters including the excitotoxin glutamate (Schwartz and Tachibana 1990), induces cell swelling (Kimelberg and Ransom 1986), and may affect cerebral blood flow (Paulson and Newman 1987). Extracellular acidosis can have direct toxic effects on both neuronal and glial membranes (Goldman et al. 1989; Kraig et al. 1987), may alter ion channel function (Chesler 1990), and blocks currents generated by activation of NMDA receptors (Tang et al. 1990).

Anoxia causes rapid changes in the extracellular concentrations of K^+ and H^+ in WM that are qualitatively similar to those seen in GM, but smaller (Ransom et al. 1992a; Kraig et al. 1985; Hansen 1985). Within 3 or 4 min of the onset of anoxia, $[K^+]_o$ in the optic nerve begins to increase and reaches a final concentration averaging 14.0 m*M*. An acid shift in pH_o develops during anoxia with a time course similar to the change in $[K^+]_o$. The average maximum acid shift in standard physiological solution is 0.31 pH unit (Ransom et al. 1992). As in GM (Kraig et al. 1985), the magnitude of the extracellular acid shift increases with higher levels of bath glucose concentration, and this is associated with smaller increases in $[K^+]_o$ and delayed loss of the CAP (Ransom et al. 1992). In the presence of 20 m*M* glucose, compared to 10 m*M*, there is more substrate for anaerobic metabolism and presumably greater generation of lactic acid and ATP. The higher levels of ATP would act to slow the deterioration of ion gradients during anoxia, accounting for the smaller increases in $[K^+]_o$ and delayed loss of the CAP (Ransom et al. 1992).

In WM, glial cells have been shown to contribute directly to anoxia-induced changes in extracellular ions, based on studies using optic nerves that (as a result of retinal ablation and resultant loss of axons) contain only glial cells (Ransom and Philbin 1992). In such "glial" nerves, anoxia causes abrupt increases in $[K^+]_o$ and decreases in pH_o that develop and recover with similar time courses to the corresponding ionic changes seen in intact rat optic nerves. Technical reasons cause the ionic changes recorded in the "glial" nerve to be smaller and more variable than ionic changes in the intact nerve (Ransom and Philbin 1992), so the quantitative degree to which glial cells contribute to the ionic changes seen in normal WM is presently not known.

The extent to which extracellular ionic changes participate in the development of anoxic injury in WM is not known. Certainly these early changes predispose to some of the other ionic events that are critical for injury (see below). Increasing $[K^+]_o$ to 30 m*M* for 60 min, under normoxic conditions, is not harmful to the optic nerve (Ransom et al. 1992) and preliminary experiments indicate that the optic nerve is also undamaged by 20 m*M* lactate buffered to a pH of 6.4. In contrast to the situation in GM (Pulsinelli et al. 1982), WM suffers less, rather than more, injury during anoxia in the presence of higher-than-usual glucose concentrations (Ransom et al. 1992), in spite of the fact that the elevated bath glucose concentration

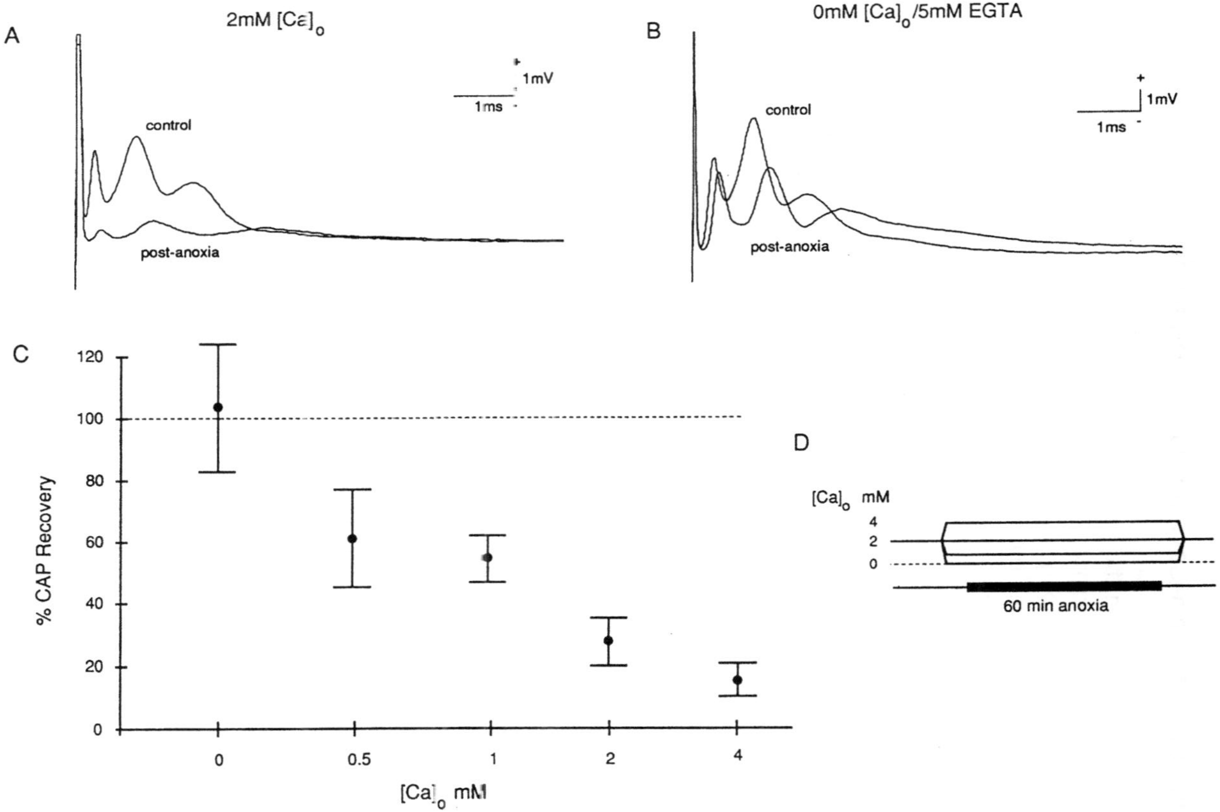

Fig. 2A–D. Effect of perfusate $[Ca^{2+}]$ on recovery of CAP from 60 min of anoxia. **A** CAPs before ("control") and 60 min after anoxia in 2 m*M* $[Ca^{2+}]$ or zero $[Ca^{2+}]$/5 m*M* EGTA are shown. CAP recovery from anoxia is enhanced in the zero Ca^{2+} solution. **B** Average (±1% S.D.) CAP recovery after a standard 60-min period of anoxia as a function of perfusate $[Ca^{2+}]$ during anoxia. Test solutions were begun 10 min before anoxia onset and continued until 10 min after the end of anoxia (see *inset*). The percent recovery of the CAP gradually diminished as perfusate $[Ca^{2+}]$ increased from 0 to 4 m*M*. The mean recovery in 2 m*M* $[Ca^{2+}]_o$ was 28.5% ± 10.6%, whereas in zero $[Ca^{2+}]$/5 m*M* EGTA the area under the CAP recovered to 103% ± 23% of the control value. All points were significantly different from the normal $[Ca^{2+}]$ of 2 m*M* ($p < 0.005$). (Modified from Stys et al. 1990a)

causes a greater acid shift, which is believed to worsen outcome in anoxic GM (Plum 1983; Pulsinelli et al. 1982).

Extracellular Ca^{2+} Is Necessary for Anoxic Injury in WM

Extracellular Ca^{2+} is critical in the production of anoxic injury in WM (Stys et al. 1990a). This was demonstrated by applying test solutions with different concentrations of Ca^{2+} for periods extending from 10 min before the onset of anoxia to 10 min after the end of anoxia (Fig. 2B, inset). As the $[Ca^{2+}]$ is decreased from 2 m*M* to zero, the degree of CAP recovery from 60 min of anoxia gradually increases. If the Ca^{2+} is omitted entirely from the perfusing fluid, the optic nerve displays essentially complete recovery, as judged by the area under the CAP, even after 60 min of anoxia; the latencies of the CAP peaks, however, do not fully recover, suggesting that although most of the axons within the nerve are protected from anoxic injury by perfusion with zero $[Ca^{2+}]$, some axons conduct more slowly after anoxia (Fig. 2A). Other experiments indicate that the deleterious effects of extracellular Ca^{2+} accumulate slowly, suggesting that Ca^{2+} influx continues throughout the entire anoxic period (Stys et al. 1990a).

Pyramidal tract responses can be recorded in vivo after 60-min periods of anoxia (Hossmann and Sato 1970). At first glance this observation seems to suggest that pyramidal tract axons in vivo can tolerate such periods of ischemia without any damage. The method used to obtain these results, however, does not allow quantitative evaluation of axonal function in the same way that it can be assessed under the conditions of our experiments on the optic nerve (Stys et al. 1991a), and thus it is unclear whether a real discrepancy exists between our in vitro observations on WM and those made in vivo. Assuming that there is less damage in vivo than in vitro, for purposes of discussion, what could be the explanation for this? The argument has been offered that under in vivo circumstances, the extracellular $[Ca^{2+}]$ falls to very low levels and in effect corresponds to the in vitro situation where bath Ca^{2+} is zero during anoxia (K.A. Hossmann, personal communication; see Fig. 2). Were this the case, the degree of in vivo axonal injury following 60 min of anoxia might be expected to be very slight or nonexistent, as in the in vitro experiments. Although $[Ca^{2+}]_o$ falls to low levels in GM during anoxia, both in vivo and in vitro (Hansen 1985), the situation in WM is less clear. Preliminary results (B.R. Ransom and D.M. Philbin, unpublished observations) suggest that $[Ca^{2+}]_o$ does not fall significantly in the rat optic nerve during anoxia (although marked changes take place in $[K^+]_o$ and pH_o, see above). Further measurements of $[Ca^{2+}]_o$ in WM during anoxia should be carried out to resolve this question.

Although it is not directly shown by these experiments, extracellular Ca^{2+} probably acted as a source for inward Ca^{2+} flux into a cytoplasmic compartment. Because it is the axons that become dysfunctional during anoxia, based on the loss of the CAP, a damaging increase in intra-axonal [Ca^{2+}] seems likely. Some support for this view comes from recent observations in our laboratory on the ultrastructural correlates of anoxia in the optic nerve indicating that 60 min of anoxia causes striking pathological alteration within axons, particularly the largest fibers (Waxman et al. 1992). Large vacuolar spaces appear between axons and their ensheathing myelin, axoplasmic mitochondria are swollen and disrupted, and neurofilaments and microtubules disappear. Some resolution of these anoxia-induced ultrastructural changes occurs during the standard 60-min recovery period although neurofilament and microtubule damage persists. Interestingly, many of the cytoskeletal alterations seen after 60 min of anoxia are prevented by perfusion with zero [Ca^{2+}] solution during anoxia (Waxman et al. 1993). The destruction of the axonal cytoskeleton seen in our experiments on the optic nerve is similar to that induced by exposure of peripheral nerve axons to the calcium ionophore A23187 (Schlaepfer 1977). Thus, it is likely that anoxia triggers an influx of Ca^{2+} into axons within the optic nerve.

Anoxia-Induced Ca^{2+} Entry in WM Is Mediated by the Na^+–Ca^{2+} Exchanger

How does the anoxia-induced Ca^{2+} influx occur in WM? At the outset it seemed likely that the mechanism would be different than in GM where anoxia-mediated Ca^{2+} influx is predominantly via glutamate activation of NMDA receptor channels and, perhaps to a lesser extent, by voltage-gated Ca^{2+} channels (Choi 1988a,b). Axons in the rat optic nerve are insensitive to excitotoxins such as glutamate (Ransom et al. 1990b) and they appear to have no voltage-gated Ca^{2+} channels (Foster et al. 1982). Not surprisingly, therefore, optic nerve axons are not damaged by high concentrations of excitotoxins (Ransom et al. 1990b) nor are they protected from anoxia by Ca^{2+} channel blocking agents (Stys et al. 1990b). Furthermore, the *N*-methyl-D-aspartate (NMDA) antagonist ketamine, at low concentrations that are relatively specific for blocking this receptor, had no protective effects on anoxia-induced rat optic nerve injury (at very high concentrations, ketamine is protective, but this effect is probably related to the drug's anesthetic-like actions on Na^+ channel permeability and not to NMDA receptor blockade; Ransom et al. 1990b).

Recent evidence from our laboratory indicates that during anoxia, Ca^{2+} influx occurs by way of reverse operation of the Na^+–Ca^{2+} exchanger, a membrane protein that normally operates to extrude cytoplasmic Ca^{2+} in exchange for Na^+ influx. This exchange mechanism does not consume ATP and is primarily driven by the transmembrane Na^+ gradient. The exchanger functions equally well in the forward or reverse direction. The stoichiometry of this process, at least in some instances, is that three Na^+ ions exchange for each Ca^{2+} ion; this exchange ratio implies that the exchange process is

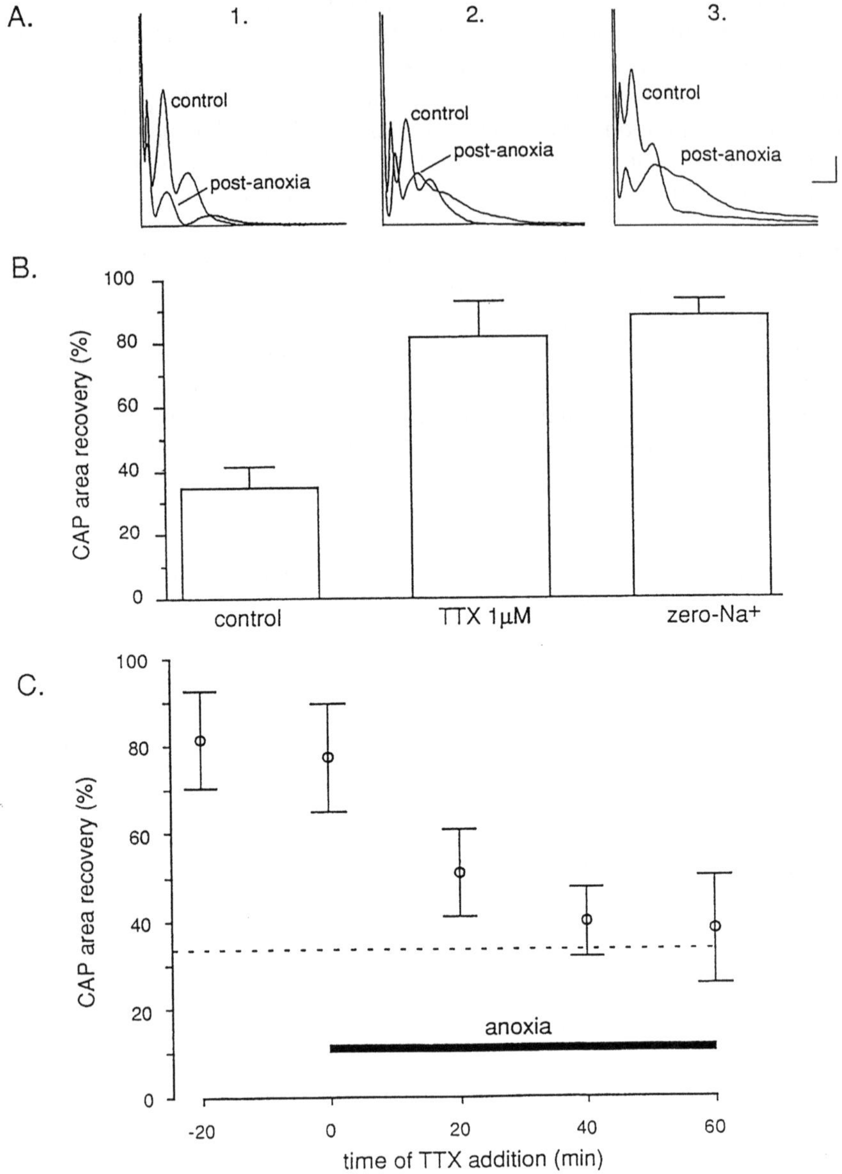
A.
1.
2.
3.
control
post-anoxia
control
post-anoxia
control
post-anoxia
B.
100
80
60
40
20
0
CAP area recovery (%)
control
TTX 1μM
zero-Na+
C.
100
80
60
40
20
0
CAP area recovery (%)
anoxia
-20
0
20
40
60
time of TTX addition (min)

electrogenic and, in fact, membrane current is generated by its operation (Lagnado et al. 1988). For this reason, the exchanger is affected by membrane potential (Blaustein 1988; Lagnado et al. 1988), with membrane depolarization favoring reverse exchange (i.e., Na^+ efflux and Ca^{2+} influx).

Anoxia in WM rapidly leads to an increase in $[K^+]_o$ which would depolarize axons and activate voltage-dependent Na^+ channels, leading to an increase in intracellular Na^+ concentration ($[Na^+]_i$). Both membrane depolarization and the increase in $[Na^+]_i$ favor reverse operation of the Na^+-Ca^{2+} exchanger which would continue until a higher, steady state intracellular $[Ca^{2+}]$ is reached that is in thermodynamic equilibrium with the new conditions; modest increases in $[Na^+]_i$, coupled with membrane depolarization, can cause $[Ca^{2+}]_i$ to rise into the micromolar range (Blaustein 1988; Stys et al. 1991b). Thus, we predicted that blocking the increase in $[Na^+]_i$ during anoxia should protect against irreversible injury.

To test this prediction we examined the effect of bathing in zero Na^+ solution and exposure to tetrodotoxin (TTX), a potent Na^+ channel-blocking agent, on functional recovery of the optic nerve from anoxia. In fact, reducing $[Na^+]_o$ during anoxia or adding TTX dramatically improved CAP recovery (Fig. 3). Intracellular Na^+ appears to accumulate slowly during anoxia, suggesting the presence of persistently open Na^+ channels (Fig. 3C; Stys et al. 1991b, 1992); one class of voltage-dependent Na^+ channels inactivates slowly or not at all (Stafstrom et al. 1985) and such channels are candidates for mediating anoxia-induced increases in $[Na^+]_i$ in WM (Stys et al. 1992).

According to the above schema, blocking the Na^+-Ca^{2+} exchanger during anoxia should improve recovery of the optic nerve after an anoxic challenge. This hypothesis was tested using two different inhibitors of this

Fig. 3A–C. Protective effect of TTX and perfusion in zero-$[Na^+]$ solution on the anoxic optic nerve. **A** Representative CAPs, before and after 60 min of anoxia, are shown in normal solution (*1*), with the addition of 1 μM TTX (*2*), and with perfusion with zero-$[Na^+]$ solution (*3*). These agents were applied beginning 15 min (TTX) or 20 min (zero-$[Na^+]$ solution) before the onset of anoxia and were continued until 15 min after the anoxic period. The TTX-treated optic nerves were perfused for 3 h in normal solution following reoxygenation to allow TTX to wash out as completely as possible before postanoxic measurements were taken. Recovery of CAP area is improved with both TTX and zero-$[Na^+]$ solution, although peak latencies were delayed. **B** Pretreatment with 1 μM TTX or zero-$[Na^+]$ significantly improves CAP recovery (81.5% ± 11%; $p < 0.0001$ and 88% ± 5%; $p < 0.0001$, respectively) compared to normal solution (CAP recovery, 36.7% ± 7%). **C** Effects of 1 μM TTX, applied at various times before or after a 60-min anoxic period, on CAP recovery. Optic nerves were washed for 3 h in normal solution after anoxia to remove as much residual TTX as possible before postanoxic measurements were taken. Control experiments showed that TTX was not completely removed even after 3 h of wash (see Stys et al. 1992). Recovery diminished gradually as the introduction of TTX solution is delayed with respect to the onset of anoxia, suggesting that Na^+ influx begins soon after the onset of anoxia and continues throughout the anoxic period. (Modified from Stys et al. 1992)

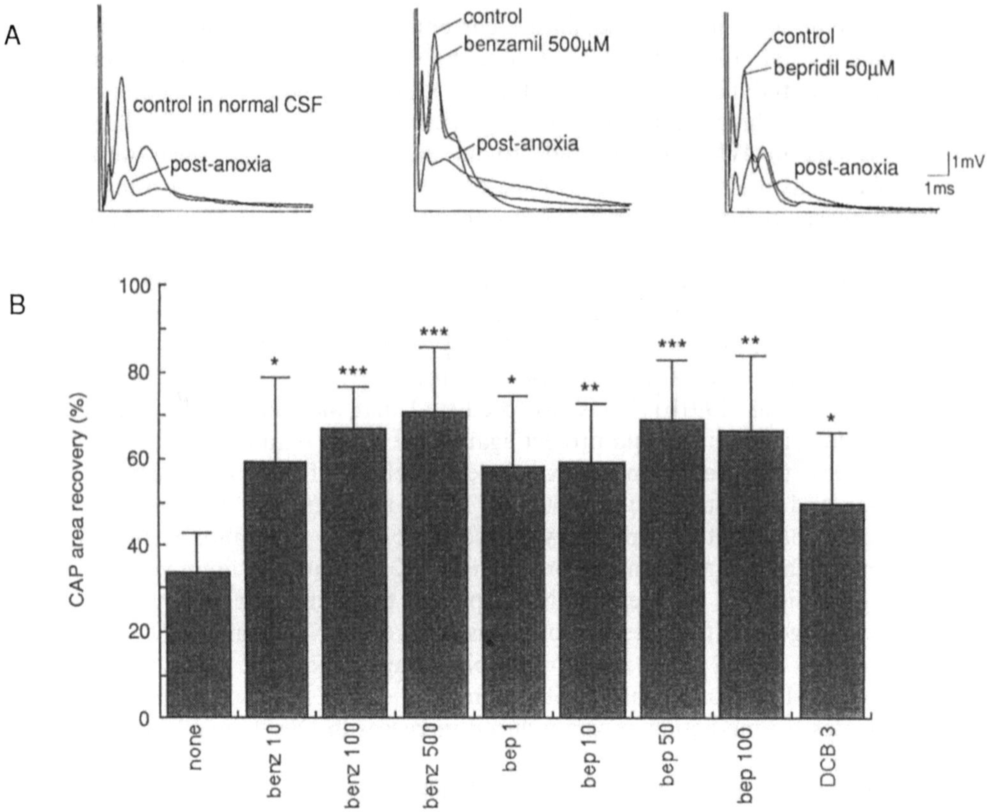

Fig. 4A,B. Effects of Na^+-Ca^+ exchange inhibitors on CAP recovery after anoxia. Benzamil (*benz*), bepridil (*bep*), or dichlorobenzamil (*DCB*) were applied 60 min before the start of anoxia and continued until 15 min after reoxygenation, when normal solution was resumed. Postanoxic activity was measured 1 h after reoxygenation. **A** Benzamil has no effect on the preanoxic response at 10 and 100 μM and reduces CAP area modestly to 61.1% ± 20% of control at 500 μM in a reversible manner. Bepridil and dichlorobenzamil have no effect on preanoxic responses at the concentrations shown. **B** Benzamil significantly improves recovery of CAP area in a dose-dependent manner over a concentration range of 10–500 μM. Bepridil (1–100 μM) and dichlorobenzamil (3 μM) also significantly improve recovery. * $p < 0.01$; ** $p < 0.001$; *** $p < 0.0001$; $n = 7-10$ for each drug and concentration. (Modified from Stys et al. 1992)

transporter, bepridil and benzamil; both drugs markedly improve post anoxic recovery (Fig. 4; Stys et al. 1991b; Stys et al. 1992). Because the amount of CAP recovery seen with blockade of the Na^+-Ca^{2+} exchanger approaches the level of recovery seen when the nerve is bathed in zero Ca^{2+} solution during anoxia (compare Figs. 2B, 4), this suggests that the majority of the damaging Ca^{2+} influx into axons during anoxia probably occurs by way of reverse operation of the exchanger.

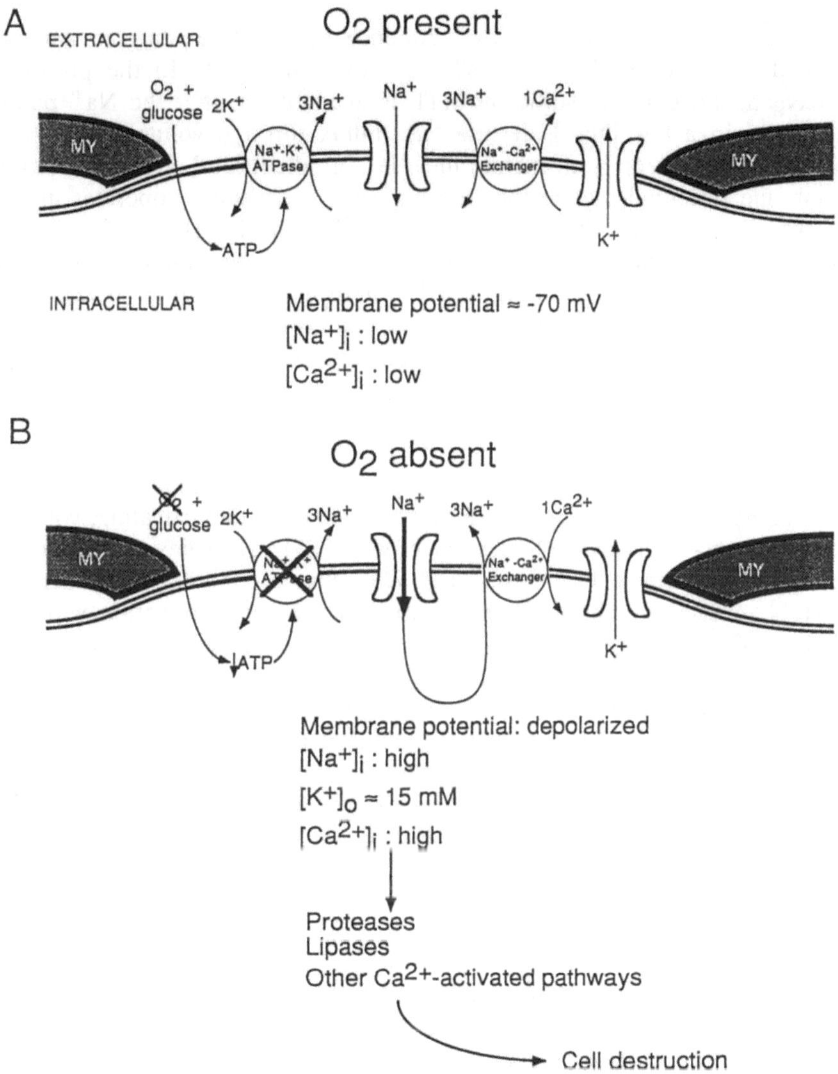

Fig. 5A,B. Key ionic events that lead to intracellular Ca^{2+} accumulation during anoxia at nodes of Ranvier of CNS axons. **A** Under normal conditions there is sufficient oxygen and glucose present to generate enough ATP to operate necessary ion pumps that are critical to maintaining excitability. If $[Na^+]_i$ increases due to action potentials, this increase is easily compensated by enhanced Na^+ pump activity. The steep Na^+ gradient produced by the Na^+ pump in conjunction with high negative membrane potential drives the high capacity Na^+-Ca^{2+} exchanger in the forward direction which helps maintain low $[Ca^{2+}]_i$. There is reason to believe that both the Na^+ pump and the Na^+-Ca^{2+} exchanger might be preferentially located at the nodes, because this is where activity-dependent ion fluxes occur in myelinated axons (*My*). **B** In the absence of oxygen the generation of ATP is seriously reduced because it is now exclusively coming from glycolsis. The shortfall of ATP causes ion gradients to deteriorate and the speed of deterioration is augmented by action potential activity which increases the work load on the Na^+ pump. As the transmembrane Na^+ gradient falls and the membrane depolarizes, the Na^+-Ca^{2+} exchanger is driven to work in reverse and begins loading the axon with Ca^{2+}. The ultimate destruction of cellular integrity is probably mediated by Ca^{2+}-activated destructive enzmyes such as proteases and lipases but these final steps have yet to be clarified. (From Ransom et al. 1993)

A schematic summary of the sequence of events leading to anoxia-induced Ca^{2+} accumulation in WM is shown in Fig. 5. In the presence of oxygen and glucose, sufficient ATP is available so that the Na^+ pump can maintain a low $[Na^+]_i$ despite Na^+ influx through voltage-gated Na^+ channels. The steep transmembrane Na^+ gradient and negative resting membrane potential dictate that the Na^+–Ca^{2+} exchanger operates in the "forward" mode, so that it acts as a Ca^{2+} extruder (Fig. 5A). Under conditions of anoxia, there is a drop in ATP and the Na^+ pump, which may consume about half of all energy used in neurons (Ames et al. 1992), is no longer able to maintain transmembrane gradients of K^+ and Na^+ (Fig. 5B). Increases in $[Na^+]_i$ secondary to influx through Na^+ channels can no longer be corrected. Myelinated axons may be especially susceptible to this cascade of events because of the very high density of Na^+ channels at the nodes of Ranvier (Waxman and Ritchie 1985). We have proposed that this would allow very high nodal Na^+ influx under pathological conditions predisposing to local increases in $[Na^{2+}]_i$ (Waxman et al. 1991). Noninactivating, voltage-dependent Na^+ channels (Stafstrom et al. 1985) are believed to mediate persistent Na^+ influx during anoxia in WM (Stys et al. 1992). The increase in $[K^+]_o$ causes membrane depolarization which, in conjunction with the diminished transmembrane Na^+ gradient, leads to reverse operation of the Na^+–Ca^{2+} exchanger and pathological increase in $[Ca^{2+}]_i$. This outline emphasizes the early ionic events that lead to an increase of $[Ca^{2+}]_i$ and ultimately to irreversible damage (Schanne et al. 1979). Although precise information is lacking, it is believed that the increase in $[Ca^{2+}]_i$ activates a set of destructive biochemical reactions, mediated by enzymes such as proteases and lipases, that are the ultimate cause of cell death (Siesjö and Wieloch 1985; Orrenius et al. 1988).

The anoxia-induced increase in $[Ca^{2+}]_i$ develops via a different mechanism and in a slower fashion in WM than in GM. In WM, Ca^{2+} entry is by a nonsynaptic route, namely, reversal of the Na^+–Ca^{2+} exchanger. The slowness of the Ca^{2+} loading process in WM (Stys et al. 1990a) derives from the fact that the $[Na^+]_i$ must increase before the exchanger runs in reverse, and this latter process appears to proceed gradually (Stys et al. 1992). The concept is emerging that therapeutic strategies for protecting WM from anoxic injury must be tailored to the unique mechanisms leading to irreversible damage in this tissue (Ransom et al. 1993). As illustrated in Fig. 5 the anoxic cascade in WM is multifactorial and complex, offering a multiplicity of sites and avenues for possible therapeutic intervention. Future investigation will focus heavily on developing protective strategies, and this effort seems highly promising based on our increasingly molecular understanding of how anoxia injures the central nervous system.

Acknowledgements. This work is supported by grants from the National Institutes of Health (BRR) and the Medical Research Service, U.S. Department of Veterans Affairs (SGW). P.K.S. was supported by fellowships from the Blinded Veterans Association and Department of Veterans Affairs and by a Centennial Traveling Fellowship from the Medical Research Council of Canada.

References

Ames A III, Li Y, Heher EC, Kimble CR (1992) Energy metabolism of rabbit retina as related to function: High cost of Na^+ transport. J Neurosci 12:840–853
Benveniste H, Drejer J, Shousboe A, Diemer NH (1984) Elevation of the extracellular concentrations of glutamate and aspartate in rat hippocampus during transient cerebral ischemia monitored by intracerebral microdialysis. J Neurochem 43:1369–1374
Blaustein MP (1988) Calcium transport and buffering in neurons. TINS 11:438–443
Chesler M (1990) The regulation and modulation of pH in the nervous system. Prog Neurobiol 34:401–427
Choi DW (1988a) Glutamate neurotoxicity and diseases of the nervous system. Neuron 1:624–634
Choi DW (1988b) Calcium-mediated neurotoxicity: relationship to specific channel types and role in ischemic damage. Trends Neurosci 11:465–469
Connors BW, Ransom BR, Kunis DM, Gutnick MJ (1982) Activity-dependent K^+ accumulation in the developing rat optic nerve. Science 216:1341–1343
Davis P, Ransom BR (1987) Anoxia and CNS white matter: in vitro studies using the rat optic nerve. Soc Neurosci Abstr 13:1634
Fisher CM (1979) Capsular infarcts: the underlying vascular lesions. Arch Neurol 36:65–73
Foster RE, Connors BW, Waxman SG (1982) Rat optic nerve: electrophysiological, pharmacological and anatomical studies during development. Dev Brain Res 3:371–386
Goldman SA, Pulsinelli WA, Clarke WY, Kraig RP, Plum F (1989) The effects of extracellular acidosis on neurons and glia in vitro. J Cereb Blood Flow Metab 9:471–477
Hansen AJ (1985) Effect of anoxia on ion distribution in the brain. Physiol Rev 65:101–148
Hossmann KA, Sato K (1970) Recovery of neuronal function after prolonged cerebral ischemia. Science 168:375–376
Kass IS, Lipton P (1982) Mechanisms involved in irreversible anoxic damage to the in vitro rat hippocampal slice. J Physiol (Lond) 332:459–472
Kimelberg HK, Ransom BR (1986) Physiological and pathological aspects of astrocytic swelling. In: Fedoroff S, Vernadakis A (eds) Astrocytes, vol 3. Academic, Orlando, pp 129–166
Kraig RP, Pulsinelli WA, Plum F (1985) Hydrogen ion buffering during complete brain ischemia. Brain Res 342:281–190
Kraig RP, Petito CK, Plum F, Pulsinelli WA (1987) Hydrogen ions kill brain at concentrations reached in ischemia. J Cereb Blood Flow Metab 7:379–386
Lagnado L, Cervetto L, McNaughton PA (1988) Ion transport by the Na-Ca exchanger in isolated rod outer segments. Proc Natl Acad Sci USA 85:4548–4552
Orrenius S, McConkey DJ, Jones DP, Nicotera P (1988) Ca^{2+}-activated mechanisms in toxicity and programmed cell death. ISI Atlas of Science: Pharmacology 2:319–324
Paulson OB, Newman EA (1987) Does the release of potassium from astrocyte endfeet regulate cerebral blood flow? Science 237:896–898
Plum F (1983) What causes infarction in ischemic brain? Neurology 33:222–233
Pulsinelli WA, Waldman S, Rawlinson D, Plum F (1982) Moderate hyperglycemia augments ischemic brain damage: a neuropathologic study in the rat. Neurology 32:1239–1246
Ransom BR, Philbin DM (1992) Anoxia-induced extracellular ionic changes in CNS white matter: the roles of glial cells. Can J Physiol Pharmacol 70:181–189
Ransom BR, Yamate CL, Connors BW (1985) Activity-dependent shrinkage of extracellular space in rat optic nerve: a developmental study. J Neurosci 5:532–535

Ransom BR, Carlini WG, Connors BW (1986) Brain extracellular space: developmental studies in rat optic nerve. Ann NY Acad Sci 481:87–105
Ransom BR, Stys PK, Waxman SG (1990a) The pathophysiology of anoxic injury in CNS white matter. Stroke 21:[Suppl III] 52–57
Ransom BR, Waxman SG, Davis PK (1990b) Anoxic injury of CNS white matter: protective effect of ketamine. Neurology 40:1399–1403
Ransom BR, Walz W, Davis PK, Carlini WG (1992) Anoxia-induced changes in extracellular K^+ and pH in mammalian central white matter. J Cereb Blood Flow Metab 12:593–602
Ransom BR, Stys PK, Waxman SG (1993) Anoxic injury of central myelinated axons: ionic mechanisms and pharmacology. In: Waxman S (ed) *Molecular and cellular approaches to the treatment of neurological disease*. Raven, New York, pp 121–151
Schanne FA, Kane AB, Young EE, Farber JL (1979) Calcium-dependence of toxic cell death: a final common pathway. Science 206:700–702
Schlaepfer WW (1977) Structural alterations of peripheral nerve induced by the calcium ionophore A23817. Brain Res 136:1–9
Schwartz EA, Tachibana M (1990) Electrophysiology of glutamate and sodium cotransport in a glial cell of the salamander retina. J Physiol (Lond) 426:43–80
Siesjö BK (1981) Cell damage in the brain: a speculative synthesis. J Cereb Blood Flow Metab 1:155–185
Siesjö BK, Wieloch T (1985) Brain injury: neurochemical aspects. In: Becker D, Povlishock JT (eds) Central nervous system trauma status report. NIH, NINCDS, Bethesda, pp 513–532
Somjen GG, Aitken PG, Balestrino M, Herreras O, Kawasaki K (1990) Spreading depression-like depolarization and selective vulnerability of neurons: a brief review. Stroke 21:III-179–III-183
Stafstrom CE, Schwindt PC, Chubb MC, Crill WE (1985) Properties of persistent sodium conductance and calcium conductance of layer V neurons from cat sensorimotor cortex. J Neurophys 53:153–170
Stys PK, Ransom BR, Waxman SG, Davis PK (1990a) The role of extracellular calcium in anoxic injury of mammalian white matter. Proc Natl Acad Sci USA 87:4212–4216
Stys PK, Ransom BR, Waxman SG (1990b) Effects of polyvalent cations and dihydropyridine calcium channel blockers on recovery of CNS white matter from anoxia. Neurosci Lett 115:293–299
Stys PK, Ransom BR, Waxman SG (1991a) Compound action potential of nerve recorded by suction electrode: a theoretical and experimental analysis. Brain Res 546:18–32
Stys PK, Waxman SG, Ransom BR (1991b) Na^+-Ca^{2+} exchanger mediates Ca^{2+} influx during anoxia in mammalian CNS white matter. Ann Neurol 30:375–380
Stys PK, Waxman SG, Ransom BR (1992) Ionic mechanisms of anoxic injury in mammalian CNS white matter: role of Na^+ channels and Na^+-Ca^{2+} exchanger. J Neurosci 12:430–439
Tang CM, Dichter M, Morad M (1990) Modulation of the N-methyl-D-aspartate channel by extracellular H^+. Proc Natl Acad Sci USA 87:6445–6449
Taylor MD, Mellert TK, Parmentier JL, Eddy LJ (1985) Pharmacological protection of reoxygenation damage to in vitro brain slice tissue. Brain Res 347:268–273
Waxman SG, Ritchie JM (1985) Organization of ion channels in the myelinated nerve fiber. Science 228:1502–1507
Waxman SG, Ransom BR, Stys PK (1991) Non-synaptic mechanisms of Ca^{2+}-mediated injury in CNS white matter. TINS 14:461–468
Waxman SG, Black JA, Stys PK, Ransom BR (1992) Ultrastructural concomitants of anoxic injury and early post-anoxic recovery in rat optic nerve. Brain Res 574:105–119
Waxman SG, Black JA, Ransom BR, Stys PK (1993) Protection of the axonal cytoskeleton in anoxic optic nerve by decreased extracellular calcium. Brain Res (in press)
Young W (1985) Blood flow, metabolic and neurophysiological mechanisms in spinal cord injury. In: Becker D, Povlishock JT (eds) Central nervous system trauma status report. NIH, NINCDS, Bethesda, pp 463–473

In Vivo Nuclear Magnetic Resonance Observation of Cerebral Metabolism

T.R. Brown

Introduction

Development of increasingly sophisticated ways to obtain noninvasive information from the brain is allowing us to probe its extracellular and intracellular environment to a greater and greater degree. Position emission tomography (PET) and single photon emission computed tomography (SPECT) have provided important information on cerebral blood flow, metabolism, and receptor distributions. Here we review the recent improvements in nuclear magnetic resonance (NMR) spectroscopy which allow an image of the energy metabolism of brain to be obtained noninvasively. First, the paper outlines the techniques used for localization and their reliability. These are then illustrated with example from brain observations.

Localization Techniques

The need for well-defined procedures for spectral localization for in vivo studies is hard to overemphasize. Unlike standard high-resolution NMR studies in chemistry, wherein the sample is homogeneous and is placed entirely within a relatively uniform radiofrequency (rf) field, studies in vivo necessarily study an inhomogeneous tissue, often with a coil which produces a spatially variable rf field. These circumstances make it very difficult to obtain a clear understanding of the relationship between observed spectra and physiology. The more precise and controlled the spectral localization obtained, the better any correlations which exist can be determined.

All methods of spectral localization depend upon a gradient in a magnetic field because the wavelength of the rf frequencies used, 10–100 MHz, are in the 3–30 m range. Thus no localization can be provided by the electromagnetic field directly, as in a light microscope; instead, all localization techniques depend upon controlled spatial variations in either the static magnetic field or the rf magnetic field.

Fox Chase Cancer Center, Department of Nuclear Magnetic Resonance and Medical Spectroscopy, Philadelphia, PA 19111, USA

Cerebral Ischemia and Basic Mechanisms
Ed. by A. Hartmann, F. Yatsu, and W. Kuschinsky

Perhaps the simplest localization scheme, that of a surface coil, was developed a number of years ago in Oxford by Ackerman et al. [1]. In this technique one relies on the localized spatial distribution of rf field from a coil, typically a circular loop of conductor, to excite and detect only those spins within a certain field strength. Since the rf field from a loop falls off rapidly beyond one radius, most of the signal comes from a volume of dimensions roughly equal to the coil. This technique has been used extensively, primarily because of its simplicity. No other gradients or complex pulse schemes are required. Its primary difficulty is the imprecision of the localization. This is particularly troublesome in brain studies as surface layers contribute significantly. Depending on the precise placement of the coil with respect to the sample, different volumes can contribute to the signal. In addition, if stronger rf pulses are used to try to excite spins deeper in the sample the volume contributing to the signal can become quite large.

Gradients in the static magnetic field can be used as well. These phase encode the positions of the spin with a short gradient pulse [2, 3] or, alternatively, use a selective rf pulse in the presence of a gradient to excite only a limited region of the sample [4–6]. By the application of multiple gradients in orthogonal directions localization can be achieved in three dimensions. Generally, the frequency-selective techniques can only obtain spectra for one or at most a few localized regions simultaneously, although complex pulse schemes do exist for multiple regions [7]. Phase-encoding techniques, on the other hand, produce localized spectra from the entire sample under study. Analogous phase-encoding techniques can be used with surface coils, although they are more complex [8].

The most widely used frequency-selective techniques are ISIS [4] and STEAM [5, 6], the former primarily for phosphorus and the later primarily for proton spectroscopy. These techniques localize a singly species of spin to a particular region very well. Because ISIS involves the addition and subtraction of different acquisitions there are some problems with motion sensitivity. However, the sizes of the localized region in phosphorus spectroscopy are generally large enough so that small in vivo motions are not a serious problem. STEAM, since it localizes the signal in a single acquisition, does not suffer from this type of problem. There is one artifact, however, from which this entire class of techniques suffers: namely that different compounds, which resonate at different frequencies, are localized to different regions in space. To understand this effect consider two compounds shifted in frequency by Δf. In a linear gradient the same frequency-selective pulse will cause the two species to be localized in two different regions, shifted in position by $\Delta f/\gamma G$ where γ is the gyromagnetic ratio of the spins in Hz/G and G is the strength of the gradient in G/cm. Using typical gradient values from human studies the shift for the proton case is relatively minor, primarily because the chemical shift range of protons is small (5–10 ppm). In the case of ^{31}P (and even more so in the case of ^{13}C), since the chemical shift range is 20–30 ppm (100–200 ppm for ^{13}C), the differential shift be-

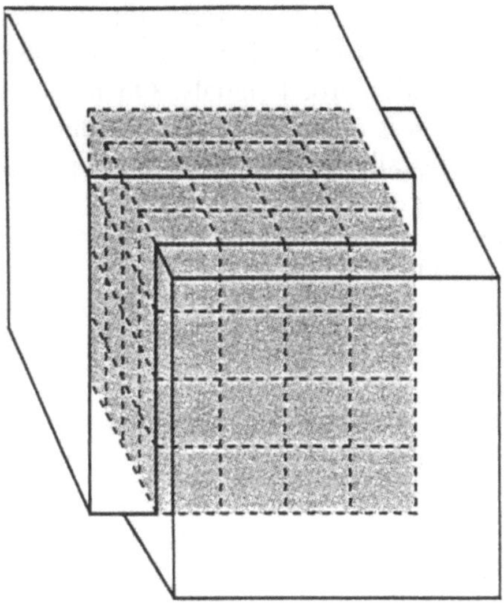

Fig. 1. Schematic illustration of the amount of common volume in three dimensions with a 20% shift in each axis

tween two compounds can be significant. For example, two compounds separated by 20 ppm will be shifted 1 cm with respect to each other in a gradient of 0.3 G/cm (3 mT/m) at a field of 1.5 T. To illustrate the consequences of this, consider Fig. 1, which schematically demonstrates two interleaved boxes coming from an ISIS or STEAM experiment in which the two spectral regions are shifted 1 cm out 5 cm with respect to one another in each dimension. This 20% linear shift causes nearly 50% of the two regions to be different (64/125 of the volumes are in common). This is a potentially serious problem if one is trying, for example, to compare the levels of inorganic phosphate to those of β-ATP; the two compounds are then coming from substantially different volumes. For smaller voxel sizes the problem is even worse; a 27-ml voxel ($3 \times 3 \times 3$), would have less than 1/3 (8/27) of the volumes in common.

Phase encoding techniques such as chemical shift imaging (CSI) [2, 3] do not suffer from this artifact because they only use the gradient as a short gradient pulse following the rf pulse to label the location of a spin in space with a phase factor. It can be shown that the mathematical description of this phase encoding is that of the Fourier transform; that is, the free induction decay (FID) of the spins in a sample at a particular time in a gradient, constant or time dependent, is the Fourier transform of the spatial spin distribution in the sample at a particular point in the transform apace [2], generally denoted as k space. An estimate of the original spatial distributions is then obtained by a finite Fourier transformations in both time and space of this dataset.

Techniques of this type suffer from artifacts as well. First, due to the gradient pulse which immediately follows the rf excitation, the early time points in the FID are unusable. This degrades broad signals. On modern machines the loss of the early timepoints is less than 1 ms, corresponding to line widths greater than 300 Hz. This is usually broader than most peaks of interest, although certain phospholipid peaks can be observed at such widths. In such cases care must be taken in quantitation. Another potential problem is the finite sampling in k space, which can produce artifacts in the reconstruction related to both the maximum spread and the interval between samples. These artifacts are well known from the standard theory of finite Fourier transforms [9] but are worth reviewing as they as so fundamental to understanding the nature of the localized spectra in CSI experiments. A finite sampling of k space produces effects on the reconstructed spatial distribution of the spins which can be represented by a point spread function (PSF), which mathematically describes how the observed spectrum at a particular location is related to the actual distribution.

This relationship takes the form of a convolution so that

$$\rho_{\mathrm{rec}}(\vec{r}) = \int \rho(\vec{s})\mathrm{PSF}(\vec{r} - \vec{s})\mathrm{d}\vec{s}$$

where ρ_{rec} is the reconstructed distribution.

For uniform sampling in k space from $-N\pi/L$ to $(N - 1)\pi/L$ in steps of $2\pi/L$, corresponding to an object of size L divided into N points, the full three-dimensional PSF is given by

$$\mathrm{PSF}(\vec{\Delta}) = \mathrm{psf}(\vec{\Delta}_x)\mathrm{psf}(\vec{\Delta}_y)\mathrm{psf}(\vec{\Delta}_z)$$

where Δ_i is the component of Δ in the ith direction. The function in one dimension is

$$\mathrm{psf}_N(\Delta) = \mathrm{e}^{-i(\pi/L)\Delta}\frac{\sin(N\pi/L)\Delta}{\sin(\pi/L)\Delta}$$

This function is plotted as a function of Δ in Fig. 2 for $N = 16$. As can be seen, it has a large peak near $\Delta = 0$ with smaller oscillations extending to the edge of the field of view L. These oscillations cause the well-known "sinc wiggles" found in finite Fourier transforms, and otherwise referred to as "Gibbs oscillations." Depending on the original distribution $\rho(\mathbf{r})$, the estimated or reconstructed distribution $\rho_{\mathrm{rec}}(\mathbf{r})$ will have varying degrees of mixing of signal from outside the region. Since the function is oscillatory these are generally small but need to be kept in mind for precise work.

Applications of CSI

It is simplest to illustrate CSI in one dimension. Figure 3 shows a schematic diagram of a simple one-dimensional experiment, consisting of a surface coil and a cylinder of inorganic phosphate aligned along the axis of the coil. In

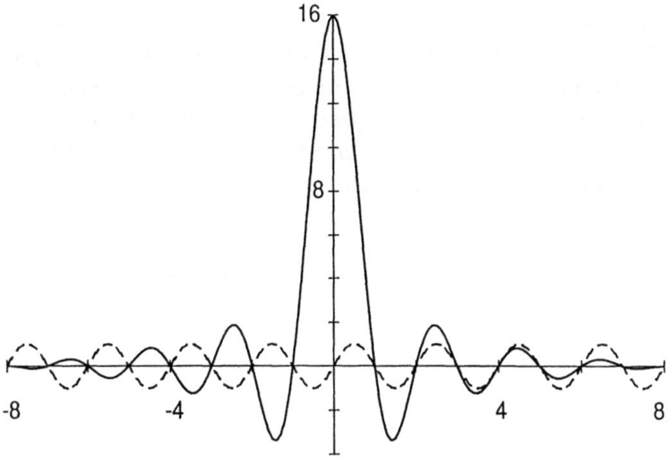

Fig. 2. Point spread function for a finite Fourier transform with $N = 16$. *Solid curve* is real part, *dotted curve* is imaginary part

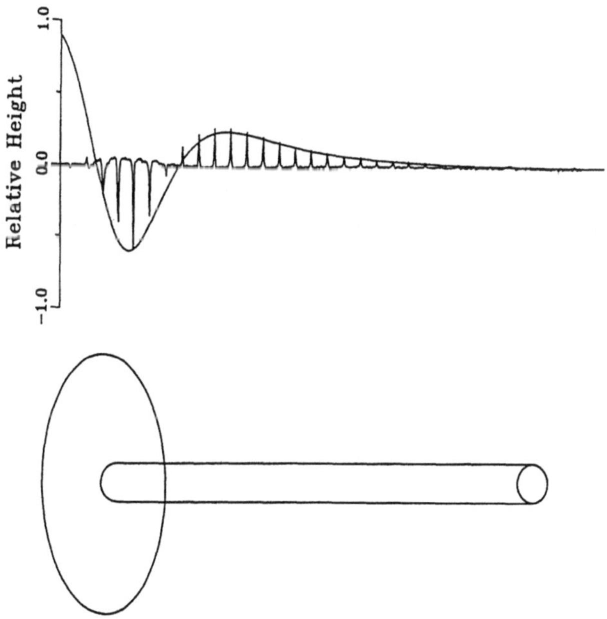

Fig. 3. *Lower*, schematic illustration of a column phantom coaxial with a loop surface coil. *Upper*, one-dimensional. CSI experiment with 32 phase encode steps of the lower phantom. The *solid curve* is the theoretically expected signal. The individual peaks are the experimentally observed signal

this experiment 32 phase-encoding steps were used with the pulse sequence consisting of a hard rf pulse (540° at the center of the coil) followed by a short gradient pulse of amplitude and width corresponding to a spacing of 5 mm for a total field of view of 16 cm. The spectra were acquired on a 1.5 T Siemens Magnetom at 25.6 Mhz. The entire data set was collected and than doubly Fourier transformed in time and space. Each spectrum was phased identically and the peak region surrounding the phosphate resonance was selected from each spectrum and plotted next to one another along the axis in Fig. 3. These signals were compared to the theoretically expected ones by plotting on the same figure $B_1(y) \sin\gamma B_1(y)T_o$ where $B_1(y)$ is the rf field strength as a function of the distance, y, along the axis, γ is the gyromagnetic ratio, and T_o is the length of the rf pulse. For the simple loop $\gamma B_1(y)T_o = 540° a^3/(y^2 + a^2)^{3/2}$. This function was calculated as a function of position (5 mm/spectrum) along the axis and scaled in the vertical direction to fit the peak heights; it is plotted as the solid line in Fig. 3. As can be seen, the agreement is excellent. The only adjustable parameter here was the vertical scaling of the theoretical curve since all the other parameters were known. These results demonstrate the ability of CSI phase-encoding measurements to provide excellent localization in space, as would be expected from their common use in the imaging field.

In two and three spatial dimensions, it becomes possible to consider alternative ways of representing this spectral information, i.e., more than simply arrays of spectra. Since we have continguous localized spectra the obvious possibility is to use an image format to represent the spatial distribution. The primary difficulty with such a procedure is the determination of the peak areas or other variables of interest from the spectra in an automatic fashion. Traditional methods of measuring area often rely on hand integration, which is unsuitable in this case since we are dealing with several hundred spectra. In order to overcome this, Dr. Sarah Nelson and I have developed a program called PIQABLE (peak identification quantification and automatic baseline estimation), which is capable of nonparametrically dividing a spectrum into peak regions, slowly varying background and high-frequency noise [10]. This procedure can be applied automatically to the CSI spectra. The results of this analysis can then be presented as images. Because there are usually only 8 × 8 or 16 × 16 steps we have found these data can be much better visualized if they have been zero filled and spatially filtered so as to provide some spatial smoothing.

These techniques can now be applied to the brain, where high-quality rf coils can be made to encompass the entire head. In fact, it is possible to design dual resonant coils capable of operation at both ^1H and ^{31}P frequencies [11]. Using such coils three-dimensional ^{31}P CSI spectra from the entire human head can be obtained with voxel sizes of less than 8 cm^3 in observation times on the order of 100 min. When lower spatial resolution can be tolerated high-quality ^{31}P spectra can be obtained from larger regions of brain tissue in 20–30 min of observation (12). This enables regional mea-

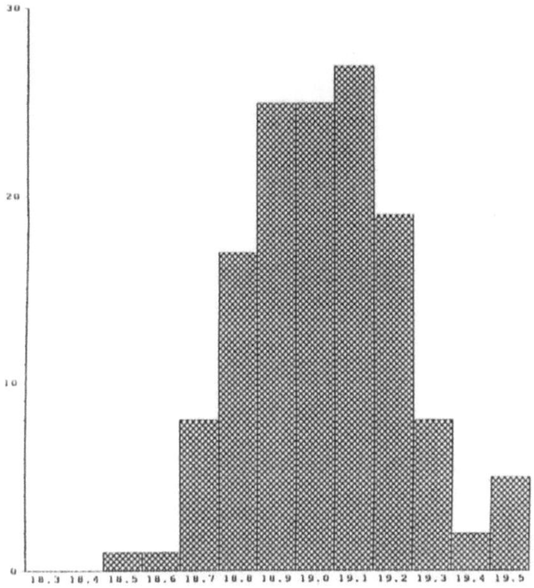

Fig. 4. A histogram of the observed chemical shift of β-ATP in human brain. The data are from five individuals and consist of measurements from 138 different voxels. The mean of the distribution is -19.02, SD = 0.20

surements of pH and pMg as well as ATP, PCr, and other phosphorylated metabolites. Free Mg^{2+} levels in the brain appear to be substantially lower than in many other tissues. Mg^{2+} levels are determined by measuring the β-ATP peak position, which is well known to titrate with Mg^{2+}. A histogram of the position of the β-ATP peak with respect to PCr (PCr = -2.52 ppm) from normal volunteers is shown in Fig. 4. The mean is approximately -19 ppm, which corresponds to a free Mg^{2+} level of about $250\,\mu M$, assuming a standard titration curve for brain tissue [14]. This concentration value is observed to increase both in brain tumors as well as cachexic patients. It is too early to assign significance to these changes.

A significant advantage can be gained by the use of a dual-tuned resonator to apply rf power at the proton-resonant frequency while observing phosphorus. This irradiation causes two major changes in the ^{31}P spectrum. First, the direct J coupling between a phosphate and directly bonded nearby protons splits the phosphorus lines into doublets, triplets, or higher order multiplets. This structure can be collapsed into a single line by sufficient irradiation power at the proton frequency during ^{31}P data acquisition. It is well known in high resolution NMR that decoupling simplifies spectra and increases sensitivity because many lines are collapsed into one. In addition to the changes in spectral shape, under certain conditions, changes in the

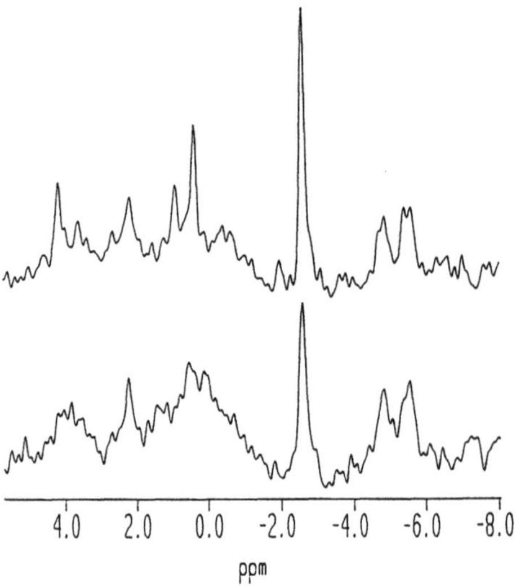

Fig. 5. A comparison of proton coupled and decoupled ^{31}P spectra from $42\,cm^3$ of human brain acquired in 35 min

steady state magnetization in the protons are reflected in the changes in steady state magnetization of phosphorus due to the dipolar spin–spin interaction between the ^1H and ^{31}P spins. This change is known as the nuclear overhauser effect (NOE) enhancement [14], and can cause a substantial increase in the phosphorus magnetization when protons are saturated.

In the human brain the effect of proton irradiation is demonstrated in Fig. 5, which compares ^{31}P spectra from a single voxel of $42\,cm^3$ acquired in 35 min with and without proton irradiation [15]. Details of the irradiation are given in the figure legend; however, it is obvious that the phosphomonoester and phosphodiester regions of the spectrum are considerably sharpened due to the elimination of the residual proton J coupling. In addition there is an NOE enhancement for some peaks (e.g., approximately 50% for PCr). The additional sensitivity that proton decoupling and NOE enhancement provides has enabled metabolic images of the human brain at a resolution less than 2 cm to be constructed. Figure 6 shows four coronal slices of the PCr intensity in a normal human volunteer taken from a dataset acquired in 115 min with a voxel size of $6\,cm^3$. As can be clearly seen, at this resolution the basal ganglia are observed as distinct anatomical objects.

Fig. 6.a MRI coronal images of a human volunteer. **b** Phosphocreatine CSI images of the same sections. Note the increased intensity in the regions of the basal ganglia, as well as the superficial muscles outside the skull. These data were acquired in 115 min at a resolution of 18 mm

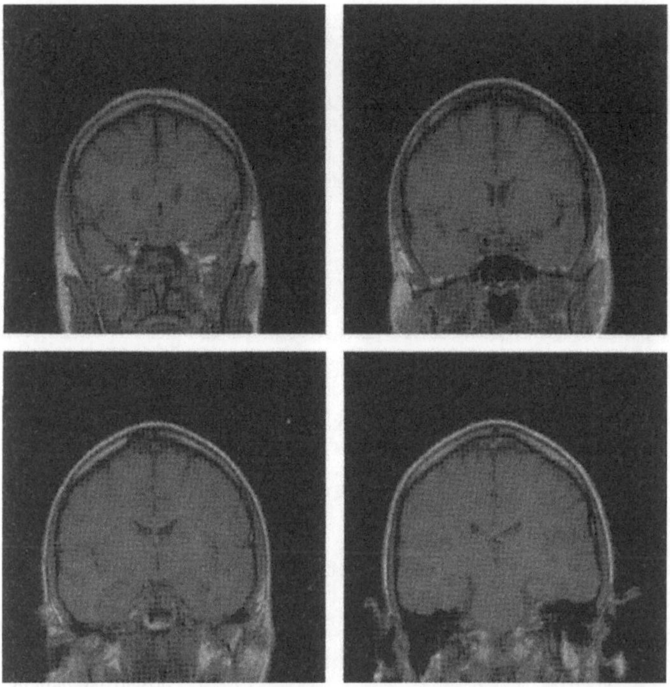

a

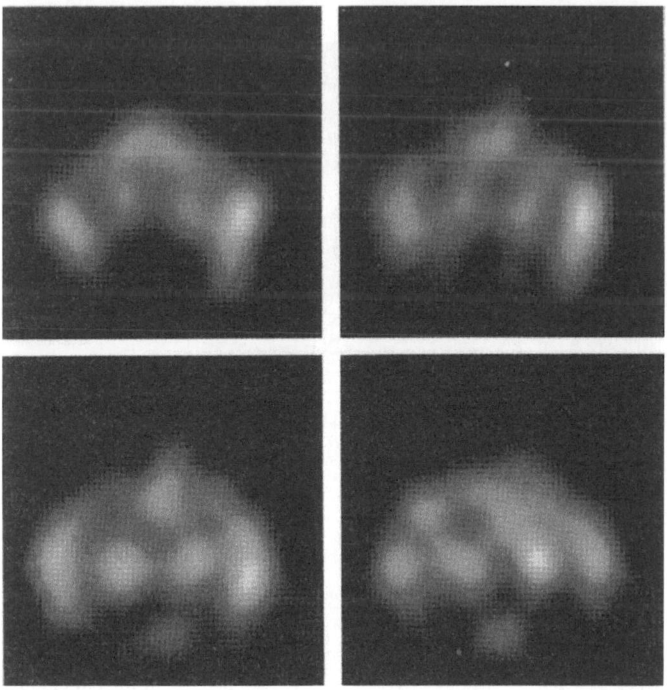

b

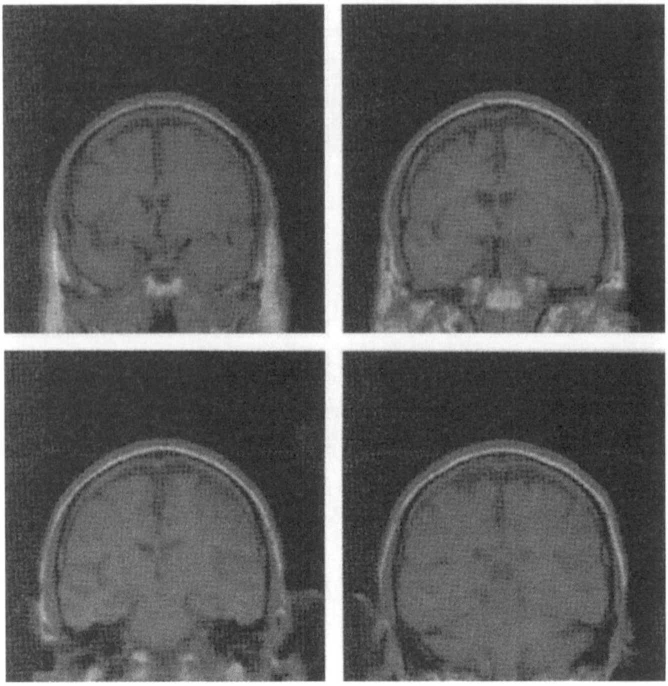

Fig. 7. T1-weighted MRI of 5-year-poststroke patient

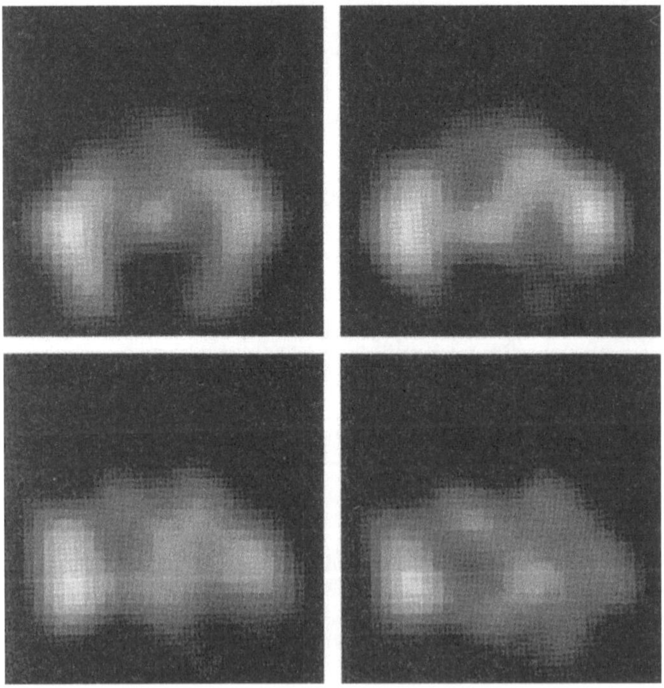

Fig. 8. PCr metabolic images of the patient in Fig. 7

In preliminary observations a patient with a 5-year-old infarct in the right hemisphere was observed twice with ^{31}P spectroscopy at a voxel size of 27 cm^3. A relatively small scar (1 × 2 cm^2) could be observed on T1-weighted magnetic resonance images (upper left panel, Fig. 7) but little other sign of the infarct was present. In the PCr metabolic image, however, a substantial deficit not centered on the scar could be observed in the right hemisphere running from the parietal to occipital lobes. The consequences and implication of this are unclear, although the observation was repeated twice with the deficit observed in the same location each time. The proton images and four coronal PCr images are shown in Figs. 7 and 8.

Conclusions

This paper has presented a brief overview of localization techniques with emphasis on CSI and some example of its use in the brain. The techniques should allow advances in observations of metabolism in different pathological conditions. Stroke and myocardial infarcts have been observed with single voxel localization techniques which are not capable of providing as broad a picture as the multivoxel CSI techniques discussed here. With the advent of clinical instrumentation capable of implementing these techniques a much wider set of applications should ensue over the next few years. The ability to pinpoint metabolic changes associated with pathological states or physiological variables should provide us with new insights into both normal and disease processes in humans.

Acknowledgements. This work was supported by NIH grants PO1-CA41708 and RO1-CA54339 and Siemens Medical Systems. I would like to thank Joseph Murphy-Boesch, Radka Stoyanova, Ravi Srinivasan, Tamela Willard, and Lucas Carvajel for their help and involvement in the work described here.

References

1. Ackerman JJH, Grove TH, Wong GG et al. (1980) Mapping of metabolites in whole animals by ^{31}P using surface coils. Nature 283:167
2. Brown TR, Kincaid BM, Ugurbil K (1982) NMR chemical shift imaging in three dimensions. Proc Natl Acad Sci 79:3523–3526
3. Maudsley AA, Hilal SK, Perman WH, Simon HE (1983) Spatially resolved high resolution spectroscopy by four-dimensional NMR. J Magn Reson 51:147–152
4. Ordidge RJ, Connelly A, Lohman JAB (1986) Image-selective in vivo spectroscopy (ISIS). A new technique for spatially selective NMR spectroscopy. J Magn Reson 66:283–294
5. Frahm J, Bruhn H, Gyngell MNL, Merboldt KD, Hanicke W, Sauter R (1989) Localized high-resolution proton NMR spectroscopy using stimulated echoes: initial applications to human brain in vivo. Magn Reson Med 9:79–93
6. Luyten PR, Marien AJ, Heindel W, van Gerwen PH, Herholz K, den Hollander JA et al. (1990) Metabolic imaging of patients with intracranial tumors: H-1 MR spectroscopic imaging and PET. Radiology 176(3):791–799
7. Goelman G, Walter G, Leigh JS (1991) Hadamard spectroscopic imaging technique on the human calf muscles. Tenth annual meeting of the Society of Magnetic Resonance in Medicine, San Francisco, p 551

8. Cox SJ, Styles P (1980) Towards biochemical imaging. J Magn Reson 40:209–212
9. Bracewell RN (1978) The fourier transform and its applications. McGraw-Hill, New York
10. Nelson SJ, Brown TR (1987) A new method for automatic quantification of 1-D spectra with low signal to noise ratio. J Magn Reson 75:229–243
11. Murphy-Boesch J, Srinivasan R, Brown TR (1990) P-31-H-1 dual-tuned quadrature bird-cage head coil for clinical imaging and proton-decouploed spectroscopy at 1.5 T. 76th Scientific Assembly and Annual Meeting, Radiological Society of North America, Chicago, Illinois, p 309
12. Vigneron DB, Nelson SB, Murphy-Boesch J, Kelley DAC, Kessler HB, Brown TR, Taylor JS (1990) Chemical shift imaging of human brain: axial, sagittal, and coronal P-31 metabolite images. Radiology 177:643–649
13. Taylor JS, Vigneron DB, Murphy-Boesch J, Nelson SJ, Kessler HB, Coia L, Curran W, Brown TR (1991) Free magnesium levels in normal human brain and brain Tumors: ^{31}P chemical-shift imaging measurements at 1.5 T. Proc natl Acad Sci 88:6810–6814
14. Noggle JH, Shirmer RE (1971) The nuclear overhauser effect: chemical applications. Academic, New York
15. Murphy-Boesch J, Stoyanbova R, Srinivasan R, Willard T, Vigneron D, Nelson SJ, Taylor JS, Brown TR (1992) Proton-decoupled P-31 chemical shift imaging of the brain. NMR Biomed (in press)

Effects of Extracellular Acidosis on Glial Cell Intracellular pH: Evidence for a Glial Spatial H^+-Buffering Mechanism?

C. Klawe, C. Volk, and O. Kempski

Introduction

Acidosis and in particular lactacidosis from anaerobic metabolism are considered of primary significance among the consequences of cerebral ischemia, seizures, and head injury (Siesjö 1981). A marked decrease in brain tissue pH has been demonstrated in cerebral ischemia – even more so in hyperglycemic subjects – where lactic acid may accumulate to 20–30 m*M* and higher concentrations (Rehncrona et al. 1980; Katsura et al. 1991). Tissue pH may drop to pH 5.5 (Chopp et al. 1988). In consequence, acidosis has since long been suspected as a mediator of brain damage (Siesjö 1981).

Using an in vitro system which permitted a close control of the extracellular (EC) environment together with a highly accurate detection of cell volume changes (Kempski et al. 1983), we have demonstrated that EC acidosis (Kempski et al. 1988) and lactacidosis (Staub et al. 1990) cause considerable volume increases in glial cells and may therefore be responsible for the early swelling of glial cells during and after cerebral ischemia, for example, cytotoxic edema. Acidosis-induced swelling was demonstrated in C6 glioma cells, used as a model cell line with glial characteristics, and in astrocytes from primary culture (Kempski et al. 1988; Staub et al. 1990). Swelling was seen at pH of 6.8 or lower, reduced by amiloride, an inhibitor of the Na^+/H^+ antiporter, and was prevented in the absence of Na^+ and HCO_3^- ions, i.e., in hydroxyethylpiperazine ethanesulfonic acid (HEPES) buffered media in which sodium ions had been repaced by choline (Kempski et al. 1988). 4-Acetamino-4′-isothiocynatostilbene-2,2′-disulfonic acid reduced acidosis-induced swelling as did pretreatment with acetazolamide, an inhibitor of carbonic anhydrase.

From these data a concept of the mechanisms involved in acidosis-induced glial swelling has been developed (Kempski et al. 1988) based on the fact that acid equivalents leaking from an ischemic focus are buffered by EC HCO_3^-, causing the production of CO_2, which can readily diffuse into glial cells. Here, carbonic anhydrase facilitates the formation of H^+ and

Institute for Neurosurgical Pathophysiology, Johannes Gutenberg-Universität Mainz, Langenbeckstr. 1, 55131 Mainz, Germany

Cerebral Ischemia and Basic Mechanisms
Ed. by A. Hartmann, F. Yatsu, and W. Kuschinsky

HCO_3^-. Hence the cell acidifies. To maintain a normal intracellular (IC) pH antiporter systems such as the Na^+/H^+- or Cl^-/HCO_3^- antiporter, are activated. During the normalization of IC pH, the cell takes up sodium ions and possibly chloride, and by this increase in IC osmolality water is drawn into the cell. As a consequence cell volume increases. A futile cycle may develop by the renewed generation of carbon dioxide. Similar mechanisms have been proposed by others (Jakubovicz and Klip 1989).

This proposed mechanism of glial swelling, which might occur in the ischemic penumbra in particular, involves the acidification of the IC compartment by CO_2 as a central parameter. In EC acidosis in the absence of carbon dioxide, IC pH should therefore be less affected. This was suggested by experiments using HEPES-buffered media in which cell swelling was inhibited. What had been missing so far was confirmation of this conclusion by measurements of IC pH changes during EC acidosis. The current data illustrate that this model, in a modified version, may explain how glial cells fulfill a spatial H^+ buffering function.

Methods

Glial Cell Culture. Experiments were performed with C6 glioma cells (Benda et al. 1968) and astrocytes from primary culture. The latter were obtained from 3-day-old neonatal rats employing a method adapted from Frangakis and Kimelberg (Frangakis and Kimelberg 1984; Staub et al. 1990). In brief, forebrains were freed from meninges and choroid plexus under a dissecting microscope. The tissue was chopped into fine pieces and repeatedly digested using dispase. Suspended cells were sedimented and seeded into petri dishes. The purity of the preparation was assessed by glial fibrillar acidic protein stained with peroxidase-antiperoxidase (DAKO). Both cell types were cultured in monolayers in petri dishes using Dulbecco's minimal essential medium supplemented with penicillin G (100 IU/ml) and streptomycin (50 μg/ml). Cell growth and differentiation was stimulated by 10% fetal calf serum. Cells were cultured at 37°C in a humidified atmosphere of 5% CO_2 and 95% room air. The culture medium was changed and C6 cells were subcultivated daily.

Measurement of Intracellular pH. Acetoxymethyl ester of bis-carboxyethyl carboxy fluorescein (BCECF) was used to measure IC pH on a quantitative basis. Cells were loaded with BCECF-AM (*4 μM*, 15 min), a hydrophobic ester which readily passes cell membranes. In the cell, BCECF-AM is split by esterases to form the active, hydrophilic, pH-sensitive dye BCECF, which is trapped intracellularly. The adherent cells were illuminated (50 Hz) alternately with two wavelengths using monochromators, a chopper disk, a xenon lamp, and a computer-controlled shutter (AMKO-LTI, Tornesch, FRG). Emitted light was collected through an inverted microscope (Leica,

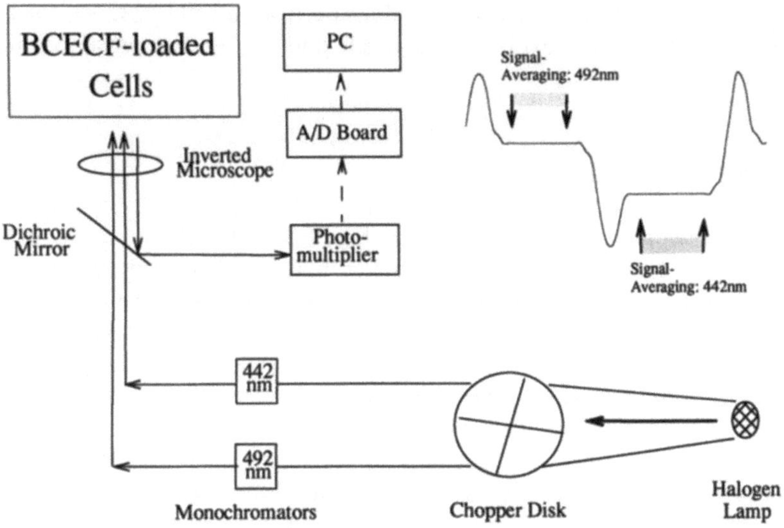

Fig. 1. Experimental setup to measure IC pH in BCECF-loaded cells. Cells are illuminated alternately with light from two monochromators allowing the passage of light with a wavelength of 442 ± 6 or 492 ± 6 nm. The emitted fluorescence (>505 nm) changes depending on the excitation wavelength. An intelligent A/D board permitting on-board peak detection and signal averaging allows discrimination between the two emission plateaus. The ratio of the two signals varies with IC pH. The system is calibrated using the nigericin method (see text)

Wetzlar, FRG) and a photomultiplier. An intelligent A/D board permitting on-board averaging (DAP, Microstar Laboratories), an AT-386SX computer, and software programmed in BASIC (Better BASIC, Summit Software Technology) were used to calculate IC pH from the ratio of the fluorescence intensities measured at over 505 nm during excitation at 442 and 492 nm (Fig. 1). The system was calibrated using the nigericin method. To do so, IC and EC pH of C6 glioma cells were equilibrated in the presence of 10 μM nigericin and 120 mM K^+ (for details see Thomas et al. 1979; Molenaar et al. 1991).

Experimental Setup. Experiments were conducted in a thermostatized incubation chamber mounted on the scanning table of the inverted microscope in HEPES-buffered (15 mM), bicarbonate-free, glucose-containing electrolyte solution. After an initial control period of 15 min at an extracellular pH of 7.4, acidosis was rapidly induced by removing the control medium and adding medium titrated to an EC pH of 6.4, with osmolality maintained at 300 mosm/l. BCECF fluorescence was monitored initially every minute and then in 5-min intervals for a total of 65 min. EC pH was then normalized again to pH 7.4. BCECF fluorescence was registered for another 25 min. IC pH was calculated from the ratio of emitted light at the two excitation wavelengths using a calibration curve established from a total of more than 200 measurements at pH levels between 6.0 and 8.2 in the presence of nigericin.

Results and Discussion

In contrast to what had been expected in HEPES-buffered media, IC pH decreased to the EC level which, with pH = 6.4, had been set to a level easily reached in cerebral ischemia. The effect was observed in C6 glioma (Fig. 2) and in astrocytes from primary culture (Fig. 3). Upon realkalization IC pH normalized rapidly in both cell types. Comparing the kinetics of acidification and pH recovery shows that acidification is rather slow whereas pH normalization is an immediate phenomenon.

The data make it necessary partially to revise the proposed scheme of events occurring during EC acidosis-induced glial swelling (see "Introduction"). In particular, other mechanisms in addition to entry of CO_2 must be considered causes of IC acidification since the present study was performed in the absence of bicarbonate in HEPES buffer; currently the only additional way of acid entry fitting the data would be through H^+-channels. Preliminary data using $ZnCl_2$, known to block proton channels, support this contention. The involvement of proton channels would make it possible not only to fit all available data but also to explain glial acidification as a pathway to (a) protect the neuronal environment from acidosis by a spatial buffering mechanism, (b) to signal enhanced neuronal oxygen demand to the microcirculation, and (c) to effectively unload acid equivalents in the vicinity of blood vessels while resupplying bicarbonate to the neuronal environment.

The improved model involves four major components: the Na^+/H^+ antiporter, Na^+/HCO_3^- cotransport, HCO_3^-/Cl^- exchange, and proton channels. It is based on the consideration that in vivo glial cells may function as intermediates between neurons and the microcirculation, and as a consequence receptors, ion channels, and transporters are likely to be unevenly distributed over the glial membrane. This has been shown elegantly for the spatial K^+-buffering mechanism. In vitro, on the other hand, glial cells are surrounded by the experimental media, and unidirectional ion fluxes may become "camouflaged." For the proposed spatial H^+-buffering mechanism to work, the Na^+/H^+ antiporer and the Na^+/HCO_3^- cotransport would have to be located more closely to the vascular component (Newman 1991), whereas HCO_3^-/Cl^- exchange (Ahmad and Loeschke 1982) and proton channels would neighbor nerve cells.

It is assumed that *under physiological conditions* glial cells extrude bicarbonate via Na^+/HCO_3^- cotransport (Deitmer and Schlue 1987; Astion and Orkand 1988; Deitmer 1991) close to blood vessels, and protons via Na^+/H^+ exchange. In addition, bicarbonate may be exchanged for chloride

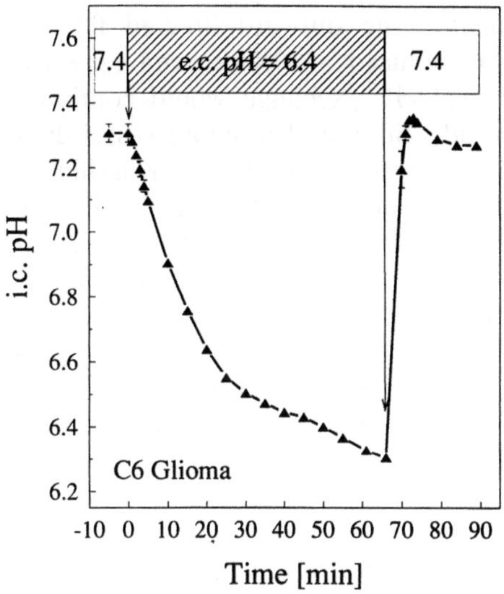

Fig. 2. Typical example of IC acidification of C6 glioma cells during lowering of the EC pH to 6.4. Upon normalization of EC pH the IC pH recovers rapidly with a slight overshoot

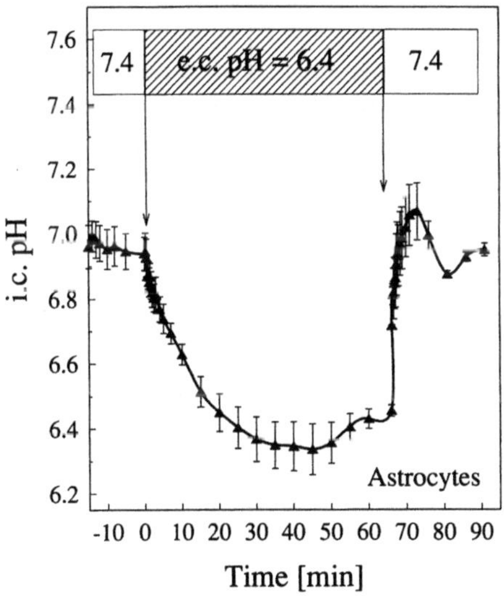

Fig. 3. IC acidification of astrocytes from primary culture during lowering of the EC pH to 6.4. As in C6 glioma, the IC pH recovers rapidly upon normalization of EC pH, again with an overshoot

(Kimelberg et al. 1979) at the glia/EC fluid interface (Ahmad and Loeschke 1982). At a normal membrane potential Na^+/HCO_3^- flux has been calculated to be outward directed (Newman 1991). CO_2 from neuronal metabolism resupplies IC bicarbonate.

With a *decreasing membrane potential* the outward-directed flux of the sodium/bicarbonate cotransporter is reduced and later even reversed (Newman 1991), while electroneutral Na^+/H^+ exchange would continue. Thereby the arteriolar environment would acidify and promote vessel dilation, i.e., the local autoregulation of local cerebral blood flow (Kuschinsky and Wahl 1978). In a depolarized state, with an inward flux of bicarbonate and continuing H^+ export, IC pH would become alkaline, as observed in vivo (Chesler and Kraig 1989; Chesler 1990). The influx of bicarbonate and sodium ions would increase IC osmolality as the cause of glial swelling. Influx of bicarbonate and alkalinization would stimulate HCO_3^-/Cl^- exchange (Reinertsen et al. 1988); bicarbonate would be extruded toward the neuronal side of the EC space. In other words, the cited mechanisms could elicit a net unidirectional flux of bicarbonate from the vascular toward the neuronal environment: spatial buffering.

With *acidosis in the neuronal environment*, H^+ ions would enter glial cells through proton channels and acidify the cell. In the presence of bicarbonate, glial acidification is faster due to the outward direction of electrogenic Na^+/HCO_3^- cotransport (Deitmer 1991). As a result the EC space close to the neurons would again be buffered. If H^+ ions continue to be extruded in the vicinity of blood vessels, the spatial buffering mechanism would be maintained by a net flux of protons from the neuronal toward the vascular environment.

In *ischemia* spatial H^+ buffering finally fails since acid is not removed via the circulation. Now, glial cells acidify through proton channels, with H^+ ions following their electrochemical gradient into the cell. This in turn – for a while at least – helps further to protect the neuronal environment from its acid load.

The ischemic situation comes close to what happens during the in vitro experiments in the absence of bicarbonate and the contiuous presence of a low EC pH: with H^+ influx through proton channels, IC pH cannot be buffered by bicarbonate influx (bicarbonate is absent, and the cell is not depolarized). Na^+/H^+ exchange should still be operative, which would explain the rather slow course of IC acidification and the rapid normalization of IC pH upon realkalization (Figs. 2, 3).

In the presence of EC bicarbonate during EC acidosis in vitro the rapid glial acidification via the carbonic anhydrase reaction would even be aggravated by the efflux of bicarbonate through electrogenic Na^+/HCO_3^- cotransport, which, however, should cease soon. Bicarbonate might be still leaving the cell in electroneutral exchange for chloride. This could explain acidosis-induced swelling as a result of a net load of sodium and chloride ions, just as suspected in the original hypothesis (Kempski et al. 1988).

In conclusion, a spatial H^+-buffering mechanism is proposed which is based on a – so far not fully verified – spatial separation of ion channels and transporters on the glial membrane. It consists of the Na^+/H^+ antiporter, the Na^+/HCO_3^- cotransporter, the HCO_3^-/Cl^- exchanger, and proton channels. Spatial pH buffering in vivo would facilitate the function of glial cells as regulating elements between neurons and the microcirculation, and hence glial homeostasis of the extracellular space of the brain.

Acknowledgements. This chapter contains material from the doctoral dissertations of C.K. and C.V. The excellent technical assistance of Barbara Kempski is gratefully acknowledged.

References

Ahmad HR, Loeschke HH (1982) Fast brain bicarbonate-chloride exchange between brain cells and brain extracellular fluid in respiratory acidosis. Pflugers Arch 395:293–299

Astion ML, Orkand RK (1988) Electrogenic Na^+/HCO_3^--cotransport in neuroglia. Glia 1: 355–357

Benda P, Lightbody J, Sato G, Levine L, Sweet W (1968) Differentiated rat glial cell strain in tissue culture. Science 161:370–371

Chesler M (1990) The regulation and modulation of pH in the nervous system. Prog Neurobiol 34:401–427

Chesler M, Kraig RP (1989) Intracellular pH transients of mammalian astrocytes. J Neurosci 9:2011–2019

Chopp M, Welch KMA, Tidwell CD, Helpern JA (1988) Global cerebral ischemia and intracellular pH during hyperglycemia and hypoglycemia in cats. Stroke 19:1383–1387

Deitmer JW (1991) Electrogenic sodium-dependent bicarbonate secretion by glial cells of the Leech central nervous system. J Gen Physiol 98:637–655

Deitmer JW, Schlue W-R (1987) The regulation of intracelluar pH by identified glial cells and neurones in the central nervous system of the Leech. J Physiol (Lond) 388:261–283

Frangakis MV, Kimelberg HK (1984) Dissociation of neonatal rat brain by dispase for preparation of primary astrocyte cultures. Neurochem Res 9:1689–1698

Jakubovicz DE, Klip A (1989) Lactic acid-induced swelling in C6 glioma cells via Na^+/H^+ exchange. Brain Res 485:215–224

Katsura K, Ekholm A, Asplund B, Siesjö BK (1991) Extracellular pH in the brain during ischemia: relationship to the severity of lactic acidosis. J Cereb Blood Flow Metab 11:597–599

Kempski O, Chaussy L, Gross U, Zimmer M, Baethmann A (1983) Volume regulation and metabolism of suspended C6 glioma cells: an in vitro model to study cytotoxic brain edema. Brain Res 279:217–228

Kempski O, Staub F, Jansen M, Schödel M, Baethmann A (1988) Glial swelling during extracellular acidosis in vitro. Stroke 19:385–392

Kimelberg HK, Biddlecome S, Bourke RS (1979) SITS-inhibitable Cl^- transport and Na^+-dependent H^+-production in primary astroglial cultures. Brain Res 173:111–124

Kuschinsky W, Wahl M (1978) Local chemical and neurogenic regulation of cerebral vascular resistance. Physiol Rev 58:656–689

Molenaar D, Abee T, Konings WN (1991) Continuous measurement of the cytoplasmic pH in Lactococcus lactis with a fluorescent pH indicator. Biochim Biophys Acta 1115:75–83

Newman E (1991) Sodium-bicarbonate cotransport in retinal Müller (glial) cells of the salamander. J Neurosci 11:3972–3983

Rehncrona S, Siesjö BK, Smith DS (1980) Reversible ischemia of the brain: biochemical factors influencing restitution. Acta Physiol Scand Suppl 492:135–140

Reinertsen KV, Tonnessen TI, Jacobsen J, Sandvig K, Olsnes S (1988) Role of chloride/bicarbonate antiport in the control of cytosolic pH. J Biol Chem 263:11117–11125

Siesjö BK (1981) Cell damage in the brain: a speculative synthesis. J Cereb Blood Flow Metab 1:155–185

Siesjö BK (1985) Acid-base homeostasis in the brain: physiology, chemistry and neurochemical pathology. Prog Brain Res 63:121–154

Staub F, Baethmann A, Peters J, Weigt H, Kempski O (1990) Effects of lactacidosis on glial cell volume and viability. J Cereb Blood Flow Metab 10:866–876

Thomas JA, Buchsbaum RN, Zimniak A, Racker E (1979) Intracellular pH measurement in Ehrlich ascites tumor cells utilizing spectroscopic probes generated in situ. Biochemistry 18:2210–2218

Intracellular Messengers

Calcium Channel Blockers Prevent Delayed Cerebral Ischemia After Intracranial Aneurysmal Subarachnoid Hemorrhage

J.S. Meyer, S. Takashima, and Y. Terayama

Introduction

Calcium entry blockers stabilize voltage-dependent calcium channels within cell membranes and prevent abnormal intracellular entry of calcium (Ca^{2+}) through endothelial and ischemic neuronal cell membranes [9, 30]. It is well known that when Ca^{2+} ions enter vascular smooth muscle, contraction results, but if calcium entry is blocked by calcium antagonists, vascular dilatation occurs. It is also well known that when extracellular Ca^{2+} moves into ischemic neurones, neuronal function initially becomes impaired, eventually followed by neuronal death as excessive intracellular Ca^{2+} interferes with intracellular messengers and mitochondrial metabolism [31]. Experimental studies of cerebral ischemia in animal models have shown that pretreatment with the calcium entry antagonist nimodipine results in protection of the brain from cerbral infarction by preventing calcium entry into ischemic neurones and glia [1].

Cerebral arterioles contract as Ca^{2+} enters vascular smooth muscle cells, which is inhibited by pretreatment with calcium antagonists, resulting in vasodilation of the small cerebral vessels and increased availability of the cerebral collateral circulation [3, 16, 20]. Unfortunately, in the clinical setting when patients have developed cerebral vasospasm following subarachnoid hemorrhage (SAH), serial angiography has failed to demonstrate reversal of cerebral vasospasm after treatment with calcium antagonists, although the capacitance of the cerebral collateral blood flow to ischemic brain via the small cerebral vessels does become increased [7].

The following clinical considerations are considered here according to basic mechanisms involved. (a) Why have clinical trials of nimodipine administered to patients *after* an acute ischemic stroke shown little or no benefit? (b) Why does nimodipine prevent delayed cerebral ischemia if administered early following intracranial SAH? (c) Why does intracranial aneurysmal rupture causing SAH *provide a predictable model of cerebral ischemia* in human subjects for testing the clinical effectiveness of a variety

Department of Neurology, Baylor College of Medicine, and Veterans Administration Medical Center, 2002 Holcombe Boulevard 151A, Houston, TX 77030, USA

Cerebral Ischemia and Basic Mechanisms
Ed. by A. Hartmann, F. Yatsu, and W. Kuschinsky

of medical interventions, such as immediate treatment with nimodipine for example? (d) After SAH, is it possible to test new forms of medical therapy in an ethically acceptable manner by carrying out controlled clinical trials *before* cerebral ischemia and stroke result among human patients in the hospital setting?

Calcium Channel Blockers and the Special Cerebral Selectivity of Nimodipine

The collective term "calcium channel antagonists" refers to a disparate group of drugs that differ structurally and pharmacologically. Diltiazem is a benzothiazepine, verapamil is a phenylalkylamine, and nimodipine, nicardipine, and nifedipine are all 1,4-dihydropyridines. It is their marked structural variation which determines their different pharmacological properties, particularly the different selectivity which they exhibit for actions on various smooth and cardiac muscle compartments, including the bundle of His and on neuronal and glial membranes as well. Nimodipine, for example is lipid soluble, passes the blood-brain barrier rapidly, and shows high sensitivity for binding with cerebral smooth muscle and neuronal and glial membrane calcium channel receptors.

There are at least three types of voltage-dependent calcium channels within cell membranes: the L (long-lasting) type, T (transient) type, and N (neither L not T but of an intermediate) type [31]. The L type of calcium channel receptors shows high sensitivity to becoming blockaded by nimodipine as the drug is selectively absorbed within their voltage-dependent channels, particularly in the presence of severe cerebral ischemia. Finally, there is an additional route of calcium entry provided by agonist-operated calcium channels, notably those gated by receptors which are activated by glutamate and other excitatory amino acids [31].

Nimodipine is much more lipophilic than other calcium antagonists and passes the blood-brain barrier more readily. As an illustration, after intravenous injection nimodipine exhibits a distribution five times greater than a similar dose of nifedipine throughout normal brain. A number of in vivo and in vitro studies have confirmed the high affinity of nimodipine for the L type channels in neuronal membranes, so that radioactively labeled nimodiine has been used in vivo as a biological marker for identifying them [14]. When considering treatment of ischemic cerebrovascular disorders, it is imperative that therapeutic agents have minimal systemic hypotensive effects. Thus, the ideal calcium antagonist for treating cerebral ischemia should have maximal activity at cerebrovascular and neuronal receptor sites and have minimal effects on peripheral vascular receptors. Fortunately, nimodipine has all of these desirable properties, and as a result significant drug-induced hypotension has rarely been reported in a large number of clinical trials.

Pharmacological Effects of Nimodipine in Experimental Models of Cerebral Ischemia

Numerous experimental studies among different animal models have demonstrated that pretreatment with nimodipine, by which we mean that the drug is administered *before* induction of experimental cerebral ischemia, reduces both the extent of cerebral infarction and accompanying neurological deficits among an impressive array of different models of experimental cerebral ischemia [19]. While beneficial effects of administration of nimodipine *before* or even *at time of* experimental ischemia have been consistent, there have not been consistent or significant benefits from administration of nimodipine some time *after* the cerebral ischemia has been initiated.

Kazda et al. [16] were the first to demonstrate that nimodipine prevents postischemically impaired cerebral reperfusion in feline models of cerebral ischemia. Later, utilizing double-label autoradiographic techniques in rats following occlusion of the middle cerebral artery, Hakim [12] showed that nimodipine ameliorated local cerebral tissue acidosis without altering cerebral blood flow in focal zones of cerebral ischemia. The observation that calcium channel blockers prevent development of ischemic cerebral tissue aciosis has important clinical implication in the treatment of human stroke since tissue acidosis is known to enhance neuronal cell death and volumes of resulting cerebral infarction in both animal experiments and human ischemic strokes. Mohamed et al. [22] demonstrated that nimodipine reduces the extent of histological damage measured 4 h after the onset of cerebral ischemia. Likewise, Jacewicz et al. [15] showed that pretreatment with nimodipine improves cerebral blood flow within the middle cerebral artery territory without evidence of any "vascular steal" and reduced the volume of both cerebral ischemic infarction and accompanying cerebral edema by 50% compared with untreated controls examined 2.5 h after the induction of ischemia. Furthermore, any mild hypotension resulting from administration of nimodipine (2 μg/kg per minute) did not aggravate the cerebral ischemia. Indeed, volumes of 2.5-h cerebral infarctions were 50%–75% less than the 24-h infarct volumes generated in similar experimental paradigms utilizing continuous nimodipine, infusion which reduced the infarct volume by 20%–30% compared to untreated controls. Recent studies in Reivich's laboratory have demonstrated that intravenous administration of nimodipine (1–2 μg/kg per minute) during or immediately after induction of cerebral ischemia attenuates in vivo measurements of Ca^{2+} ion influx into feline cerebral neurones and hastens electroencephalographic recovery [11, 34].

Controlled Clinical Trials of Nimodipine Treatment in Threatened Acute Ischemic Stroke Among Human Subjects

Results of clinical trials concerned with the effects of nimodipine treatment administered to human patients *after* the onset of acute ischemic stroke have been inconsistent. As shown in Table 1, seven controlled trials of nimodipine in acute ischemic stroke have been published to date. The first study by Gelmers et al. [10] concluded from a controlled clinical trial among 186 patients with ischemic stroke that those receiving oral nimodipine in daily doses of 120 mg exhibited early improvements in their neurological outcome and improved levels of consciousness compared with the control group receiving placebo. These authors also reported significant reductions in the 6-month mortality rate ($p < 0.05$).

Paci et al. [26] also reported improved neurological outcome, as judged by the Mathew Scale, among a nimodipine-treated group of patients with stroke compared with a placebo-treated group. However, Bogousslavsky et al. [6] reported negative results in their clinical trials of nimodipine treatment of acute strokes because both the Mathew neurological outcome scores and mortality rates showed no difference between the nimodipine-and placebo-treated groups.

Martínez-Vila et al. [21] and Krämer et al. [17] likewise reported no differences in neurological outcomes and mortality rates between nimodipine- and placebo-treated groups in their clinical trials of acute stroke.

Table 1. Controlled clinical trials of nimodipine in acute ischemic stroke

Reference	*n*	Time of treatment	Nimodipine treatment	Follow-up period	Neurological outcome	Mortality
Gelmers et al. (1988) [10]	186	<24 h	120 mg	28 days 6 months	Effective Effective	Significantly improved Significantly improved
Paci et al. (1989) [26]	41	<12 h	120 mg	28 days	Effective	NS
Bogousslavsky et al. (1990) [6]	52	<48 h	120 mg	4 months	NS	NS
Martínez-Vila et al. (1990) [21]	123	<48 h	120 mg	28 days	NS (effective in a subgroup)	NS
Krämer et al. (1990) [17]	356	<48 h	120 mg	21 days	NS (effective in a subgroup)	NS
Trust Study Group (1990) [33]	1215	<48 h	120 mg	6 months	NS	NS
The American Nimodipine Study Group (1992) [32]	1064	<48 h	60 mg 120 mg 240 mg	21 days	NS (effective in a subgroup)	NS

However, Martínez-Vila et al. [21] stated that when the patients "most likely to benefit" from therapeutic intervention with nimodipine, having marked neurological deficits with Mathew Scale scores of 65 or lower at baseline, were analyzed separately as a post hoc defined subgroup, this group did show some improved neurological outcome.

Krämer et al. [17] likewise reported among a similar post hoc selected subgroup of patients with the more severe neurological deficits from strokes and having the lowest Mathew scores that nimodipine treatment produced significantly better results, particularly among women.

Recently, there have been consistent but negative results reported from two extensive and carefully controlled collaborative clinical trials also testing nimodipine treatment *after* acute stroke. These are the Trust Study Group [33] and the American Study Group of Nimodipine in Stroke [32]. Neither of these extensive studies found significant differences between placebo- and nimodipine-treated groups. The Trust Study Group in the United Kingdom found no differences either in 6-month mortality rate or in neurological outcome after 3 weeks of treatment with 120 mg nimodipine daily compared with placebo-treated controls among 1215 patients with ischemic stroke aged 35–97 years, all of whom were treated some time *after* the ischemic event. The American Nimodipine Study Group directed by Mohr likewise reported that the nimodipine-treated group showed no overall benefits compared to the placebo-treated group when treatment was begun within 48 h *after* the ischemic stroke in a large series of patients. However, post hoc subgroup analysis did show some reductions in neurological deterioration among those treated with nimodipine shortly after the stroke by 30% compared with the placebo-treated group. Significantly better neurological outcome scores were found among patients treated with 120 mg nimodipine daily *only* if treatment was started *within 18 h* after the stroke. In this group, neurological improvement as judged by the Toronto neurological scale was statistically significant ($p < 0.005$).

Rationale for Clinical Trials of Calcium Antagonists After SAH Because SAH Provides a Predictable Human Model of Delayed Cerebral Ischemia

Cerebral arterial vasospasm is a common and predictable complication of ruptured intracranial aneurysm associated with SAH, which occurs in at least 30%–40% of patients [8]. Among these patients cerebral vasospasm has been confirmed by cerebral angiography showing either diffuse or focal arterial spasm. Likewise, cerebral blood flow has been shown to become decreased among patients with cerebral vasospasm, and both these measures of focal cerebral ischemia correlate with the associated delayed ischemic neurological deficits [35]. These delayed neurological deficits usually occur within 2–10 days after rupture of an intracranial aneurysm and are always associated with vasospasm. They have been termed delayed ischemic neurological deficits since they clearly result from delayed vasospasm of the appropriate cerebral arteries responsible for the neurological deficits demonstrated by arteriography.

Over one-third of patients surviving SAH from rupture of intracranial aneurysm, provided that their state of consciousness permits recognition of worsening of neurological status, develop clinically responsible signs of delayed ischemic neurological deficits. Whatever the nature of the cerebral vasospasm, it is clearly brought about by the presence of blood and blood products in the subarachnoid space surrounding and contacting the major cerebral arteries, so that entry of Ca^{2+} ions is certain to act as the important intracellular messenger which activates contractile proteins within vascular smooth muscle cells.

Likewise, the intracellular accumulations of Ca^{2+} ions in neurones and glia play a major role in ischemic cell damage. Taking these considerations into account, there was the probability that treatment with calcium antagonists at the onset of SAH might well prevent or ameliorate the incidence of delayed ischemic neurological deficits. Clinical trials which are discussed here support this hypothesis. In other words, SAH due to ruptured intracranial arterial aneurysm now *provides us with a predictable model of human cerebral ischemia* in which the clinical effectiveness of treatment by other medical interventions, such as the administration of excitatory amino acids antagonists or antioxidants, can be combined in two groups of patients with SAH, one with nimodipine treatment and the other with nimodipine plus placebo, utilizing double-blind controlled clinical trials before predictable cerebral ischemic events occur. We suggest this as a valid and ethical approach for clinical evaluations of treatment of acute ischemic strokes since it is virtually impossible to evaluate the use of such promising agents in any other way *before* the onset of an ischemic stroke.

Results of Controlled Trials of Nimodipine Therapy Among Patients with SAH

Six controlled clinical trials have been published to date which establish the clinical effectiveness of nimodipine therapy among patients with SAH *before* the onset of delayed neurological deficits due to ruptured intracranial aneurysm (Table 2). The first of the controlled clinical trials of nimodipine in SAH due to ruptured aneurysm was organized by Allen et al. [2] representing five collaborating centers throughout North America. These authors reported that persistent and severe neurological deficits or death resulting from cerebral arterial spasm at the end of a 21-day trial of treatment occurred in 8 of 60 patients given placebo but in only 1 of 56 given nimodipine ($p = 0.03$, Fisher's exact test). These authors concluded that

Table 2. Controlled clinical trials of nimodipine among patients with SAH

Reference	*n*	Grade (Hunt)	Nimodipine treatment	Follow-up period	Study endpoints	Results
Allen et al. (1983) [2]	116	I–II	Oral (21 days)	21 days	Deficits (all causes)	NS
					Severe deficits due to vasospasm	Significantly decreased
					Side effects	NS
Phillippon et al. (1986) [28]	81	I–III	Oral (21 days)	21 days	Deficits (all causes)	NS
					Deficits due to vasospasm	Significantly decreased
					Angiographic vasospasm	NS
Neil-Dwyer et al. (1987) [23]	75	I–V	Intrathecal	3 months	Poor outcome (all cases)	NS
			Oral (21 days)		Poor outcome (eligible cases)	Significantly improved
					Mortality (eligible cases)	Significanlty reduced
					Side effects	NS
Petruk et al. (1988) [27]	188	III–V	Oral (21 days)	3 months	Good outcome	Significantly improved
					Deficits due to vasospasm	Significantly decreased
					Angiographic vasospasm	NS
Pickard et al. (1989) [29]	554	I–V	Oral (21 days)	3 months	Delayed ischemia	Significantly decreased
					Poor outcome	Significantly improved
					Adverse reactions	NS
Öhman et al. (1991) [25]	213	I–III	Intravenous (7–10 days)	1–3 years	Overall outcome	NS
			Oral (to 21 days)		Death due to ischemia	Significantly decreased
					Infarction on CT	Significantly decreased

nimodipine should be given to *all* patients, including those who were neurologically normal, immediately after SAH to reduce delayed neurological deficits brought about by cerebral arterial spasm.

Philippon et al. [28] reported among patients whose neurological deficits were attributable to vasospasm that those treated with nimodipine showed improved neurological outcomes ($p < 0.05$). Ten of the 39 treated with placebo suffered severe neurological deficits or died, but only 2 of the 31 treated with nimodipine showed poor neurological outcomes. Neil-Dwyer et al. [23] also reported from their controlled trial of 50 patients that only one patient who received nimodipine, compared with six treated with placebo, died within 3 months of treatment.

In the cooperative Canadian study [27] 154 patients with SAH were entered within 96 h after aneurysmal rupture; 72 received nimodipine and 82 placebo. Nimodipine treatment was associated with significantly improved outcome ($p < 0.001$). Of the nimodipine-treated patients 29% had good neurological outcome 3 months after SAH compared to only 11% in the placebo-treated group. Following nimodipine treatment there was a significantly larger number of patients surviving with good neurological outcome compared with placebo, and the incidence of delayed ischemic deficits due to vasospasm was also reduced. These Canadian authors concluded that the proven beneficial effects of nimodipine must occur by mechanisms other than prevention of large cerebral arterial spasm visualized by angiography since the angiographic incidence of vasospasm did not differ between nimodipine- and placebo-treated groups.

In Pickard's definitive British collaborative nimodipine treatment trial of SAH due to ruptured aneurysm [29], a total of 554 patients were entered into the study within 96 h following SAH. The diagnosis was confirmed by clinical signs, lumbar puncture and computed tomography (CT), but arteriography was optional in order to start treatment early. Placebo was assigned to 276 patients, and nimodipine (240 mg daily) was administered to 278 patients. Patients were followed for 3 months and occurrences of cerebral infarction and ischemic neurological deficits were carefully reporded. The incidence of cerebral infarction was 22% in the group receiving nimodipine and 33% in the group receiving placebo. The significant reduction in delayed neruological deficits was by 34%. Poor neurological outcomes (moderate to severe disability) were reduced significantly by 40% in the group receiving nimodipine. Adverse reactions were reported among 14 patients receiving nimodipine and 10 given placebo.

The study of Öhman et al. [25] was conducted within a single center in Helsinki. Of 213 patients admitted to the study, 58 underwent operation for clipping the aneurysm early (within 72 h after the SAH; days 0–3), 69 underwent surgery subacutely (between days 4 and 7), and 74 had late surgical interventions (on day 8 or later). Follow-up examination with CT performed 1–3 years after the SAH (mean 1.4 years) revealed no significant differences in overall outcome between groups. However, the nimodipine-treated group had a significantly lower mortality rate caused by delayed cerebral ischemia ($p < 0.01$) and a significantly lower incidence of cerebral infarctions visualized by CT ($p < 0.05$). Neurological benefits were most evident among patients who did not exibit associated intracerebral hemorrhage detected by admission CT scans ($p = 0.03$).

Statistically significant improvements in neurological outcome were observed among nimodipine-treated patients with SAH compared with those treated with placebo in all six of these controlled double-blind trials. Results can be summarized as follows: (a) nimodipine treatment does not decrease the incidence of angiographic vasospasm as judged by angiography, (b) the mortality rate, incidence, and severity of delayed ischemic neurological deficits are significantly reduced among those treated with nimodipine, and (c) hypotension and other side effects during nimodipine treatment are minimal.

Possible Mechanisms for Nimodipine's Therapeutic Benefits After SAH

Beneficial effects of calcium entry blockers appear to be brought about by improving cerebral blood flow at the onset of cerebral ischemia and by direct protection of nerve and glial cells from the effects of ischemia. Possible mechanisms of calcium antagonists include the following [4]:

- Hemodynamic effects
  - Improvement of ischemic flow
  - Attenuation of postischemic hypoperfusion
  - Prevention of vasospasm
  - Improvement of rheological properties
- Cellular protection
  - Prevention of enhanced proteolysis
  - Prevention of formation of free fatty acids and free radicals
  - Prevention of oxidation by free radicals
  - Protection against energy failure
  - Protection against acidosis
- Inhibition of secondary brain damage
  - Protection against serotonin, histamine, kinin, glutamate, and other excitatory amino acids

It has been shown by different investigators utilizing experimental models of ischemia that calcium entry blockers improve cerebral blood flow under ischemic conditions [9, 15, 16], attenuate postischemic hypoperfusion [13, 16], and may attenuate cerebral vasospasm [20]. However, the clinically protective effects of nimodipine against delayed cerebral ischemia after SAH occur despite consistent evidence that it does not prevent or reverse cerebral "vasospasm" or improve overall tissue perfusion. Nevertheless, whether or not nimodipine treatment reduces the incidence and severity of vasospasm, it certainly minimizes its ischemic consequences and also selectively increases local perfusion in zones of reduced blood flow [3] so that beneficial effects are, in part, due to its vasoactive properties.

Direct protection of ischemic neurones and glia may be attributed to inhibition of calcium entry, preventing intracellular overloading by Ca^{2+} ions [1] and reduced tissue acidosis [12]. Enhanced cellular proteolysis and lipolysis are thereby prevented, and release of free fatty acids and free radicals is reduced [18]. Calcium entry blockers also possess antioxidant properties and protect against cellular energy failure. Effective antagonism of the release of highly active compounds including free fatty acids, kinins, glutamate, and other excitatory amino acids from ischemic brain also appears to be an important aspect of neuronal cell protection. Evidence has also been provided that Ca^{2+} entry blockers interfere with pathogenetic mechanisms caused by intracellular mediator compounds [4].

However, Ca^{2+} entry blockers do not enhance deficient respiratory activity of subcellular particles obtained from ischemic gerbil brain [24]. There were significant reductions of cerebral blood flow in the forebrain ischemia model, but treatment with verapamil was unable to induce powerful cerebroprotective effects in models of severe global forebrain ischemia [5].

Conclusions

In conclusion, patients with ruptured intracranial aneurysms and SAH treated immediately with nimodipine suffer fewer and milder neurological deficits and decreased mortality with improved quality of life. Presumed mechanisms for therapeutic benefits brought about by pretreatment with nimodipine are protection of neuronal and glial membranes against ischemia by blockade of Ca^{2+} entry and improved collateral capacitance of the cerebral circulation.

Ruptured aneurysm with SAH now offers a predictable model of cerebral ischemia in patient populations in which controlled double-blind trials of other medical interventions (such as blockers of excitatory amino acid and antioxidants) may be tested by comparing groups treated with the new drugs to be tested combined with nimodipine versus placebo combined with nimodipine.

References

1. Abe K, Kogure K, Watanabe T (1988) Prevention of ischemic and postischemic brain edema by a novel calcium antagonist (PN200-110). J Cereb Blood Flow Metab 8:436–439
2. Allen GS, Ahn HS, Preziosi TJ, Battye R, Boone SC, Chou SN, Kelly DL, Weir BK, Crabbe RA, Lavik PJ, Rosenbloom SB, Dorsey FC, Ingram CR, Mellitis DE, Bertsch LA,

Boisvert DPJ, Hundley MB, Johnson RK, Strom JoA, Transou CR (1983) Cerebral arterial spasm – a controlled trial of nimodipine in patients with subarachnoid hemorrhage. N Engl J Med 308:619–624
3. Auer LM (1982) Cerebrovascular activity of the Ca-antagonist nimodipine. J Cereb Blood Flow Metab 2:268
4. Baethmann A, Jansen M (1986) Possible role of calcium entry blockers in brain protection. Eur Neurol 25 Suppl 1:102–114
5. Berger JR, Busto R, Ginsberg MD (1984) Verapamil: failure of metabolic amelioration following global forebrain ischemia in the rat. Stroke 15:1029–1032
6. Bogousslavsky J, Regli F, Zumstein V, Köbberling W (1990) Double-blind study of nimodipine in non-severe stroke. Eur Neurol 30:23–26
7. Findlay JM, Macdonald RL, Weir BKA (1991) Current concepts of pathophysiology and management of cerebral vasospasm following aneurysmal subarachnoid hemorrhage. Cerebrovasc Brain Metab Rev 3:336–361
8. Fisher CM (1975) Clinical syndroms in cerebral thrombosis, hypertensive hemorrhage, and ruptured saccular aneurysm. Clin Neurol 22:117–147
9. Gelmers HJ (1983) The role of calcium and the significance of calcium antagonists in some neurological and neurosurgical disease. Clin Neurol Neurosurg 85:213–224
10. Gelmers HJ, Gorter K, Weerdt CJ, Wiezer HJA (1988) A controlled trial of nimodipine in acute ischemic stroke. N Engl J Med 318:203–207
11. Greenberg JH, Uematsu D, Araki N, Hickey WF, Reivich M (1990) Cytosolic free calcium during focal cerebral ischemia and the effects of nimodipine on calcium and histologic damage. Stroke 21 Suppl 4:IV72–IV77
12. Hakim AM (1986) Cerebral acidosis in focal ischemia. II. Nimodipine and verapamil normalize cerebral pH following middle cerebral artery occlusion in the rat. J Cereb Blood Flow Metab 6:676–683
13. Hakim AM, Evans AC, Berger L, Kuwabara H, Worsley K, Marchal G, Biel C, Pokrupa R, Diksic M, Meyer E, Gjedde A, Marrett S (1989) The effect of nimodipine on the evolution of human cerebral infarction studied by PET. J Cereb Blood Flow Metab 9:523–534
14. Hakim A, Hogan M, Berger L, Gjedde A (1989) Activation of the calcium channels in focal cerebral ischemia. J Cereb Blood Flow Metab 9 Suppl 1:s14
15. Jacewicz M, Brint S, Tanabe J, Wang XJ, Pulsinelli WA (1990) Nimodipine pretreatment improves cerebral blood flow and reduces brain edema in conscious rats subjected to focal cerebral ischemia. J Cereb Blood Flow Metab 10:903–913
16. Kazda S, Garthoff B, Krause HP, Schloßmann D (1982) Cerebrovascular effects of the calcium antagonistic dihydropyridine derivative nimodipine in animal experiments. Arzneimittelforschung 32:331–338
17. Krämer G, Tettenborn B, Rothacher G, Hacke W, Busse O, Hornig CR, Aichner F, Ladurner G (1990) Nimodipine German Austrian stroke trial. Neurology 40 Suppl 1:415–416
18. Kubo K, Yoshitake I, Kumada Y, Shuto K, Nakamizo N (1984) Radical scavenging action of flunarizine in rat brain in vitro. Arch Int Pharmacodyn Ther 272:283–295
19. Langley MS, Sorkin EM (1989) Nimodipine: a review of its pharmacodynamic and pharmacokinetic properties, and therapeutic potential in cerebrovascular disease. Drugs 37:669 2699
20. Leblanc R, Feindel W, Yamamoto L, Milton JG, Frojmovic MM (1983) Reversal of acute experimental cerebral vasospasm by calcium antagonism with verapamil. Can J Neurol Sci 11:42–47
21. Martínez-Vila E, Guillén F, Villanueva JA, Matías-Guiu J, Bigorra J, Gil P, Carbonell A, Martínez-Lage JM (1990) Placebo-controlled trial of nimodipine in the treatment of acute ischemic cerebral infarction. Stroke 21:1023–1028
22. Mohamed AA, Gotoh O, Graham DI, Osborne KA, McCulloch J, Mendelow AD, Teasdale GM, Harper AM (1985) Effect of pretreatment with the calcium antagonist nim-

odipine on local cerebral blood flow and histopathology after middle cerebral artery occlusion. Ann Neurol 18:705–711
23. Neil-Dwyer G, Mee E, Dorrance D, Lowe D (1987) Early intervention with nimodipine in subarachnoid haemorrhage. Eur Heart J 8 Suppl K:41–47
24. Nowicki JP, MacKenzie ET, Young AR (1982) Brain ischemia. Calcium and calcium antagonists. Pathol Biol 30:282–288
25. Öhman J, Servo A, Heiskanen O (1991) Long-term effects of nimodipine on cerebral infarcts and outcome after aneurysmal subarachnoid hemorrhage and surgery. J Neurosurg 74:8–13
26. Paci A, Ottaviano P, Trenta A, Iannone G, Santis LD, Lancia G, Moschini E, Carosi M, Amigoni S, Caresia L (1989) Nimodipine in acute ischemic stroke: a double-blind controlled study. Acta Neurol Scand 80:282–286
27. Petruk KC, West M, Mohr G, Weir BKA, Benoit BG, Gentili F, Disney LB, Khan MI, Grace M, Holness RO, Karwon MS, Ford RM, Cameron GS, Tucker WS, Purves GB, Miller JDR, Hunter KM, Richard MT, Durity FA, Chan R, Clein LJ, Maroun FB, Godon A (1988) Nimodipine treatment in poor-grade aneurysm patients. Results of a multicenter double-blind placebo-controlled trial. J Neurosurg 68:505–517
28. Philippon J, Grob R, Dagreou F, Guggiari M, Rivierez M, Viars P (1986) Prevention of vasospasm in subarachnoid haemorrhage. A controlled study with nimodipine. Acta Neurochir(Wien) 82:110–114
29. Pickard JD, Murray GD, Illingworth R, Shaw MDM, Teasdale GM, Foy PM, Humphrey PRD, Lang DA, Nelson R, Richards P, Sinar J, Bailey S, Skene A (1989) Effect of oral nimodipine on cerebral infarction and outcome after subarachnoid haemorrhage: British aneurysm nimodipine trial. Br Med J 298:636–642
30. Robinson MJ, Teasdale GM (1990) Calcium antagonists in the management of subarachnoid haemorrhage. Cerebrovasc Brain Metab Rev 2:205–226
31. Siesjö BK, Bengtsson F (1989) Calcium fluxes, calcium antagonists, and calcium-related pathology in brain ischemia: a unifying hypothesis. J Cereb Blood Flow Metab 9:127–140
32. The American Nimodipine Study Group (1992) Clinical trial of nimodipine in acute ischemic stroke. Stroke 23:3–8
33. Trust Study Group (1990) Randomised, double-blind, placebo-controlled trial of nimodipine in acute stroke. Lancet 336:1205–1209
34. Uematsu D, Greenberg JH, Hickey WF, Reivich M (1989) Nimodipine attenuates both increase in cytosolic free calcium and histologic damage following focal cerebral ischemia and reperfusion in cats. Stroke 20:1531–1537
35. Yamamoto M, Meyer JS, Naritomi H, Sakai F, Yamaguchi F, Terry S (1979) Noninvasive measurement of cerebral vasospasm in patients with subarachnoid hemorrhage. J Neurol Sci 43:301–311

Monosialoganglioside GM_1 in Experimental Models of Stroke

M.S. SEREN, A. LAZZARO, M.C. COMELLI, R. CANELLA, R. ZANONI, D. GUIDOLIN, and H. MANEV

It is now well established that the neurotransmitter glutamate plays a prominent role in the pathogenesis of hypoxic-ischemic brain damage [1, 2] High levels of glutamate accumulate in the interstitial fluid during an ischemic/traumatic episode [3] and, through the activation of ionotropic glutamate receptors, initiate a cascade of cellular events which ultimately lead to irreversible cell damage. These cellular events contributing to excitotoxicity appear to be triggered largely by destabilization of calcium homeostasis [4].

Glutamate receptor antagonists have been proposed as a strategy to protect neurons from glutamate-triggered neuronal death, but an indiscriminate blockade of glutamate receptor-mediated neurotransmission negatively affects numerous physiological functions, including neuroplastic responses (for a review see [5]). An alternative approach is the use of gangliosides which do not interfere with glutamate receptor function but limit events occurring downstream from the receptor [6, 7].

The Model of the Middle Cerebral Artery Occlusion in the Rat

Several reports have shown that the monosialoganglioside GM_1 can improve outcome in animals subjected to both global and focal ischemic insults. Moreover, encouraging data are obtained with clinical application of GM_1 in the acute traumatic or ischemic brain pathology (for a review see [8]). We have used the model of permanent middle cerebral artery occlusion (MCAo) according to Tamura et al. [9]. Sprague-Dawley rats (250–300 g) were subjected to proximal occlusion of the MCA. This procedure induces a large damage to striatum and cortex. The animals, in which the body temperature was kept at 37°C throughout the experiment, received GM_1 (30 mg/kg intravenously) soon after the MCAo was completed, then 3 h

Fidia Research Laboratories, Via Ponte della Fabbrica, 3/A 35031 Abano Terme (PD), Italy

Cerebral Ischemia and Basic Mechanisms
Ed. by A. Hartmann, F. Yatsu, and W. Kuschinsky

Table 1. Effects of GM_1 treatment on the infarct volume, 7 days after MCAo

	Saline ($n = 13$)	GM_1 ($n = 9$)
Hemisphere	142.37 ± 17.68	73.38 ± 10.67*
Cortex	97.45 ± 18.57	43.81 ± 7.24*
Sriatum	37.96 ± 3.01	28.59 ± 3.57

Seven days following MCAo, the animals were perfusion-fixed with Bouin's solution. Paraffin-embedded coronal sections were taken throughout the extent of the lesion and stained with cresyl violet–Luxol fast blue. The area was measured in eight selected sections, spaced at 800-μm intervals. The cross-sectional areas of infarction were measured separately in striatum, cortex, and each hemisphere using a semiautomatic image analysis system (VIDS, AMS, UK). The volume of infarction was calculated by integrating the areas with the distance between them. The extent of ischemic damage is expressed in cubic millimeters.
*$p < 0.05$ compared to saline-treated animals (Student's unpaired t test).

Table 2. Effect of GM_1 treatment on the MCAo-induced decrease in cortical choline acetyltranferase (ChAT) activity

Saline ($n = 8$)	−49.7% ± 5.6%
GM_1 ($n = 8$)	−34% ± 4.7%*

Seven days after MCAo the cortex was dissected on ice and immediately frozen on dry ice. ChAT activity, a marker of cholinergic neurons projecting to the cortex from the nucleus basalis magnocellularis, was measured according to a slightly modified method of Fonnum [14]. Values are expressed as percentage of ChAT activity compared to the contraleral hemisphere.
*$p < 0.05$ (Student's t test).

later, and daily for 6 days. At the end of this period the extent of brain infarction was measured (for the method, see Table 1).

As shown in Table 1, ganglioside administration reduced the magnitude of infarct volume. The protective effect was also evident when the biochemical parameters were measured. Namely, the loss of cortical choline acetyl transferase activity was significantly reduced by GM_1 treatment (see Table 2). These results extend previous observations obtained in another model of focal cerebral ischemia (distal occlusion of MCA and occlusion of common carotid arteries) in which the administration of monosialoganglioside GM_1 reduced the acute increase of water content (i.e., edema) and the acute accumulation of Ca^{2+} in areas affected by the ischemic episode

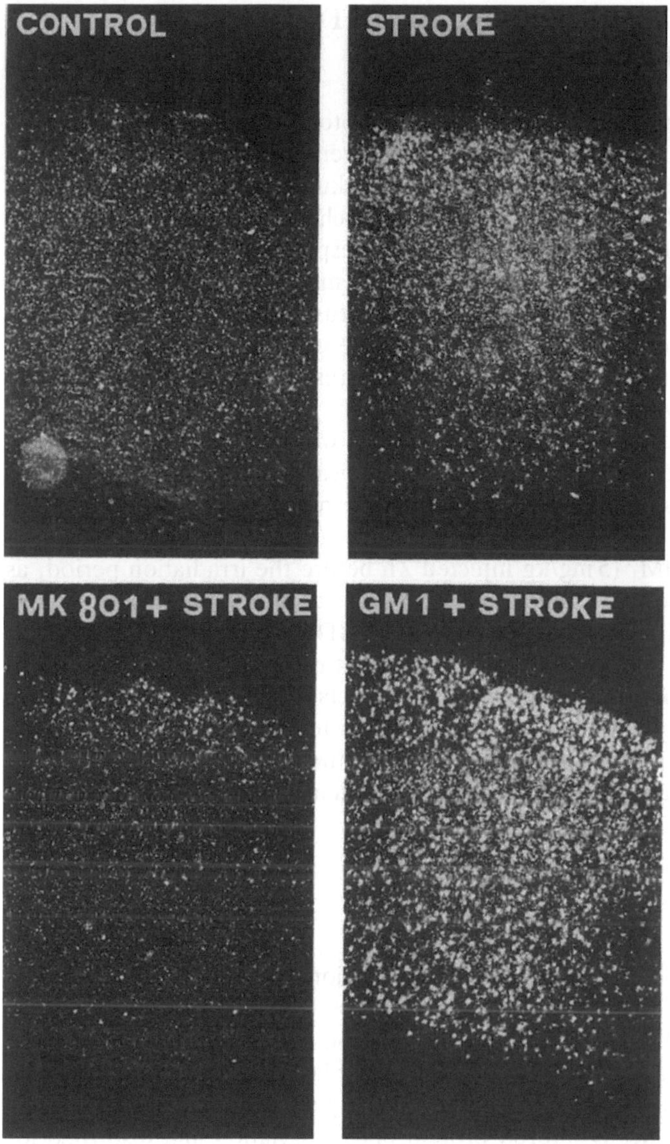

Fig. 1. Dark field photomicrographs of emulsion-dipped slides showing examples of BDNF mRNA expression in the cortical area, ipsilateral to the lesion, of control, photothrombotic, MK-801 and GM_1 pretreated photothrombotic rats (for details see [12])

[10]. Hence, our results indicate that the acute biochemical improvement observed in GM_1-treated animals is indeed associated with a reduction in brain infarct volume.

The Model of Photochemically Induced Focal Cortical Thrombosis in the Rat

A focal thrombotic lesion of the sensorimotor cortex was produced in Sprague-Dawley rats (280–330 g) by intravenously injecting the light-sensitive dye rose bengal and exposing the skull to a controlled beam of light, as previously described [11, 12]. Photochemically induced unilateral focal cortical stroke is associated with a widespread ipsilateral stimulation of brain-derived neurotrophic factor (BDNF) mRNA expression in cortical and hippocampal regions, as assessed by in situ hybridization and northern blot techniques (see Fig. 1) [12]. The increase in BDNF mRNA was maximal at 4 h following the light exposure. The response was completely prevented by the systemic administration of the noncompetitive glutamate receptor blocker dizolcipine (MK-801), which was administered 1 h prior to the induction of stroke. Hence, the results indicate that the observed BDNF mRNA increase is mediated via the glutamate receptor activation (Fig. 1).

On the other hand, the intracerebroventricular administration of monosialoganglioside GM_1 (5 mg/kg injected 2 h before the irradiation period, as previously described by Cuello et al. [13]), did not inhibit the stroke-induced increase in BDNF mRNA (Fig. 1). As the BDNF mRNA expression was increased in areas that did not show significant morphological alterations, it can be assumed that the BDNF response reflects the compensatory response mechanism through which the brain responds to acute injury. If so, GM_1, unlike the glutamate receptor antagonists, can maintain the capability of the brain to react to an insult by enhancing the neuronal expression of trophic factors.

Conclusion

The mechanism of the neuroprotective action of gangliosides must be viewed not only as an antineurotoxic process but also with regard to its potential role in the regulation and stimulation of structural and functional compensations via the process of neuronal plasticity. The antiexcitotoxic action of gangliosides, which is based on the ganglioside-mediated normalization of pathologically altered protein phosphorylation and disturbed ion homeostasis, prevents the development of neurodegeneration. On the other hand, the neuroplastic action of gangliosides, which might be seen as the ganglioside-mediated preservation or facilitation of autoprotective responses of the brain to injury (for example, the increased expression of BDNF), allows the appropriate expression of functional or structural compensatory processes. The dual antiexcitotoxic and neuroplastic action of gangliosides distinguishes these molecules from glutamate receptor antagonists and suggests their pharmacological potential for the treatment not only of acute stroke but also of chronic degenerative brain pathologies.

References

1. Choi DW (1988) Glutamate toxicity and diseases of the nervous system. Neuron 1:623–634
2. Rothman SM, Olney JW (1986) Glutamate and the pathophysiology of hypoxic/ischemic brain damage. Ann Neurol 19:105–111
3. Benveniste H, Jorgensen MB, Sandberg M (1989) Ischemic damage in hippocampal CA_9 is dependent on glutamate release and intact innervation from CA_3. J Cereb Blood Flow Metab 9:629–639
4. Manev H, Favaron M, Guidotti A, Costa E (1989) Delayed in crease of Ca^{2+} influx elicited by glutamate: role in neuronal death. Mol Pharmacol 36:106–112
5. Buchan AM (1990) Do NMDA antagonists protect against cerebral ischemia: are clinical trials warranted? Cerebrovasc Brain Metab Rev 2:1–26
6. Favaron M, Manev H, Alho H, Bertolino M, Ferret B, Guidotti A, Costa E (1988) Gangliosides prevent glutamate and kainate neurotoxicity in primary neuronal cultures of neonatal rat cerebellum and cortex. Proc Natl Acad Sci USA 85:7351–7355
7. Manev H, Costa E, Wroblewski JT, Guidotti A (1990) Abusive stimulation of excitatory amino acid receptors: a strategy to limit neurotoxicity. FASEB J 4:2789–2797
8. Carolei A, Fieschi C, Bruno R, Toffano G (1991) Monosialoganglioside GM_1 in cerebral ischemia. Cerebrovasc Brain Metab Rev 3:134–157
9. Tamura A, Graham DI, McCulloch J, Teasdale GM (1981) Focal cerebral ischemia in the rat. I. Description of technique and early neuropathological consequences following middle cerebral artery occlusion. J Cereb Blood Flow Metab 1:53–60
10. Karpiak SE, Mahadik SP, Wakade CC (1990) Ganglioside reduction of ischemic injury. CRC Crit Rev Neurobiol 5:221–237
11. Watson BD, Dietrich WD, Busto R, Wachtel MS, Ginsberg MD (1985) Induction of reproducible brain infarction by photochemically initiated thrombosis. Ann Neurol 17: 497–504
12. Comelli MC, Seren MS, Guidolin D, Manev RM, Favaron M, Rimland JM, Canella R, Negro A, Manev H (1992) Photochemical stroke and brain-derived neurotrophic factor (BDNF) mRNA expression. Neuroreport 3(6):473–476
13. Cuello AC, Garofalo L, Kenigsberg RL, Maysinger D (1989) Gangliosides potentiate in vivo and in vitro effects of nerve growth factor on central cholinergic neurons. Proc Natl Acad Sci USA 86:2056–2060
14. Fonnum F (1975) A rapid radiochemical method for the determination of choline acetyltranferase. J Neurochem 24:407–409

What Are We Learning from Clinical Trials in Acute Cerebral Ischemia?

C. Fieschi, M. Frontoni, M. Fiorelli, M.L. Sacchetti, D. Toni, C. Argentino, and M.C. Gori

At the beginning of the 1990s we can make a long list of studies on various pharmacological tools and approaches to the treatment of acute cerebral ischemia, such as osmotics, steroids, vasodilators, hemodilution, which have all systematically failed. We would rather not elaborate that discouraging list but think about how to make use of these negative experiences as examples of methodological errors which should not be repeated. Thus, studies with a poor diagnostic characterization of patients due to the lack of a computed tomography (CT) scan and trials carried out using insufficient sample sizes should never be performed.

The lack of knowledge of pathophysiological mechanisms existing at the beginning of these experiences can no longer be considered an overwhelming barrier. In fact, over 10 years have passed since we first discovered the existence of an ischemic penumbra around the infarcted area in which neurons are functionally blocked but liable to recover if the blood flow is restored within the first few hours after the onset of ischemia. This brings us the concept of a therapeutic window for at least part of the ischemic tissue, but the treatment must begin very early. Animal experiments (Astrup et al. 1981), confirmed later by neuroimaging tecniques (Lenzi et al. 1982), provided this information and also defined the temporal limits of 6–8 h for the window (Weinstein et al. 1980). Therefore, trials designed without considering these facts should be avoided as they could yield further negative results because of the delay in treatment rather than the inefficacy of the drugs.

As our experience suggests, logistical problems related to the need for early treatment can be overcome. In fact, the creation of a Stroke Unit in Rome has enabled us over 10 years of activity to reduce the interval between stroke onset and treatment from 12 h in the 1988 study on hemodilution (Italian Acute Stroke Study Group 1988) to 6 h in half the patients in the GM_1 trial (Argentino et al. 1989), and recently to 5 h in the Early Stroke Trial (EST). Correct information to the public health service, 24-h neurooncall service, strict cooperation with the Emergency Room, noninvasive examinations, and neuroradiology services represent the cornerstones which have supported these experiences.

Department of Neurological Sciences, University of Rome, La Sapienza, Viale dell' Universita 30, 00174 Rome, Italy

Cerebral Ischemia and Basic Mechanisms
Ed. by A. Hartmann, F. Yatsu, and W. Kuschinsky

The concerted effort of an adequate number of efficient neurological centers is essential for reliable proof of drug efficacy. Indeed, the process of testing a drug may pass through exploratory studies which assess any positive effect on biological parameters but do not require the observation of many cases. To demonstrate clinical benefits, however, "pragmatic" trials of a more advanced phase will be requested, in which several hundreds of patients need to be enrolled to reach statistically significant figures. For this reason, in pragmatic trials the crucial problem of obtaining a good accrual rate needs to be resolved by means of a strict coordination of multiple participating centers.

While methodological shortcomings of nonrandomized or nonblinded trials speak for themselves, the question of endpoints deserves some discussion. Death rates should never be the sole parameter of judgement on drug efficacy since naturally occurring mortality is so low in stroke that only an extremely effective drug would yield significant results. Besides stroke scales or activities of daily life (ADL) indexes can also be utilized, although it must be remembered that both have certain limitations. The former provide no information on the patient's functional state while the latter cannot be used to evaluate subjects in the acute phase. A careful choice should be made, which also takes into account the need for an adequate follow-up to assess the functional outcome of stabilized patients. This follow-up should not last more than a few months to avoid the occurrence of possible invalidating events not related to the initial stroke.

All these considerations, when put together, give an idea of how difficult planning a good clinical trial in acute cerebral ischemia can be. This fact is confirmed by the recently published American Nimodipine Study (American Nimodipine Study Group 1992) which satisfied all the multicenter, randomized, doubleblind, multidose, adequately sized design rules and also included a complete set of evaluation scales, but had a major drawback – early treatment, since patients had been given the drug within the first 48 h from onset. As a result, only a small subset of patients seemed to have some benefit from the treatment, casting doubts – should there be any – on the overall usefulness of nimodipine in acute cerebral ischemia. We do not yet know how effective this drug can be when given promptly.

Previous methodological errors and positive acquirements prompted and supported the planning of the already mentioned EST study. This trial, which was designed to assess the safety and efficacy of monosialoganglioside GM_1 in patients with ischemic stroke in the middle cerebral artery territory, was carried out in a randomized, placebo-controlled, double-blind, multicenter manner (Rocca et al. 1992). Sixteen centers in Europe and the United States participated in the study, which allowed the randomization of 805 out of 8781 recruited subjects (9.2%). Each patient underwent evaluation (including physical examinations, blood tests, EKG, Doppler sonography of the neck vessels, and brain CT) and treatment within 5 h of the onset of stroke. Neurological and disability scales used were the Canadian

Table 1. Distribution of EST patients according to age, sex, and severity of stroke at entry

	Men				Women			
	<65 years		>65 years		<65 years		>65 years	
CNS score	*n*	%	*n*	%	*n*	%	*n*	%
Severe (1.5–5.0)	58	7.3	120	15.2	37	4.7	148	18.7
Less severe (5.5–8.5)	118	14.9	142	17.9	56	7.1	113	14.3
Total	176	22.2	262	33.1	93	11.7	261	33.0

Neurological Score (CNS), the Toronto scale, and the Barthel index. Follow-up lasted 4 months in survivors. A specific schedule of GM_1 monosialoganglioside or identical-appearing placebo, was administered for 21 days. To balance the treatment and control groups for known prognostic variables within each center, patients were stratified according to sex, age (<65, >65 years), and clinical severity on admission (<5, >5 on CNS). Data on patient distribution are shown in Table 1.

As the 36-month recruitment period ended only in April 1990, the statistical evaluation of drug efficacy is still in progress. Observations can, however, be made on some aspects emerging from the baseline results. The relatively low percentage of randomization (9.2) was due mainly to the fact that treatment was delayed over the protocol limits. Hospital referral to the centers was generally rapid, as shown by the fact that approximately 46% of the patients were admitted within 1 h. The delay between stroke onset and treatment was therefore basically related to baseline assessment at the centers. Mortality rates ranged widely, from 0% in Lausanne and Pavia to 45.9% in Madrid. This may reflect some variability in the interpretation of selection criteria in different centers. While evidence of drug efficacy is still being assessed, logistical and protocol adaptation problems still appear to be major issues and are a long way from being completely resolved. It can nevertheless be said that the feasibility of a well-designed, large-scale trial has finally been demonstrated. What experiences could reliable clinical studies in future address? The unsatisfying results obtained so far from calcium antagonists and the achievements with gangliosides have not exhausted all pharmacological possibilities.

Besides the information on the penumbra, animal experiments have provided us with an increasing amount of knowledge on the cascade events which lead from ischemia to neuronal death (Siesjö 1981). In this process the central role of an excessive intracellular calcium influx is generally accepted, but a complex system of ion channel regulation has been identified

(Siesjö and Bengtsson 1989). Voltage-sensitive Ca^{2+} channels directly involved in Ca^{2+} influx modulation are thought to represent the site of action of nimodipine and similar compounds. Different, receptor-operated Ca^{2+} channels are activated by *N*-methyl-D-aspartate (NMDA), an analogue of the excitatory neurotransmitter glutamate, whose extracellular concentration is thought to reach cytotoxic levels in the ischemic area (Rothman and Olney 1986). Glutamate release actually involves multiple subtypes of postsynaptic receptors. Among them, AMPA seem to play a primum movens role, activating voltage- and receptor-operated calcium channel regulation in turn (Siesjo, personal communication). The progressive discovery of these mechanisms has moved the focus of pharmacological attempts from the final common event (i.e., excessive calcium entry) back to antecedent pathophysiological mechanisms. Competitive NMDA receptor blocking agents such as MK-801 did not overcome the initial phases of clinical studies, mainly because of their hallucinogenic properties (Albers et al. 1989). Noncompetitive blockers, devoid of relevant side effects, have been introduced more recently. Following a pilot study in which our group also participated, SL 82.0715 was tested in a multicenter European trial, the results of which are being eagerly awaited. Human studies on AMPA receptor active drugs may represent the future of this research area.

A different therapeutic approach is related to the potential cytotoxic effects of some free radicals. Superoxide ($O_2^{-\cdot}$) and hydroxyl ($OH^{\cdot}$), two of these molecules with a "half" bond available, seem to be involved in cerebral ischemia, on account of their damaging properties to membrane lipids (Cord 1985). A low concentration of molecular O_2 in the ischemic is thought to be responsible for the accumulation of free radicals, thus leading to a chemical disruption of membranes which concerns mainly the polyunsaturated fatty acid component. Intrinsic molecules, including vitamins A and E as well as enzymes such as superoxide dismutase, catalases, and peroxidases are known to act as scavengers against the accumulation of free radicals. A recently synthesized class of steroid derivatives, obtained by aminosubstitution in the 21 position to eliminate the classical hormone activity, demonstrated good scavenger properties in animals (Hall et al. 1988; Young et al. 1988; Hall and Pazara 1989). Studies which have not yet been published on tolerance and safety have been conducted on the 21-aminosteroid U-74006F. This compound is now ready for a phase III study, in which we will cooperate.

All the pharmacological approaches that we have so far considered pertain to brain tissue protection from cytotoxic damage. Another fundamental mainstream of therapy addresses the recanalization of occluded cerebral arteries by means of thrombolytic agents. Although the concept of early fibrinolysis is far from new, its application to acute cerebral ischemia is still under experimentation. Nevertheless, previous experiences in an analogous treatment of myocardial infarction have been encouraging (GISSI 1987), and recent pathophysiological results also support this approach

(Fieschi et al. 1989). We observed 80 ischemic stroke patients, evaluated within 6 h of onset with a protocol including all the instrumental exams of the EST plus angiography in each case. Intracranial arterial occlusions, very likely embolic, were detected in 58 cases (66%). Of 40 occluded patients followed up using transcranial Doppler, none of 11 in whom spontaneous recanalization had been observed, improved clinically. A good collateral circulation was attributed to subjects in whom vessels distal to the occlusion were fully visualized within 5 s of the end of contrast injection. As expected, patients with poor or absent collateral filling had the worst clinical outcome.

Since previous studies on the prevalence of cerebral embolism in stroke had reported lower figures (Fisher 1954; Mohr et al. 1978), we concluded that: (a) early angiography allowed us to detect a higher rate of embolic lesions; (b) spontaneous recanalization may occur; and (c) early fibrinolysis represents a rational intervention as spontaneous late reperfusion does not result in clinical improvement.

Past attempts at local intra-arterial injection (Del Zoppo et al. 1986) have been replaced by newer intravenously administered agents, such as rt-PA and scu-PA. A number of exploratory, angiographically controlled studies on posterior and anterior circulation strokes performed over the past few years with different intra-arterially administered fibrinolytics have been reported (Del Zoppo et al. 1988; Mori et al. 1988; Theron et al. 1989; Matsumoto and Satoh 1991; Henze et al. 1987; Hacke et al. 1987; Buteux et al. 1988). These observations were aimed at evaluating the balance between the occurrence of recanalization and the prevalence of hemorrhagic complications. Intravenous administration of the same agents seemed to be less risky and could reduce the delay in treatment, but doubts arose concerning its efficacy as regards both recanalization and clinical results. To clarify these issues, further trials were therefore carried out, some of which focused on angiographic findings and others on the patients' outcome (Abe et al. 1990; Brott et al. 1988; rt-PA Acute Stroke Study Group 1991; Mori et al. 1991; Yamaguchi 1991; von Kummer 1991).

Although a number of studies have been completed, many questions still remain unanswered. We now know that recanalization occurs following of fibrinolytic drug action, and that the risk/benefit ratio is positive. What we do not yet know is which drug is most effective and at what dosage, how often reocclusion occurs, how useful anticoagulants can be against this phenomenon, and how dangerous early reperfusion is for previously ischemic tissue (Schmidley 1990). However, pragmatic studies are essential if we are to answer the crucial question of clinical efficacy. At present, five randomized, placebo-controlled trials are in progress. Researchers appear oriented in different directions: a very short delay versus a longer interval between onset and treatment and rt-PA versus streptokinase administration. We are looking forward to data on prevalence and prognosis of hemorrhagic complications as much as we are to clinical improvements; the results of the trials are expected by the end of 1993.

Depending on these results, a four-arm trial should be planned at that time which combines the best fibrinolytic with the most promising anti-cytotoxic drug. That is in our opinion the rationale needed to obtain from brain protection agents a broader therapeutic window for revascularization. Early intervention, in the context of a correctly designed trial, is the basis on which good results will be achieved.

References

Abe T, Terashi A, Tohgi H et al. (1990) Clinical efficacy of intravenous administration of SM-9527 (t-PA) in cerebral thrombosis. Clin Eval 18:39–69

Albers GW, Goldberg MP, Choi DW (1989) N-Methyl-D-aspartate antagonists: ready for clinical trial in brain ischemia? Ann Neurol 25:398–403

American Nimodipine Study Group (1992) Clinical trial of nimodipine in acute ischemic stroke. Stroke 23:3–8

Argentino C, Sacchetti ML, Toni D et al. (1989) GM1 ganglioside therapy in acute ishemic stroke. Stroke 20:899–903

Astrup J, Siesjö BK, Symon L (1981) The state of penumbra in the ischemic brain: viable and lethal threshold in cerebral ischemia. Stroke 12:723–725

Brott T, Haley EC, Levy DE et al. (1988) The investigational use of t-PA for stroke. Ann Emerg Med 17:1202–1205

Buteux G, Jubault V, Suisse A (1988) Local recombinant tissue plasminogen activator to clear cerebral artery thrombosis developing soon after surgery. Lancet 1:1143–1144

Cord JM (1985) Oxygen-derived free radicals in postischemic tissue injury. N Engl J Med 312:159–163

Del Zoppo GJ, Zeumer H, Harker LA (1986) Thrombolytic therapy in stroke: possibilities and hazards. Stroke 17:595–607

Del Zoppo GJ, Ferbert A, Otis S et al. (1988) Local intra-arterial fibrinolytic therapy in acute carotid territory stroke: a pilot study. Stroke 19:307–313

Fieschi C, Argentino C, Lenzi GL et al. (1989) Clinical and instrumental evaluation of patients with ischemic stroke within the first six hours. J Neurol Sci 91:311–322

Fisher M (1954) Occlusion of the carotid arteries. Further experiences. Arch Neurol Psychiatry 72:187–204

GISSI (1987) Long-term effects of intravenous thrombolysis in acute myocardial infarction: final report. Lancet 1:871–874

Hacke W, Zeumer H, Ferbert A et al. (1988) Intra-arterial thrombolitic therapy improves outcome in patients with acute vertebro basilar occlusive disease. Stroke 19:1216–1222

Hall ED, Pazara KE (1989) Effects of novel 21-aminosteroid antioxidants on postischemic neuronal degeneration. In: Ginsberg MD, Dietrich WD (eds) Cerebrovascular diseases. Raven, New York, pp 387–391 (16th Princeton Conference)

Hall ED, Pazara KE, Braughler JM (1988) 21-Aminosteroid antioxidant peroxidation inhibitor U74006F protects against cerebral ischemia in gerbils. Stroke 19:997–1002

Henze TH, Boeer A, Tebbe U, Romatowski J (1987) Lysis of basilar artery occlusion with tissue plasminogen activator. Lancet 1:1391

Italian Acute Stroke Study Group (1988) Hemodilution in acute stroke: results of the Italian hemodilution trial. Lancet 1:318–321

Lenzi GLL, Frackowiack RSJ, Jones T (1982) Cerebral oxygen metabolism and blood flow in human cerebral ischemic infarction. J Cereb Blood Flow Metab 2:321–335

Matsumoto K, Satoh K (1991) Topical intraarterial urokinase infusion for acute stroke. In: Hacke W, del Zoppo GJ, Hirschberg M (eds) Thrombolytic therapy in acute ischemic stroke. Springer, Berlin Heidelberg New York

Mohr JP, Caplan LR, Melski JW et al. (1978) The Harvard Cooperative Stroke Registry: a prospective registry. Neurology (Minneap) 28:754–762

Mori E, Tabuchi M, Yoshida T, Yamadori A (1988) Intracarotid urokinase with thromboembolic occlusion of the middle cerebral artery. Stroke 19:802–812

Mori E, Yoneda Y, Ohkawa S et al. (1991) Double-blind placebo-controlled trial of intravenous recombinant tissue plasminogen activator (rt-PA) in acute carotid stroke. Neurology 41 Suppl 1:347

Rocca WA, Dorsey FC, Grigoletto F et al. (1992) Baseline results of the monosialoganglioside Early Stroke Trial. Stroke 23:519–526

Rothman SM, Olney JW (1986) Glutamate and the pathophysiology of hypoxic-ischemic brain damage. Ann Neurol 19:105–111

rt-PA Acute Stroke Study Group (1991) An open safety-efficacy trial of rt-PA in acute thromboembolic stroke: final report. Stroke 22:153

Schmidley JW (1990) Free radicals in central nervous system ischemia. Stroke 21:1086–1090

Siesjö BK (1981) Cell damage in the brain. A speculative synthesis. J Cereb Blood Flow Metab 1:155–185

Siesjö BK, Bengtsson F (1989) Calcium fluxes, calcium antagonists and calcium related pathology in brain ischemia, hypoglicemia and spreading depression: a unifying hypothesis. J Cereb Blood Flow Metab 9:127–140

Theron J, Courtheoux P, Casaseo A et al. (1989) Local intraarterial fibrinolysis in the carotid territory. AJNR 10:753–765

Von Kummer R (1991) Intravenous tissue plasminogen activator in acute stroke. In: Hacke W, del Zoppo GJ, Hirschberg M (eds) Thrombolytic therapy in acute ischemic stroke. Springer, Berlin Heidelberg New York, pp 161–167

Weinstein PR, Anderson GG, Telles DA (1980) Neurological deficit and cerebral infarction after temporary middle cerebral artery occlusion in unanesthetized cats. Stroke 17:318–324

Yamaguchi T (1991) Intravenous rt-PA in acute embolic stroke. In: Hacke W, del Zoppo GJ, Hirschberg M (eds) Thrombolytic therapy in acute ischemic stroke. Springer, Berlin Heidelberg New York, pp 168–174

Young W, Wojak JC, De Crescito V (1988) 21-Aminosteroid reduces ion shifts and edema in the rat middle cerebral artery occlusion model of regional ischemia. Stroke 19:1013–1019

Threshold of Calcium Influx After Global and Focal Ischemia: Implications for the Window of Therapeutic Opportunity

T.J. De Graba and J.C. Grotta

Introduction

Extensive experimental evidence has confirmed that the early events of ischemia-induced neuronal damage are characterized by excessive influx of calcium [13, 14]. Activation of calcium-dependent enzymes such as protein kinase C, calmodulin (CaM) kinase II, calpain I, calcineurin, and phospholipase A_2 is believed to play a major role in neuronal excitotoxicity [2, 4, 5, 16]. In addition, therapeutic intervention with agents which target either voltage- or receptor-mediated ion channels are found to have neuroprotective affects in animal models of focal and global ischemia [6, 7, 12, 14]. Despite these findings, the results of clinical therapeutic trials using these agents have thus far been inconclusive in human stroke patients [1]. Possible explanations for these results include differences in vulnerability between human and animal neurons, lower dose tolerance due to coexisting cardiovascular disease, and insensitivity in methods of detecting modest neuronal protection in humans.

However, another important reason for the negative results may be that neuroprotective therapy has been started too late. This hypothesis is supported by the improvement in efficacy seen with preischemic versus postischemic neuroprotective agent treatment in animal models [7]. In addition, it has been shown that maximal infarct volume [9], activation of calcium dependent enzymes [2], and alterations in neuronal protein synthesis [10] occurs early after focal ischemia.

Understanding of the sequence of events leading to the ultimate destruction of ischemic neurons is essential in instituting neuroprotective agents in a timely fashion. The purpose of recent work in our laboratory has been to explore the time course of disturbed calcium homeostasis as determined by measurements of Ca-CaM binding after global and focal ischemia, and by correlating these data with subsequent permanent histological damage to establish theoretical time windows of opportunity beyond which irreversible calcium-mediated damage has occurred. The present study evaluates the effects of variable durations of global and focal ischemia on Ca-

Department of Neurology, University of Texas, 6431 Fannin, B 20708, Houston, TX 77225, USA

Cerebral Ischemia and Basic Mechanisms
Ed. by A. Hartmann, F. Yatsu, and W. Kuschinsky

CaM binding using immunohistochemical staining as previously described [11]. A secondary objective was to explore the effects of reperfusion itself on reversing disruption of calcium homeostasis since neuroprotective therapies may ultimately be combined with reperfusion therapy (i.e., thrombolysis).

Methods

Global Ischemia

Male Wistar rats (250–300 g) were subjected to global ischemia using an adaptation of the four-vessel occlusion model. Coagulation of the vertebral arteries was performed under chloral hydrate anesthesia, and loose ligatures were placed on the carotid arteries for future access. Twenty-four hours later, after a few seconds of ether inhalation, nontraumatic clips were placed on the common carotid arteries, and the cervical muscles were ligated for 5, 10, or 20 min. A heating lamp and warming blankets were used to maintain rectal and scalp temperature at 37°C. Electroencephalographic (EEG) activity was monitored to ensure duration and effectiveness of the global ischemia. Rats that did not display a complete loss of EEG rhythm or had a postischemic seizure were not included in the analysis.

Preparation of brain slices (40-μm thickness) was carried out for CaM immunohistochemical staining as previously described [11]. Brain sections in groups of six rats exposed to 5, 10, or 20 min of ischemia after 2 and 24 h of reperfusion were incubated in sheep-induced CaM antibody, which binds specifically to free CaM (i.e., the CaM not bound to Ca^{2+} and target protein). The sections were incubated with rabbit anti-sheep peroxidase-conjugated secondary antibody, which binds and stains only CaM-antibody complex. Thus, only normal cells with free CaM stain, and ischemic cells with Ca^{2+}-activated CaM-target protein complex do not stain. Sets of normal and postischemic brain were processed simultaneously to ensure comparable staining. Intensity of staining in the dorsal CA1, lateral CA3, endal limb of dentate, and layers I–VI of the parietal cortex was graded on a 4-point scale (4, extensive staining of normal neuronal soma; 3, some staining of soma; 2, some staining but soma not distinguishable; 1 minimal staining; 0, no staining). The evaluator was blinded to the duration of ischemia and the period of reperfusion. Statistical analysis was performed using Kruskal-Wallis one-way analysis of variance (ANOVA) by rank.

Seventy-two hours after 5, 10, or 20 min of ischemia, while under ether anesthesia, six rats from each group and six control rats were perfused with phosphate-buffered 10% formalin, decapitated, and the brains embedded in paraffin to be sectioned and stained with hematoxylin and eosin (H&E). Ischemic histological neuronal changes were quantitated in CA1, CA3, dentate, and parietal cortex by light microscopy. Regions were scored from

0 to 4 based on the percentage of neurons with shrunken eosinophilic cytoplasm and pyknotic nuclei (0, 0%; 1, 25%; 2, 50%; 3, 75%; 4, 100%). Evaluation was again blinded as to the duration of ischemia, and group differences were evaluated by Kruskal-Wallis one-way ANOVA.

Focal Ischemia

Male spontaneously hypertensive rats (250–300 g) were fasted overnight with free access to water and anesthetized with chloral hydrate intraperitoneally prior to surgery. The femoral artery was cannulated with a polyethylene catheter for blood pressure monitoring and serial blood gas sampling. A microprobe (IT-18 flexible implantable) was used to keep skull temperature at 37° ± 0.5°C, and rectal temperature was monitored with a YSI Tele-thermometer. The right common carotid artery was isolated through a midline ventral cervical incision and tagged with 3.0 silk. An incision was made perpendicular to a line between the external auditory ear canal and the lateral canthus of the right eye through the temporalis muscle which was retracted with 4.0 silk. Under direct visualization with a Zeiss surgical microscope, the right middle cerebral artery (MCA) was exposed through a 1 × 3 mm burr hole 2–3 mm rostral to the fusion of the zygomatic arch with the squamosal bone. A continuous 0.9% saline drip was maintained over the drilling site to prevent thermal injury to the underlying brain tissue. A 23-G hypodermic needle was used to pierce and open the dura along the entire length of the burr hole.

The method for producing reversible ischemia was a modification of the procedure described by Kaplan et al. [9] and Brint et al. [3]. A 10-0, BV130-5 Ethicon Neurosurgical suture was placed under the MCA rostral to the rhinal fissure, proximal to the major bifurcation of the MCA but distal to the lenticulostriate arteries. The suture was rotated clockwise so that one end pointed toward the frontal bone forming a right angle to the burr hole. Verification of the absence of blood flow was made visually with the surgical microscope. Immediately after MCA occlusion, the right common carotid artery was occluded using two atraumatic Heifetz aneurysm clips to prevent blood flow. Reperfusion was achieved by turning the suture counter clockwise and removing it. Aneurysm clips were removed and flow in both the common carotid artery and MCA were confirmed visually with the operating microscope. In rats exposed to 24 h of focal ischemia, permanent ligation of the right common carotid artery and MCA was performed. Again, absence of blood flow was verified visually.

Head and rectal temperature was maintained between 36.5° and 37°C. This technique produces a nearly exclusive cortical infarction sparing the deep white matter and basal ganglia.

Rats exposed to 2 min (n = 11), 2 h (n = 9), 4 h (n = 10), or 24 h (n = 7) of focal ischemia were anesthetized and sacrificed without reperfusion.

Groups of rats were exposed to 1 h of occlusion with 1 h of reperfusion (n = 8) and 2 h of occlusion with 22 h of reperfusion. Animals were anesthetized with ethyl ether (purified grade) and sacrificed.

Brain sections were prepared as above, and CaM staining assays were performed. The ischemic region of each animal exposed to focal ischemia was graded on a scale of 0–15 (0, normal; 15, largest area of severe loss of CaM staining). H&E staining of rats exposed to 2 h of occlusion with 22 h of reperfusion and rats exposed to 24 h of MCA occlusion was performed. The area of infarction based on computer-assisted image analysis was recorded.

The specificity of the CaM antibody to free CaM was demonstrated by in vitro immunohistochemical dot staining of CaM incubated in ethylene glycol tetraacetic acid, Ca^{2+}, and Ca^{2+} with brain homogenate. Staining occurred only in the absence of Ca^{2+} and target protein [11]. Overlays of sodium dodecyl sulfate polyacrylamide gel electrophoresis confirm the presence of target protein and CaM in ischemic brain homogenate up to 4 h of focal ischemia.

Results

No significant difference in physiologic parameters, including blood pressure, pO_2, pCO_2, pH, and temperature, were noted among all comparable control and ischemic animal groups.

Loss of CaM staining, representing Ca^{2+} influx and Ca-CaM binding, correlated with increasing durations of ischemia across all brain regions (Fig. 1). CaM staining 24 h after 5 min versus 10 min of ischemia was 3.2 versus 1.9, respectively ($p < 0.05$), and after 10 min versus 20 min of ischemia was 1.9 versus 1.0, respectively ($p < 0.001$). The CA1 region and the cortex displayed the greatest reduction of CaM staining in response to the increasing durations of ischemia. Similar but much less severe reduction was found after 2 h of reperfusion (Fig. 2).

The histological score demonstrated worse damage after 20 min versus 10 min of ischemia across all brain regions at 72 h of reperfusion (2.4 versus 1.4, respectively; $p < 0.005$) (Table 1). Again, the most significant neuronal damage was seen in the CA1 region, with over 85% ischemic neurons after 20 min of ischemia (Table 1, Fig. 3). In comparison with other neuronal populations, the CA1 showed significantly more damage after shorter durations of ischemia. After 10 min of ischemia, the CA3, dentate, and cortex displayed under 25% ischemic neurons, whereas the CA1 had over 65% (Table 1). Histological damage correlated very closely with CaM staining, best seen in the CA1 region (Fig. 3). An increase in the percentage of ischemic neurons seen after 72 h of reperfusion corresponds to a decrease in CaM staining.

An apparent threshold of Ca^{2+} influx to ischemic duration is represented by the change in CaM staining in all regions seen at 5 versus 10 min

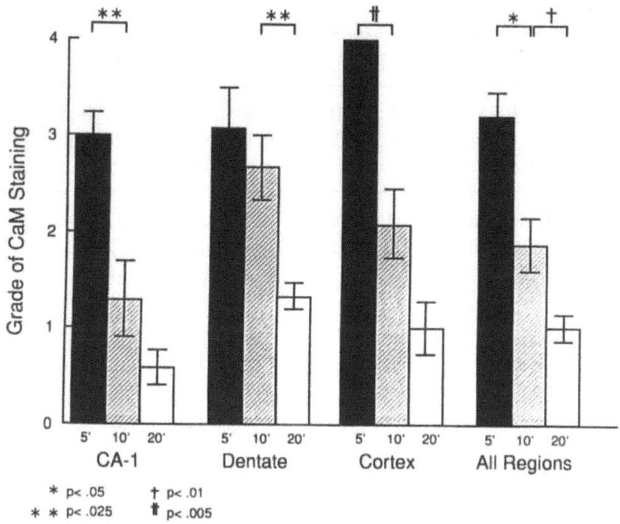

Fig. 1. Global ischemia. CaM staining after 24 h of reperfusion in all regions, CA1, CA3, dentate, and cortex, demonstrating a graded response to increasing durations of ischemia (5, 10, or 20 min). CA1 and cortex show the greatest vulnerability to ischemia, with significantly decreased staining between 5 and 10 min of ischemia ($p < 0.025$). Grading score of 4, normal staining; 0, no staining

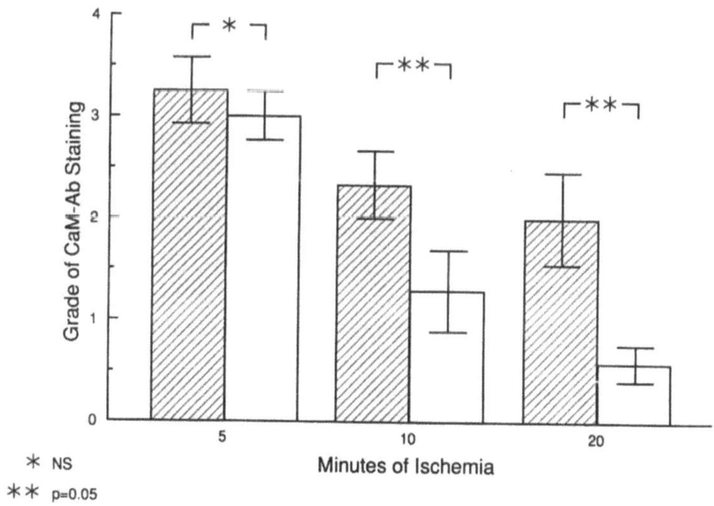

Fig. 2. Global ischemia. Graded response of Ca-CaM binding to increasing durations of ischemia after 24 h of reperfusion (*open bars*) in all regions combined. Note a similar pattern but significantly less staining after 2 h of reperfusion (*hatched bars*)

Table 1. Histological score (72-h reperfusion)

Ischemia	All regions	CA-1	CA-3	DENT	CTX
5 min	1.05 ± 0.35	1.86 ± 0.22	0.42 ± 0.18	0.50 ± 0.20	1.00 ± 0.20
10 min	1.43 ± 0.38	2.75 ± 0.28	0.50 ± 0.20	0.58 ± 0.14	0.83 ± 0.19
20 min	2.38 ± 0.32	3.53 ± 0.11	1.83 ± 0.55	1.75 ± 0.47	1.58 ± 0.25

Values are mean ± SEM Score determined by hematoxylin and eosin staining, based on percentage of cells with shrunken eosinophilic cytoplasm and pyknotic nuclei (0, 0%; 1, 25%; 2, 50%; 3, 75%; 4, 100%)

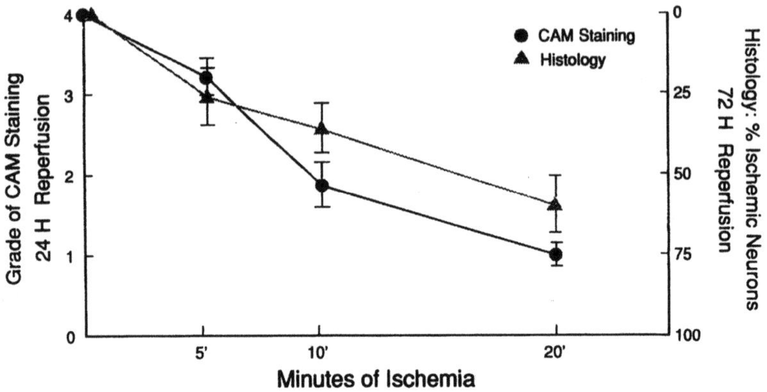

Fig. 3. Histologic damage, represented by percentage of ischemic neurons seen at 72 h of reperfusion, compared with CaM staining at 24 h in all regions. Increase in percentage of ischemic neurons correlates closely with increasing Ca^{2+} influx, represented by loss of calmodulin staining seen after 5, 10, and 20 min of global ischemia. Grading score (CaM staining) of 4, normal staining; 0, no staining. Histologic score, see text

of ischemia (3.2 versus 1.9, $p < 0.05$; Fig. 1). This threshold phenomenon can be seen by examining CA1 CaM staining in individual rats (Fig. 4). In all rats, neurons revealed nearly normal staining after 5 min of ischemia and severe loss of CaM staining after 20 min of ischemia. A variable response, however, was seen after 10 min of ischemia; some rats displayed nearly normal staining while others showed severe loss of staining, suggesting a threshold of Ca-CaM binding to target protein in vulnerable neurons at 10 min of ischemia.

In focal MCA ischemia, significant loss of CaM staining was noted by 1 h and was maximal by 4 h. Again, Ca-CaM was seen in a time-dependent, graded pattern with progressive decrease in staining between immediate versus 1-h occlusion 1-h reperfusion (0.57 versus 8.13, $p < 0.005$), 1-h versus 2 h (8.13 versus 11.28, $p < 0.05$), 2 h versus 4 h (11.28 versus 13.8, $p < 0.01$), and 4 h versus 24 h (13.8 versus 12.83, NS; Fig. 5). Of note, was an apparent threshold of calcium influx seen at 1 h of ischemia, demonstrated

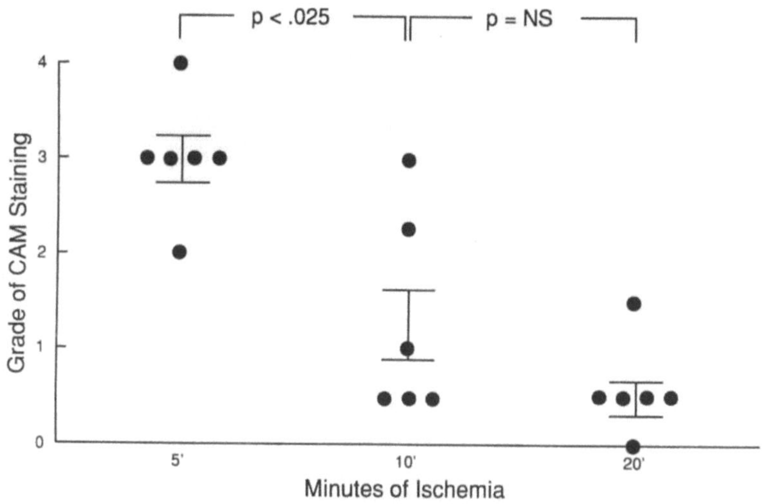

Fig. 4. Global ischemia. CaM staining of CA1 region after 24 h of reperfusion, represented as individual scores and means ± S.E. Nearly normal staining is seen in all rats exposed to 5 min of ischemia, and servere loss of staining is seen in all rats after 20 min of ischemia. A variable response is seen after 10 min, suggesting a threshold of Ca^{2+} influx at this duration of ischemia

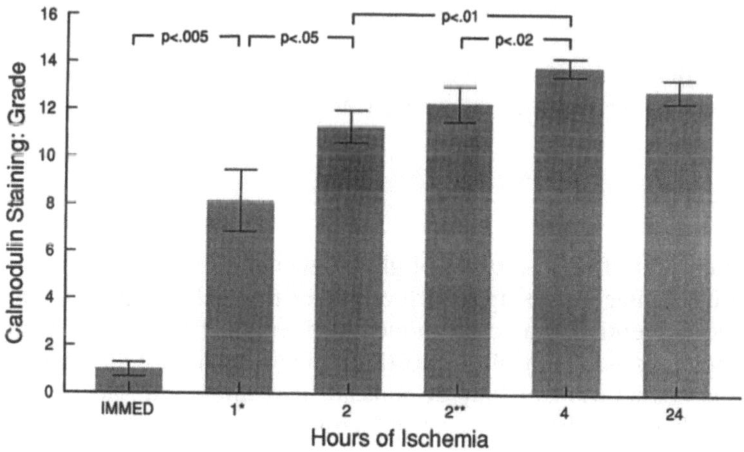

Fig. 5. CaM staining after increasing durations of focal MCA occlusion. Significant decrease in staining is noted at 1 h of occlusion with 1 h of reperfusion, becoming maximal at 4 h of occlusion. *, 1 h reperfusion; **, 22 h reperfusion

by plotting individual scores (Fig. 6). A variable response to ischemia was seen, as was noted after 10 min of global ischemia.

H&E staining demonstrated infarcts comparable in size to the size of the "central core" of CaM staining loss seen in the ischemic area 24 h after

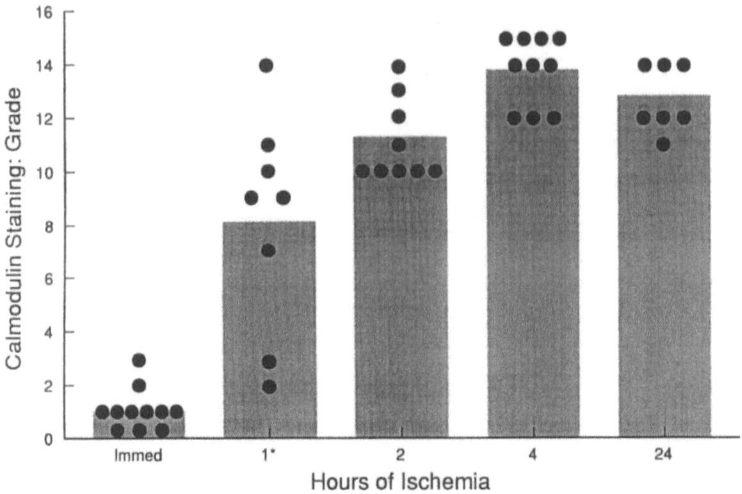

Fig. 6. Individual score of each animal CaM staining and ischemic region. Nearly normal staining is seen in brief (2-min) ischemia, and severe loss of staining is seen at 2, 4, and 24 h of ischemia. A variable response is seen at 1 h of occlusion/1 h of reperfusion, suggesting a threshold at this duration of ischemia. *, 1 h reperfusion

ischemia. Note that areas of mild loss of CaM staining (considered to be a border zone) did not show up as infarct by H&E staining.

Discussion

The observation that after 10–20 min of global ischemia followed by reperfusion, neuronal damage takes up to 72 h or longer to reach maximal extent has led to the perception that calcium-mediated events triggered by such brief but complete ischemic insults occur throughout this time. Our data do not refute this possibility, but they do indicate that after both global and focal ischemia there is substantial Ca-CaM target protein activation by 2 h even in the presence of reperfusion. In the global model we found that more than 10 min of ischemia followed by reperfusion results in irreversible Ca-CaM target protein activation and subsequent maximal cellular damage in vulnerable regions (CA1).

In the present study focal MCA ischemia lasting more than 1 h was associated with abnormal Ca-CaM binding in the entire cortical region supplied by the occluded artery, and maximum Ca-CaM binding occupying this entire region was found after 2 h of MCA occlusion. Of particular interest is that reperfusion after only 1 h of occlusion did not reverse Ca-CaM binding, indicating that reperfusion therapies must be coupled with neuroprotective therapies to provide optimal reduction of calcium mediated ischemic injury. The inability of reperfusion to rectify abnormal calcium homeostasis and the complimentary effect of reperfusion and neuronal protective therapies have recently been demonstrated by Uematsu et al. [13] and Ziven and Mazzarella [15].

The regions of maximal Ca-CaM binding found in our study correlated almost exactly with the extent of cortical necrosis seen by light microscopy after 12 h. Less than 2 h of ischemia was associated with a border zone of limited Ca-CaM binding surrounding a central core of maximal binding. The border zone was not included in the area of necrosis seen on light microscopy at 24 h. Longer periods of MCA occlusion, even if followed by reperfusion, resulted in maximal Ca-CaM binding and necrosis in the entire MCA territory at 24 h. Therefore, reperfusion occurring by 2 h can salvage threatened border zones having less severe Ca-CaM binding. While we have not directly evaluated the ability of calcium antagonist drugs to reverse this process, other investigators have found progressively decreasing protection using calcium or glutamate antagonists started up to 2 h after the onset of MCA occlusion.

Our data may help explain why calcium and glutamate antagonists are more effective in focal than in global ischemic models. After only 10 min of global ischemia there is maximal Ca-CaM binding and neuronal destruction in CA1 where histologic damage is usually measured. However, more than 2 h of MCA occlusion is required to produce maximal Ca-CaM binding and neuronal destruction in the entire cortex supplied by the occluded MCA. Although cerebral blood flow measurements were not carried out in this study, Brint et al. [3] shown that except in a central core of cortex supplied by a permanently occluded MCA, cerebral blood flow does not fall in collaterally supplied border zones to levels as low as those occurring after four-vessel occlusion. Thus it is likely that damage in less severely ischemic border zones after MCA occlusion is due to partial, and therefore presumably more therapeutically reversible, calcium disturbances.

It is worth noting that our studies are confined only to disturbances of calcium, in particular its binding with CaM. We have carried out parallel studies of CaM kinase II activity [2, 5] and have demonstrated that calcium-activated enzyme systems are also profoundly disturbed very early after the onset of ischemia. However, our data do not preclude the possibility that other "rescue therapies" such as growth factors, which target later events in the ischemic cascade, might be effective if started more than 2 h after ischemia.

We have demonstrated that CaM and its target proteins appear to remain present at least after 2 h of focal ischemia. It must be considered that the loss of CaM staining at later durations of ischemia may be due to CaM binding proteins which are preferentially preserved during ischemic insult [8]. This however, is more likely to be a late occurrence in comparison to early maximal Ca-CaM binding.

In conclusion, to the extent that our animal models reflect conditions of ischemia in humans, our data indicate that if left untreated, irreversible calcium-mediated damage occurs after 10 min of global ischemia and 1 h of MCA occlusion, even if reperfusion occurs. After MCA occlusion, maximal calcium-mediated damage occurs by 4 h. These data establish the windows of opportunity for therapeutic intervention aimed at limiting early events activated by Ca-CaM binding.

References

1. American Nimodipine Study Group (1992) Clinical trial of nimodipine in acute ischemic stroke. Stroke 23:3–8
2. Aronowski J, Grotta JC, Waxham MN (1992) Ischemia-induced translocation of Ca/calmodulin-dependent protein kinase II: potential role in neuronal damage. J Neurochem 58:1743–1753
3. Brint S, Jacewicz M, Kiessling M, Tanabe J, Pulsinelli W (1988) Focal brain ischemia in the rat: method for reproducible neocortical infarction using tandem occlusion of the distal middle cerebral and ipsilateral common carotid arteries. J Cereb Blood Flow Metab 8:474–485
4. Choi DW (1990) Cerebral hypoxia: some new approaches and unanswered questions. J Neurosci 10(8):2493–2501
5. DeGraba TJ, Aronowski J, Waxham N, Grotta JC (1991) Calcium influx and CaM-kinase II activity as predictors of neuronal damage. J Cereb Blood Flow Metab 11 Suppl 2:S-521
6. Grotta JC (1987) Current medical and surgical therapy for cerebrovascular disease. N Engl J Med 317:1505–1515
7. Grotta JC, Picone CM, Ostrow PT, Strong RA, Rhoades HM, Earls RM, Yao LP, Dedman JR (1990) CGS-19755, a competitive NMDA receptor antagonist, reduces calcium-calmodulin binding and improves outcome after global cerebral ischemia. Ann Neurol 27:612–619
8. Jacewicz M, Kiessling M, Pulsinelli WA (1986) Selective gene expression in focal cerebral ischemia. J Cereb Blood Flow Metab 6:263–272
9. Kaplan B, Brint S, Tanabe J, Jacewicz M, Wang XJ, Pulsinelli W (1991) Temporal thresholds for neocortical infarction in rats subjected to reversible focal cerebral ischemia. Stroke 22:1032–1039
10. Mies G, Ishimaru S, Xie Y, Soe K, Hossmann K-A (1991) Ischemic thresholds of cerebral protein synthesis and energy state following middle cerebral artery occlusion in rat. J Cereb Blood Flow Metab 11:753–761
11. Picone CM, Grotta JC, Earls R, Strong R, Dedman J (1989) Immunohistochemical determination of calcium-calmodulin binding predicts neuronal damage after global ischemia. J Cereb Blood Flow Metab 9:805–811
12. Simon R, Shiraishi K (1990) N-Methyl-D-aspartate antagonist reduces stroke size and regional glucose metabolism. Ann Neurol 27:606–611
13. Uematsu D, Araki N, Greenberg JH, Sladky J, Reivich M (1991) Combined therapy with MK-801 and nimodipine for protection of ischemic brain damage. Neurology 41:88–94
14. Wieloch T, Siesjö BK (1982) Ischemic brain injury: the importance of calcium, lipolytic activities, and free fatty acids. Pathol Biol (Paris) 20:269–277
15. Zivin JA, Mazzarella (1991) Tissue plasminogen activator plus glutamate antagonist improves outcome after embolic stroke. Arch Neurol 48:1235–1238
16. Zivin JA, Kochhar A, Saitoh T (1990) Protein phosphorylation during ischemia. Stroke 21 Suppl 3:117–121

Non-NMDA Glutamate Receptors in the Regulation of Neuronal Ca^{2+} and Excitotoxicity

R.J. Miller[2], J.R. Brorson[1,2], D. Bleakman[2], and P.S. Chard[2]

The excitatory amino acid glutamate mediates excitatory synaptic transmission and causes elevations in the intracellular Ca^{2+} concentration ($[Ca^{2+}]_i$) by acting on a variety of receptor types. Under normal physiological conditions, this happens in a controlled manner [28]. However, in pathological situations such as ischemia, status epilepticus, or trauma excessive glutamate release is thought to lead to prolonged exposure of the receptors to their agonist, causing excessive accumulation of intracellular Ca^{2+} [28]. This uncontrolled elevation of $[Ca^{2+}]_i$ can lead to neuronal death, with the time delay providing a potential window of opportunity for therapeutic intervention. Thus, understanding these processes seems to offer hope for the more effective treatment of brain ischemia and stroke [13].

Several sources can contribute to the increase in $[Ca^{2+}]_i$. Ca^{2+} can enter the cell through voltage-gated Ca^{2+} channels or through ligand-gated Ca^{2+} channels such as the ionotropic glutamate receptors. In addition, Ca^{2+} can enter the cytoplasm by release from intracellular Ca^{2+} stores, such as those activated by inositol trisphosphate (IP_3) or by Ca^{2+}-induced Ca^{2+} release, associated with ryanodine receptors [8]. Once in the cytoplasm, Ca^{2+} participates in the regulation of a number of processes. First, it can directly affect the electrical properties of the cell membrane by its regulation of Ca^{2+}-activated K^+ or Cl^- channels. Second, it acts as a trigger for neurotransmitter release. Finally, Ca^{2+} can activate a number of intracellular enzymes such as Ca^{2+}-activated protein kinases, phosphatases, lipases, and proteases. These enzymatic processes allow Ca^{2+} to control the remodeling of cell structural elements as well as providing a mechanism by which excessive elevations in $[Ca^{2+}]_i$ can cause long-term structural damage leading to neuronal cell death [39].

Glutamate receptors can be divided into ionotropic receptors, which are directly linked to cationic ion channels, and metabotropic receptors, where glutamate leads to the activation of an effector through the participation of a G protein. The ionotropic receptors have been further divided into

[1] Department of Neurology, University of Chicago, Chicago, IL 60637, USA
[2] Department of Pharmacology and Physiological Sciences, University of Chicago, Chicago, IL 60637, USA

Cerebral Ischemia and Basic Mechanisms
Ed. by A. Hartmann, F. Yatsu, and W. Kuschinsky

subclasses identified by the specific synthetic agonist acting at each subclass. The NMDA receptors, activated by *N*-methyl-D-aspartate, are known to be highly permeable to Ca^{2+} [29], so that they directly contribute to the intracellular Ca^{2+} load. Thus they have been considered likely participants in processes of neuronal plasticity and Ca^{2+}-mediated excitotoxic damage. On the other hand, the non-NMDA glutamate receptors, activated by kainate (KA) and AMPA, have conventionally been thought to be Ca^{2+} impermeable [28, 29] and to contribute to the $[Ca^{2+}]_i$ signal only indirectly, following the Na^+-dependent depolarization of the cell membrane and the opening of voltage-gated Ca^{2+} Channels [32].

NMDA receptors have received the greatest amount of attention in relation to glutamate-mediated neuronal death. Work from a number of laboratories has implicated NMDA receptors in in vitro models of glutamate excitotoxicity [38, 39], and NMDA receptor activation seems to be an essential mediator of neuronal damage in several in vivo models of ischemia and trauma [16, 35]. However, there is also reason to believe that the NMDA receptor does not completely account for glutamate-mediated neuronal toxicity. In certain models of global ischemia, specific antagonists of non-NMDA receptors alone [41] or in combination with NMDA antagonists [24] are neuroprotective. Also, domoic acid poisoning, resulting from the ingestion of contaminated mussels, is an example of a clinical disease in which activation of non-NMDA glutamate receptors is the likely culprit [45]. Second, recent evidence has also suggested that in some physiological systems non-NMDA receptors also exhibit a high permeability to Ca^{2+}. Murphy and Miller [32] showed that in cultured striatal neurons the Ca^{2+} influx induced by KA could not be entirely blocked by antagonists of voltage-gated Ca^{2+} channels. Iino et al. [22, 34] showed that a small fraction of cultured hippocampal neurons express glutamate receptors which when activated by KA are Ca^{2+} permeable, giving rise to an inwardly rectifying current-voltage (I–V) relationship. They denoted these Ca^{2+}-permeable KA receptors as type II KA receptors; in distinction to the typical type I KA receptors which are Na^+-permeable, Ca^{2+}-impermeable, and have an I–V relationship that is approximately linear. Other authors have also reported evidence for Ca^{2+}-permeable KA/AMPA receptors in various types of cultured neurons [20, 33, 37]. Additionally, the cloning of several non-NMDA glutamate receptor subunits has revealed that these proteins, when expressed in various combinations in oocytes or cell lines, can give rise to glutamate receptors which are either Ca^{2+}-permeable or Ca^{2+} impermeable [18].

To study the various aspects of Ca^{2+} signaling in normal physiology and in excitotoxicity, we sought an in vitro neuronal system in which these various types of Ca^{2+}-mediated processes might play important roles. We chose to study cerebellar Purkinje neurons because several of their features indicate that Ca^{2+}-mediated processes might be particularly important in these cells. First, they have among the highest levels in the brain of several of the Ca^{2+} binding proteins, such as calbindin D-28k [2, 12], thought to be involved in Ca^{2+} buffering. Second, they bear markers of intracellular Ca^{2+} stores such as the IP_3 and ryanodine receptors [8]. Finally, they are known to have profuse glutamatergic synaptic inputs, to exhibit Ca^{2+}-dependent action potentials, and to be extremely sensitive to hypoxic and ischemic neuronal death [26]. Interestingly, most studies have found that in adult tissues, Purkinje neurons lack NMDA receptors [1]. Thus it would seem that excitotoxicity mediated by pathways other than the NMDA receptor might be particularly important to the vulnerability of Purkinje cells.

We made use of the embryologic data which suggest that Purkinje neurons are produced early in cerebellar development, on embryonic day 15 in the rat, preceded only by the production of the deep cerebellar nuclear neurons in order to obtain cultured neurons enriched in Purkinje cells. Thus, by taking cerebellar from gestational day 16 rat embryos, we obtained dissociated neuronal cultures in which about one-half of the neurons bore markers of Purkinje cells, including calbindin, IP_3 receptors, and ryanodine receptors [8]. We utilized whole-cell patch clamp and Fura-2 based $[Ca^{2+}]_i$ microfluorimetric methods, either singly or in combination, to study these cultured neurons. First, using Fura-2 microfluorimetry, we screened the cultured neurons for glutamate receptors. We found that all of the viable neurons responded to KA (10–100 μM) with increases in $[Ca^{2+}]_i$. In contrast, some neurons did not respond at all to NMDA (50 μM) even under conditions maximizing NMDA responsiveness, i.e., Mg^{2+}-free, glycine-supplemented solutions. Other cells had small responses to NMDA that seemed to be synaptically mediated in that they could be blocked by 6-cyano-7-nitroquinoxaline-2,3-dione (CNQX), an antagonist of non-NMDA glutamate receptors, indicating that the cells did not themselves bear NMDA receptors. Only a minority of the cultured neurons, about one-third, responded directly to NMDA with increases in $[Ca^{2+}]_i$ which were not blocked by CNQX but could be blocked by the NMDA antagonist aminophosphovalerate (APV) [8]. Thus all cells responded to KA with Ca^{2+} influxes, but most did not show direct responses to NMDA.

According to the original model Ca^{2+} influx observed upon activation of non-NMDA receptors is due to Na^+ influx through the receptor ionophore, causing depolarization and Ca^{2+} influx through voltage-gated Ca^{2+} channels. If this were the explanation for the KA-induced rise in $[Ca^{2+}]_i$ in cultured cerebellar neurons, it should not occur if the external Na^+ were replaced by an impermeant cation such as *N*-methyl-D-glucamine (NMDG). In some cells, this indeed seemed to be the case, in that there was little or no Ca^{2+} influx in Na^+ free solutions, despite large $[Ca^{2+}]_i$ rises in response to KA in Na^+-containing solutions. However, in most of the cells studied, a substantial Ca^{2+} influx was observed even in Na^+-free conditions [9]. Some 77% of cells responded in a solution of 2 mM Ca^{2+}, 140 mM NMDG (2 Ca/NMDG) to 30 μM KA, with increases in the $[Ca^{2+}]_i$ of up to 1 μM. In cells exposed to KA in both Na^+-containing and Na^+-free solutions, the $[Ca^{2+}]_i$ increases in Na^+-free solutions were about 60% of those in Na^+-containing solutions. The concentration dependence of this KA-induced Ca^{2+} influx showed a threshold concentration of between 3 and 10 μM KA; AMPA could also induce a Na^+-independent Ca^{2+} influx. The effects of KA in either Na^+-containing or Na^+-free media were not reduced by the NMDA receptor antagonist APV (50 μM) but could be completely blocked by the non-NMDA antagonist CNQX (10 μM; Fig. 1). These data suggested that the KA/AMPA receptors might be directly Ca^{2+} permeable.

In order to confirm the conclusion that KA causes Ca^{2+} influx in these cerebellar neurons that does not involve voltage-sensitive Ca^{2+} channels we made use of both pharmacological and physiological approaches. We attempted to block Ca^{2+} influx using the nonselective organic voltage gated Ca^{2+} channel antagonist TA-3090 (clentiazem) [5]. TA-3090 effectively inhibited both depolarization-induced whole-cell Ca^{2+} currents and 50 mM K^+-induced Ca^{2+} influxes. In contrast, when KA-induced Ca^{2+} influxes were examined in 2 Ca/NMDG-containing solutions, TA-3090 was ineffective at blocking the $[Ca^{2+}]_i$ rise. Similarly, under voltage clamp conditions in 2 Ca/NMDG solutions, KA-induced inward currents were unaffected by TA-3090, nor were they inhibited by 100 μM Cd^{2+}, an inoganic blocker of voltage-gated Ca^{2+} channels.

Another way to eliminate contributions of voltage-gated Ca^{2+}currents to Ca^{2+} influx is to prevent depolarization of the cell membrane by voltage

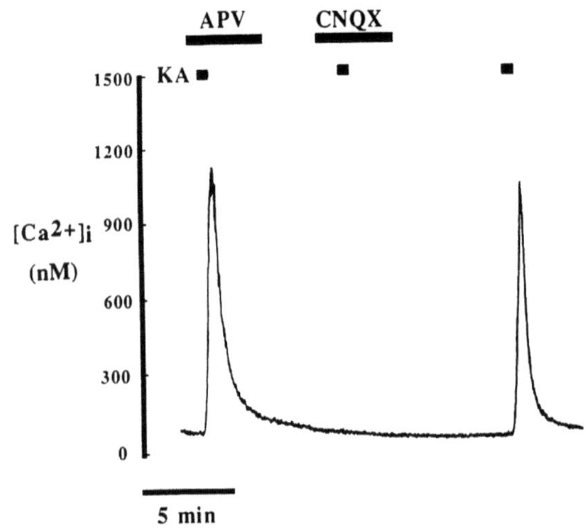

Fig. 1. Na^+-independent KA-induced $[Ca^{2+}]_i$ responses in cultured cerebellar neurons. $[Ca^{2+}]_i$ was monitored in a single cerebellar neuron, using Fura-2 based $[Ca^{2+}]_i$ microfluorimetry, during exposure to KA (30 μM) in a Na^+-free solution (2 mM Ca^{2+}, 140 mM NMDG). KA produced a large rise in $[Ca^{2+}]_i$ which was not inhibited by the NMDA antagonist APV (50 μM), but fully blocked by the non-NMDA antagonist CNQX (10 μM)

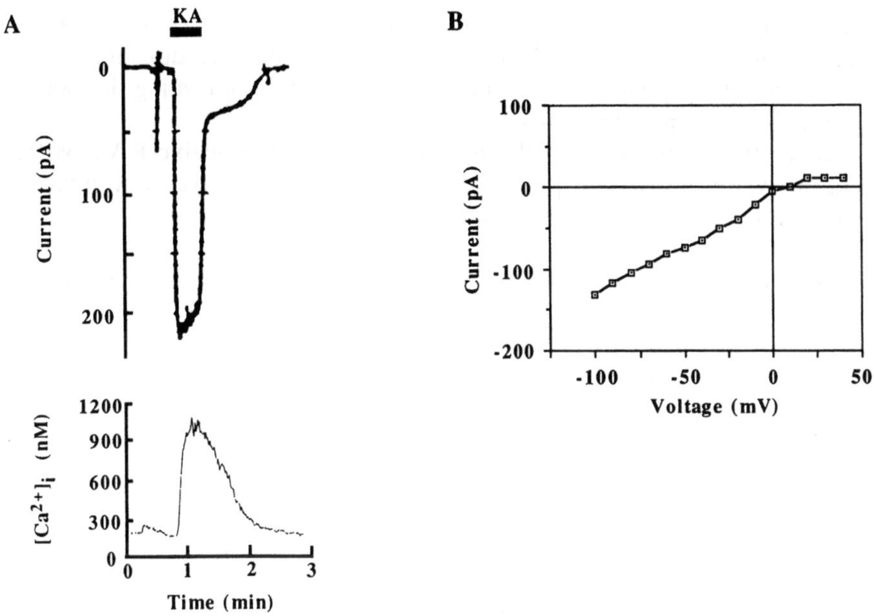

Fig. 2A,B. KA-induced Ca^{2+} currents in voltage-clamped cerebellar neurons. **A** In simultaneous whole-cell voltage-clamp and $[Ca^{2+}]_i$ microfluorimetric recordings, KA (30 μM) induced an inward current and an increase in $[Ca^{2+}]_i$ in a neuron which was held at −80 mV, preventing activation of voltage gated Ca^{2+} channels. (Intracellular solution: 140 mM K^+ gluconate, 100 μM Fura-2; extracellular solution: 2 mM Ca^{2+} 140 mM NMDG). **B** Measurement of the I–V relationship under conditions eliminating contributions from monovalent cationic currents and voltage-gated Ca^{2+} currents revealed an inwardly rectifying Ca^{2+} current induced by KA (30 μM). (Intracellular solution: 140 mM NMDG Fl^-; extracellular solution: 2 mM Ca^{2+} 140 mM NMDG and 100 μM Cd^{2+})

clamping. KA also produced clear Ca^{2+} influx in Na^+-free medium into the majority of neurons that were voltage clamped at −80 mV, a potential at which high threshold Ca^{2+} currents are not activated in these cells (Fig. 2A). However, there were also cells that did not respond in this way. In these cells not showing a Ca^{2+} influx nor an inward current in Na^+-free medium, large KA-induced currents could still be observed in Na^+-containing medium, without concomitant increases in $[Ca^{2+}]_i$ [9].

Current-voltage relationships for the KA-induced currents were measured both in Na^+-containing and Na^+-free media, using 100 μM Cd^{2+} to block the voltage-gated Ca^{2+} currents. In Na^+-containing media, the KA-induced large inward currents had approximately linear I/V relationships, similar to results previously reported for KA-induced currents [29]. In contrast, the KA-induced currents recorded in Na^+-free external solutions were much smaller in magnitude, and displayed inward rectification (Fig. 2B) as was described for the type II KA receptors in hippocampal neurons [22] and for the Ca^{2+}-permeable KA/AMPA receptors in oocytes injected with glutamate subunit messenger RNA [18]. The I/V curves for the KA-induced currents recorded in NMDG shifted in the positive direction when the external [Ca^{2+}] was raised from 2 to 10 m*M*, again indicating the opening of a Ca^{2+}-permeable channel [9].

Given that in these cerebellar neurons the Ca^{2+}-permeable KA/AMPA receptors were always coexpressed with a substantially larger KA-induced Na conductance, it could be postulated that this relatively small KA-induced Ca^{2+} current matters little to the Ca^{2+} homeostasis of these neurons. We therefore also wished to obtain an indication of the relative contributions of voltage-gated Ca^{2+} channels and Ca^{2+}-permeable KA/AMPA receptors to Ca^{2+} influx in a more physiological medium containing Na^+. In "responding" cells (those with a Ca^{2+} influx when stimulated with KA in Na^+-free solutions), TA-3090, while substantially reducing the 50 m*M* K^+-induced increase in $[Ca^{2+}]_i$, produced only about 13% inhibition of the KA-induced peak $[Ca^{2+}]_i$ responses in Na^+-containing solutions. These data are consistent with the idea that when neurons possess Ca^{2+}-permeable KA/AMPA receptors, KA-stimulated Ca^{2+} influx occurs substantially via this pathway even in the presence of Na^+.

Ca^{2+}-permeable KA/AMPA receptors constructed in oocytes are very nonselective with respect to cation permeability and are quite permeable to Co^{2+} [30], which does not permeate voltage-sensitive Ca^{2+} channels, NMDA receptors, or Ca^{2+}-impermeable KA/AMPA receptors. Using a histochemical silver-staining method to identify Co^{2+} uptake, Pruss et al. [37] have shown that stimulation with KA causes Co^{2+} influx through Ca^{2+}-permeable KA/AMPA receptors in neurons. Following this method, we found that 65% of the cerebellar neurons in culture showed Co^{2+} uptake staining following incubation with 100 μM KA. This effect was blocked by 20 μM CNQX but not by 50 μM APV. The morphology of neurons stained for Co^{2+} uptake included both those with Purkinje cell morphology and also some non-Purkinje cells.

These data clearly demonstrate that KA can activate a substantial Ca^{2+} influx in Purkinje cells and other cerebellar neurons which does not require cell depolarization and influx via voltage-sensitive Ca^{2+} channels. It is likely that these Ca^{2+}-permeable, KA-activated channels are formed by the association of KA receptor binding subunits similar or identical to those shown to produce Ca^{2+}-permeable channels in oocytes [18]. What seems particularly significant is that this phenomenon does not only occur in a minority of neurons but in the majority of cultured cerebellar cells where these receptors seem to coexist with traditional types of non Ca^{2+}-permeable KA receptors. Moreover, judging by Co^{2+} uptake staining data [37], these results are not just confined to cell culture but are also observed in the mature rat brain. Indeed, several of the non-NMDA receptor subtypes have been shown by in situ hybridization to be expressed particularly heavily in cerebellar Purkinje cells in adult rat brain sections as well as in a variety of other cell types throughout the brain [25]. This Ca^{2+}-permeable pathway may be very important in mediating excitatory amino acid induced phenomena in various neuronal cell types, particularly in instances where NMDA receptors do not appear to operate.

We have now started to investigate whether such Ca^{2+}-permeable KA receptors might be responsible for excitotoxicity in these cultured cerebellar neurons. Preliminary evidence [10] suggests that KA can indeed induce excitotoxic cell death through Na^{+}-independent pathways in these neurons, and that this is blocked fully by CNQX and partially by removal of extracellular Ca^{2+}. These results suggest that in these neurons Ca^{2+} permeation through the KA receptor plays an important role in glutamate-mediated excitotoxicity. Thus the role played by glutamate receptors in clinical situations might well include Ca^{2+} influx through non-NMDA receptors as well as NMDA receptors, with different pathways having different relative importance in various neuronal cell types.

The picture of glutamate receptor physiology has been illuminated recently by the cloning of a number of non-NMDA glutamate receptor subunits [3, 4, 14, 18, 25] as well as an NMDA receptor [31] and four metabotropic receptors [27, 44]. Hollman et al. [19] cloned the first non-NMDA receptor subunit (gluR1) by expression of brain RNA fractions in oocytes and functional screening for glutamate responses. Subsequently three other members of this family (gluR2–4) have been cloned, three members of a related family (gluR5–7) of non-NMDA receptor subunits have also been identified, and a third group of related KA binding proteins has also come to light [40, 48] (Fig. 3). The receptor subunits have been found to form AMPA and KA selective receptors when expressed in oocytes or cell lines [19, 25]. Among combinations of the first four subunits, it has been found that gluR2 confers Ca^{2+} impermeability and linear or outward rectification to the receptors formed, and this property has been specifically ascribed to a single amino acid residue position, which is filled by an arginine in gluR2 but by glutamine in the other subunits [7, 21, 46]. This arginine residue in gluR2 has been found to be necessary and sufficient to confer the usual Ca^{2+} impermeability of KA/AMPA receptors. Thus, unlike combinations including native gluR2 which form Ca^{2+} impermeable ionophores, receptor subunit combinations not including gluR2, or including an altered gluR2 with glutamine replacing the arginine, form Ca^{2+} permeable, inwardly rectifying receptor ionophores. Furthermore, Sommer et al. [42] showed that this crucial arginine residue in gluR2 is present only because of posttranscriptional editing of a single nucleotide of the mRNA; at the genomic level the codon for gluR2 encodes a glutamine residue as for the other glutamate receptor subunits. This suggests a possible point of regulation of the properties of non-NMDA glutamate receptors, which are involved in most fast excitatory transmission in the brain, and may be the crux of pathophysiological breakdown of glutamate receptor physiology in ischemia as well as developmental and plastic regulation of synaptic properties.

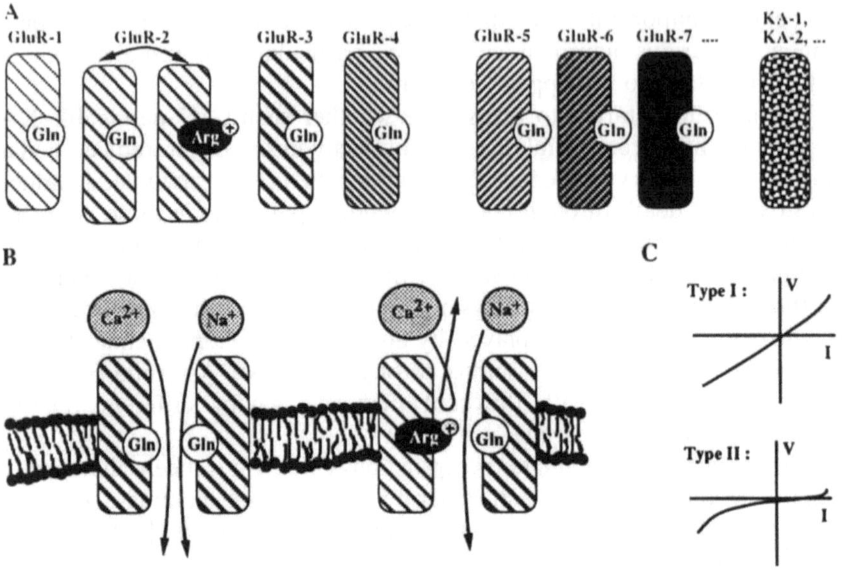

Fig. 3A–C. Cloned non-NMDA glutamate receptor subunits. **A** At least nine different non-NMDA receptor subunit genes have been cloned, coding for related subunits which fall into three homologous families. The first group, gluR1–4, bind AMPA with high affinity and have been shown to form functional AMPA/KA receptors when expressed in oocytes or cell lines. Each subunit contains a glutamine residue in one of the transmembrane spanning regions (at the "Q/R" site) which has been found to be crucial to the characteristics of the receptor channel formed; this glutamine codon is edited to one for arginine in the gluR2 subunit message. The members of the gluR5–7 homologous group also bear a glutamine codon in this position, which is partially edited to arginine [11]. These subunits seem to bind KA more specifically and with intermediate affinity. KA-1 and KA-2 (or gamma-2) form a third group, the KA binding proteins, which bind KA specifically and with high affinity. These subunits do not form functional homomeric channels, but KA-2 can form channels in combination with other subunits. **B** When channels are formed by gluR subunits including only subunits with glutamine residues at the Q/R site, a "type II" receptor forms which is permeable to Ca^{2+} as well as Ca^{2+}. Receptors including a subunit with the arginine residue at this position are type I receptors, impermeable to Ca^{2+}. **C** Type I receptors exhibit linear or outwardly rectifying I–V curves, whereas type II receptors are inwardly rectifying

With at least nine non-NMDA receptor subunits, forming receptors which are probably pentameric [47], the possible number of KA/AMPA receptor subtypes is enormous, and it will be of great interest to begin to catalogue the specific neuroanatomic expression patterns of each subtype [25, 36] and to relate the expression pattern in an individual neuronal type to the receptor physiology expressed. Preliminary work, for instance, suggests that some glial cells express gluR1 and gluR4, and this may relate to growing evidence of glutamate responses and $[Ca^{2+}]_i$ increases in these "nonexcitable" brain cells [17, 23]. Regional variations in the patterns of neuronal expression of the gluR subunits have also been noted and the functional consequences of these expression patterns need to be explored.

A great diversity of glutamate receptors is involved in the normal physiology of the central nervous system and potentially in excitotoxic processes as well. The importance of NMDA receptor to many forms of delayed neuronal damage is well established, but it is also becoming clear that at least in some neurons and in some types of injury, the non-NMDA receptors play a major or predominant role as well. The relevance of these different classes of glutamate receptors to pathophysiological processes needs to be further defined. Another area which remains wide open for exploration, given this complexity of glutamate receptors, is that of the interactions of sets of these receptors. The most commonly described form of long-term potentiation involves the simultaneous activation of NMDA receptors and non-NMDA glutamate receptors, causing long-term increases in the strength of the synapses involving the non-NMDA receptors. What might be the consequences of simultaneous activation of type II and type I non-NMDA receptors? One hypothesis is that this interaction is responsible for those types of synaptic plasticity which do not involve NMDA receptors. Additionally, what about the simultaneous activation of metabotropic and ionotropic receptors? This combination has been implicated as the mechanism of long-term depression of parallel fiber–Purkinje's cell synapses in the cerebellum, and evidence from our laboratory suggests that simultaneous metabotropic receptor activation can also markedly enhance the responsiveness of other glutamate receptors in a reversible fashion[6].

The potential diversity of glutamate receptors at synapses adds an additional dimension of complexity to excitatory neural networks over and above the marvelous anatomic complexity of the nervous system. The daunting nature of trying to understand the physiology of this intricacy should not obscure the fact that such multiplicity of receptor types empowers us with greater potential specificity and variety in our therapeutic interventions.

References

1. Audinat E, Knöpfel T, Gahwiler BH (1990) Responses to excitatory amino acids of Purkinje cells and neurones of the deep nuclei in cerebellar slice cultures. J Physiol (Lond) 430:297–313
2. Baimbridge KG, Miller JJ, Parkes CO (1982) Calcium-binding protein distribution in the rat brain. Brain Res 239:519–525
3. Bettler B, Boulter J, Hermans-Borgmeyer I, O'Shea-Greenfield A, Deneris ES, Moll C, Borgmeyer U, Hollmann M, Heinemann S (1990) Cloning of a novel glutamate receptor subunit, gluR5: expression in the nervous system during development. Neuron 5:583–595
4. Bettler B, Egebjerg J, Sharma G, Pecht G, Hermans-Borgmeyer I, Moll C, Stevens CF, Heinemann S (1992) Cloning of a putative glutamate receptor: a low affinity kainate-binding subunit. Neuron 8:257–265

5. Bleakman D, Chard PS, Foucart S, Miller RJ (1991) Block of neuronal Ca^{2+} influx by the antiischemic agent TA-3090. J Pharmacol Exp Ther 259:430–438
6. Bleakman D, Rusin KI, Chard PS, Glaum SR, Miller RJ (1992) Metabotropic glutamate receptors potentiate ionotropic glutamate responses. Mol Pharmacol 42:192–196
7. Bochet P, Dutriaux A, Lambolez B, Nalivaiko E, Rossier J, Prado de Carvalho L (1991) A chimeric glutamate receptor subunit: discrete changes modify the properties of the channel. Biochem Biophys Res Commun 177:1183–1187
8. Brorson JR, Bleakman D, Gibbons SJ, Miller RJ (1991) The properties of intracellular Ca^{2+} stores in cultured rat cerebellar neurons. J Neurosci 11:4024–4043
9. Brorson JR, Bleakman D, Chard PS, Miller RJ (1992) Calcium directly permeates kainate/alpha-amino-3-hydroxy-5-methyl-4-isoxazolepropionic acid receptors in cultured cerebellar Purkinje neurons. Mol Pharmacol 41:603–608
10. Brorson JR, Manzolillo PA, Miller RJ (1992) Kainate-induced excitotoxicity in cultured cerebellar neurons by a Na^+-independent mechanism (Abstr). Neurology 42:196
11. Burnashev N, Monyer H, Seeburg PH, Sakmann B (1992) Divalent ion permeability of AMPA receptor channels is dominated by the edited form of a single subunit. Neuron 8:189–198
12. Celio MR (1990) Çalbindin D-28k and parvalbumin in the rat nervous system. Neuroscience 35:375–475
13. Choi DW (1987) Ionic dependence of glutamate neurotoxicity. J Neurosci 7:369–379
14. Egebjerg J, Bettler B, Hermans-Borgmeyer I, Heinemann S (1991) Cloning of a cDNA for a glutamate receptor subunit activated by kainate but not by AMPA. Nature 351:745–748
15. Ellisman MH, Deerinck TJ, Ouyang Y, Beck CF, Tanksley SJ, Walton PD, Airey JA, Sutko JL (1990) Identification and localization of ryanodine binding proteins in the avian central nervous system. Neuron 5:135–146
16. Faden AI, Simon RP (1988) A potential role for excitotoxins in the pathophysiology of spinal cord injury. Ann Neurol 23:623–626
17. Glaum SR, Holzwarth J, Miller RJ (1990) Metabotropic glutamate receptors activate Ca^{2+} mobilization and Ca^{2+} influx in astroglial cells. Proc Natl Acad Sci USA 87:3454–3458
18. Hollman M, Hartley M, Heinemann S (1991) Ca^{2+} permeability of KA-AMPA gated glutamate channels depends on subunit composition. Science 252:851–853
19. Hollman M, O'Shea-Greenfield A, Rogers SW, Heinemann S (1989) Cloning by functional expression of a member of the glutamate receptor family. Nature 342:643–648
20. Holopainen I, Enkvist MOK, Akerman KEO (1989) Glutamate receptor agonists increase intracellular Ca^{2+} independently of voltage gated Ca^{2+} channels in rat cerebellar granule cells. Neurosci Lett 98:57–62
21. Hume RI, Dingledine R, Heinemann SF (1991) Identification of a site in glutamate receptor subunits that controls calcium permeability. Science 253:1028–1031
22. Iino M, Ozawa S, Tsuzuki K (1990) Permeation of calcium through excitatory amino acid receptor channels in cultured rat hippocampal neurones. J Physiol (Lond) 424:151–165
23. Jensen AM, Chiu SY (1991) Differential intracellular calcium responses to glutamate in type 1 and type 2 cultured rat brain astrocytes. J Neurosci 11:1674–1684
24. Kaku DA, Goldberg MP, Choi DW (1991) Antagonism of non-NMDA receptors augments the neuroprotective effect of NMDA receptor blockade in cortical cultures subjected to prolonged deprivation of oxygen and ischemia. Brain Res 554:344–347
25. Keinanen K, Wisden W, Sommer B, Werner P, Herb A, Verdoorn TA, Sakmann B, Seeburg PH (1990) A family of AMPA-selective glutamate receptors. Science 249:556–560
26. Llinas R, Sugimori M (1980) Electrophysiological properties of in vitro Purkinje cell somata in mammalian cerebellar slices. J Physiol (Lond) 305:171–195
27. Masu M, Tanabe Y, Tsuchida K, Shigemoto R, Nakanishi S (1991) Sequence and expression of a metabotropic glutamate receptor. Nature 349:760–765
28. Mayer Ml, Miller RJ (1990) Excitatory amino acid receptors: regulation of neuronal $[Ca^{2+}]_i$ and other second messengers. Trends Pharmacol Sci 11:254–260

29. Mayer ML, Westbrook GL (1987) Permeation and block of N-methyl-D-aspartic acid receptor channels by divalent cations in mouse cultured central neurons. J Physiol (Lond) 394:501–527
30. McGurk JF, Roginski RS, Zukin Rs, Bennett MVL (1991) Divalent ion permeability and pharmacology of cloned kainate/AMPA receptors expressed in oocytes. Soc Neurosci Abstr 17:335
31. Moriyoshi K, Masu M, Ishii T, Shigemoto R, Mizuno N, Nakanishi S (1991) Molecular cloning and characterization of the rat NMDA receptor. Nature 354:31–37.
32. Murphy SN, Miller RJ (1989) Regulation of Ca^{2+} influx into striatal neurons by kainic acid. J Pharmacol Exp Ther 249:184–193
33. Ogura A, Akita K, Kudo Y (1990) Non-NMDA receptor mediates cytoplasmic Ca^{2+} elevation in cultured hippocampal neurones. Neurosci Res 9:103–113
34. Ozawa S, Iino M, Tsuzuki K (1991) Two types of kainate response in cultured rat hippocampal neurons. J Neurophysiol 66:2–11
35. Park CK, Nehls DG, Graham DI, Teasdale GM, McCulloch J (1988) The glutamate antagonist MK-801 reduces focal ischemic brain damage in the rat. Ann Neurol 24:543–551
36. Petralia RS, Wenthold RJ (1992) Light and electron immunocytochemical localization of AMPA-selective glutamate receptors in the rat brian. J Comp Neurol 318:329–354
37. Pruss RM, Akeson RC, Racke MM, Wilburn JL (1991) Agonist activated cobalt uptake identifies divalent cation permeable kainate receptors on neurons and glial cells. Neuron 7:509–518
38. Regan RF, Choi DW (1991) Glutamate neurotoxicity in spinal cord cell culture. Neuroscience 43:585–591
39. Rothman SM, Olney JW (1987) Excitotoxicity and the NMDA receptor. TINS 10:299–302
40. Sakimura K, Morita T, Kushiya E, Mishina M (1992) Primary structure and expression of the gamma2 subunit of the glutamate receptor channel selective for kaninate. Neuron 8:267–274
41. Sheardown MJ, Nielsen EØ, Hansen AJ, Jacobsen P, Honore T (1990) 2,3-Dihydroxy-6-nitro-7-sulfamoyl-benzo(F)quinoxaline: a neuroprotectant for cerebral ischemia. Science 247:571–574
42. Sommer B, Keinanen K, Verdoorn TA, Wisden W, Burnashev N, Herb A, Kohler M, Takagi T, Sakmann B, Seeburg PH (1990) Flip and flop: a cell-specific functional switch in glutamate-operated channels of the CNS. Science 249:1580–1584
43. Sommer B, Køhler M, Sprengel R, Seeburg PH (1991) RNA editing in brain controls a determinant of ion flow in glutamate-gated channels. Cell 67:11–19
44. Tanabe Y, Masu M, Ishii T, Shigemoto R, Nakanishi S (1992) A family of metabotropic glutamate receptors. Neuron 8:169–179
45. Teitelbaum JS, Zatorre RJ, Carpenter S, Gendron D, Evans AC, Gjedde A, Cashman NR (1990) Neurologic sequelae of domoic acid intoxication due to ingestion of contaminated mussels. N Engl J Med 322:1781–1787
46. Verdoorn TA, Burnashev N, Monyer H, Seeburg PH, Sakmann B (1991) Structural determinants of ion flow through recombinant glutamate receptor channels. Science 252: 1715–1718
47. Wenthold RS, Yokotani N, Doi K, Wada K (1992) Immunochemical characterization of the non-NMDA glutamate receptor using subunit specific antibodies. J Biol Chem 267: 501–507
48. Werner P, Volgt M, Keinanen K, Wisden W, Seeburg PH (1991) Cloning of a putative high-affinity kainate receptor expressed predominantly in hippocampal CA3 cells. Nature 351:742–744

Phenylalkylamine Calcium Antagonists as Cerebroprotective Agents

L. Szabo, H.P. Hofmann, and L. Unger

Calcium Antagonists and Cerebral Ischemia

Ion channels which are selectively permeable to Ca^{2+} and regulated primarily by changes in membrane potential are present in virtually every excitable cell. The physiological role and pharmacological relevance of these voltage-sensitive Ca^{2+} channels was first recognized in the heart, where they were shown to be the site of action of cardiodepressant drugs described functionally as calcium antagonists.

Although recent progress in electrophysiology has led to the recognition and characterization of several voltage-dependent Ca^{2+} channel subtypes (Nowycky et al. 1985), the term calcium antagonist refers in general to compounds showing activity on the "slow" myocardial Ca^{2+} channel (Kohlhardt et al. 1972), now termed the L-type Ca^{2+} channel.

Shortly after the discovery of calcium antagonism the hypothesis of Ca^{2+}-mediated ischemic myocardial damage was formulated (see Fleckenstein 1983 for references). The postulate stated that ischemic cell damage is causally related to mitochondrial Ca^{2+} overload leading to irreversible changes in the function and structure of the mitochondria. Subsequently the hypothesis was extended to other tissues including the brain (Siesjö 1981), and Ca^{2+}-dependent mechanisms such as the activation of intracellular proteases and lipases were also recognized as important factors contributing to ischemic cell death (Orrenius et al. 1989). On these grounds, compounds that block the entry of Ca^{2+} into the cell are potential therapeutic agents for the treatment of ischemia.

In brain tissue the L-type Ca^{2+} channel possesses distinct binding sites for three chemically disparate groups of molecules. These binding sites are allosterically coupled to each other and to the Ca^{2+} pore, and each of them can mediate inhibitory effects by prolonging the inactivated state of the channel. Named after the selective compounds at the respective binding site, blockers of the L-type Ca^{2+} channel are customarily subdivided into verapamil-like, nifedipine-like and diltiazem-like calcium antagonists, corresponding structurally to phenylalkylamines, 1,4-dihydropyridines, and benzothiazepines (Vanhoutte and Paoletti 1987).

Knoll AG, Department of Neuropharmacology, 67061 Ludwigshafen, Germany

Cerebral Ischemia and Basic Mechanisms
Ed. by A. Hartmann, F. Yatsu, and W. Kuschinsky

Since both 1,4-dihydropyridines and phenylalkylamines bind with high affinity to saturable binding site populations in the brains of different species, including humans (Quirion 1985), the use of calcium antagonists for the treatment of cerebral ischemia is a rationally justified approach if the compounds in question possess appropriate pharmacokinetic properties. Obviously, calcium antagonists that are intended for treating diseases of the central nervous system must reach therapeutically relevant concentrations at the site of action in order to achieve efficacy.

While dihydropyridines pass the blood-brain barrier quite readily, this is not the case for other chemical groups. For example, the negative results encountered in experimental studies using verapamil for the treatment of cerebral ischemia are probably related primarily to the inferior acute brain availability of the compound and not to an inefficacy of the phenylalkylamines. The recognition of this general problem has led to the development of compounds with improved cerebral availability. As a result, the brain availability of the phenylalkylamine emopamil is approximately 20 times higher than that of the parent compound verapamil (Szabo 1989), and the follow-up 1,4-dihydropyridine nimodipine exhibits a three times higher distribution volume in the brain than the prototype nifedipine (van den Kerckhoff and Drewes 1985).

In the past decade the effect of calcium antagonists was tested in different experimental models of cerebral ischemia employing a large number of parameters of varying clinical relevance. In view of the often conflicting data, it is difficult to make a general statement regarding the efficacy of calcium antagonists. The situation is illustrated by two of the rare pharmacological studies investigating different calcium antagonists under identical conditions. The neuroprotective effects of nicardipine, another 1,4-dihydropyridine calcium antagonist, and nimodipine were compared in the gerbil model of transient cerebral ischemia employing a combined pre- and posttreatment protocol. Based on the survival of hippocampal CA_1 neurons, nicardipine provided considerable neuronal protection while nimodipine remained without effect (Alps et al. 1988). In a comparative study of focal cerebral ischemia in the mouse, that employed changes in the peripheral type (ω_3) benzodiazepine binding sites as a measure of infarct size, nimodipine treatment initiated a few minutes after vessel occlusion resulted in tissue protection while diltiazem and verapamil displayed no effect (Gotti et al. 1990). However, when the first drug administration was performed prior to the occlusion, diltiazem was found to be highly protective, and the efficacy of nimodipine was also improved.

In essence, the preclinical data suggest that calcium antagonists can limit the necrosis of selectively vulnerable neurons induced by transient ischemic episodes and reduce the development of infarct in focal ischemia under certain conditions. In general, however, significant effects are seen only if the compounds are administered before or immediately after the induction of ischemia.

Large-scale clinical investigations in cerebral ischemic disorders have been performed almost exclusively with nimodipine. Several trials have demonstrated that the compound effectively reduces the development of infarction after subarachnoid hemorrhage and improves clinical outcome when given prophylactically to patients in good neurological condition (Pickard et al. 1989). This effect was not associated with changes in the frequency of vasospasm development, suggesting a direct cytoprotective effect. On the other hand, inspite of the large number of patients treated with nimodipine, the compound has not been shown to be useful for the treatment of acute ischemic stroke (Trust Study Group 1990; American Nimodipine Study Group 1992). This discrepancy in clinical efficacy correlates quite well with differences observed between pre- and posttreatment protocols in preclinical studies.

It should be mentioned, however, that even within the same chemical group the effects of calcium antagonists can be quite disparate. Nicardipine exhibits a clinical profile opposite to that of nimodipine. When given after acute aneurysmal hemorrhage, nicardipine prevents the development of both symptomatic and angiographic vasospasm without influencing neurological outcome or mortality at 3 months after the initial hemorrhage (Findlay et al. 1991).

Basically, both clinical and preclinical data imply that calcium antagonists differ substantially in their cerebroprotective efficacy, but the underlying causes are unclear.

Anti-ischemic Effects of Phenylalkylamines

As mentioned above, all three binding sites of the L-type Ca^{2+} channel can mediate the blockade of Ca^{2+} influx. Nevertheless, the exploration of phenylalkylamines in neurological disorders, with the notable exception of migraine, was given up rather quickly after the first negative findings with verapamil had been published (Berger et al. 1984). In contrast to verapamil, the anti-ischemic properties of the brain-penetrating phenylalkylamine levemopamil [(*S*)-emopamil] have been shown repeatedly. The protective effect of levemopamil against ischemic neuronal cell death was investigated in several models of transient ischemia. Neuronal survival in the selectively vulnerable CA_1 region of the hippocampus was found to be promoted by levemopamil pretreatment both in the gerbil, using bilateral carotid artery occlusion (Szabo and Hofmann 1989), and in the rat, employing either bilateral carotid occlusion combined with systemic hypotension (Block et al. 1990) or the simultaneous occlusion of carotid and vertebral arteries, known as four-vessel occlusion (Bielenberg et al. 1989).

In a further rat study of transient global ischemia (Lin et al. 1990) the efficacy of pre- and postischemic application of levemopamil was compared. While prophylactically administered levemopamil resulted in increased neuronal survival, no significant protection was observed when treatment was begun 30 min after the onset of the ischemic insult. Similar conclusions were reached in a cat model of cardiac arrest evoked by ventricular fibrillation and terminated by standard cardiopulmonary resuscitation (Fleischer et al. 1992). Neuropathological examination of the occipital cortex and the CA_1 region of animals pretreated with levemopamil showed less severe damage than the same regions in the placebo group. However, when treatment was delayed for 5 min after resuscitation, no cellular protection was seen.

Under the conditions of transient global ischemia, levemopamil exhibits the typical features of a calcium antagonist. If substance application is started before onset of the ischemic event, the compound prevents neuronal degeneration to a significant degree. In contrast to other compounds, this effect has been shown unequivocally in all species examined. However, postischemic levemopamil therapy – just as therapy with other calcium antagonists – does not lead to any improvement in neuronal survival.

A quite different picture emerges with regard to the efficacy of levemopamil in focal ischemia. The compound's effect on the histopathological outcome following permanent middle cerebral artery (MCA) occlusion was evaluated after 3 days of survival in three groups of rats (Nakayama et al. 1988). Levemopamil was administered either 30 min prior to, immediately after, or 1 h after MCA occlusion. Twice daily treatment was continued until the end of the study. At the 3-day endpoint cortical infarct size was significantly reduced by 55%–65% in all treatment groups, even when the first levemopamil injection was given 1 h after vessel occlusion.

A follow-up study (Morikawa et al. 1991) explored the postischemic therapeutic window for levemopamil, defined as the maximal possible delay between the induction of ischemia and the commencement of an eventually successful therapy. These experiments corroborated the compound's postischemic therapeutic efficacy with regard to permanent MCA occlusion. The administration of levemopamil reduced cortical infarct volume by 48% when therapy was initiated 1 h after ischemia. When treatment was deferred for 2 h, mean cortical infarct volume was still reduced by 34%, but this difference did not attain statistical significance. Treatment delayed for 3 h was without any effect. Accordingly, the therapeutic window of levemopamil after MCA occlusion in the rat lies between 1 and 2 h.

The effect of levemopamil on permanent MCA occlusion was also investigated in the cat. In this study (Szabo and Urbanics 1991), intravenous levemopamil infusion was started 30 min after the onset of ischemia, and the size of the resulting infarct was determined after a maturation period of 16 h. Levemopamil significantly altered the relationship between ischemic severity (defined as the degree of EEG reduction observed before the initiation of treatment) and final infarct volume, corresponding to a pronounced reduction of tissue damage at comparable degrees of ischemia. The average volume of cerebrocortical infarcts was 60% smaller in levemopamil-treated animals than in the placebo group.

While a therapeutic window between 1 and 2h is probably at the lower limit of the clinically useful range, the delayed postischemic efficacy of levemopamil nevertheless represents a unique finding since no comparable effect has been reported with any other compound as yet.

Receptor Pharmacology of Emopamil Enantiomers

As discussed above, the protective effects of levemopamil in models of global cerebral ischemia are similar to those of other L-type Ca^{2+} channel antagonists. The compound's postischemic efficacy in focal ischemia, however, is difficult to explain by its calcium antagonistic property alone, seeing that potent 1,4-dihydropyridines are without effect under the same conditions. Since phenylalkylamines are known to possess relevant affinities to a number of receptors present in the brain in addition to the L-type Ca^{2+} channel (DeFeudis 1987), it is conceivable that levemopamil activates additional neuroprotective mechanisms which contribute substantially to its anti-ischemic efficacy. In order to clarify this issue, the receptor binding properties of levemopamil and its enantiomer (*R*)-emopamil were investigated in several assays using standard procedures. Table 1 shows the equilibrium dissociation constant (K_i) for the inhibition of ligand binding by the emopamil enantiomers for all receptors with $K_i < 1\,\mu mol/l$ together with the same data for nimodipine. The receptor binding results clearly show that nimodipine is a highly selective antagonist of the L-type Ca^{2+} channel while the emopamil enantiomers exhibit a rather complex receptor profile. In particular, levemopamil possesses appreciable affinity to several serotonin (5-HT) receptor subtypes ($5\text{-}HT_{1A}$, $5\text{-}HT_{1C}$, $5\text{-}HT_2$) as well as to some adrenoceptors (α_1, α_2), to certain subtypes of the dopamine (D_2) and histamine (H_2) receptors, to the muscarinic receptor, and the σ binding site. While levemopamil is an agonist at the $5\text{-}HT_{1A}$ receptor (R. Binder, personal communication), the compound's intrinsic activity at most other targeted receptors is of antagonistic nature.

Interestingly, not only the L-type Ca^{2+} channel, but several other receptors interacting with levemopamil have been implicated in the ischemic process, albeit to different degrees. Several compounds possessing agonistic activity at the $5\text{-}HT_{1A}$ receptor have been shown to be neuroprotective in both global and focal cerebral ischemia (Bielenberg and Burkhardt 1990; Bode-Greuel et al. 1990; Prehn et al. 1991). The protective effect was attributed to the hyperpolarizing action of $5\text{-}HT_{1A}$ agonists, which possibly counteracts the hyperexcitation produced by the ischemic activation of the glutamatergic system.

The anti-ischemic effect of $5\text{-}HT_2$ receptor blockade is supported by experiments with the cerebral vasodilator naftidrofuryl that reportedly protects neurons from the effects of transient ischemia (Krieglstein et al. 1989; Fujikura et al. 1989) and is undergoing therapeutic trials in stroke patients

Table 1. Receptor binding profile of the phenylalkylamine emopamil and the 1,4-dihydropyridine nimodipine

Receptor site	Tissue (species)	Radioligand	Blank	Agent (K_i, nmol/l)		
				Levemopamil	(R)-Emopamil	Nimodipine
Ca^{2+} channel	Skeletal muscle (guinea pig)	[^{3}H](S)-D888 [^{3}H]PN200-110	(*S*)-D888 PN200-110	73 –	66 –	– 4.5
Serotonin 5-HT_{1A}	Hippocampus (rat)	[^{3}H]8-OH-DPAT	Serotonin	37	30	>1000
Serotonin 5-HT_{1C}	Choroid plexus (pig)	[^{3}H]Mesulergine	Serotonin	4.2	149	>1000
Serotonin 5-HT_2	Cerebral cortex (rat)	[^{3}H]Ketanserin	Cyproheptadine	4.4	58	>1000
α_1-Adrenoceptor	Cerebrum (rat)	[^{3}H]Prazosin	Phentolamine	157	54	>1000
α_2-Adrenoceptor	Cerebral cortex (bovine)	[^{3}H]Rauwolscine	Phentolamine	253	49	>1000
Dopamine D_2	Nucleus caudatus (bovine)	[^{3}H]Spiperone	Haloperidol	20	85	>1000
Histamine H_2	Brain cortex (guinea pig)	[^{3}H]Tiotidine	Histamine	264	246	>1000
Muscarinic	Cerebrum (rat)	[^{3}H]QNB	Atropine	329	64	>1000
σ Site	Cerebrum (rat)	[^{3}H]DTG	Haloperidol	132	68	>1000

(Capon et al. 1990). Implicated mechanisms include ischemic changes in blood-brain barrier permeability, edema development, and platelet aggregation.

The noradrenergic neurons originating in the locus coeruleus represent another inhibitory system that is supposed to exert a protective effect by preventing postischemic hyperactivity. This system can be stimulated by α_2-antagonists that act on autoreceptors and thereby increase the extracellular level of noradrenaline. Indeed, the administration of the α_2-antagonist idazoxan was shown to significantly reduce neuronal damage in the hippocampus and in the vulnerable parts of the neocortex (Gustafson et al. 1990). Similar inhibition-mediated protection has been achieved experimentally by the α_1-adrenoceptor agonist methoxamine as well (Araki et al. 1989). Idazoxan also reduces the size of infarcts elicited by the occlusion of the MCA (Maiese et al. 1992) although this result was traced back pharmacologically to a possible interaction with what is known as the imidazole receptor.

The involvement of dopaminergic neurotransmission in the development of ischemic tissue damage has been demonstrated in the striatum. In global cerebral ischemia induced by four-vessel occlusion, preischemic lesioning of the substantia nigra not only prevented the massive release of dopamine into the extracellular space (Globus et al. 1988a) but also markedly attenuated the degree of neuronal damage in the striatum (Globus et al. 1988b). These findings suggest that similar protection may be achieved by the blockade of dopamine receptors.

Recent reports about the affinity of the neuroprotective agents ifenprodil and dextromethorphan to the σ site have led to speculations about the cerebroprotective potential of σ ligands in general (Rao et al. 1990), but due to the interaction of these compounds with the NMDA receptor complex the verification of this concept requires further work.

The complexity of levemopamil's receptor binding profile obscures the possible causality between individual mechanisms and the compound's neuroprotective effects. The present evaluation of the relative importance of tentative protective mechanisms incorporated into the levemopamil molecule originates from the observation that the two emopamil enantiomers differ not only with regard to their receptor profile – as shown in Table 1 – but also regarding their in vivo anti-ischemic efficacy in simple screening models (Hofmann et al. 1989). In an attempt to correlate pharmacological properties with anti-ischemic efficacy, the neuroprotective effects of levemopamil and (*R*)-emopamil were tested comparatively in global and focal cerebral ischemia.

Neuroprotective Mechanism of Emopamil Enantiomers in Global and Focal Ischemia

Transient Global Ischemia in the Gerbil

The experiments on global ischemia were carried out in male Mongolian gerbils which had free access to food and tap water until the beginning of the experimental procedure and after its completion. The animals were anesthetized with halothane in a mixture of 70% nitrous oxide and 30% oxygen delivered through a loose-fitting face mask. Cerebral ischemia was produced by bilateral carotid occlusion for 7 min with the aid of micro-aneurysm clips. The animals were killed 7 days later, and the brains were processed according to standard procedures (Szabo and Hofmann 1989). Coronal sections of 5 μm thickness at approximately 1.8 mm posterior to the bregma were prepared with a microtome and stained with cresyl fast blue.

Quantitative evaluation of neuronal survival in the CA_1 sector of the hippocampus was based on relative optical density measurements. On each individual section the optical density of CA_1, dentate gyrus (DG), and stratum radiatum (RAD) was determined. Since neither DG nor RAD are affected by a 7-min ischemic episode, their values were used as internal standards, hence compensating for differences in staining. As a measure of neuronal survival, the relative neuronal density of CA_1 was calculated by the formula $(CA_1 - RAD)/(DG - RAD)$, where all symbols denote the optical density of the corresponding region.

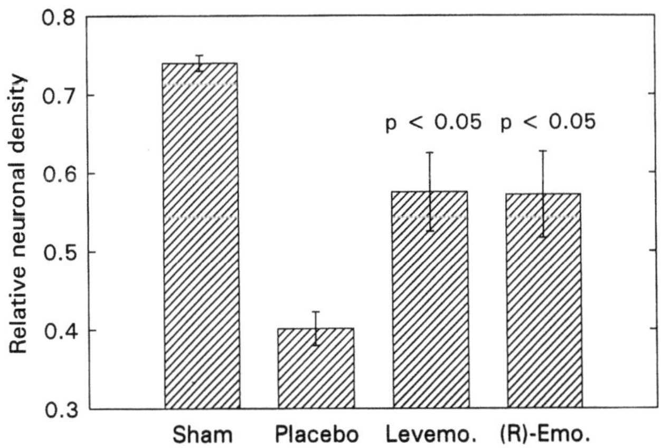

Fig. 1. Effect of the emopamil enantiomers levemopamil and (*R*)-emopamil on the ischemic survival of hippocampal CA_1 neurons in the gerbil expressed as relative neuronal density (mean ± SEM). Treated animals received 31.6 mg/kg intraperitoneally of either compound 30 min before the induction of a 7-min ischemic episode in the forebrain. Statistical significance indicated refers to comparisons between treatment and placebo (uncorrected *t* test)

Treated animals received an intraperitoneal injection of 31.6 mg/kg levemopamil or (*R*)-emopamil in saline 30 min before carotid occlusion. Placebo-treated animals were injected with the same amount of saline only, whereas sham-operated animals served as controls. By using the definition of relative neuronal density as given above, the value found for the CA_1 sector of the hippocampus in normal animals was 0.74 ± 0.01 (n = 9). The 7-min ischemia employed in these experiments was associated with a relative neuronal density of 0.44 ± 0.04 (n = 10). Pretreatment with either emopamil enantiomer significantly increased the number of surviving neurons (Fig. 1), resulting in density values of 0.58 ± 0.05 and 0.57 ± 0.05 (n = 12) for levemopamil and (*R*)-emopamil, respectively. When compared with placebo, the administration of emopamil protected approximately 50% of the neurons that were destined to die without treatment. Obviously, there was no statistically significant difference between the two emopamil treatments.

Permanent Focal Ischemia in the Mouse

This investigation was performed in female mice with free access to food and water before and after surgery. Under halothane/nitrous oxide anesthesia the skull was opened over the right MCA leaving the dura intact. With the aid of a bipolar device the MCA as well as all visible branches were coagulated permanently, the wound was closed, and the animals were permitted to recover. The brains were removed 24 h later and incubated in

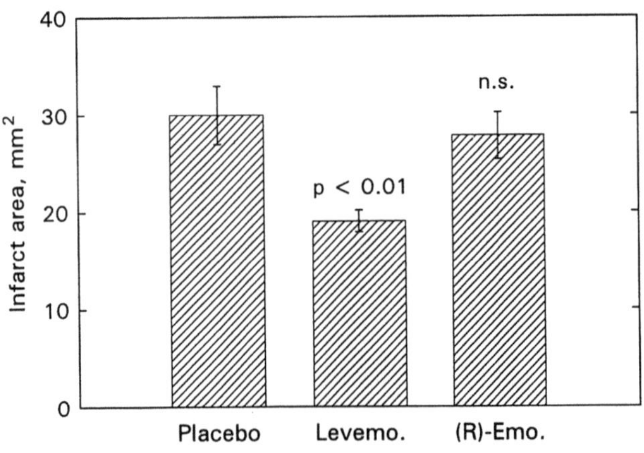

Fig. 2. Effect of the emopamil enantiomers levemopamil and (*R*)-emopamil on infarct size in the mouse determined 24 h after permanent occlusion of the middle cerebral artery (mean ± SEM). The animals in the treatment groups received 30 and 15 mg/kg intraperitoneally of either compound 30 min before and 3 h after vessel occlusion. Statistical significance given refers to comparisons between treatment and placebo (uncorrected *t* test)

2% 2,3,5-triphenylatetrazolium chloride solution at 37°C for 30 min. Since under the described conditions ischemic damage is restricted to the cortex, the size of infarction visible by the absence of staining was quantified by planimetry of the brain surface.

In the treatment groups, animals received an intraperitoneal injection of 30 mg/kg levemopamil or (*R*)-emopamil in distilled water 30 min before occlusion and a second dose of 15 mg/kg 3 h after occlusion. Animals in the placebo group were given distilled water.

The experimental results are shown in Fig. 2. In the untreated animals the cortical infarct area amounted to $30 \pm 3\,mm^2$ ($n = 16$). Treatment with levemopamil reduced the average infarct size as compared to the placebo group by 36% to $19 \pm 1\,mm^2$ ($n = 15$, $p < 0.01$). In contrast, administration of (*R*)-emopamil remained without a statistically significant effect on the infarcted area ($28 \pm 2\,mm^2$, $n = 16$).

Anti-ischemic Mechanism of Levemopamil

With regard to the protection of selectively vulnerable hippocampal neurons in global ischemia, the anti-ischemic potency of (*R*)-emopamil closely resembles that of levemopamil. Accordingly, the pharmacological mechanism responsible for this effect must be of equal potency in both enantiomers. As shown in Table 1, this condition is met by the affinities to the calcium channel, the 5-HT_{1A} receptor, and the histamine H_2 receptor. In view of the relatively weak affinity of the enantiomers to the histamine receptor, however, calcium antagonism and/or 5-HT_{1A} agonism are more likely to be responsible for the neuroprotective effect observed under global ischemia.

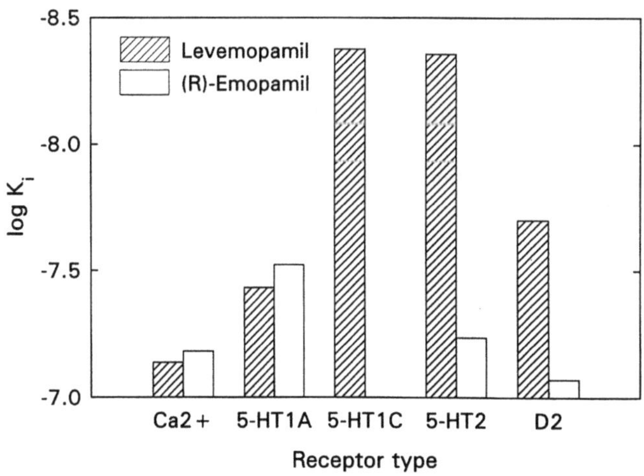

Fig. 3. Principal receptor affinities of levemopamil in comparison with (*R*)-emopamil. Only affinities with $K_1 < 100\,nmol/l$ are shown. See text for details

This conclusion is in good agreement with the published findings of other groups discussed above.

Under the conditions of focal ischemia, the significant infarct-reducing effect of levemopamil was not matched by (*R*)-emopamil. Consequently, levemopamil must activate a pharmacological mechanism that is either not present or much weaker in the (*R*) configuration. As illustrated by Fig. 3, which shows the receptor affinities with $K_i < 100\,\text{nmol/l}$, this requirement is fulfilled by the affinities to the 5-HT_{1C}, 5-HT_2, and D_2 receptors. It can therefore be concluded that antagonism at the 5-HT_{1C}, 5-HT_2, and/or D_2 receptor is the most plausible mechanism of action responsible for the protection offered by levemopamil in focal ischemia. It is unclear at present how the blockade of one of these receptors leads to an infarct reduction, and whether there is a synergism with calcium antagonism. It is not unlikely, however, that vascular mechanisms play a substantial role. This assumption is supported by the observation that levemopamil significantly increases local cerebral blood flow by up to 100% whereas (*R*)-emopamil exerts no effect on cerebral hemodynamics (Szabo 1989).

The involvement of 5-HT_{1A} receptors in hippocampal neuronal protection and the dominant role of 5-HT_2 receptors in the reduction of cortical infarcts – as exemplified by levemopamil's mode of action – appears to be plausible in view of autoradiographic studies investigating cerebral receptor distribution. 5-HT_{1A} receptors are most abundant in the hippocampus and hardly present in the cortex. In contrast, 5-HT_2 receptors are found predominantly in cerebrocortical regions (Hoyer et al. 1986a; Hoyer et al. 1986b). The relatively sparse presence of D_2 receptors in the cortex makes their contribution to infarct reduction less likely (Charuchinda et al. 1987).

The data presented above corroborate the notion that the amelioration of ischemic damage requires the activation of separate pharmacological mechanisms in global and focal ischemia. Consequently, compounds preventing the ischemia-induced death of selectively vulnerable neurons cannot be regarded a priori as drug candidates for the treatment of focal ischemic events such as stroke. On the other hand, a single chemical molecule, such as levemopamil, can activate more than one neuroprotective mechanism, resulting in improved therapeutic efficacy.

References

Alps BJ, Calder C, Hass WK, Wilson AD (1988) Comparative protective effects of nicardipine, flunarizine, lidoflazine and nimodipine against ischaemic injury in the hippocampus of the Mongolian gerbil. Br J Pharmacol 93:877–883

American Nimodipine Study Group (1992) Clinical trial of nimodipine in acute ischemic stroke. Stroke 23:3–8

Araki T, Kogure K, Izumiyama K (1989) Prevention of ischemic neuronal damage by α1-adrenoceptor agonist (methoxamine). Acta Neurol Scand 80:451–454

Berger JR, Busto R, Ginsberg MD (1984) Verapamil: failure of metabolic amelioration following global forebrain ischemia in the rat. Stroke 15:1029–1032

Bielenberg GW, Burkhardt M (1990) 5-Hydroxytryptamine$_{1A}$ agonists: a new therapeutic principle for stroke treatment. Stroke 21 [Suppl IV]:161–163

Bielenberg GW, Sauer D, Nuglisch J, Beck T, Rossberg C, Mennel HD, Krieglstein J (1989) Effects of emopamil on postischemic blood flow and neuronal damage in rat brain. Naunyn Schmiedebergs Arch Pharmacol 339:230–235

Block F, Jaspers RMA, Heim, C, Sontag K-H (1990) S-Emopamil ameliorates ischemic brain damage in rats: histological and behavioural approaches. Life Sci 47:1511–1518

Bode-Greuel KM, Klisch J, Horváth E, Glaser T, Traber J (1990) Effects of 5-hydroxytryptamine$_{1A}$-receptor agonists on hippocampal damage after transient forebrain ischemia in the Mongolian gerbil. Stroke 21 [Suppl IV]:164–166

Capon A, Lehert P, Opsomer L (1990) Naftidrofuryl in the treatment of subacute stroke. J Cardiovasc Pharmacol 16 [Suppl 3]:62–66

Charuchinda C, Supavilai P, Karobath M, Palacios JM (1987) Dopamine D_2 receptors in the rat brain: autoradiographic visualization using a high-affinity selective agonist ligand. J Neurosci 7:1352–1360

DeFeudis FV (1987) Interactions of Ca^{2+} antagonists at 5-HT_2 and H_2 receptors and GABA uptake sites. Trends Pharmacol Sci 8:200–201

Findlay JM, MacDonald RL, Weir BKA (1991) Current concepts of pathophysiology and management of cerebral vasospasm following aneurysmal subarachnoid hemorrhage. Cerebrovasc Brain Metab Rev 3:336–361

Fleckenstein A (1983) History of calcium antagonists. Circ Res 52:3–16

Fleischer JE, Nakakimura K, Drummond JC, Scheller MS, Zornow MH, Grafe MR, Shapiro HM (1992) Effects of levemopamil on neurologic and histologic outcome after cardiac arrest in cats. Crit Care Med 20:126–134

Fujikura H, Kato H, Nakano S, Kogure K (1989) A serotonin S_2 antagonist, naftidrofuryl, exhibited a protective effect on ischemic neuronal damage in the gerbil. Brain Res 494: 387–390

Globus MYT, Busto R, Dietrich WD, Martinez E, Valdes I, Ginsberg MD (1988a) Effect of ischemia on the in vivo release of striatal dopamine, glutamate, gamma-aminobutyric acid studied by intracerebral microdialysis. J Neurochem 51:1455–1464

Globus MYT, Busto R, Dietrich WD, Martinez E, Valdes I, Ginsberg MD (1988b) Intraischemic extracellular release of dopamine and glutamate is associated with striatal vulnerability to ischemia. Neurosci Lett 91:36–40

Gotti B, Benavides J, MacKenzie ET, Scatton B (1990) The pharmacotherapy of focal cortical ischaemia in the mouse. Brain Res 522:290–307

Gustafson I, Westerberg E, Wieloch T (1990) Protection against ischemia-induced neuronal damage by the α_2-adrenoceptor antagonist idazoxan: influence of time of administration and possible mechanisms of action. J Cereb Blood Flow Metab 10:885–894

Hofmann HP, Raschack M, Unger L (1989) (S)-Emopamil, a novel calcium and serotonin antagonist for the treatment of cerebrovascular disorders. 1st communication: pharmacological profile. Arzneimittelforschung 39:304–308

Hoyer D, Pazos A, Probst A, Palacios JM (1986a) Serotonin receptors in the human brain. I. Characterization and autoradiographic localization of 5-HT_{1A} recognition sites. Apparent absence of 5-HT_{1B} recognition sites. Brain Res 376:85–96

Hoyer D, Pazos A, Probst A, Palacios JM (1986b) Serotonin receptors in the human brain. II. Characterization and autoradiographic localization of 5-HT_{1C} and 5-HT_2 recognition sites. Brain Res 376:97–107

Kerckhoff W van den, Drewes LR (1985) Transfer of the Ca-antagonists nifedipine and nimodipine across the blood-brain barrier and their regional distribution in vivo. J Cereb Blood Flow Metab 5:S459–S460

Kohlhardt M, Bauer B, Krause H, Fleckenstein A (1972) Differentiation of the transmembrane Na and Ca channels in mammalian cardiac fibres by the use of specific inhibitors. Pflugers Arch 335:309–322

Krieglstein J, Sauer D, Nuglisch J, Rossberg C, Beck T, Bielenberg GW, Mennel H-D (1989) Naftidrofuryl protects neurons against ischemic damage. Eur Neurol 29:224–228

Lin B, Dietrich WD, Busto R, Ginsberg MD (1990) (S)-Emopamil protects against global ischemic brain injury in rats. Stroke 21:1734–1739

Maiese K, Pek L, Berger SB, Reis DJ (1992) Reduction in focal cerebral ischemia by agents acting at imidazole receptors. J Cereb Blood Flow Metab 12:53–63

Morikawa E, Ginsberg MD, Dietrich WD, Duncan RC, Busto R (1991) Postischemic (S)-emopamil therapy ameliorates focal ischemic brain injury in rats. Stroke 22:355–360

Nakayama H, Ginsberg MD, Dietrich WD (1988) (S)-Emopamil, a novel calcium channel blocker and serotonin S_2 antagonist, markedly reduces infarct size following middle cerebral artery occlusion in the rat. Neurology 38:1667–1673

Nowycky MC, Fox AP, Tsien RW (1985) Three types of neuronal calcium channel with different calcium agonist sensitivity. Nature 316:440–443

Orrenius S, McCondey DJ, Bellomo G, Nicotera P (1989) Role of Ca^{2+} in toxic cell killing. Trends Pharmacol Sci 10:281–285

Pickard JD, Murray GD, Illingworth R, Shaw MDM, Teasdale GM, Foy PM, Humphrey PRD, Lang DA, Nelson R, Richards P, Sinar J, Bailey S, Skene A (1989) Effect of oral nimodipine on cerebral infarction and outcome after subarachnoid haemorrhage: British aneurysm nimodipine trial. BMJ 298:636–642

Prehn JHM, Backhauss C, Karkoutly C, Nuglisch J, Peruche B, Rossberg C, Krieglstein J (1991) Neuroprotective properties of 5-HT_{1A} receptor agonists in rodent models of focal and global cerebral ischemia. Eur J Pharmacol 203:213–222

Quirion R (1985) Characterization of binding sites for two classes of calcium channel antagonists in human forebrain. Eur J Pharmacol 117:139–142

Rao TS, Cler JA, Mick SJ, Ragan DM, Lanthorn TH, Contreras PC, Iyengar S, Wood PL (1990) Opipramol, a potent sigma ligand, is an anti-ischemic agent: neurochemical evidence for an interaction with the N-methyl-D-aspartate receptor complex in vivo by cerebellar cGMP measurements. Neuropharmacology 29:1199–1204

Siesjö BK (1981) Cell damage in the brain: a speculative synthesis. J Cereb Blood Flow Metab 1:155–185

Szabo L (1989) (S)-Emopamil, a novel calcium and serotonin antagonist for the treatment of cerebrovascular disorders. 2nd communication: brain penetration, cerebral vascular and metabolic effects. Arzneimittelforschung 39:309–314

Szabo L, Hofmann HP (1989) (S)-Emopamil, a novel calcium and serotonin antagonist for the treatment of cerebrovascular discorders. 3rd communication: effect on postischemic cerebral blood flow and metabolism, and ischemic neuronal cell death. Arzneimittelforschung 39:314–319

Szabo L, Urbanics R (1991) Levemopamil posttreatment reduces cerebrocortical infarct size after permanent middle cerebral artery occlusion in the cat. J Cereb Blood Flow Metab 11:S710

Trust Study Group (1990) Randomised, double-blind, placebo-controlled trial of nimodipine in acute stroke. Lancet 336:1205–1209

Vanhoutte PM, Paoletti R (1987) The WHO classification of calcium antagonists. Trends Pharmacol Sci 8:4–5

Intracellular Messengers After Ischemia

H. Onodera, H. Aoki, and K. Kogure

Introduction

Cerebral ischemia depresses brain energy metabolism and neuronal function. Although most neurons can tolerate short-term ischemia, the CA1 pyramidal cells in the hippocampus are selectively vulnerable to even 5 min of transient forebrain ischemia. Neuronal death of CA1 pyramidal neurons occurs after an interval of 1–2 days following recirculation, during which time no energy crisis or morphological changes are observed (delayed neuronal death). However, recent reports have indicated that nucleic acid metabolism and protein synthesis are disrupted preceding the death of the CA1 pyramidal cells. Various mechanisms have been suggested to account for the CA1 neuronal death. These include the excitatory neurotransmitter theory, calcium overload, and alteration of gene expression. Recently, attention has been focused on the role of intracellular signal transduction systems in the pathogenesis of ischemic CA1 pyramidal cell death.

We report here the changes in postischemic intracellular signal transduction systems and assess the contribution of various agents which can modify intracellular messenger systems. We use Pulsinelli's four-vessel occlusion model of transient forebrain ischemia (20 min) in introduce delayed death of CA1 pyramidal cells [7].

Second Messenger Systems and Calcium Channels: In Vitro Autoradiographic Study

Protein Kinase C (Phorbol Ester Binding)

Protein kinase C (PKC; Fig. 1) [7] is a calcium- and phospholipid-dependent enzyme highly concentrated in the brain. The activation of PKC plays a pivotal role in neurotransmitter release and synaptic plasticity. The tumor promoter phorbol ester binds PKC with nanomolar affinity, and mapping of PKC obtained by in vitro [^{3}H]phorbol 12,13-dibutyrate (PDBu)

Institute of Brain Diseases, 1-1 Seiryon-machi, Sendai 980, Japan

Cerebral Ischemia and Basic Mechanisms
Ed. by A. Hartmann, F. Yatsu, and W. Kuschinsky

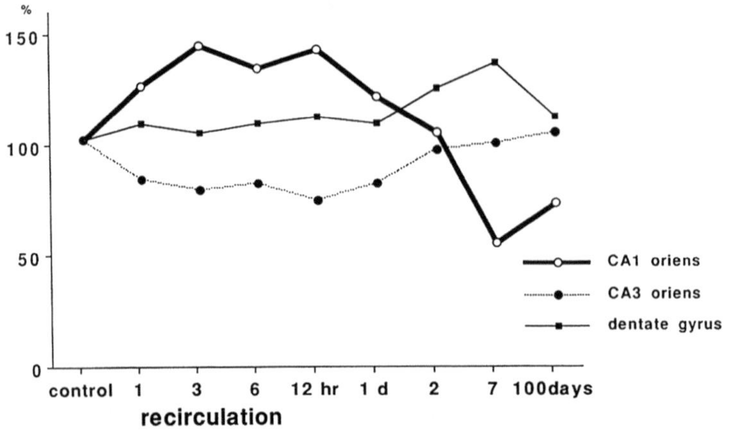

Fig. 1. Phorbol ester (PDBu) binding activity (percentage of control) in the hippocampal formation after 20 min of transient forebrain ischemia in the rat four-vessel occlusion model

autoradiography is identical to the distribution of PKC. The distribution of PDBu binding in the rat brain is heterogeneous, and hippocampal formation exhibits high binding activity. Dendritic fields of the CA1 have the highest grain density. The CA3 and dentate gyrus have moderate grain densities with laminar distribution. PDBu binding in postischemic hippocampus changes not only in the CA1 but also in the histologically intact CA3 and dentate gyrus. The CA1 subfield shows a gradual increase in the PDBu binding during early recirculation (1–12 h after reperfusion), when minimal damage of the CA1 pyramidal cells is seen. Grain density in the CA1 is maximum 6–12 h after recirculation. Thereafter the grain density in the CA1 subfield decreases to show a marked drop 7 days after ischemia. Thus, the loss of PDBu binding of the dendritic fields in the CA1 reflects a close association of phorbol ester binding sites with CA1 pyramidal cells. In contrast, the stratum pyramidale of the CA1 retains a higher grain density than dendritic fields. In the molecular layer of the dentate gyrus, PDBu binding is elevated 2–7 days after ischemia. Interestingly, the inner region of the molecular layer has a higher grain density than the outer region.

We previously found that ipsilateral excitatory input from CA3 pyramidal cells (Schaffer collaterals) plays a critical role in CA1 pyramidal cell death [5]. Since the CA3 pyramidal cells have excitatory input to the CA1 pyramidal cells, repetitive stimulation of CA3 neurons may trigger epileptiform activity of CA1 pyramidal cells. Thus, the mechanism of CA1 neuronal death is analogous to that of long-term potentiation (LTP). Under the condition of unilateral prevention of ischemic CA1 pyramidal cell death by unilateral lesioning of Schaffer collaterals, we observed marked alteration in the 2-deoxyglucose uptake pattern in the CA1 and in subiculum with intact Schaffer collaterals 20 h after ischemia, compared to the side deprived of Schaffer collaterals [6]. PKC plays a role in controlling the release of neurotransmitters and may be involved in the generation of LTP. Translocation of PKC activity may mediate hippocampal LTP since PKC activity is increased in the membrane and decreased proportionally in the cytosol after LTP. Thus, transient increase in the PDBu binding in the CA1 may reflect the translocation of PKC to the membrane. We also observed no significant change of PKC mRNA after ischemia (data not shown). Enhancement of PKC activity and/or translocation of the enzyme play an important role in the postischemic modulation of synaptic efficacy in the hippocampal formation and neuronal death of CA1 neurons. This hypothesis is partly verified by the protective effect of the protein kinase inhibitor staurosporine (discussed in the following section).

Inositol 1,4,5-Trisphosphate Receptors

Inositol 1,4,5-trisphosphate (IP_3) receptors (Fig. 2) [3] are located predominantly on the microsomal membrane and play a pivotal role in the regulation of intracellular calcium concentration. Distribution of IP_3 receptors are heterogeneous and are highly concentrated in the CA1 subfield of the hippocampus, although the CA3 and dentate gyrus exhibit trace IP_3 receptors. The dendritic fields of the CA1 (strata oriens, radiatum, and lacunosum moleculare) in the CA1 show a 20% loss in IP_3 binding activity 3 h after ischemia, without obvious morphological abnormalities. Thereafter, grain density in these layers decrease, and half of the binding sites are lost 2 days after ischemia. By contrast, the stratum pyramidale of the CA1

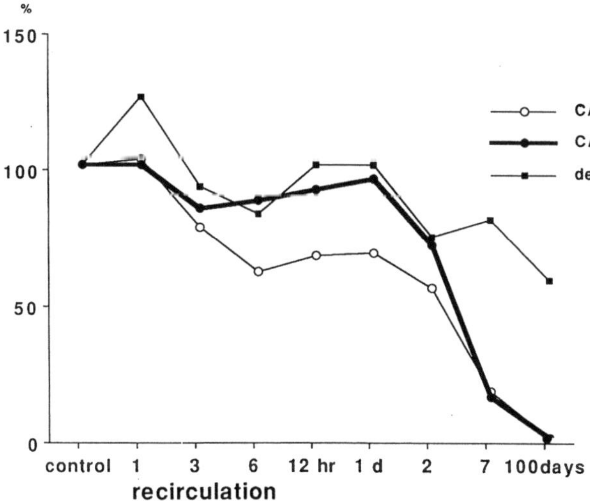

Fig. 2. IP_3 binding activity (percentage of control) in the hippocampal formation after 20 min of transient forebrain ischemia in the rat four-vessel occlusion model

shows no significant change until 2 days after recirculation. Seven days after ischemia all layers in the CA1 have lost 85% of IP_3 receptors. If reduced IP_3 binding activity in the dendritic fields during early recirculation means microsomal dysfunction, disruption of the intracellular calcium buffering system may lead to lethal damage to neurons. We cannot determine whether the postischemic reduction in IP_3 receptors in dendritic fields is directly related to the damage of the CA1 pyramidal cells, or whether it is the reflection merely of cellular dysfunction. Further investigation is needed to clarify the contribution of IP_3 receptors to CA1 pyramidal cell death.

GTP Binding Sites

In normal animals [^{3}H]GTP binding (Fig. 3) [1] in the hippocampal formation is highly concentrated in the dendritic fields of the CA1 and in the molecular layer of the dentate gyrus. Binding is moderate in the dendritic fields of the CA3 and the hilus. The pyramidal cell layer exhibits very low binding. At 1–48 h after ischemia GTP binding remains unaltered in the hippocampal formation. Seven days after ischemia [^{3}H]GTP binding in the dendritic fields of the CA1 subfield decreases by 30%. No significant alteration of GTP binding is noticed in the CA3 or dentate gyrus throughout the reperfusion period studied. Thus, a significant decrease in GTP binding is not observed until delayed CA1 pyramidal cell death. Our [^{3}H]GTP binding study indicated a minor contribution of the CA1 pyramidal neurons to the total GTP binding sites in the CA1 subfield (ca. 30%). Residual GTP binding sites in the CA1 may be located on glial cells, presynaptic fibers, and GABAergic interneurons.

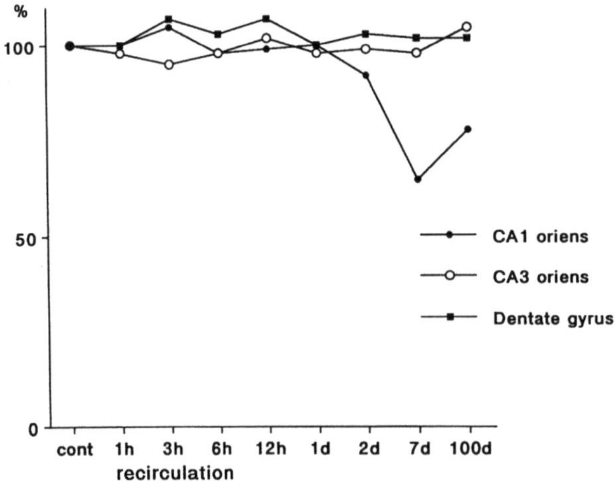

Fig. 3. GTP binding activity (percentage of control) in the hippocampal formation after 20 min of transient forebrain ischemia in the rat four-vessel occlusion model

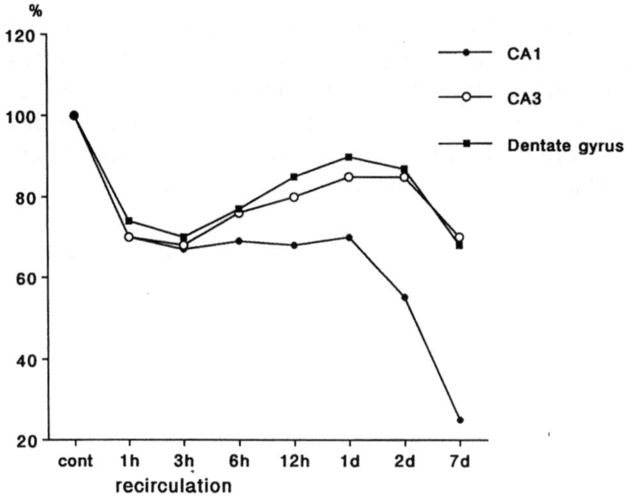

Fig. 4. PN200-100 binding activity (percentage of control) in the hippocampal formation after 20 min of transient forebrain ischemia in the rat four-vessel occlusion model

L-Type Calcium Channels (PN200-110 Binding)

L-type calcium channels visualized by [^{3}H]PN200-110 (PN; Fig. 4) [4] is highly concentrated in the dentate gyrus and is moderate in the CA3 subfield. The CA1 has relatively low PN binding. After ischemia a decrease in PN binding precedes delayed neuronal death of CA1 pyramidal cells. In the stratum oriens of the CA1, PN binding decreases significantly after 3 h of recirculation. Thus, neuronal activity of the CA1 pyramidal cells may be altered despite minimal morphological damage early during recirculation. Seven days after ischemia approximately 75% of PN binding is lost in the CA1. Thus, calcium antagonist binding sites in the CA1 are located predominantly on the CA1 pyramidal cells. Postischemic calcium homeostasis is markedly modulated immediately after recirculation. Thus, many studies have been designed to estimate the effect of calcium antagonists on ischemic CA1 neuronal death. Since PN binding was reduced immediately after reperfusion in the CA1, postischemic administration of calcium antagonists may be much less effective than preischemic administration.

Immediate Early Genes After Ischemia

Immediate early genes [8] are implicated in cell growth and differentiation. The c-*fos* and c-*jun* belong to a class of genes encoding nuclear proteins. Although their basal levels for mRNA are very low, expression of the genes is increased to high transcript levels by factors that stimulate neuronal activity and cell differentiation. Transient and massive increase of c-*fos* and c-*jun* mRNAs is observed in the hippocampal formation and cerebral cortex after ischemia (0.5–3 h after recirculation). In contrast, another unclear proto-oncogene, c-*myc*, was not induced by ischemic insult alone. During the early recirculation period, transcription levels of most proteins are markedly depressed. Many factors affecting c-*fos* and c-*jun* expression are recognized during and after ischemia (e.g., membrane depolarization, calcium influx, neurotransmitter release, and disruption of the membrane structure). Transient elevation of cyclic AMP concentration during early recirculation may also play a role. The induction of a nuclear proto-oncogene may be involved in the cellular processes that modulate genomic events and neuronal activity, although the significance of these increased proto-oncogenes is still unknown. The products of *fos* and *jun* form a complex that associate with transcriptional control elements containing AP-1 sites and mediates the long-term responses to signals that regulate differentiation. Fos and Jun may serve as a sort of master switch for turning on other genes in response to ischemic insult. Although the target proteins are not known at present, proteins required for the recovery process and neurotransmission could be the preferential targets of Jun-Fos complex after ischemic insult.

Protective Effect of Agents Which Can Modulate Intracellular Signaling

Pertussis Toxin

Pertussis toxin (PTX) [1] was stereotaxically injected into the unilateral CA1 subfield (2 μg) or into the unilateral lateral ventricle (4 μg) 3 days before ischemic insult. As previously reported, such injection inactivates G proteins in the hippocampus 3 days after injection. Six days after ischemia the number of surviving CA1 pyramidal neurons were counted in each specimen.

Direct injection of PTX into the CA1 itself causes no neuronal damage. Both PTX injection into the CA1 and the lateral ventricle fail to prevent CA1 pyramidal cell death. Postischemic modulation of G_i and/or G_o by means of PTX depends on preserving the signal transduction system mediated by G proteins. A significant decrease in GTP binding is not observed until delayed CA1 pyramidal cell death. Certain types of G proteins that couple with cell surface receptors regulate activity of the enzyme phospholipase C and control the level of IP_3 and diacylglycerol, both of which convey intracellular signals. Therefore, we tested the hypothesis that PTX treatment can protect against ischemic CA1 damage by suppressing the increase in intracellular calcium concentration and the activation of PKC through the reduction of both IP_3 and diacylglycerol production. Ischemic change in the neurotransmission from CA3 neurons to CA1 neurons may be caused by a mechanism similar to the generation of LTP, and lesioning to Schaffer collaterals prevents CA1 pyramidal cell death. PTX is known to prevent LTP in both the stratum radiatum-CA1 and mossy fiber-CA3 synaptic systems, indicating the involvement of PTX-sensitive G protein in LTP. However, PTX treatment preceding ischemic insult, which is able to block the G_o- and G_i-mediated transduction system during and after ischemia, is not beneficial for the survival of CA1 neurons. Rather, it aggravates postischemic CA1 neuronal damage when administered into the lateral ventricle. One possible cause of increased CA1 damage is the blockade of inhibitory systems involved in synaptic transmission. Since PTX is known to suppress the actions of inhibitory systems via G_i-linked proteins such as adenosine A1 receptors and gamma-aminobutyric acid B receptors, increased neuronal damage may be explained as being due partially to the blockade of these inhibitory systems.

Protein Kinase Inhibitors

In a study of protein kinase inhibitors [2] the CA1 neuronal cell density in control animals was 165 cells/mm. In the vehicle-treated ischemia group, neuronal cell density of the CA1 ipsilateral to the injection was 28/mm, and the density of the contralateral CA1 was 22/mm. The unilateral injection of staurosporine at a dose of 10 ng 30 min before induction of ischemia reduced the damage of the CA1 pyramidal cells ipsilateral to the injection, and the CA1 pyramidal cell density ipsilateral to the staurosporine injection was 78/mm. In contrast to the protective effect of staurosporine administered 30 min before induction of ischemia, staurosporine treatments 10 or 180 min after recirculation failed to prevent the loss of CA1 pyramidal cells.

In the gerbil model, staurosporine (0.1–10 ng), KT5720 (cyclic AMP dependent protein kinase inhibitor, 10 ng), KT5822 (cyclic GMP dependent protein kinase inhibitor), and W-7 (calmodulin antagonist, 10 ng) were stereotaxically injected unilaterally into the CA1 on the day of ischemia. Interestingly, only the side with staurosporine was protected against CA1 pyramidal cell loss in a dose-dependent manner. On the other hand, KT-5720, KT5822, and W-7 had no protective effect.

Staurosporine is a nonspecific protein kinase inhibitor acting at the ATP binding site of protein kinases. Since A and G kinase inhibitors and calmodulin antagonist were not effective, staurosporine in the present study plays a protective role by the inhibition of PKC. As stated above, phorbol ester binding in the CA1 increases during early reperfusion period. This phenomena reflects translocation of PKC molecules from cytosol to membrane fraction. The target phospho-proteins that are critical for neuronal death remains to be clarified. The delayed neuronal death of CA1 neurons cannot be explained solely by the activation of PKC because staurosporine cannot protect all CA1 neurons. In the rat model half the pyramidal cells are damaged even when staurosporine is administered before the ischemic insult. Furthermore, we cannot rule out the possibility that other actions of staurosporine, such as inhibition of tyrosine protein kinase, play a role in the amelioration of neuronal damage. This suggests that mechanisms other than PKC activation are involved in the generation of CA1 neuronal death. Postischemic administration has little protective effect, even if administered immediately after recirculation. This suggests that PKC-mediated processes that occur during and/or shortly after recirculation lead to subsequent events that ultimately determine the fate of pyramidal cells.

Conclusion

Transient forebrain ischemia induces marked changes in inter- and intracellular signal transduction systems. In the CA1 subfield of the hippocampus, IP_3 metabolism and PKC activity are altered preceding delayed-type CA1 pyramidal cell death. Although selective loss of CA1 pyramidal cells is protected by the depletion of major excitatory inputs, which clearly indicates the contribution of excitatory amino acids to neuronal death, critical intracellular events that induce neuronal cells are still unknown. Failure of PTX to protect neuronal death suggests the contribution of other factors than GTP binding proteins (G_o, G_i) and phosphoinositides cycles. Preischemic administration of staurosporine partly protected CA1 neuronal death, although postischemic administration failed to prevent neuronal death. The failure of postischemic staurosporine administration and change of IP_3 and phorbol ester binding sites early during reperfusion period suggests that triggering events leading to the death of CA1 pyramidal cells occur during the early recirculation period. The mechanism that finally determines the fate of pyramidal cells remains to be clarified.

References

1. Aoki H, Onodera H, Yamasaki Y, Yae T, Jian Z, Kogure K (1992) The role of GTP binding proteins in ischemic brain damage: autoradiographic and histopathological study. Brain Res 570:144–148
2. Hara H, Onodera H, Yoshidomi M, Matuda Y, Kogure K (1990) Staurosporine, a novel protein kinase C inhibitor, prevents postischemic neuronal damage in the gerbil and rat. J Cereb Blood Flow Metab 10:646–653
3. Onodera H, Kogure K (1989) Mapping second messenger systems in the rat hippocampus after transient forebrain ischemia: in vitro [^{3}H]forskolin and [^{3}H]inositol 1,4,5-trisphosphate binding. Brain Res 487:343–349
4. Onodera H, Kogure K (1990) Calcium antagonist, adenosine A1, and muscarinic binding in rat hippocampus after transient ischemia. Stroke 21:771–776
5. Onodera H, Sato G, Kogure K (1986) Lesions to Schaffer collaterals prevent ischemic death of CA1 pyramidal cells. Neurosci Lett 68:169–174

6. Onodera H, Tobita M, Kogure K (1988) The effect of transient ischemia on 2-[^{14}C]deoxyglucose metabolism in the rat hippocampus pretreated with kainic acid. Neurosci Lett 89:90–96
7. Onodera H, Araki T, Kogure K (1989) Protein kinase C activity in the rat hippocampus after forebrain ischemia: autoradiographic analysis by [^{3}H]phorbol 12,13-dibutyrate. Brain Res 481:1–7
8. Onodera H, Kogure K, Ono Y, Igarashi K, Kiyota Y, Nagaoka A (1989) Proto-oncogene c-fos is transiently induced in the rat cerebral cortex after forebrain ischemia. Neurosci Lett 98:101–104

Influence of NMDA Receptor Antagonists on Glycine Receptor Isoform Expression in Spinal Cord Cultures

C.-M. Becker[1,2], W. Hoch[2,5], M. Schramm[3], I. Wolters[2,4], and H. Betz[4]

Introduction

The physiology of the amino acid transmitter glycine is well characterized in spinal cord, where it is involved in both mutual and feedback inhibition by recurrent axon collaterals of spinal motoneurons. Glycinergic inhibition is mediated by small interneurons within the spinal gray matter, including the Renshaw cells (Aprison and Daly 1978). Upon its release by the inhibitory neuron, glycine binds to postsynaptic receptors where it causes an intrinsic chloride channel to open. The resulting chloride influx produces a postsynaptic hyperpolarization and drastically decreases neuronal firing (Bormann et al. 1987). Glycinergic inhibition is also found in spinal nociceptive, brain stem auditory, and other CNS systems as well as in the retina (reviewed by Becker 1992). The convulsant alkaloid strychnine competetively antagonizes receptor binding of glycine, and, conversely, glycine-displaceable [^{3}H]strychnine binding has been established as a pharmacological probe of inhibitory glycine receptors. Heterogeneity is a widespread phenomenon of neurotransmitter receptors in the mammalian CNS (Betz 1990). Different isoforms have also been identified of the inhibitory glycine receptor (reviewed by Betz and Becker 1988; Langosch et al. 1990; Becker 1992). In contrast, high-affinity binding sites for [^{3}H]glycine are distinct from inhibitory glycine receptors and are thought to reflect a regulatory domain of the *N*-methyl-D-aspartate (NMDA) receptor (Betz and Becker 1988; Monaghan et al. 1989; Moriyoshi et al. 1991).

[1] Neurologische Universitätsklinik, Universität Heidelberg, Im Neuenheimer Feld 400, 69120 Heidelberg, Germany
[2] Zentrum für Molekulare Biologie, Universität Heidelberg, Im Neuenheimer Feld 282, 69120 Heidelberg, Germany
[3] Department of Biological Chemistry, Hebrew University of Jerusalem, Jerusalem 91 904, Israel
[4] Max-Planck-Institut für Hirnforschung, Deutschordenstr. 46, 60528 Frankfurt, Germany
[5] Howard Hughes Medical Institute, Department of Molecular and Cellular Physiology, Beckman Center, Stanford University, Stanford, CA 94305, USA (present address)

Cerebral Ischemia and Basic Mechanisms
Ed. by A. Hartmann, F. Yatsu, and W. Kuschinsky

Table 1. Glycine receptor protein isoforms

	Adult isoform	Neonatal isoform
Subunits	α_1 (48 kDa) β (58 kDa)	α_2 (49 kDa)
Structure of transmembrane complex	Heterooligomer (pentamer $\alpha_3\,\beta_2$)	Homooligomer (as analyzed in spinal cord culture)
Binding affinity for strychnine	High ($K_D \approx 10\,nM$)	Low
Regional expression	Adult spinal cord, adult brain stem	Neonatal spinal cord, adult and neonatal cerebral cortex

Structure and Isoforms of the Inhibitory Glycine Receptor

During postnatal development, two protein isoforms of the inhibitory glycine receptor are sequentially expressed in rodent spinal cord (Table 1). The adult isoform ($GlyR_A$) binds strychnine with high affinity and behaves as a complex glycoprotein of molecular weight of 250 kDa or more (Pfeiffer et al. 1982; Langosch et al. 1990). It comprizes three polypeptides of 48, 58, and 93 kDa (Pfeiffer et al. 1982; Becker et al. 1986). The strychnine binding site resides on the 48-kDa (α1) subunit (Pfeiffer et al. 1982) which coassembles with the 58-kDa (β) polypeptide to form a pentameric chloride channel (Langosch et al. 1990). The 93-kDa protein (gephyrin) residing at the

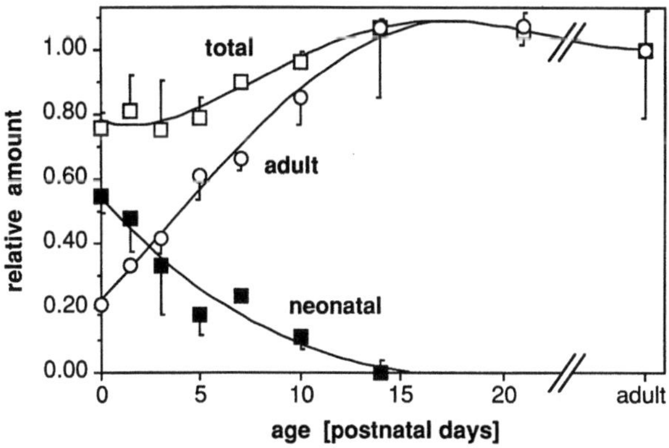

Fig. 1. Postnatal development of glycine receptor isoforms in rat spinal cord. Total glycine receptor content (□) was assessed by MAb 4a immunoreactivity, $GlyR_A$ levels (○) were determined by high-affinity [^{3}H]strychnine binding. The fraction of low-affinity antagonist binding receptors, i.e., $GlyR_N$ (■), was derived from these values (from Becker et al. 1988, 1992; by permission of Cell Press)

cytoplasmic face of the postsynaptic membrane has been implicated in synaptic anchoring of the receptor channel (Triller et al. 1985; Prior et al. 1992).

A distinct neonatal isoform of the glycine receptor ($GlyR_N$) prevails in spinal cord of newborn rodents. Within 2 weeks postnatally it is fully replaced by the adult receptor (Becker et al. 1988, 1992) (Fig. 1). Its low antagonist binding affinity precludes a detection by conventional [^{3}H]strychnine binding techniques. $GlyR_N$ contains a ligand binding $\alpha2$ subunit which in both molecular weight (49 kDa) and antigenic epitopes differs from the $\alpha1$ polypeptide of $GlyR_A$ (Becker et al. 1988; Hoch et al. 1989; Table 1). A $GlyR_N$-like protein has also been identified in cerebral cortex of rat and man (Naas et al. 1991; Becker et al., submitted).

Regulation of Glycine Receptor Isoform Expression

Glycine receptor heterogeneity reflects a sequential expression of homologous α subunit variants identified by cDNA cloning (Grenningloh et al. 1987; Kuhse et al. 1990a,b, 1991; Malosio et al. 1991). These α subunit variants are highly homologous to each other (Grenningloh et al. 1987, 1990; Kuhse et al. 1990a,b, 1991) and to the deduced primary structure of the β subunit (Grenningloh et al. 1990). By hydropathy profiles, the same transmembrane topology is predicted for all of these polypeptides with four hydrophobic segments spanning the postsynaptic membrane. The ligand binding sites of the α subunits are localized on a large extracellular N-terminal domain (reviewed by Becker 1992). Upon recombinant expression in eukaryotic cells, α subunits assemble into functional glycine receptor channels (Schmieden et al. 1989; Sontheimer et al. 1989).

As expected for the ligand binding subunit of the neonatal glycine receptor protein, $\alpha2$ transcripts are abundant in spinal cord around birth but display a drastic decrease thereafter (Becker et al. 1988; Kuhse 1990a, 1991). Consistent with biochemical studies on $GlyR_N$ (Becker et al. 1988; Hoch et al. 1989), expression in *Xenopus laevis* oocytes of the $\alpha2$ subunit leads to the formation of glycine-gated chloride channels of low affinity for the antagonist strychnine. In contrast, tissue contents of $\alpha1$ subunit mRNA significantly increase during postnatal development (Kuhse et al. 1990a). Additional glycine receptor α subunit variants ($\alpha3$, $\alpha4$) have been identified but are transcribed at low levels only and, thus, are not likely to account for a major fraction of $GlyR_A$ (Kuhse et al. 1990b; Malosio et al. 1991). In agreement with the conclusion of cortical glycine receptors resembling $GlyR_N$, in situ hybridization reveals $\alpha2$ transcripts in the deeper layers of the developing and adult rat cortex (Malosio et al. 1991). At the protein level, the sequential expression of gluycine receptor α subunit variants can be monitored by monoclonal antibody (MAb) binding. While MAb 4a defines an epitope common to all GlyR α subunits known, MAb 2b identifies the $\alpha1$

subunit of $GlyR_A$ (Pfeiffer et al. 1984; Becker et al. 1988; Schröder et al. 1991).

The differential expression of glycine receptor subunit variants is not understood. Primary cultures of neural and muscle cells are widely employed to investigate regulation and surface distribution of neuro-transmitter receptors during cellular maturation. In particular, primary cultures of fetal rodent spinal cord have been used to characterize functional properties of glycine receptor channels (Hamill et al. 1983; Bormann et al. 1987). However, cultured spinal neurons predominantly accumulate $GlyR_N$ while $GlyR_A$ accounts for less than 10% of the total glycine receptor protein. Even after extended periods of cultivation, the switch of glycine receptor isoforms characterizing postnatal development of rodent spinal cord is not seen under culture conditions (Hoch et al. 1989).

NMDA Receptor Antagonists Increase Glycine Receptor Levels in Culture

The influence of NMDA receptor antagonists on glycine receptor expression was analyzed in cultured mouse spinal neurons. NMDA receptor antagonists efficiently inhibit glutamate-mediated neurotoxicity (Choi 1990) and promote long-term survival in culture of highly vulnerable cerebellar granule cells (Schramm et al. 1990). In cultures of spinal cord, long-term blockade (days 3–21 after plating) of NMDA receptors by the noncompetitive channel blocker MK 801 increased glycine receptor contents per culture dish more than 15-fold (Fig. 2) as determined by the common probe of all α subunit variants known, MAb 4a (Hoch et al. 1992). These changes were accompanied by an about 2.5-fold increase in the total membrane protein content per dish (not shown). Similar alterations were produced by treatment with the competitive antagonist 2-amino-5-phosphonovalerate (APV; Fig. 2A). Furthermore, spinal cord cultures subjected to NMDA receptor blockade developed a significant reactivity with the selective marker of $GlyR_A$, MAb 2b (Fig. 2B). Induction by MK 801 of $GlyR_A$ was confirmed by immunoprecipitation of the $\alpha 1$ polypeptide from extracts of [^{35}S]methionine-labeled cultures; the immunoprecipitation was inhibited by a peptide corresponding to the MAb 2b epitope of the $\alpha 1$ subunit (Hoch et al. 1992). As detailed above, $GlyR_A$ is expressed in postnatal rodent spinal cord but not under standard conditions of primary culture (Hoch et al. 1989). Thus, NMDA receptor antagonists induced $GlyR_A$ causing a biochemical maturation of cultured spinal cord neurons. Blockade of NMDA receptors also increased levels of different presynaptic marker proteins (not shown) including synaptophysin, a major integral membrane protein of synaptic vesicles (Wiedenmann and Franke 1985) and the neurotransmitter-synthesizing enzyme choline acetyltransferase (ChAT; Fonnum 1975; Geyer et al. 1987). Likewise, NMDA-receptor antagonists altered culture morphology.

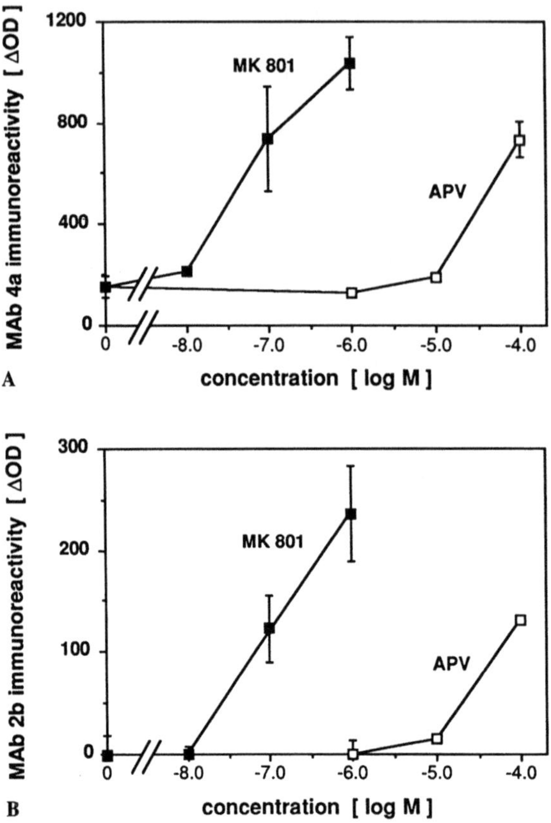

Fig. 2A,B. Effects of NMDA receptor antagonists on glycine receptor expression. Contents of total glycine receptor protein as defined by MAb 4a (**A**) and $GlyR_A$ as assessed by MAb 2b immunoreactivity (**B**) were determined in crude membranes from 21-day-old cultures (from Hoch et al. 1992; by permission of Oxford University Press)

Untreated spinal neurons established a dispersed network of cells interconnected by numerous processes. Upon application of MK 801, neurons arranged in large clusters interconnected by thick axon bundles (not shown; see Hoch et al. 1992).

To define the mechanism of NMDA receptor antagonist action on glycine receptor expression in spinal cord cultures, α subunit mRNA levels were investigated by the polymerase chain reaction. Amplification of $\alpha1$ and $\alpha2$ transcripts using synthetic oligonucleotide primers (Kuhse et al. 1990a) indicated that levels of $\alpha1$ mRNA were low in untreated spinal neurons throughout the entire culture period. Treatment by MK 801 increased the amount of $\alpha1$ subunit mRNA by four- to eightfold after 2 weeks in culture (Fig. 3) while $\alpha2$ subunit transcripts displayed a different accumulation pattern; contents increased to high levels at day 7 after plating irrespective

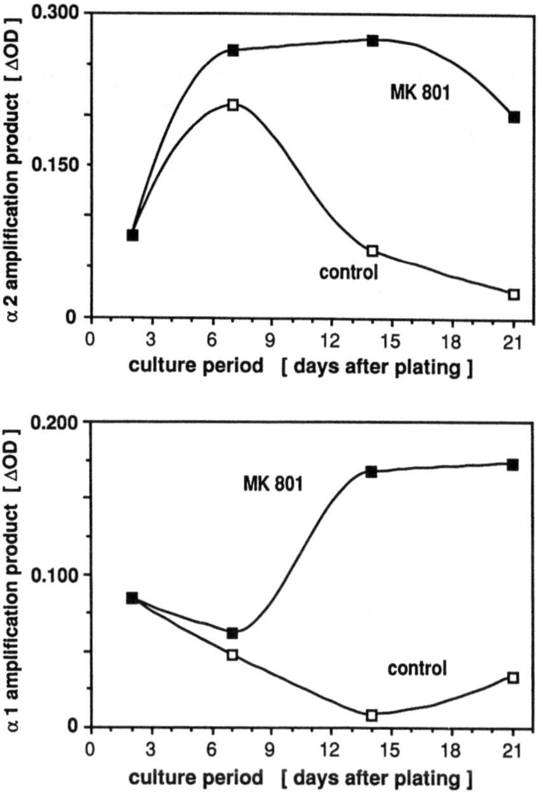

Fig. 3. Effect of MK 801 on the accumulation of glycine receptor $\alpha2$ and $\alpha1$ transcripts in cultures. Spinal neurons were cultured for 2, 7, 14, and 21 days in the presence or absence of MK 801. RNA was isolated from the cells, and polymerase chain reaction was performed on corresponding cDNAs. Semiquantitative analysis of $\alpha1$ and $\alpha2$ amplification products was obtained by scanning of autoradiographs (from Hoch et al. 1992)

of antagonist treatment. Thereafter, high expression of $\alpha2$ mRNA was continued only upon NMDA receptor blockade. Without treatment, transcript levels declined rapidly (Fig. 3). In the same samples, ChAT expression was high throughout the culture period and not influenced by MK 801 (Hoch et al. 1992).

Even small amounts of glutamate present in serum-supplemented culture medium or generated by conversion of glutamine, a major metabolite of most media formulations, severely affect neuronal survival in vitro. In cultures of cerebellar granule cells, NMDA receptor antagonists promote long-term survival by efficiently inhibiting glutamate-mediated excitotoxicity (Schramm et al. 1990). In primary cultures of mouse spinal cord, pharmacological blockade of NMDA receptors increased the levels of glycine receptor and other neuronal markers. Apparently, treatment with MK

801 and that with APV exerted similar effects on neuronal survival in our spinal cord cultures by preventing glutamate-mediated neuronal lesions. Alterations in cell culture morphology induced by MK 801 treatment were generally more pronounced when neurons were grown in small rather than large culture dishes (not shown). Possible, a higher fluid stress in smaller plates during medium changes aggravated glutamate-induced cell damage. Indeed, mechanical irritation of cultured neurons causes degenerative changes which are alleviated by NMDA receptor antagonists (Telcoma et al. 1989).

Expression of $GlyR_A$ was not previously observed in primary cultures (Hoch et al. 1989). Both, the appearance of $\alpha 1$ transcripts and the MAb 2b-reactive polypeptide of 48 kDa characterizing $GlyR_A$, indicate that cultured neurons expressed novel proteins upon treatment by NMDA receptor antagonists. This considerable cellular maturation is further evident from extensive cell adhesion and process formation as well as increases in the presynaptic vesicle protein, synaptophysin (Wiedenmann and Franke 1985) while the key enzyme of acetylcholine synthesis (Geyer et al. 1987) was largely unaffected. Thus, effects of NMDA receptor blockade initiated cellular differentiation and extended beyond pure numerical changes by increased cell survival. While NMDA receptor blockade appeared to prevent cell death, it is not known whether it enhanced the survival of a subpopulation of differentiated cells highly vulnerable to glutamate, or whether it induced the maturation of glycinoceptive neurons. Whereas, in rodent spinal cord, postnatal $GlyR_A$ accumulation coincides with the repression of $GlyR_N$ synthesis (Becker et al. 1988; Kuhse et al. 1990a), treated spinal cultures never showed a complete switch of glycine receptor isoforms; $GlyR_N$ persisted even with MK 801 present. This may reflect a lack in the culture system of neuronal differentiation factors or patterned electrical activity.

Sensitive Period of Protection by NMDA Receptor Antagonists

Treatment of spinal cord cultures at different periods of time after plating indicated that the influence of MK 801 on glycine receptor expression apparent during the second week in vitro reflected maturational changes in sensitivity to NMDA receptor antagonists. Glycine receptor levels determined by MAb 4a were enhanced only when MK 801 was present during the second week in culture (Fig. 4). Even short-term treatment restricted to an interval covering days 10–13 after plating produced a significant induction of receptor (Hoch et al. 1992). Thus, NMDA receptor blockade produced its effects on glycine receptor expression during a critical period. Presently, little is known about the development of NMDA responsive ion channels in cultured spinal cord neurons. Glutamate-induced currents in dissociated spinal neurons increase during the second week of culture

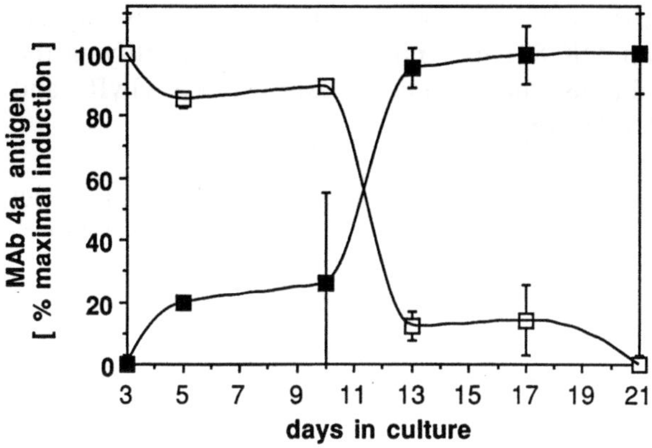

Fig. 4. Critical period of MK 801 treatment of primary cultures from fetal mouse spinal cord. Spinal cultures were treated with $10^{-6}\,M$ MK 801 for different intervals within the 21-day culture period: ■, termination of drug treatment was varied with MK 801 present in the medium from day 3 after plating to the day indicated; □, initiation of drug treatment was varied with MK 801 present in the medium from the day indicated until day 21 after plating. On day 21, all cultures were harvested, and total glycine receptor content was determined in membrane fractions using MAb 4a. Values represent means ± SEM of two culture dishes for each interval of treatment, and are expressed as percentage of maximal glycine receptor induction over untreated controls (0%). Maximum expression (100%) corresponds to values obtained after cultivation in the continuous presence of MK 801 from day 3 to day 21 (from Hoch et al. 1992).

as revealed by electrophysiology (Ransom et al. 1977). It is tempting to speculate that neuronal sensitivity to glutamate is acquired during that period by incorporation of NMDA receptors into the plasma membrane. Although detailed studies on glutamate receptor expression in cultured neurons are not available, high-affinity binding of [^{3}H]MK 801 indicates that NMDA receptors are transiently expressed during a corresponding maturational period (second postnatal week) in neonatal rat spinal cord (Becker, unpublished).

Conclusion

Primary cultures are considered valuable tools for studying receptor regulation and metabolism at the cellular level. Our data, however, demonstrate that dissociated cell cultures do not necessarily reproduce the ontogenesis of neurons in the intact CNS. Accumulation in the medium of toxic metabolites such as glutamate or the absence of essential growth factors may severely disturb a normal differentiation program. Here, suppression of glutamate-induced neurotoxicity allowed $GlyR_A$ to accumulate in spinal cord cultures. Even upon NMDA receptor blockade, however, our cultures did not mimic the expression pattern of glycine receptor isoforms encountered in postnatal spinal cord as $GlyR_N$ was not fully replaced by $GlyR_A$. Spinal cultures treated with MK 801 may now be used for identifying the factors and/or growth conditions which terminate expression of $GlyR_N$ and other early markers of CNS development.

Acknowledgements. This work was supported by the Deutsche Forschungsgemeinschaft (SFB 317, Leibniz Program to H.B. and Heisenberg Program to C.-M.B.) and the German-Israeli Foundation for Scientific Research.

References

Aprison MH, Daly EC (1978) Biochemical aspects of transmission at inhibitory synapses: the role of glycine. Adv Neurochem 3:203–294

Becker C-M (1992) Selective neurotoxicity: convulsants acting at the inhibitory glycine receptor. In: Herken H, Hucho F (eds), Handbook of experimental pharmacology, vol 102. Springer, Berlin Heidelberg New York pp 539–575

Becker C-M, Hermans-Borgmeyer I, Schmitt B, Betz H (1986) The glycine receptor deficiency of the mutant mouse spastic: evidence for normal glycine receptor structure and localization. J Neurosci 6:1358–1364

Becker C-M, Hoch W, Beta H (1988) Glycine receptor heterogeneity in rat spinal cord during postnatal development. EMBO J 7:3717–3726

Becker C-M, Schmieden V, Tarroni P, Strasser U, Betz H (1992) Isoform-selective deficit of glycine receptors in the mouse mutant spastic. Neuron 8:283–289

Betz H (1990) Ligand-gated ion channels in the brain: the amino acid receptor family. Neuron 5:383–392

Betz H, Becker C-M (1988) The mammalian glycine receptor: biology and structure of a neuronal chloride channel protein. Neurochem Int 13:137–146

Bormann J, Hamill OP, Sakmann B (1987) Mechanism of anion permeation through channels gated by glycine and γ-aminobutyric acid in mouse cultured spinal cord neurons. J Physiol (Lond) 385:243–286

Choi D (1990) The role of glutamate in hypoxic-ischemic neuronal death. Annu Rev Neurosci 13:171–182

Fonnum F (1975) A rapid radiochemical method for the determination of choline acetyl transferase. J Neurochem 24:407–409

Geyer SW, Gudden W, Betz H, Gnahn H, Weindl A (1987) Colocalization of choline acetyltransferase and postsynaptic glycine receptors in motoneurons of rat spinal cord demonstrated by immunocytochemistry. Neurosci Lett 82:11–15

Grenningloh G, Rienitz A, Schmitt B, Methfessel C, Zensen M, Beyreuther K, Gundelfinger ED, Betz H (1987) The strychnine-binding subunit of the glycine receptor shows homology with nicotinic acetylcholine receptors. Nature 328:215–220

Grenningloh G, Pribilla I, Prior P, Multhaup G, Beyreuther K, Taleb O, Betz H (1990) Cloning and expression of the 58 kd beta subunit of the inhibitory glycine receptor. Neuron 4:963–970

Hamill OP, Bormann J, Sakmann B (1983) Activation of multiple-conductance state chloride channels in spinal neurones by glycine and GABA. Nature 305:805–808

Hoch H, Betz H, Becker C-M (1989) Primary cultures of mouse spinal cord express the neonatal isoform of the inhibitory glycine receptor. Neuron 3:339–348

Hoch W, Betz H, Schramm M, Wolters I, Becker C-M (1992) Modulation by NMDA-receptor antagonists of glycine receptor isoform expression in cultured spinal cord neurons. Eur J Neurosci 4:389–395

Kuhse J, Schmieden V, Betz H (1990a) A single amino acid exchange alters the pharmacology of neonatal rat glycine receptor subunit. Neuron 5:867–873

Kuhse J, Schmieden V, Betz H (1990b) Identification and functional expression of a novel ligand binding subunit of the inhibitory glycine receptor. J Biol Chem 265:22317–22320

Kuhse J, Kuryatov A, Maulet Y, Malosio M-L, Schmieden V, Betz H (1991) Alternative splicing generates two isoforms of the $\alpha 2$ subunit of the inhibitory glycine receptor. FEBS Lett 283:73–77

Langosch D, Becker C-M, Betz H (1990) The inhibitory glycine receptor: a ligand-gated chloride channel of the central nervous system. Eur J Biochem 194:1–8

Malosio M-L, Marquèze-Pouey B, Kuhse J, Betz H (1991) Widespread expression of glycine receptor subunit mRNAs in the adult and developing rat brain. EMBO J 10:2401–2409

Monaghan DT, Bridges RJ, Cotman CW (1989) The excitatory amino acid receptors: their classes, pharmacology, and distinct properties in the function of the central nervous system. Annu Rev Pharmacol Toxicol 29:365–402

Moriyoshi K, Masu M, Ishii T, Shigemoto R, Mizuno N, Nakanishi S (1991) Molecular cloning and characterization of the rat NMDA receptor. Nature 354:31–37

Naas E, Zilles K, Gnahn H, Betz H, Becker C-M, Schröder H (1991) Glycine receptor immunoreactivity in cerebral cortex of man and rat. Brain Res 561:139–146

Pfeiffer F, Graham D, Betz H (1982) Purification by affinity chromatography of the glycine receptor of rat spinal cord. J Biol Chem 257:9389–9393

Pfeiffer F, Simler R, Grenningloh G, Betz H (1984) Monoclonal antibodies and peptide mapping reveal structural similarities between the subunits of the glycine receptor of rat spinal cord. Proc Natl Acad Sci USA 81:7224–7227

Prior P, Schmitt B, Grenningloh G, Pribilla I, Multhaup G, Beyreuther K, Maulet Y, Werner P, Langosch D, Kirsch J, Betz H (1992) Primary structure and alternative splice variants of gephyrin, a putative glycine receptor-tubulin linker protein. Neuron 8:1161–1170

Ransom B, Bullock PN, Nelson PG (1977) Mouse spinal cord in cell culture. III. Neuronal chemosensitivity and its relationship to synaptic activity. J Neurophysiol 40:1163–1177

Schmieden V, Grenningloh G, Schofield PR, Betz H (1989) Functional expression in Xenopus oocytes of the strychnine binding 48 kd subunit of the glycine receptor. EMBO J 8:695–700

Schramm M, Eimerl S, Costa E (1990) Serum and depolarizing agents cause acute neurotoxicity in cultured cerebellar granule cells: role of glutamate receptor responsive to N-methyl-D-aspartate. Proc Natl Acad Sci USA 87:1193–1197

Schröder S, Hoch W, Becker C-M, Grenningloh G, Betz H (1991) Mapping of antigenic epitopes on the $\alpha 1$ subunit of the inhibitory glycine receptor. Biochemistry 30:42–47

Sontheimer H, Becker C-M, Pritchett DP, Schofield PR, Grenningloh G, Kettenmann H, Betz H, Seeburg PH (1989) Functional chloride channels by mammalian cell expression of rat glycine receptor subunit. Neuron 2:1491–1497

Telcoma ES, Monyer H, Goldberg MP, Choi DW (1989) Traumatic neuronal injury is attenuated by NMDA antagonists in vivo. Neuron 2:1541–1545

Triller A, Cluzeaud F, Pfeiffer F, Betz H, Korn H (1985) Distribution of glycine receptors at central synapses: an immunoelectron microscopy study. J Cell Biol 101:683–688

Wiedenmann B, Franke WW (1985) Identification and localization of synaptophysin, an integral membrane glycoprotein of M_r 38,000 characteristic of presynaptic vesicles. Cell 41:1017–1028

Effects of Calcium Antagonists on Experimentally Induced Carotid Intimal Proliferations

E. Betz and D. Roth

Atherosclerotic proliferates causing occlusion or stenoses of carotid arteries are a frequent cause of stroke or transient ischemic attacks (TIA). It is not possible directly and systematically to follow the cellular processes in a patient's carotid artery wall leading to the final disaster of a stroke. We therefore sought experimental approaches to inhibiting atherogenetic carotid occlusions and secondary stenoses.

The multifactorial causality of atherogenesis (Fig. 1) cannot be simulated completely in a single experimental model. The left side of the chain of causation shown in the scheme became the basis for the lipid theory of atherogenesis. As early as 1908 Ignatowski had shown that rabbits responded with aortic and coronary atheromas when they received a diet enriched with egg yolk and fat. Anitschkow and Chalatow (1913) showed that cholesterol in whatever form added to the food of rabbits was the decisive component. Later it was confirmed in patients with abnormally high blood cholesterol that high serum levels of low-density lipoproteins combined with low levels of high-density lipoproteins is a risk factor, and the lipid theory of atherogenesis became widely accepted also for the development of atheromas in extracerebral brain supplying arteries. By the feeding of a high cholesterol diet it is not easy quantitatively to study the efficiency of drugs in inhibiting the development of atherosclerosis in brain-supplying arteries because these arteries often respond to hypercholesterinemia considerably less than do the aortas or coronary arteries.

In 1963 Baumgartner and Studer reported that after dilatation of a carotid artery with an inflatable balloon catheter and endothelial denudation by the balloon, the distended, deendothelialized artery section responded with proliferation of smooth muscle cells (SMC) into the intima of the lesioned artery. Platelets adhered on the luminal artery surface denuded of endothelium. Ross and Glomset (1976) found that platelet-derived growth factor (PDGF) is released from the adhering platelets, and since in cultures of rabbit smooth muscle cells PDGF increases the mitotic rate, they proposed the injury hypothesis of atherogenesis. In contrast to the arteries of animals fed a high cholesterol diet, in rabbits that received normal food the

Institute of Physiology I, University of Tübingen, Gmelinstr. 5, 72076 Tübingen, Germany

Cerebral Ischemia and Basic Mechanisms
Ed. by A. Hartmann, F. Yatsu, and W. Kuschinsky

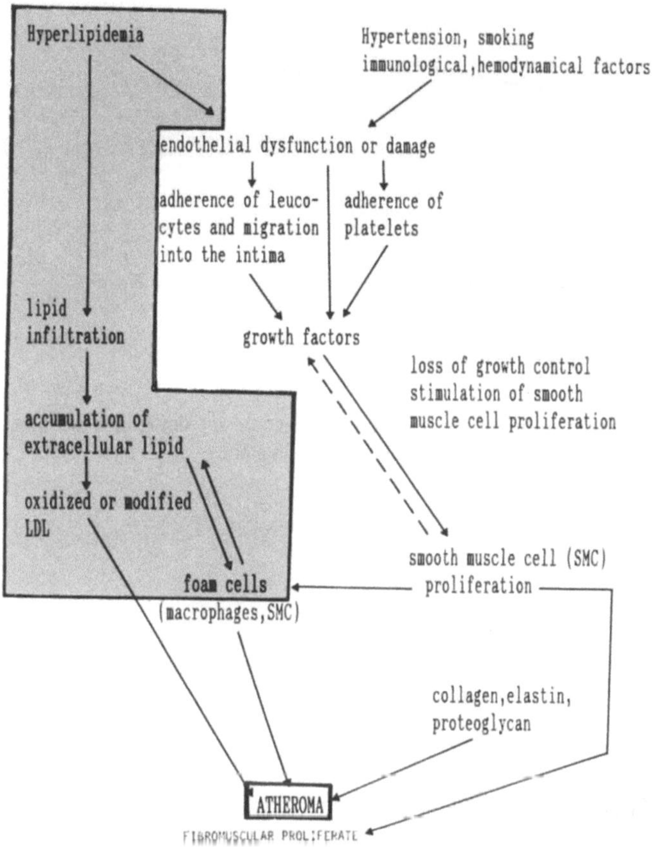

Fig. 1. Network of factors involved in the development of atheromas or fibromuscular proliferates. If no hyperlipidemia (*left*) exists, the proliferation is a fibromuscular intimal thickening

circular proliferate contains no lipids. By sheating an artery with a rigid cuff, a similar proliferate is induced, but the endothelium is maintained hence endothelial denudation is only one possibility of inducing an intimal proliferate in normocholesterinemic animals. Using the ballooning model, drug effects on SMC growth in carotid artery regions free of endothelium have been studied quantitatively. We have standardized the ballooning procedure in the following way: In rats a balloon catheter was inserted into a common carotid artery. After inflation of the balloon the catheter was pulled back and then deflated. This procedure was repeated three times. To control whether the ballooning procedure deendothelializes the artery wall completely, Evans blue stain was injected intravenously. This substance stains deendothelialized intimal regions blue, and this was always seen after the procedure. After various time intervals both carotid arteries were excised

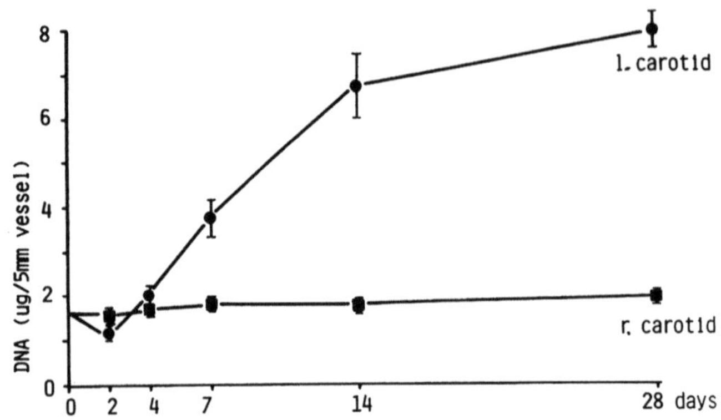

Fig. 2. Time course of intimal lesions in the left carotid artery of rabbits after ballooning, determined by measuring the DNA content per 5-mm carotid segment

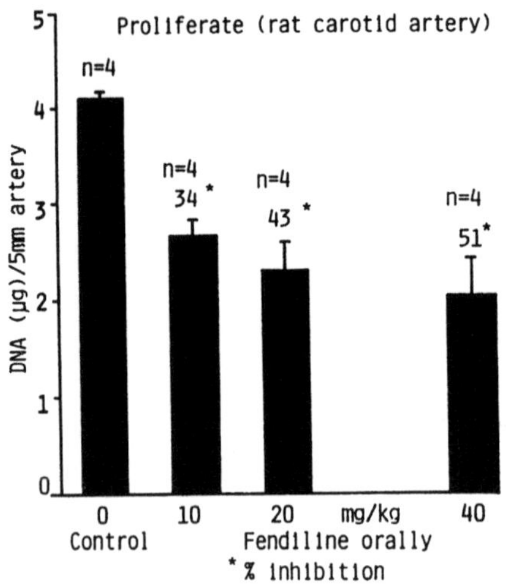

Fig. 3. Dose-response inhibition of fendiline of intimal thickenings 14 days after ballooning. Each column represents the mean of four animals; *vertical bars*, SD

and the DNA content of 5-mm-long sections of both arteries was measured (Labarca and Paigen 1980).

Since the amount of DNA is the same for each cell nucleus, the DNA content can be used as a quantitative measure for the amount of cells in an artery wall section. The comparison of the DNA content of equally long sections of a ballooned carotid artery with that of the contralateral artery (Fig. 2) in rat carotids shows an initially steep increase in cell proliferation that returns toward normal after about 2–3 weeks. Comparison of the proliferation of ballooned carotids of drug-treated animals with untreated ones allows the study of the effect of proliferation inhibitors. Figure 3 depicts how the calcium-calmodulion antagonist fendiline inhibits intimal cell proliferation in a dose-dependent manner, and Table 1 shows the effect of sulfated polyanoins, including a low molecular weight heparin (MW 4000 mass units). To study the mitotic rate of the proliferating cells we subcutaneously injected the thymidin analog substance 5-bromodeoxyuridine (BrdU) 18 h and again 8 h prior to the excision of the artery and stained the microscopical sections with a fluorescent BrdU antibody. BrdU is incorporated into the DNA of dividing cells in their S phase and is therefore an indicator of the mitotic activity of cells. In accordance with the DNA study, the mitotic rate was lower when proliferation inhibiting drugs were given after ballooning.

For the study of calcium antagonists we mainly used the electrostimulation model for the production of fibromuscular proliferates or atheromas. Two fine graphite-coated gold electrodes are chronically implanted on the adventitia of carotid arteries in such a way that weak electrical DC stimuli (0.1 mA 10 Hz, 15 ms/impulse) pass transmurally through a 5-mm-long section of the artery (for details of the technique see Betz and Schlote 1979; Betz and Hämmerle 1984). Leads from the electrodes are conducted subcutaneously to a socket fixed in the skull. They can be connected to an electrical stimulator in such a way that stimulations can be applied to the unrestrained, conscious animal repetitively. With this model all components of atherogenesis shown in Fig. 1 can be taken into consideration.

In stimulated arteries a massive proliferation appears at the anode side that causes a stenosis of the artery. If an animal is exposed to a single 30min stimulation period with weak impulses no proliferate develops. However, the transendothelial transport of macromolecules and ions toward the anode increases. This has been checked with a 30-min period of stimulation with the above stimulation parameters measuring the effect on the transendothelial transport of intravenously injected peroxidase of a molecular weight of 40 000 mass units. When both arteries were excised 6 min

Table 1. Inhibition in the development of proliferations in ballooned carotid arteries of rats with sulfated polyanions of the heparin type and dextransulfate

Substance	DNA (g/5 mm artery)	Inhibition	Number of animals
Controls	3.82 ± 0.32	0%	4
Heparin, type II (Sigma)	2.11 ± 0.08	45%	4
Low molecular weight heparin (4300 dalton)	2.30 ± 0.39	40%	4
Dextransulfate	1.67 ± 0.32	56%	4

In comparison with normal heparin (Sigma) the anti-Xa activity of low molecular weight heparin is low, so that high concentrations are tolerated by the animals. All substances were injected subcutaneously twice daily in a dose of 1.5 mg/kg.

after the injection of 10000 U peroxidase, in the subendothelial space of the contralateral nonstimulated carotid artery almost no reaction products of peroxidase can be seen. Using a densitometric technique, it was possible to quantify the accumulation of peroxidation products in the subendothelial space of the carotid arteries. By this method it was shown that the increase in permeation through the endothelial lining is transient and can be completely inhibited by the calcium antagonist flunarizine. Various other calcium antagonists possess the ability to inhibit transendothelial transport of macromolecules (Betz et al. 1985).

With the technique of repetitive standardized transmural stimulation of carotid arteries it is also possible to study the inhibitory action of calcium antagonistic substances on the proliferation of cells in the stimulated region. Table 2 demonstrates that no relationship exists between the inhibitory effect of the drugs on SMC proliferation or the concentration of serum cholesterol in the animals that received 1% cholesterol in their food during the 4-week stimulation time. The serum triglycerides also showed no correlation to the extent of proliferation. In vivo tests do not enable one to decide clearly whether drugs inhibit proliferation as they are toxic in the concentration used; it is also not possible to test in vivo whether a drug specifically inhibits the proliferation on SMC only or in addition the growth of endothelial cells, blood cells in the bone marrow, cells in the intestine etc. The in vivo experiments must therefore be supplemented by in-vitro studies. With cell cultures it is also possible directly to demonstrate differences in responses to drugs between animal cells, and cells obtained from human artery walls.

The effect of the calcium antagonists demonstrated in the in-vivo experiments (Table 2) was also checked in mass cultures of SMC and endothelial cells of various species and of fibroblasts of rabbits (Table 3). There exist

Table 2. Effect of various calcium antagonists on the proliferation of smooth muscle cells in the intima of electrically stimulated rabbit carotid arteries

Substance	Dose (mmol/kg)	Number of animals	Controls (number of animals)	Percentage inhibition of proliferation (comparison with controls)	Percentage change of serum cholesterol of treated animals (comparison with untreated controls)
Flunarizine	0.063 (o)	10	30	−68*	−37*
Verapamil	0.046 (o + sc)	12	30	−48*	−6
Nimodipine	0.24 (o)	10	30	±0	+2
Fendiline	0.12 (o)	7	10	−5	normocholesterinemic

Arteries were stimulated daily with DC impulses of 0.1 mA, 10 Hz, 15 ms/impulse, 30 min in the morning and 15 min in the evening for 28 days. The rabbits received 0.5% cholesterol in their food during the 28-day experiments. As can be seen, there is no clear correlation between the changes in serum cholesterol or trigyceride levels and drug-induced reduction of the proliferation. o, Oral; sc, administration subcutaneous administration of the drug.
$^{*}p \leqslant 0.05$.

Table 3. Effect of calcium antagonists (percentage inhibition) on the proliferation of vascular smooth muscle cells, endothelial cells, and fibroblasts in mass cultures of cells

	10^{-6} mol/l	10^{-5} mol/l	10^{-4} mol/l	Species
Smooth muscle cells				
Verapamil	12	14	100 (toxic)	Rabbit
Flunarizine	11	23	95	Rabbit
Fendiline	0	27	Toxic	Rabbit
Nimodipine	30	110 (toxic)	Toxic	Rabbit
Verapamil	10	12	Toxic	Human
Flunarizine	11	21	85	Human
Fendiline	0	50	Toxic	Human
DHP	0	6	Toxic	Human
DHP Servier L+	7	70	100 (toxic)	Human
DHP Servier D−	3	93	100 (toxic)	Human
Diltiazem	7	0	28	Human
Endothelial cells				
Verapamil	14			Bovine
Flunarizine	5	18	95	Pig
Nimodipine	35			Bovine
Fendiline	0	52	Toxic	
Verapamil	15			Human
Flunarizine	10	14	82	Human
Fendiline	−12	10	Toxic	Human
Felodipine	0	0	Toxic	Human
DHP Servier L+	6	59	Toxic	Human
DHP Servier D−	17	60	Toxic	Human
Fibroblasts				
Verapamil	0	17	100	Rabbit
Flunarizine	14	32	98	Rabbit
Nimodipine	44	100 (toxic)	Toxic	Rabbit

DHP, Dihydropyridine (felodipine). SMC and endothelial cells were obtained from arteries, fibroblasts from subcutaneous tissue. As one can see from Table 2, part of the substances have been used for the inhibition of intimal proliferation in in vivo experiments. There are differences in the responses to some of the drugs (e.g., nimodipine).

differences in the inhibitory activities of some calcium antagonists between animal experiments and cell cultures. It also can be seen that none of the substances used inhibits only SMC proliferation or the proliferation of fibroblasts. When human arterial SMC and arterial endothelial cells were exposed to increasing concentrations of verapamil, both cell species were inhibited by the drug. With the calcium-calmodulin antagonist fendiline mentioned above, the inhibitory action on human endothelial cells is somewhat less pronounced than its action on human SMC. However, the action of this drug is also not specific for SMC.

Mass cultures and clone cultures of single cell species (SMC, endothelial cells, or fibroblasts) of human or animal origin are often used to screen for antiproliferative drugs. However, in comparison with in vivo systems these cultures have the disadvantage that cellular or at least humoral contacts to the other cell species of artery walls are absent. Attempts have therefore been made to establish cocultures which imitate the cellular composition of an artery wall. In our research group, transfilter cocultures have been developed. In these cultures, a collagen-coated polycarbonate filter (thickness 10 μm) imitates the porous elastic internal lamina. On the upper side of the filter endothelial cells are seeded in a high density so that they become confluent after a short time. After turning the filter upside down, SMC are seeded on the other filter side, or a piece of media is laid on the filter. If a second filter is laid on the SMC, and if this filter contains such fine pores that SMC cannot pass, the second filter is used as a base for adventitial cells, so that in this sandwichlike arrangement the structure of a carotid artery wall is imitated (Fig. 4). If the endothelial lining is then destroyed, SMC migrate through the pores in a similar way as they do in vivo after removal of the endothelium with a balloon catheter, and the SMC form an in vitro

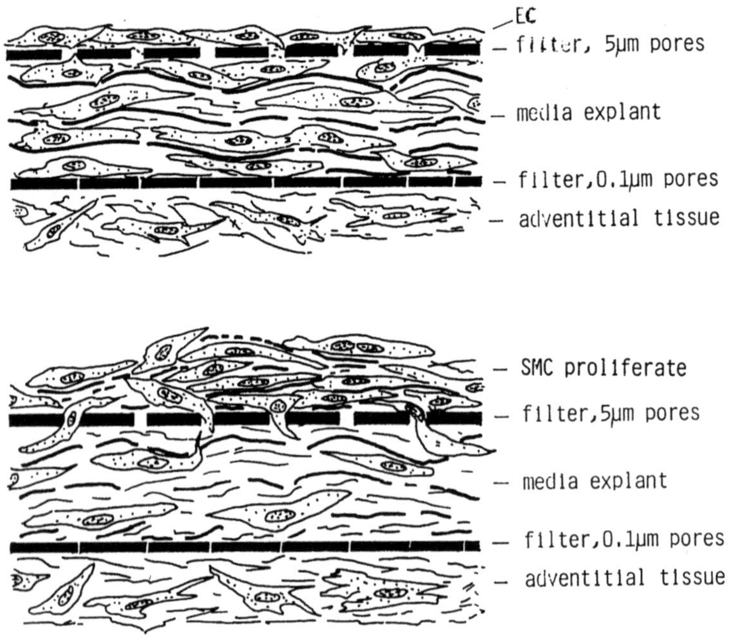

Fig. 4. Schematic drawing of a coculture system that imitates the morphology of a rabbit carotid artery wall. *Above*, an imitate of an artery is depicted in which a confluent endothelial layer inhibits the migration of smooth muscle cells from the media explant into the intima. *Below*, the effect 14 days after removal of the endothelium. The pore size of the filter between adventitia and media is so small that cells cannot migrate through this filter, whereas the pores between endothelium and media are so wide that cells can migrate. The cells form a fibromuscular proliferate. This model has been used to study the effect of various drugs to inhibit migration and proliferation of SMC

proliferate. This coculture system has already been used to test the effect of drugs on SMC proliferation. Comparison with in vivo effects showed that the time course of inhibition was similar. Calcium antagonists were also effective. However, there do not yet exist dose-response curves obtained with these drugs in these models. The system of transfilter cocultures has also been used to produce in vitro atheroma by addition of oxidized low-density lipoproteins and monocytes to the culture medium. Recently, it became possible to product in vitro atheromatous proliferates not only with vessel wall cells obtained from animals but also those of human arterial cells, and these new techniques promise to become an excellent tool in therapeutic research on the inhibition of stenosing processes in brain-supplying arteries.

References

Anitschkow N, Chalatow S (1913) Über experimentelle Cholesterinsteatose und ihre Bedeutung für die Entstehung einiger pathologischer Prozesse. Z Allg Pathol Pathol Anat 24:1–14

Baumgartner HR, Studer A (1963) Gezielte Überdehnung der Aorta abdominalis am normo- und hypercholesterinämischen Kaninchen. Pathol Microbiol 26:129–148

Betz E, Hämmerle H (1984) Arterienwandproliferate und Zellkulturen als Indikatoren für Hemmstoffe der Atherogenese. Funkt Biol Med 3:46–55

Betz E, Schlote W (1979) Responses of vessel walls to chronically applied electrical stimuli. Basic Res Cardiol 74:10–20

Betz E, Hämmerle H, Strohschneider T (1985) Inhibitory actions of calcium entry blockers on experimental atheromas. In: Goodfraind T, Vanhoutte PM, Gowoni S, Paoletti R (eds) Calcium entry blockers and tissue protection. Raven, New York, p 117

Ignatowski AC (1908) Influence of animal food on the organism of rabbits. S Peterb Izv Imp Voyenno Med Akad 16:154–173

Labarca C, Paigen K (1980) A simple, rapid and sensitive DNA assay procedure. Ann Biochem 102:344–352

Ross R, Glomset JA (1976) The pathogenesis of atherosclerosis. N Engl J Med 295:420–425

Excitatory Amino Acids and Excitotoxicity

Unsettled Role of Glutamate/Glutamate Receptors in Ischemic Excitotoxic Neuronal Damage

A. Sugimoto, A. Takeda, H. Onodera, S. Nakanishi, and K. Kogure

It has been a widely accepted hypothesis that the vulnerability of the brain cells to brief periods of ischemia is caused by the excitotoxicity of glutamate. As a matter of fact, almost all of these ischemia-susceptive neurons are densely innervated by glutamate fibers. In the case of pyramidal cells in the hippocampal CA1 sector, the glutamate fibers innervate the dendritic and the somal cell layers through multisynaptic, feed-foreward networks from the ipsilateral and the contralateral entorhinal cortex.

Onodera et al. disconnected the major input to the CA1 neurons by destroying the pyramidal neurons in the CA3 sector by means of unilateral cerebroventricular injection of kainic acid and found that the denervated pyramidal cells in the ipsilateral CA1 subfield were markedly spared for 5 days after 20 min of forebrain ischemia in the rat, while the contralateral CA1 neurons innervated by intact Schaffer collaterals were nearly completely destroyed. The difference between the denervated and the innervated neurons after ischemia was also clear in glutamate receptor autoradiography. Non-N-methyl-D-aspartate (NMDA) receptors are evenly distributed in the pyramidal cell layer of the CA1 sector as well as the dendritic layers. In contrast, competitive and noncompetitive NMDA receptors are densely distributed in the dendritic layers, while the pyramidal cell layer lacks these receptor sites. According to our observations, postischemic elevation in the intracellular calcium ion activity, succeeding alteration in the immunoreactivity of MAP2, and other diverse abnormalities observed in the CA1 subfield begin in the dendritic fields but not in the pyramidal cell layer. For instance, [^{3}H]2-deoxyglucose molecules are selectively lodged into the thin dendrite 30 min after 10-min forebrain ischemia in the rat. A rather simple deduction from these experimental data is that the quisqualate receptors contribute little to the excitotoxic neuronal injury.

Kainate receptor density in the CA1 is low and ^{3}H-kainate binding in the CA1 was reduced after ischemia. In the case of competitive NMDA receptors, a noticeable decrease in the binding activity can be seen only after the beginning of autolysis of the affected CA1 cell somata. Seven

Tohoku University, Institute of Brain Diseases, 1-1-Seiryou-machi, Sendai 980, Japan

Cerebral Ischemia and Basic Mechanisms
Ed. by A. Hartmann, F. Yatsu, and W. Kuschinsky

days after the insult, the activity decreased more than 50% even in the morphologically intact CA3 and the dentate gyrus. The postischemic change in the CA1 was also delayed in non-competitive NMDA receptors.

In the face of disintegration, do the remaining receptors transmit the glutamate signal? We assume that the data simply reflect the fact that the catabolic process of the receptor protein is interrupted. Northern blot of the NMDA mRNA in the CA1 area also indicated that the mRNA remains unchanged until 24 h after the ictus, and a noticeable fall of NMDA receptor mRNAs was seen only 2 days after ischemia. We also assume that the result represents not a preserved transcription process but a tardy decomposition of the molecule.

In our laboratory Kato et al. unveiled the fact that a second 3-min ischemia, which was given 60 min after the first 2-min insult, exhibited a cumulative effect and accentuated the vulnerability of CA1 neurons. However, a second ischemia given several days after the initial one markedly prevents delayed neuronal death. These two conditions are then named as the vulnerability stage and the tolerance stage. Interestingly, the amount of glutamate released during and after ischemia, as measured by a microdialysis technique, does not differ among these three conditions, namely 3-min single ischemia – i.e., lethergic; 3-min ischemia 60 min after a 2-min insult – i.e., vulnerability stage; and 3-min ischemia given 4 days after the first 2-min ischemia – i.e., tolerance stage. This phenomenon means that the amount of glutamate per se does not settle the final outcome of ischemia-induced excitotoxic neuronal change. Presumably, diverse intra-and extracellular responses elicited from the glutamate/glutamate receptor activation influence the progression of postischemic reactions, respectively. Thus, virtually no part of a single step can play a pivotal role in the fate of affected neurons.

Glutamate/NMDA receptor interaction induces heat shock proteins (HSP) and therefore the stage of tolerance obtained from the initial conditioning ischemia is thought to be acquired at least in part by induced stress proteins. MK-801 prevents HSP production and limits the cellular ability to resist a second ischemia in the tolerance stage. Incidentally, however, we should point out that anisomycin, which is a prominent inhibitor of protein synthesis at the translational step, also blocks HSP production but does not alter viability of the brain cells.

At present, experimental data on cross-talk reactions among each different signal transduction system corresponding to each different glutamate receptor after ischemia are scanty, and we cannot initiate reasonable discussions about the role of glutamate receptors in excitotoxic neuronal death by ischemia.

References

Kato H, Liu Y, Araki T and Kogure K (1991) Temporal profile of the effects of pretreatment with brief cerebral ischemia on the neuronal damage following secondary ischemic insult in the gerbil: cumulative damage and protective effects. Brain Res 553:238–242

Onodera H, Sato G and Kogure K (1986) Lesions to Schaffer collaterals prevent ischemic death of CA1 pyramidal cells. Neurosci Lett 68:169–174

Neuroprotective Effects of 5-HT_{1A} Receptor Agonists

B. Peruche, F. Ausmeier, C. Backhauss, J. Nuglisch, J.H.M. Prehn, and J. Krieglstein

Introduction

During recent years a multiplicity of 5-HT receptor subtypes has been characterized by pharmacological studies as well as by molecular cloning of the receptors. Serotonin (5-HT) is a neurotransmitter detectable in most organisms, ranging from highly specialized humans to species with a primitive CNS [1]. This neurotransmitter is responsible for a wide diversity of physiological and behavioral modulation such as, for example, the control of blood pressure, emotional behavior, endocrine functions, perception of pain, as well as sleep and wakefulness [2]. This great palette of different serotonergic functions is due to the fact that 5-HT-containing neurons are centralized in the raphé nuclei of the brainstem, while the 5-HT-releasing fibers project to virtually all parts of the CNS [1, 2]. As the majority of these fibers release 5-HT into the extracellular space instead of establishing synaptic contact with target cells via the synaptic cleft, it is suggested that, in addition to synaptic transmission, 5-HT also induces a so-called volume transmission, which is slower but more prolonged than synaptic signaling [1].

The mammalian 5-HT receptors were originally divided into two families: 5-HT_1 and 5-HT_2, due to the different affinity for 5-HT. Very soon, further classes of 5-HT receptors were identified (e.g., 5-HT_3 and 5-HT_4) [3]. Additionally, the 5-HT_1 receptor, the receptor family with the highest 5-HT binding affinity, was subdivided into 5-$HT_{1A,1B,1C,1D}$, which can be differentiated according to their physiological and pharmacological properties (see Table 1) as well as their subcellular localization. For example, 5-HT_{1A} and 5-HT_{1B} receptors are both expressed by 5-HT-containing raphé neurons and both inhibit adenylyl cyclase; however, the 5-HT_{1A} receptor subtype is localized on the soma and dendrites of the neurons (i.e., somadendritic 5-HT autoreceptor), while the 5-HT_{1B} subtype is concentrated on their terminals. Furthermore, these G-protein-coupled receptors are linked to different second messenger systems. Thus, activation of the 5-$HT_{1A,1B,1D}$ receptors inhibits adenylyl cyclase, whereas 5-HT_{1C}, 5-HT_2 receptors activate

Institut für Pharmakologie und Toxikologie im Fachbereich Pharmazie und Lebensmittelchemie, Philipps-Universität Marburg, Ketzerbach 63, 35037 Marburg/Lahn, Germany

Cerebral Ischemia and Basic Mechanisms
Ed. by A. Hartmann, F. Yatsu, and W. Kuschinsky

Table 1. Putative functional correlates of 5-HT receptor subtypes. (Modified from [34])

Receptor subtype	Functional correlate
5-HT_{1A}	Adenylate cyclase modulation Raphé cell and CA1 hippocampal inhibition Canine basilar artery contraction Forepaw treading, tremor, head-weaving Thermoregulation Hypotensive effects
5-HT_{1B}	Autoreceptor
5-HT_{1C}	Phosphatidylinositol turnover
5-HT_{1D}	? Rat kidney perfusion ? Rat stomach fundus
5-HT_2	Phosphatidylinositol turnover Contraction of vascular smooth muscle 5-HT or mescaline-induced head twitches Tryptamine-induced seizures Rat forepaw edema Discriminative cue properties Contraction of bronchial smooth muscle Platelet shape changes and aggregation Smooth muscle prostacyclin synthesis
5-HT_3	Depolarization of postganglionic autonomic neurons Contraction of ileal smooth muscle Pain, wheal, flare reaction
5-HT_4	Stimulation of adenylyl cyclase, increase in cAMP, closure of K^+ and opening of voltage-sensitive Ca^{2+} channels Facilitates acetylcholine release Facilitatory role in the peristaltic reflex Smooth muscle contractile effects Positive inotropic effects Steroid release from adrenocortical cells

phospholipase C [1]. Recently, an adenylyl cyclase-independent coupling of 5-HT_{1A} receptors to K^+ channels was found (Fig. 1) [2, 4, 5]; this is suggested to have a potential role in the pathogenesis of neuronal damage during cerebral ischemia.

Neuronal damage induced by cerebral ischemia is related to a massive increase in excitatory amino acids (EAA) and the intracellular Ca^{2+} level as well as free oxygen radicals. In particular, the selectively vulnerable brain regions, such as, for example, the CA1 subfield of the hippocampus, reveal impressive neuronal necrosis when exposed to short-term ischemic periods. An EAA-induced overexcitation of the neurons seems to play a pivotal role in the pathogenesis of postischemic neuronal damage. Predominantly gluta-

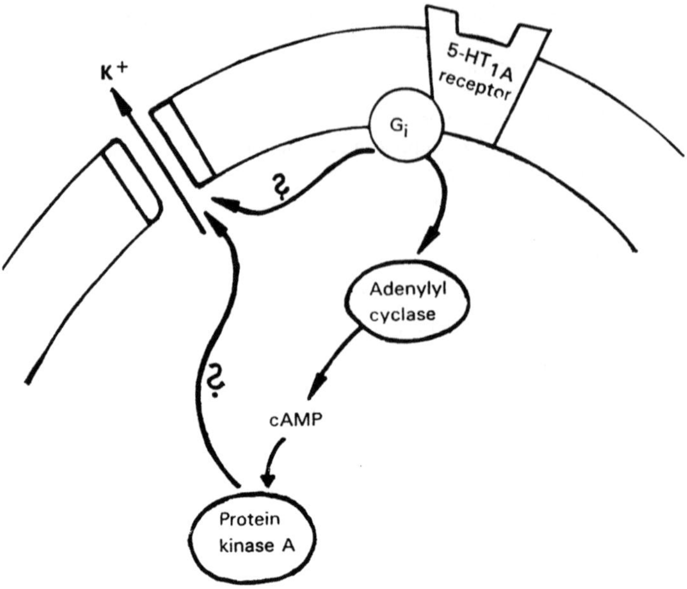

Fig. 1. 5-HT_{1A} receptor and their coupled effector systems. 5-HT_{1A} receptors are suggested to open a receptor-linked K^+ channel either directly mediated by the receptor-associated G_i protein or indirectly via G_i protein-induced inhibition of adenylyl cyclase. According to [1]

mate causes an excessive influx of Na^+ and Ca^{2+} into the neuronal cell via the ionotropic *N*-methyl-D-aspartate (NMDA) and non-NMDA receptors. The neuronal membrane depolarizes and, in addition, Ca^{2+} enters the cell through the voltage-operated Ca^{2+} channels (VOCC). This intracellular Ca^{2+} accumulation is neurotoxic and induces neuronal death by a deterioration of cellular metabolism (for a review, see [6–9]).

During ischemia an eight- to tenfold increase in 5-HT concentration has been detected in the brain, especially in the hippocampal extracellular space [10]. 5-HT has repeatedly been shown to exert both excitatory as well as inhibitory actions in the CNS [5, 11]. While excitation is due to the activation of 5-HT_2 receptors, evidence has accumulated that the inhibitory compound on neuronal excitation is due to an agonist activity of 5-HT at the 5-HT_{1A} receptor [12]. Furthermore, it has been shown that the 5-HT_{1A} receptor is expressed in the ischemia-sensitive brain regions [13], whereas the 5-HT_2 receptor is preferably located in the cerebral cortex [14]. Pazos and Palacios [13, 14] showed a high density of 5-HT_{1A} receptors in the external layers of the entorhinal cortex. Layer 2 of the entorhinal cortex projects to the hippocampus via the perforant path. Additionally, there are commissural projections of this cortical area [15]. In the hippocampus, a high density of the 5-HT_{1A} receptor was identified in the molecular layer of

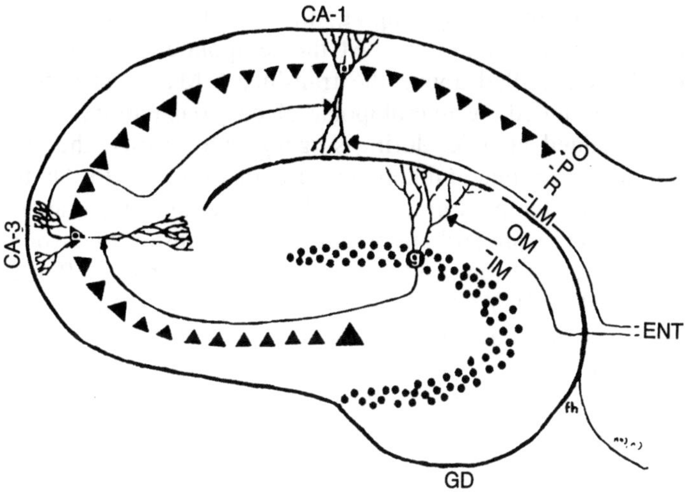

Fig. 2. Schematic representation of the trisynaptic pathway in the hippocampus. *ENT*, entorhinal cortex; *g*, granule cells; *p*, pyramidal cells; *O*, stratum oriens; *P*, stratum pyramidale; *R*, stratum radiatum; *LM*, stratum lacunosum moleculare; *OM*, outer molecular layer; *IM*, inner molecular layer; *GD*, gyrus dentatus; *fh*, hippocampal fissure. According to [7]

the dentate gyrus, which contains all the dendrites of the granule cells as well as in the CA1 subfield (Fig. 2). 5-HT, which is released during cerebral ischemia, may stimulate 5-HT_{1A} receptors within the hippocampus and may inhibit ischemia-induced neuronal excitation in this selectively vulnerable brain area. The inhibitory effect of 5-HT is related to the activation of a Ca^{2+}-independent K^+ current via postsynaptic 5-HT_{1A} receptors. This K^+ efflux from the neuronal cell induces hyperpolarization, decreases the input resistance, which finally leads to an increased firing threshold as well as to a reduced cell discharge [5]. Substances which activate this receptor subtype (5-HT_{1A} agonists), mimic the hyperpolarizing action of 5-HT [5, 16, 17]. Thus, the question arises whether these substances can provide protection of neurons against ischemic damage by a reduction of the neuronal firing rate. In the present study different models of cerebral ischemia in vivo and neuronal injury in vitro were used to characterize the potential neuroprotective action of the 5-HT_{1A} agonists in vivo as well as in vitro.

Models of Cerebral Ischemia and Neurotoxicity

Focal Ischemia in Mice

Permanent occlusion of the middle cerebral artery (MCA) was performed according to Welsh et al. [18] using male NMRI mice. The animals were anesthetized with tribromoethanol (600 mg/kg, i.p.), and an incision was made between the left orbit and the ear. At the tempolateral surface of the skull a small hole was drilled above the translucent MCA. The stem of the left MCA was occluded by microbipolar electrocoagulation. Body temperature was maintained at 37°C during surgery. Afterwards, the animals were kept at an environmental temperature of 30°C up to 4 h to prevent a decrease of body temperature. After 48 h the animals were anesthetized again and perfused transcardially with a fine suspension of carbon black (0.5 ml). The brains were removed and fixed in a phosphate-buffered formalin solution (4%, pH 7.4). As the infarct is restricted to the neocortex [18], the infarct size was determined planimetrically as infarcted surface area (mm^2), measured as the area of the carbon black-unperfused tissue by means of an image-analyzing system. Backhauss et al. [19] found a significant correlation between the carbon black-unperfused area and the infarct volume that had been determined by standard histology.

Focal Ischemia in Rats

Occlusion of the MCA was performed in male Fischer 344 rats (250–300 g) according to Tamura et al. [20]. Animals were anesthetized with a nitrous oxide/oxygen mixture (70%/30%) containing 1.5% halothane. A vertical skin incision was made midway between the left orbit and the ear, and the skull was exposed. The stem of the left MCA was approached by a subtemporal craniotomy beside the foramen ovale without removing the zygomatic arch. The stem of the MCA as well as its lenticulostriate branches were permanently occluded by microbipolar electrocoagulation. Body temperature was controlled as mentioned above.

Two days after occlusion the animals were again anesthetized and the brains were perfusion-fixed transcardially with a phosphate-buffered formalin solution (4%, pH 7.4). The brains were removed and stored in the fixative for at least 24 h . Then the brains were embedded in Paraplast, and coronal slices of 5 μm were taken every 0.5 mm, these were stained with 0.5% cresyl violet. The infarcted area was determined planimetrically as unstained tissue calculated by means of the above-mentioned image-analyzing system, separately for cortical, striatal, and total infarct. The infarct volume (mm^3) was calculated from the infarcted area of each slice and the distance between succeeding slices.

Global Ischemia in Rats

Transient forebrain ischemia was performed in male Wistar rats (250–300 g) by reversible occlusion of both common carotid arteries and additional lowering of arterial blood pressure to 40 mmHg for 10 min (modified from

[21]). Arterial pH, PaO_2, $PaCO_2$, arterial blood pressure, and plasma glucose were determined 20 min before and 10 min after ischemia. To prevent a decrease of intra- and postischemic cerebral temperature, environmental temperature was adjusted to 30°C during and up to 4 h after ischemia.

Seven days after ischemia, the rats were anesthetized again and the brains were perfusion-fixed as described above. For histological evaluation of the ischemic damage, coronal slices of 5-μm thickness were taken from each brain in a plane 6 mm rostral to the tentorial incision. The slices were stained with a mixture of 1% celestine blue and 1% acid fuchsin. Both the intact (unstained) and damaged (stained) pyramidal cells of the CA1, CA3, and CA4 hippocampal subfield were counted. Neuronal injury was expressed as the percentage of damaged neurons in each subfield.

Glutamate-Induced Excitotoxicity in Primary Neuronal Cultures of Rat Hippocampus

Primary neuronal cultures were prepared from 1-day-old rat hippocampi according to Huettner and Baughman [22]. The suspension of hippocampal cells was seeded into poly-L-lysine-coated 35-mm Petri dishes (Falcon) and the cultures were maintained in minimum essential medium (MEM) supplemented with 10% NU serum (Serva, Heidelberg, FRG), 100 units/ml penicillin, 100 μg/ml streptomycin, and 2 m*M* L-glutamine. After 5–6 days in culture, 1 μ*M* cytosine arabinoside was added for 48 h to inhibit glial growth. The medium was exchanged every 3–4 days. After a total cultivation period of 14 days, cells were switched to serum-free culture medium and 1 m*M* L-glutamate was added for 24 h, comparable to the procedure described by Koh et al. [23]. Cell viability was determined 18 h after glutamate exposure with the trypan blue exclusion method [24]. With a phase contrast microscope, damaged and nondamaged neurons were counted in three randomized subfields per culture dish (2×10^5 cells/cm^2). Only dark-stained neurons that were not phase bright were considered damaged. The percentage of damaged neurons was calculated from each dish by adding the counts of the three subfields. Control cultures received buffer without glutamate. Drug effects were evaluated on sister cultures from the same plating.

Results

Effects of 5-HT_{1A} Agonists in the Mouse Model of Focal Ischemia

In the mouse model of focal ischemia different drugs were tested over a wide range of doses. Results are presented in Table 2. All the drugs tested revealed neuroprotective effects. Additionally, CM 57493 (10 mg/kg, i.p.)

Table 2. The effects of various 5-HT_{1A} receptor agonists on the infarct area after MCA occlusion in the mouse

Drug	Drug administration (min) After[a] or before[b] ischemia	Drug concentration (mg/kg)	Infarct area (mm^2)	Reference
Ipsapirone	15 b	0	22.5 ± 3.2	
	15 b	10 i.p.	17.2 ± 2.8***	[19]
CM 57493	30 b	0	26.1 ± 2.9	
	30 b	5 i.p.	21.5 ± 3.6**	
	30 b	10 i.p.	18.8 ± 4.2***	
	30 b	20 i.p.	21.6 ± 3.6**	[19]
	2 a + 60 a	0	26.8 ± 4.4	
	2 a + 60 a	2 × 10 i.p.	22.4 ± 2.2**	[25]
Urapidil	30 b	0	19.6 ± 5.5	
	30 b	20 p.o.	17.7 ± 5.1	
	30 b	0	20.4 ± 2.6	
	30 b	80 p.o.	16.3 ± 6.3*	[19]
Roxindole	30 b	0	22.2 ± 3.2	
	30 b	0.3 s.c.	19.2 ± 2.7	
	30 b	1 s.c.	17.4 ± 4.2**	
	30 b	3 s.c.	15.7 ± 3.8***	

[a] a, after ischemia.
[b] b, before ischemia.
Data are presented as means ± SD of 10–12 experiments; different from control: $^{*}p < 0.05$; $^{**}p < 0.01$; $^{***}p < 0.001$ (Kruskal-Wallis H test and Mann Whitney U test).

showed neuroprotective activity when given postischemically (2 min and 60 min after occlusion).

Effects of 5-HT_{1A} Agonists in the Rat Model of Focal Ischemia

Data are presented in Table 3. CM 57493, urapidil, as well as roxindole were tested in this model. Drug treatment induced a significant reduction of infarct volume in nearly all the doses applied. It is surprising that urapidil reveals neuroprotection in this model at a low-dose range of 2 and 20 mg/kg, while this effect was exclusively measurable in the mouse model only when the drug was given in an extremely high dose of 80 mg/kg. The neuroprotective effect of roxindole (1 mg/kg, s.c.) was moderately expressed in

Table 3. The effects of various 5-HT_{1A} receptor agonists on the infarct volume after MCA occlusion in the rat

Drug	Drug administration (min before ischemia)	Drug concentration (mg/kg)	Total infarct volume (mm^3)	Cortical infarct volume (mm^3)	References
CM 57493	30	0	127.8 ± 27.7		
	30	1 i.p.	81.6 ± 12.2*		
	30	5 i.p.	94.1 ± 33.7*		
	30	0	120.6 ± 33.3	106.5 ± 35.8	
	30	10 i.p.	79.0 ± 35.2	62.3 ± 30.4*	[25]
Urapidil	30	0	128.9 ± 25.2		
	30	2 p.o.	71.1 ± 31.2***		
	30	20 p.o.	78.6 ± 38.2***		[19]
Roxindole	30	0	105.5 ± 33.9	88.9 ± 28.0	
	30	1 s.c.	70.3 ± 35.6	54.7 ± 30.1*	
	30	3 s.c.	93.7 ± 33.4	80.9 ± 28.9	

Data are presented as means ± SD from 8–13 experiments; different from control: $^{*}p < 0.05$; $^{***}p < 0.001$ (Kruskal-Wallis H test and Mann Whitney U test).

a significant reduction of cortical infarct volume. Total infarct volume remained unchanged.

Effects of 5-HT_{1A} Agonists in a Rat Model of Global Ischemia

CM 57493 was tested in this model in several dosages. The results are shown in Tables 4 and 5. The dorsal hippocampal section (Table 4) reveals severe neuronal damage of pyramidal neurons in the CA1 subfield of solvent- and drug-treated animals after ischemia. However, the histological quantification shows a 10% reduction of the damaged CA1 neurons after treatment with 1 mg/kg CM 57493 ($p < 0.01$); the effect of 5 mg/kg CM 57493 was less pronounced, but the reduction of neuronal damage was still significant ($p < 0.05$). Treatment with 10 mg/kg CM 57493 showed no neuroprotection (for review see [25]).

The ventral hippocampal section (Table 5) revealed moderate neuronal damage in the CA1 subfield in control rats. Rats treated with 1 mg/kg CM 57493 showed a more than 50% reduction of ischemia-induced neuronal damage. Six out of 13 rats showed no neuronal damage. Treatment with 5 mg/kg CM 57493 did not cause a significant decrease in the mean value of hippocampal CA1 damage [25].

Table 4. The effects of 5-HT_{1A} receptor agonists on neuronal damage after 10-min transient forebrain ischemia in the rat dorsal hippocampus [25]

Drug	Drug administration (min before ischemia)	Drug concentration (mg/kg)	Damaged neurons (%)
CM 57493	30	0	88.31 ± 3.76
	30	1	78.96 ± 16.28**
	30	5	83.54 ± 8.43*
	30	0	75.07 ± 19.50
	30	10	71.41 ± 19.80

Values are given as means ± SD of 11–15 experiments; different from control: $^{*}p < 0.05$; $^{**}p < 0.01$ (Kruskal-Wallis H test and Mann Whitney U test).

Table 5. The effects of 5-HT_{1A} receptor agonists on neuronal damage after 10 min transient forebrain ischemia in the rat ventral hippocampus [25]

Drug	Drug administration (min before ischemia)	Drug concentration (mg/kg)	Damaged neurons (%)
CM 57493	30	0	30.3 ± 18.7
	30	1	13.4 ± 10.5**
	30	5	18.8 ± 16.2

Values are given as means ± SD of 13 experiments; different from control: $^{**}p < 0.01$ (Kruskal-Wallis H test and Mann Whitney U test).

Table 6. The effects of various 5-HT_{1A} receptor agonists on cellular viability after glutamate-induced excitotoxicity on primary neuronal cultures of rat hippocampus [25]

Treatment	Damaged neurons (%)
Control	12.6 ± 1.6
1 m*M* Glutamate	58.2 ± 2.6
1 m*M* Glutamate + 10 μM CM 57493	39.8 ± 1.7***
+ 10 μM Roxindole	40.9 ± 3.4***
+ 10 μM Ipsapirone	46.6 ± 3.4***
+ 10 μM Buspirone	54.4 ± 1.4*
+ 10 μM Urapidil	52.1 ± 1.8*

Data are presented as means ± SD of four experiments; $^{*}p < 0.05$, $^{***}p < 0.001$ (one-way analysis of variance and Duncan's test). Drugs were added to the cultures during exposure to glutamate.

Effects of 5-HT_{1A} Agonists in a Primary Neuronal Cell Culture System

Exposure of the neuronal cultures to 1 m*M* glutamate for 24 h resulted in an increased number of damaged, trypan blue-stained neurons. There were approximately four times as many injured neurons in glutamate-exposed cultures compared to untreated controls. Addition of 5-HT_{1A} agonisits (Table 6) induced neuroprotection, indicated by a preserved viability of hippocampal cells after exposure to glutamate.

Discussion

The results demonstrate neuroprotective activity of 5-HT_{1A} agonists against ischemic damage. The substances were tested in different models representing different pathological aspects of cerebral ischemic injury.

The mouse model of focal cerebral ischemia was used to get a first insight into the neuroprotective activity of 5-HT_{1A} agonists in postischemic neuroprotection. Backhauss et al. [19] showed that the model with its simple technique of measuring the infarct size is suitable for screening purposes. They proved that the infarct area on the brain surface after occlusion of the MCA is representative of the infarct volume. Thus, the measurement of the infarct area has been evaluated for the quantification of the infarct size. Furthermore, the animal model of focal ischemia resembles the pathological situation of stroke in humans.

Preischemic application of all the drugs tested showed neuroprotection by reduction of postischemic infarct size. These results are in line with a study of Bielenberg and Burkhardt [26], who demonstrated neuroprotection of the 5-HT_{1A} agonists 8-OH-DPAT, BAY R 1531, buspirone, ipsapirone, and gepirone using the same model. They also administered the drugs preischemically. Ipsapirone was the only substance which was additionally successfully tested postischemically. In the latter experiment, a 50% reduction of infarct size was measurable. However, their study was performed without temperature control. Nuglisch et al. [27] demonstrated neuroprotection of CM 57493, given 15 min preischemically, in a temperature-controlled study. Using the same experimental conditions, Prehn et al. [25] demonstrated a reduction of infarct size with CM 57493 when applied immediately after occlusion of the MCA. Furthermore, preischemic application of urapidil (80 mg/kg) and roxindole (1 and 3 mg/kg) diminished infarct size. While CM 57493 and roxindole revealed their neuroprotective effects in a dose range of 5–10 mg/kg and 1–3 mg/kg, respectively, urapidil showed significant neuroprotection at a dose of 80 mg/kg [25]. This different neuroprotective potency seems to be related to a different affinity of the substances to the 5-HT_{1A} receptors, as shown in the studies of Adrien et al. [28] and Gross et al. [29] using in vitro binding assays with rat hippocampal membranes.

The results obtained in this screening model of focal ischemia in mice were reproduced in the rat model of focal ischemia. In this study the neuroprotective effect of the drugs (CM 57493, urapidil, roxindole) was restricted to the cortical tissue and was not measurable in the striatum. This could be related to the profound reduction of cerebral blood flow in the lateral striatum after occlusion of the MCA and its lenticulostriate branches [30].

To obtain further support of the neuroprotective effect, CM 57493 was tested in the model of transient forebrain ischemia in rats with extensive neuronal damage of the hippocampal CA1 subfield [31]. Smith et al. [32] demonstrated a dorsal-ventral gradient in the selective vulnerability of the hippocampus using this model. As the ventral hippocampus contains a higher density of 5-HT_{1A} receptors in comparison to the dorsal hippocampus, it was not surprising that CM 57493 revealed better neuroprotective effects in the ventral hippocampal region [25]. The fact that the neuroprotective effect is more pronounced in areas with high 5-HT_{1A} receptor density supports the assumption that neuroprotection is mediated by an inhibitory action on neurons, as already mentioned above.

In general, the neuroprotective effect of CM 57493 in this model was less pronounced compared to that obtained after occlusion of the MCA. It is interesting to note that the 5-HT_{1A} agonist revealed neuroprotective effects exclusively in small doses, whereas in high doses no effect was demonstrable. This has been shown also by Bode-Greuel et al. [33] with, for example, ipsapirone. An explanation for this phenomenon can be an increasing affinity of the substances to other receptor systems, e.g., dopaminergic receptors in high doses [28], or hypotensive effects in these dose ranges (for review see [34, 35]).

To exclude circulatory effects of the drugs and to clearly demonstrate their activity directly at the neuronal cell, drugs were tested in a primary neuronal cell culture system. Neuronal cells were generally protected against glutamate-induced excitotoxicity by the preexcitotoxic administration of the drugs. As 5-HT-induced hyperpolarization of the neuronal membrane has been directly demonstrated by several electrophysiological studies on cortical and hippocampal neurons [5, 16, 36], those results obtained in vitro support the hypothesis that, under pathological conditions, the activation of especially the 5-HT_{1A} receptor subtype by 5-HT_{1A} agonists is responsible for the inhibitory and, thereby, protective effect on neurons.

In summary, the study shows pronounced neuroprotective effects of 5-HT_{1A} agonists in different models of focal and global ischemia in vivo as well as in an in vitro system of glutamate-induced neuronal degeneration. In these studies the drugs were usually applied before the induction of neuronal injury. Neuroprotection was also demonstrable by the postischemic application of CM 57493 in the mouse model of focal ischemia. The protective effect is suggested to be mediated by the hyperpolarization of the neuronal membrane via 5-HT-induced opening of 5-HT_{1A} receptor-linked K^+ channels.

Thus, the 5-HT_{1A} agonists may be of interest for the treatment of ischemic brain diseases.

References

1. Hen R (1992) Of mice and flies: commonalities among 5-HT receptors. TIPS 13(4):160–166
2. Göthert M, Schlicker E (1990) Identification and classification of 5-HT_1 receptor subtypes. J Cardiovasc Pharmacol 15 Suppl 7:51–57
3. Humphrey PPA, Hartig P, Hoyer D (1993) A proposed new nomenclature for 5-HT receptors. TIPS 14:233–236
4. Aghajanian GK, Sprouse JS, Rasmussen K (1988) Electrophysiology of central serotonin receptor subtypes. In: Sanders-Bush E (ed) Serotonin receptors. Humana, Clifton, pp 225–252
5. Colino A, Halliwell JV (1987) Differential modulation of three separate K^+-conductances in hippocampal CA1 neurons by serotonin. Nature 328:73–77
6. Siesjö BK, Bengtsson F (1989) Calcium fluxes, calcium antagonists and calcium-related pathology in brain ischemia, hypoglycemia and spreading depression – a unifying hypothesis. J Cereb Blood Flow Metab 9:127–140
7. Jorgensen MB, Diemer NH (1982) Selective neuronal loss after cerebral ischemia in the rat: possible role of transmitter glutamate. Acta Neurol Scand 66:536–546
8. Choi DW (1988) Glutamate neurotoxicity and diseases of the nervous system. Neuron 1:623–634
9. Peruche B, Krieglstein J (1993) Mechanisms of drug actions against neuronal damage caused by ischemia – an overview. Prog Neuro Psychopharmacol Biol Psychiatry 17:21–70
10. Sarna GS, Obrenovitch TP, Matsumoto T, Simon L, Curzon G (1990) Effect of transient ischemia and cardiac arrest on brain extracellular dopamine and serotonin as determined by in vivo dialysis in the rat. J Neurochem 55:937–940
11. Nedergaard S, Engberg I, Flatman JA (1986) Serotonin facilitates NMDA response of cat neocortical neurons. Acta Physiol Scand 128:323–325
12. Clarke WP, DeVivo M, Beck SG, Maayani S, Goldfarb J (1987) Serotonin decreases population spike amplitude in hippocampal cells through a pertussis toxin substrate. Brain Res 410:357–361
13. Pazos A, Palacios JM (1985) Quantitative autoradiographic mapping of serotonin receptors in the rat brain. I. Serotonin-1 receptors. Brain Res 346:205–230
14. Pazos A, Palacios JM (1985) Quantitative autoradiographic mapping of serotonin receptors in the rat brain. II. Serotonin-2 receptors. Brain Res 346:231–242
15. Prehn JHM, Krieglstein J (1992) Potential role of 5-hydroxytryptamine$_{1A}$ receptors in cerebral ischemia. In: The role of neurotransmitters in brain injury. In: Globus MYT, Dietrich WD (eds) Proceedings of the official satellite symposium of Brain '91. Plenum, New York, pp 137–146
16. Andrade R, Nicoll RA (1987) Novel anxiolytics discriminate between postsynaptic serotonin receptors mediating different physiological responses on single neurons of rat hippocampus. Naunyn Schmiedebergs Arch Pharmacol 336:5–10
17. Rowan MJ, Anwyl R (1987) Neurophysiological effects of buspirone and ipsapirone in hippocampus: comparison with 5-hydroxytryptamine. Eur J Pharmacol 132:93–96
18. Welsh FA, Sakamoto T, McKee AE, Sims RE (1987) Effect of lactacidosis on pyridine nucleotide stability during ischemia in the mouse brain. J Neurochem 49:846–851

19. Backhauss C, Karkoutly C, Welsch M, Krieglstein J (1992) A mouse model of focal ischemia for screening neuroprotective drug effects. J Pharmacol Methods 27:27–32
20. Tamura A, Graham DJ, McCulloch J, Teasdale GM (1981) Focal cerebral ischemia in the rat. I. Description of technique and early neuropathological consequences following middle cerebral artery occlusion. J Cereb Blood Flow Metab 1:53–60
21. Smith ML, Bendek G, Dahlgren M, Roosen J, Wieloch T, Siesjö BK (1984) Models for studying long-term recovery following forebrain ischemia in the rat. II. A 2-vessel occlusion model. Acta Neurol Scand 69:385–401
22. Huettner JE, Baughman RW (1986) Primary culture of identified neurons from the visual cortex of postnatal rats. J Neurosci 6:3044–3068
23. Koh J, Goldberg MP, Hartley DM, Choi DW (1980) Non-NMDA receptor-mediated neurotoxicity in cortical culture. J Neurosci 10:693–705
24. Paul J (1975) Cell and tissue culture, 5th edn. Churchill Livingstone, Edinburgh
25. Prehn JHM, Backhauss C, Karkoutly C, Nuglisch J, Peruche B Krieglstein J (1991) Neuroprotective properties of 5-HT_{1A} receptor agonists in rodent models of focal and global ischemia. Eur J Pharmacol 203:213–222
26. Bielenberg GW, Burkhardt M (1990) 5-Hydroxytryptamine$_{1A}$ agonists. A new therapeutic principle for stroke treatment. Stroke 21 Suppl 4:IV161–IV163
27. Nuglisch J, Karkoutly C, Peruche B, Prehn JHM, Welsch M, Mennel HD, Roßberg C, Krieglstein J (1990) Effect of the 5-HT_{1A} agonist CM 57493 on infarct area, infarct volume and hippocampal neuronal damage after focal and global cerebral ischemia in mice and in rats. In: Krieglstein J, Oberpichler H (eds) Pharmacology of cerebral ischemia 1990. Wissenschaftliche Verlagsgesellschaft, Stuttgart, pp 493–501
28. Adrien J, Lanfumey L, Gozlan H, Fattaccini CM, Hamon M (1989) Biochemical and electrophysiological evidence for an agonist action of CM 57493 at pre- and postsynaptic 5-hydroxytryptamine$_{1A}$ receptors in brain. J Pharmacol Exp Ther 248:1222–1230
29. Gross G, Hanft G, Kolassa N (1987) Urapidil and some analogues with hypotensive properties show high affinities for 5-hydroxytryptamine (5-HT) binding sites of the 5-HT_{1A} subtype and for α_1-adrenoceptor binding sites. Naunyn Schmiedebergs Arch Pharmacol 336:597–607
30. Tamura A, Graham DI, McCulloch J, Teasdale GM (1981b) Focal cerebral ischemia in the rat. II. Regional cerebral blood flow determined by [^{14}C]iodoantipyrine autoradiography following middle cerebral artery occlusion. J Cereb Blood Flow Metab 1:61–68
31. Kirino T (1982) Delayed neuronal death in the gerbil hippocampus following ischemia. Brain Res 510:57–69
32. Smith ML, Auer RN, Siesjö BK (1984) The density and distribution of ischemic brain injury in the rat following 2–10 min of forebrain ischemia. Acta Neuropathol (Berl) 64:319–322
33. Bode-Greuel KM, Klisch J, Horvath E, Glaser T, Traber J (1990) Effects of 5-hydroxytryptamine$_{1A}$ agonists on hippocampal damage after transient forebrain ischemia in the mongolian gerbil. Stroke 21 Suppl 4:IV164–IV167
34. Peroutka SJ (1988) 5-Hydroxytrytamine receptor subtypes. Annu Rev Neurosci 11:45–60
35. Doods HN, Kalkman HO, de Jonge A, Thoolen M, Wilffert B, Timmermans P, van Zwieten PA (1985) Differential selectivities of RU 24969 and 8-OH-DPAT for the purposed 5-HT_{1A} and 5-HT_{1B} binding site. Correlation between 5-HT_{1A} affinity and hypotensive activity. Eur J Pharmacol 112:363–370
36. Beck SG, Clarke WP, Goldfarb J (1985) Spiperone differentiates multiple 5-hydroxytryptamine responses in rat hippocampal slices in vitro. Eur J Pharmacol 116:195–201

In Vitro Ischemia in the Hippocampal Slice

M.N. Waxham, S.A. Westgate, and M.D. Mauk

Introduction

Our current concepts relating ischemia to irreversible neuronal damage are based on data from multiple in vivo and in vitro model systems. Results from these models are helping to define the temporal sequence of events that lead to neuronal damage, and a complete understanding of these events will help optimize strategies for neuroprotective therapies. One particularly interesting type of neuronal damage after an ischemic episode appears to be initiated by sustained increases in intracellular Ca^{2+}, and defining the mechanisms that lead to these increases is being intensively investigated. In some ischemia models, excessive glutamate receptor stimulation appears to be an early and causal event in producing the increased intracellular Ca^{2+} levels (Abele et al. 1990; Choi 1990; Manev et al. 1989; Siesjö and Bengtsson 1989) and is referred to as glutamate excitotoxicity. Glutamate receptors can be classified into two major types: *N*-methyl-D-aspartate (NMDA) and non-NMDA (Mayer and Miller 1990). Non-NMDA receptors can be further subdivided into ionotropic and metabotropic depending on whether they open ion channels directly or activate intracellular processes through G-protein interactions, respectively. NMDA receptors permit Na^+ and K^+ flux and are also an important portal for Ca^{2+} entry in neurons (Mayer and Miller 1990; Yoneda and Ogita 1991). The ionotropic non-NMDA receptor mediates the majority of the excitatory postsynaptic potential (EPSP) and is permeable mainly to Na^+ and K^+; however, recently they were also shown to permit Ca^{2+} influx in some circumstances (for discussion see Choi 1990). The metabotropic non-NMDA receptors induce Ca^{2+} release from intracellular stores by activating phospholipase C generating the intracellular messenger IP_3. The postsynaptic effects of excitatory amino acid neurotoxicity induced by ischemia are mediated through these glutamate receptors. Since each receptor subtype has the potential to affect the intracellular concentration of Ca^{2+}, either directly or indirectly, they all have roles in determining the outcome of glutamate-mediated neurotoxicity due to sustained intracellular Ca^{2+}. Because of these findings, glutamate receptor

The University of Texas Medical School at Houston, Department of Neurobiology and Anatomy, P.O. Box 20708, Houston, TX 77225, USA

Cerebral Ischemia and Basic Mechanisms
Ed. by A. Hartmann, F. Yatsu, and W. Kuschinsky

antagonists have gained prominence in the search for neuroprotective agents.

Several models are available to investigate the mechanisms of ischemia-induced neuronal damage. Work in neuronal cell culture provided us with much of the experimental evidence for glutamate-mediated excitotoxicity in relation to ischemia and has also provided the most convincing evidence that antagonists of glutamate receptors can provide protection from a toxic dose of glutamate (Abele et al. 1990; Choi 1990). These culture studies have emphasized the role of the NMDA receptor, since antagonists specific for this receptor provide complete neuroprotection (Choi 1990). The NMDA receptor's role in this process relates to its ability to permit Ca^{2+} influx into the cell, since removal of extracellular Ca^{2+} completely protects neurons in culture (Choi 1990). Unfortunately, as encouraging as these studies seem, NMDA antagonists have shown variable effects in protecting neurons from ischemia in animal models (Choi 1990; Nellgard and Wieloch 1992; Siesjö and Bengtsson 1989). This suggests that the neuronal culture model has limited predictive ability in determining the outcome of neuroprotective agents in vivo and suggests other models should be pursued in addition to neuronal cultures to help provide more accurate predictive value for neuroprotective agents.

Hippocampal slices are an attractive adjunct model for studying ischemia and they might provide a more realistic prototype than neuronal cell culture to investigate certain aspects of ischemic damage. Because the synaptic architecture remains intact, ischemically induced changes should initiate a similar, if not identical cascade of events to that which occurs in vivo. Observations made in hippocampal slices are also not complicated by changes in cerebral blood flow or temperature which accompany pharmacologic investigations in animal models. Some other advantages of hippocampal slices include: (a) precise control of the composition of the extracellular environment; (b) precise temporal control for the delivery of compounds both extracellularly and intracellularly; and (c) the presence of differentially susceptible neurons in the same preparation (CA1 pyramidal cells vs. dentate granule cells). Results from ischemic studies in hippocampal slices also have their own set of limitations. For example, the most convenient determination for irreversible loss of neuronal function in this model is an irreversible loss of synaptic activity. Whether this relates to histopathologic neuronal cell death (the normal end point in in vivo models of ischemia) has not been thoroughly investigated. However, in some ways, the irreversible loss of synaptic transmission is probably a more relevant and sensitive indicator for functionally significant damage in nerve cells.

In this report, we attempt to describe the series of events that occur when hippocampal slices are exposed to in vitro ischemic conditions (the lack of both oxygen and glucose). In addition, the efficacy of several pharmacologic agents in blocking the irreversible loss of synaptic transmission is described.

Materials and Methods

Hippocampal slices were prepared from 60–90-day-old male Sprague-Dawley rats. Rats were anesthetized with a lethal injection of sodium pentobarbital and then decapitated. The hippocampi were dissected from each cortex in ice-cold oxygenated Ringer's solution containing 2 m*M* $MgCl_2$ and 2 m*M* $CaCl_2$; 400 μM transverse slices were prepared using a McIlwaine tissue chopper at 4°C. Slices were then incubated in oxygenated Ringer's solution for at least 1 h at 30°C before use. An alternative method for cutting slices used Ringer's solution containing 10 m*M* $MgCl_2$ and no $CaCl_2$ also at 4°C and then adding Ca^{2+} to 2 m*M* before incubating at 30°C. The slices were then transferred to oxygenated Ringer's containing 2 m*M* $MgCl_2$ and 2 m*M* $CaCl_2$ for at least an additional 30 min before use. The latter technique for preparing slices was shown previously to provide significant protection of pyramidal neurons presumably by preventing Ca^{2+}-mediated excitotoxicity (Feig and Lipton 1990). We found that slices produced using this protocol improved the quality and consistency of recordings.

Field potential recordings were obtained in the molecular layer of CA1 pyramidal cells or the molecular layer of dentate granule cells. Extracellular recording electrodes (1–3 MΩ) were filled with Ringer's solution without Ca^{2+} or Mg^{2+}. Recordings were made in a submersion chamber at 37°C. The flow rate was at least 0.5 ml/min. EPSP were obtained by placing tungsten-stimulating electrodes in either Schaffer collaterals for CA1 or axons of the perforant path for dentate granule cells (see Fig. 1). Iontophoretic applications were accomplished using 3-MΩ electrodes filled with 10 m*M* NMDA or 10 m*M* α-amino-3-hydroxy-5-methylisoxazole-4-propionic acid (AMPA). Every 5 min, agonist was applied by injecting 2.5 μA of current for 100 ms through the electrode. In vitro ischemia was induced by superfusing the slice with Ringer's without glucose which was first boiled for 5 min and then equilibrated with 95% N_2/5% CO_2. After various periods of time in ischemic buffer, the slices were returned to oxygenated Ringer's containing glucose. The maximum slope of the initial phase of the EPSP and the amplitude of the presynaptic fiber volley were plotted in real time using an on-line program developed by one of us (M.D.M.). Unless otherwise noted, drug protection experiments were accomplished by washing-on the experimental drug for 5–10 min before inducing ischemia while monitoring synaptic transmission. The drug was then present at the same concentration during ischemia and removed when the slice was returned to normal oxygenated Ringer's.

Agonists and Antagonists. Dextrorphan-tartrate, ketamine-HCl, 8-phenyltheophylline (8-PT), 8-cyclopentyl-1, 3-dimethylxanthine (8-CPT), 1,3-di(2-tolyl)guanidine (DTG), and AMPA were purchased from Research Biochemicals, Inc. NMDA was purchased from Sigma. Dextrorphan and ketamine are noncompetitive NMDA receptor antagonists which at higher

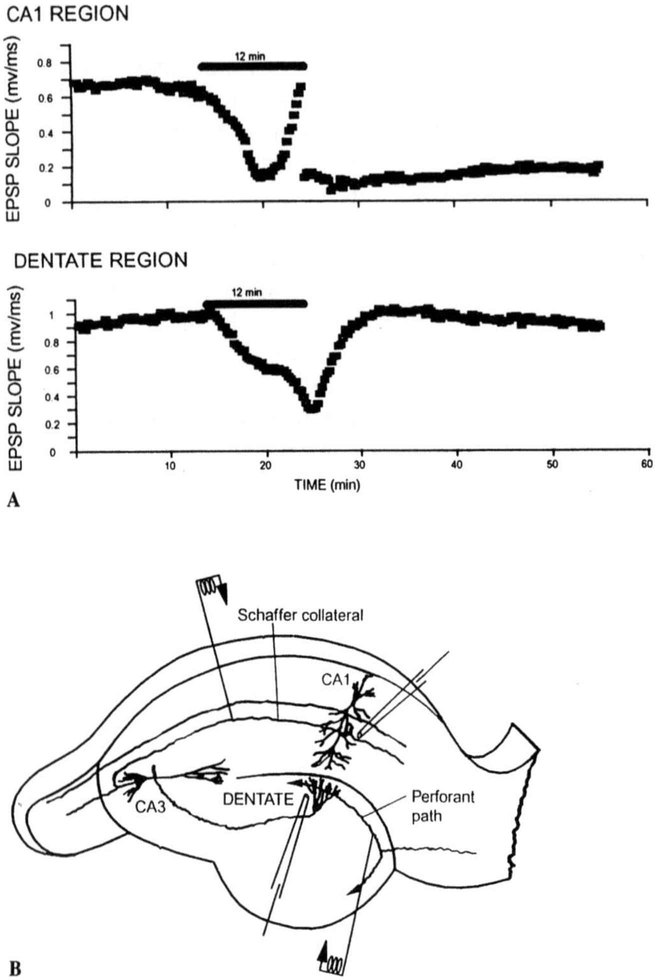

Fig. 1A,B. Electrophysiologic effects of ischemia in the hippocampal slice. **B** Positions of the recording electrodes in the CA1 and dentate regions. Stimulating electrodes were placed in the Schaffer collaterals or perforant path for establishing recordings in CA1 and dentate, respectively; stimulus frequency was 0.06 Hz. **A** Time course of changes (measured as the maximum slope of the intital phase of the EPSP) induced by applying ischemia buffer (no glucose, no oxygen) for 12 min (*bar*) in the CA1 and dentate regions

concentrations also block sigma receptors, and DTG is an antagonist presumably specific for sigma receptors. 8-phenyltheophylline (K_i = 400 n*M*) and 8-CPT (K_i = 10.9 m*M*) are adenosine antagonists with 10- to 100-fold greater specificity to the A_1 than A_2 receptor subtype and were solubilized in dimethyl sulfoxide and 0.1 *N* NaOH as 100× stocks, respectively.

Results

Extracellular field potential recordings were established in both CA1 and dentate granule cell dendrites (see Fig. 1). Figure 1 also shows the time course of synaptic responses before, during, and after perfusing the slice with Ringer's without glucose and equilibrated with N_2/CO_2. This is termed in vitro ischemia. The synaptic responses in CA1 disappear completely within 2–6 min after applying ischemic buffer, while responses in dentate decrease over a longer period and do not disappear completely during this 12-min period of ischemia. In later experiments this is referred to as the early depression in synaptic transmission. In almost every experiment, evoked responses in CA1 returned after about 10 min while still in ischemia buffer (this spontaneous return in synaptic transmission is referred to as the burst or burst phase). Shortly after the burst phase, the EPSP disappeared along with the presynaptic fiber volley. If oxygenated Ringer's containing glucose was returned before the disappearance of the burst phase, the synaptic responses returned to normal. If the responses and presynaptic fiber volley disappeared before the slice was returned to oxygenated Ringer's, the EPSPs were less likely to return, while the presynaptic fiber volley always returned. A consistent finding was that if the EPSP did not return after reoxygenation, the presynaptic fiber volley returned potentiated and had a longer latency to peak. Recordings in dentate normally disappeared at a slower rate, only occasionally exhibited the return of evoked responses, and survived longer periods of ischemia. Typically, dentate recordings returned to normal if perfused with oxygenated Ringer's within 13 min. These findings suggest that dentate granule cells are inherently more resistant than CA1 pyramidal cells to ischemia in this slice model. An important technical note is that the differences between irreversible changes in CA1 and dentate could be compressed dramatically by increasing the buffer flow rate over the slice. At flow rates of 4 ml/min and above the difference between the irreversible loss of recordings in CA1 and dentate was reduced to 30 s and the time of irreversible changes was shortened to 6–8 min in CA1 (data not shown). At these faster flow rates, the burst phase in CA1 during ischemia was much smaller and occasionally absent. Varying the flow rate might be one method to induce ischemia of different severity, and we are currently evaluating the electrophysiologic changes in CA1 and dentate by systematically altering the flow rate.

To ascertain why CA1 synaptic responses disappeared in the early depression of synaptic transmission, we first determined whether the postsynaptic cell could respond to neurotransmitter during in vitro ischemia. These experiments were accomplished by iontophoretically applying glutamate receptor agonists to the slices before, during, and after ischemia. As seen in Fig. 2A, postsynaptic responses to iontophoretically applied AMPA, an agonist that specifically opens non-NMDA receptors, were unaffected during the first 10 min of ischemia even though electrically evoked responses

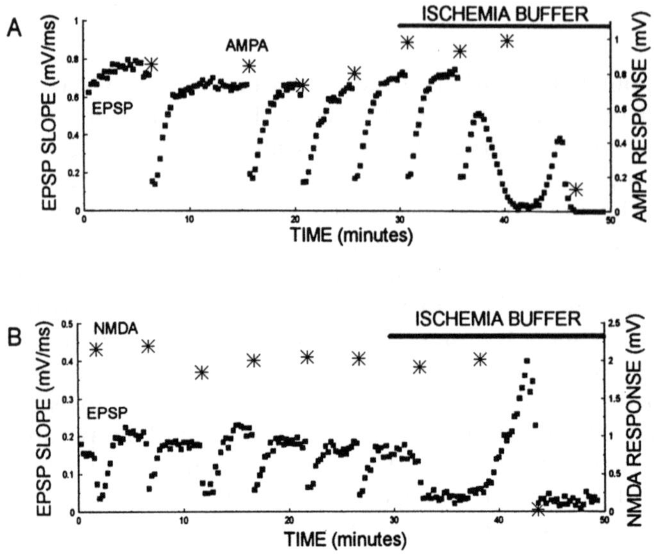

Fig. 2. A Effects of ischemia on the EPSP and iontophoretic response to AMPA in area CA1. EPSPs were evoked at 0.6 Hz. AMPA responses were obtained by iontophoretic application (10 m*M* AMPA, 2.5 μA, 100 ms) every 5 min. The amplitude of the individual responses to AMPA (in mV) are indicated on the plot as *asterisks*. **B** Effects of ischemia on the EPSP and iontophoretic response to NMDA in area CA1. Responses were obtained as described in **A** using iontophoresis electrodes filled with 10 m*M* NMDA. The amplitude of individual responses to NMDA (in mV) are indicated by *asterisks*

in the CA1 region disappeared (about 42 min from the start of this experiment). The depression seen in evoked responses after each AMPA application is due to sustained depolarizations caused by the slow clearance of agonist. Iontophoretic responses remained unchanged until the evoked responses disappeared along with the presynaptic fiber volley which occurred after the spontaneous return of activity around 45 min into this experiment. Both AMPA and electrically evoked responses disappeared post-burst. Similar response patterns were obtained when NMDA was used to induce the iontophoretic responses (Fig. 2B). Again NMDA responses were present after the electrically evoked activity disappeared (32–40 min into this experiment). After the burst, NMDA responses could not be evoked. These results indicate that both NMDA- and non-NMDA-type glutamate receptors are intact during ischemia and that the postsynaptic neuron can still support ionic responses to neurotransmitter. This finding emphasizes the fact that the postsynaptic cell remains sensitive to glutamate, or its agonists, through the early phases of the ischemic period. This also suggests that the early disappearance of electrically evoked synaptic responses (pre-burst) is probably not due to a change in the postsynaptic cell and implicates a possible alteration of presynaptic transmitter release.

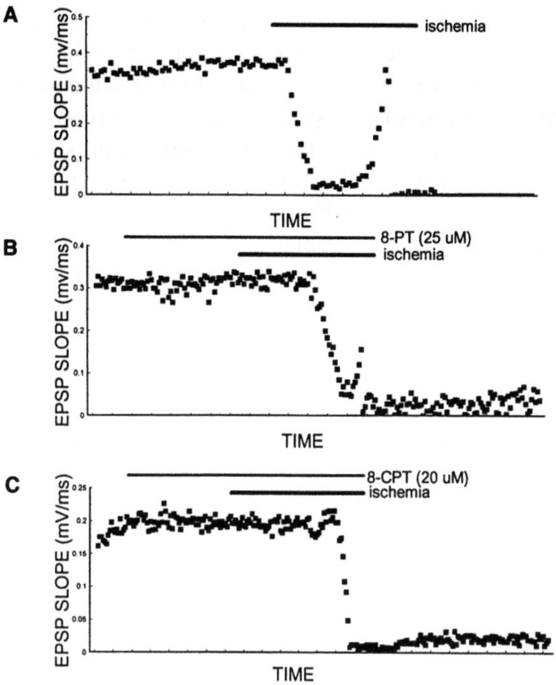

Fig. 3A–C. Effect of ischemia in the presence of adenosine antagonists on the EPSP in area CA1. Recordings were established as described in Fig. 1. **A** Time course of EPSP changes during ischemia (*bar*). **B** Time course of changes in the presence of 25 μM 8-PT (*thin bar*) before and during (*thick bar*) ischemia. **C** Time course of changes in the presence of 20 μM 8-CPT (*thin bar*) before and during ischemia (*thick bar*)

During ischemia, ATP levels drop rapidly with the ultimate formation of adenosine as a metabolite (Hagberg et al. 1985; Phillis et al. 1987). In fact, extracellular adenosine detected by microdialysis is elevated by fivefold in in vivo models of ischemia. Adenosine is a potent inhibitor of presynaptic neurotransmitter release, acting primarily through A1 receptors, and we postulated that the early depression of synaptic transmission might be explained by increased extracellular adenosine. To test this hypothesis we applied adenosine antagonists to the slices before and during ischemia. As seen in Fig. 3B no significant alteration was seen in synaptic responses with the application of adenosine antagonist (8-PT) in normal buffer. However, during ischemia, in the continued presence of 8-PT, the disappearance of synaptic responses was significantly delayed (compare Fig. 3A and 3B), indicating that the adenosine antagonist counteracted the loss of synaptic transmission. A more potent adenosine antagonist (8-CPT; 20 μM) was even more effective at preventing the disappearance of synaptic transmission (Fig. 3C). Interestingly, the burst noted in the absence of adenosine antagonists can be seen attenuated in the presence of 8-PT and as only a small increase in the slope of the EPSP in the presence of 8-CPT just before synaptic transmission disappears. We did not determine whether adenosine antagonists shortened the time to irreversible loss of synaptic transmission.

To test the protective effects of various pharmacologic agents, we established a time of ischemia that produced irreversible loss of synaptic transmission in the CA1 region in more than 90% of slices (12 min; see also Fig. 1). Various drugs were then applied to populations of individual slices to determine their potential protective effects. A drug was considered protective if synaptic transmission returned to normal after 13 min of ischemia. In some experiments, drugs were applied for 10 min before and during ischemia and in others the drug was present only during ischemia. As seen in Fig. 4A, 1 m*M* ketamine (a reported noncompetitive NMDA antagonists) applied to the slice 10 min before ischemia produced a marked suppression in synaptic transmission. Ischemia in the continued presence of 1 m*M* ketamine provided complete protection, evidenced by the return of

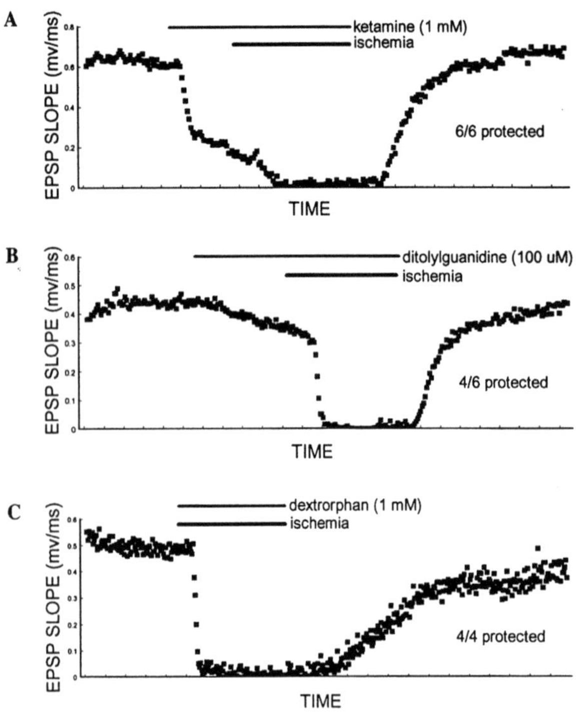

Fig. 4A–C. Protective effects of NMDA and sigma receptor antagonists in area CA1. Recordings were established as described in Fig. 1. **A** Effects of 1 m*M* ketamine (*thin bar*) on the synaptic responses before and during ischemia (*thick bar*). **B** Effects of 100 μM DTG (*thin bar*) on the synaptic responses before and during ischemia (*thick bar*). **C** Effects of 1 m*M* dextrorphan (*thin bar*) on the synaptic responses during ischemia (*thick bar*). Each drug was removed after ischemia the EPSP. Ketamine at a concentration of 1 m*M* protected six out of six slices; however, at 250 μ*M* ketamine protected none of four slices. Because of the high dose required for protection, and that synaptic transmission was also suppressed at these doses, it seems likely that ketamine's protective effects are not due solely to blocking NMDA receptors. Since high-dose ketamine is also thought to block sigma receptors, we tested whether sigma receptor antagonists would also provide protection in this model. DTG (100 μ*M*) produced small but consistent effects on baseline synaptic transmission (Fig. 4B) and provided protection from ischemia when applied 10 min before and during the ischemic period. These results were reproduced in four out of six slices, suggesting that sigma receptor antagonists provide protection in this model of ischemia. Higher concentrations (0.5 m*M* and above) of DTG produced an irreversible block in synaptic transmission and, therefore, could not be evaluated for protective effects. Finally, we tested another noncompetitive NMDA antagonist, dextrorphan. At 1 m*M*, dextrorphan protected four out of four slices, however, lowering the concentration by fourfold (250 μ*M*) provided no protection in an additional four slices suggesting that, like ketamine, dextrorphan's protection was by blocking processes in addition to NMDA receptors. Dextrorphan was the only drug tested in our model that consistently protected slices when applied with the ischemia buffer, without a 10 min preincubation (Fig. 4C). Dextrorphan at 1 m*M* also abruptly and completely, but reversibly, abolished synaptic transmission (data not shown). Dextrorphan also washed out of the slice more slowly than ketamine (often taking 1 h or more; compare the rate of synaptic transmission return in Fig. 4A and 4C). At the concentrations shown in Fig. 4, all three of these compounds eliminated the burst activity. In fact, in almost all experiments protection (return of the EPSP) correlated with an absence of the burst phase.

Discussion

We initiated studies using hippocampal slices for two reasons. The first was to understand the sequence of electrophysiologic changes that occurs in pyramidal cells and dentate granule cells during ischemia. The second was to establish a model that attempts to accurately reflect changes seen in in vivo models of ischemia and then use the model to test various compounds as neuroprotective agents. Previous hippocampal slice studies have shown that dentate granule cells appear more resistant to ischemia than CA1 pyramidal cells (Aitken and Schiff 1986). We found that this was also the case in our model. Recordings in the dendritic regions of dentate granule cells consistently survived longer periods of ischemia than recordings in CA1. Future studies will be directed at determining the electrophysiologic and biochemical differences between these two cell populations to determine the basis of their differential susceptibility.

In the CA1 region of the hippocampus, one of the first changes after inducing ischemia is that synaptic transmission is blocked. Our results with iontophoretic application of NMDA and AMPA suggest that this early depression is not due to alterations in the postsynaptic cells response and further suggest that ionic imbalances are unlikely to explain the disappearance in synaptic transmission. Since adenosine antagonists prevent this disappearance in synaptic transmission, a likely explanation is that adenosine levels accumulate extracellularly activating presynaptic adenosine receptors that attenuate neurotransmitter release. The adenosine antagonist results presented here are very similar to those reported by others (Fowler 1990; Lipton and Whittingham 1984). Apparently, extracellular accumulation of adenosine provides a natural mechanism for neuroprotection that prevents excess release of excitotoxic neurotransmitters. One possibility, that we have not yet evaluated, is that application of adenosine antagonists will block this protective mechanism and shorten the period of ischemia that the neurons will survive.

The next phase of electrophysiologic change during ischemia in the CA1 region is the spontaneous return of synaptic activity. This result might be explained by the wash-out of endogenous adenosine; however, this burst still occurs in the presence of adenosine antagonists suggesting the situation is likely more complex. One possibility, supported by other electrophysiologic findings (Hansen 1984; Kass and Lipton 1982; Lipton and Whittingham 1984; Sick et al. 1987), is that extracellular K^+ begins to build up quite rapidly at about this time, sometimes exceeding 30 m*M*. The reasons for extracellular K^+ accumulations are probably numerous and include rundown of the ATP-dependent Na^+/K^+-ATPase, opening of ATP-dependent K^+ channels and opening of Ca^{2+}-dependent K^+ channels. Failure of the Na^+/K^+-ATPase also causes increased accumulation of intracellular Na^+, reversing the Na^+-dependent glutamate transporter, pumping glutamate out of the cell from the metabolic pool (Ikeda et al. 1989; Kauppinen et al. 1988). This would lead to progressive accumulation of extracellular glutamate and would be augmented by impaired glutamate uptake. Since the postsynaptic cell remains sensitive to glutamate-induced depolarization up to this point (Fig. 2), a feed-forward cascade is likely established where extracellular K^+ induces more glutamate release which activates postsynaptic glutamate receptors leading to increased extracellular K^+, and the cycle continues. Once extracellular K^+ reaches a critical concentration the membranes depolarize completely, evidenced by the loss of evoked responses including the disappearance of the presynaptic fiber volley and the loss of iontophoretic responses. Once this massive depolarization occurs, very few slices will recover when returned to normal buffer. Other studies have suggested that intracellular Ca^{2+} increases are actually due to a failure of Ca^{2+}-ATPases that normally extrude Ca^{2+} and are not due to increased Ca^{2+} permeability (Lipton and Lobner 1990). In either case, we have found that removal of extracellular Ca^{2+} and lower concentrations of NMDA receptor antagonists (100 μM) prevent the burst (data not shown), suggesting Ca^{2+} is playing a central role.

Our attempts to protect slices pharmacologically from irreversible loss of synaptic transmission were partially successful. High doses of dextrorphan and ketamine were the most promising; however, at these concentrations (1 mM) they likely had numerous effects including activation of sigma receptors. Concentrations of these compounds where some specificity to the NMDA receptor could be assumed (250 μM or less) were ineffective in protecting the slices, consistent with earlier reports showing that NMDA antagonists were not effective neuroprotective agents in hippocampal slices (Aitken et al. 1988). Others have shown that low-dose ketamine (25 μM) and other NMDA antagonists were effective in protecting CA1 neurons from irreversible loss of synaptic transmission (Clark and Rothman 1987; Rothman et al. 1987). These latter reports used anoxia alone to induce electrophysiologic change and likely represents a milder form of ischemia than we have used (anoxia and aglycemia). Just as found in in vivo models, the method and severity of ischemia must be taken into account when evaluating neuroprotective agents. Interestingly, the sigma receptor antagonist DTG provided significant protection in our model. Lobner and Lipton (1990) reported similar findings in hippocampal slices and, in addition, showed that DTG blocked ischemia-induced glutamate release. While still rather poorly defined (for review see Ferris et al. 1991), sigma receptors might provide additional targets for neuroprotective therapies. As noted above, if the predominant reason for increased intracellular $Ca^{2+}$is a failure of the Ca^{2+}-ATPase (Lipton and Lobner 1990), then one would expect little protection from these NMDA antagonists unless they also slow the rate of ATP utilization within the neurons. Interestingly, NMDA application to slices increases the rate that intracellular ATP falls (Raley-Susman and Lipton 1990). In this scheme, any treatment that prevents ATP utilization during ischemia would have some beneficial effect and may be a major reason that hypothermia is protective in various models (Busto et al. 1989; Churn et al. 1990; Okada et al. 1988). Whether the hippocampal slice will provide a useful predictive model for neuroprotective agents in vivo awaits the evaluation of a larger spectrum of compounds.

Acknowledgments. This research was supported in part by the National Institute of Health NINDS-NS26086 (M.N.W.). S.A.W. is supported by a scholarship from the American Heart Association, M.D.M. is supported by the McKnight Foundation, and M.N.W. is supported by a research career development award NINDS-NS01509.

References

Abele AE, Scholz KP, Scholz WK, Miller RJ (1990) Excitotoxicity induced by enhanced excitatory neurotransmission in cultured hippocampal pyramidal neurons. Neuron 2: 413–419

Aitken PG, Schiff SJ (1986) Selective neuronal vulnerability to hypoxia in vitro. Neurosci Lett 67:92–96

Aitken PG, Balestrino M, Somjen GG (1988) NMDA antaonists: lack of protective effect against hypoxic damage in CA1 region of hippocampal slices. Neurosci Lett 89:187–192

Busto R, Dietrich WD, Globus MY-T, Ginsberg MD (1989) The importance of brain temperature in cerebral ischemic injury. Stroke 20:1113–1114

Choi DW (1990) Cerebral hypoxia: some new approaches and unanswered questions. J Neurosci 10:2493–2501

Churn SB, Taft WC, Billingsly MS, Blair RE, DeLorenzo RJ (1990) Temperature modulation of ischemic neuronal death and inhibition of calcium/calmodulin-dependent protein kinase II in gerbils. Stroke 21:1715–1721

Clark GD, Rothman SM (1987) Blockade of excitatory amino acid receptors protects anoxic hippocampal slices. Neuroscience 21:665–671

Feig S, Lipton P (1990) N-methyl-D-aspartate receptor activation and Ca2+ account for poor pyramidal cell structure in hippocampal slices. J Neurochem 55:473–483

Ferris CD, Hirsch DJ, Brooks BP, Snyder SH (1991) Sigma receptors: from molecule to man. J Neurochem 57:729–737

Fowler JC (1990) Adenosine antagonists alter the synaptic response to in vitro ischemia in the rat hippocampus. Brain Res 509:331–334

Hagberg H, Anderson P, Lazarewicz J, Jacobsen I, Butcher S, Sandberg M, Lehmann A, Hamberger AX (1985) Extracellular adenosine, inosine, hypoxanthine, and xanthine in relation to tissue nucleotides and purines in rat striatum during transient ischemia. J Neurochem 49:227–231

Hansen AJ (1985) Effects of anoxia on ion distribution in the brain. Physiol Rev 65:101–148

Ikeda M, Nakazawa T, Abe K, Kanekon T, Yamatsu K (1989) Extracellular accumulation of glutamate in the hippocampus induced by ischemia is not calcium-dependent-in vitro and in vivo evidence. Neurosci Lett 96:413–419

Kass IS, Lipton P (1982) Mechanisms involved in irreversible anoxic damage to the in vitro hippocampal slice. J Physiol (Lond) 332:459–472

Kauppinen RA, McMahon HT, Nicholls DG (1988) Ca2+-dependent and Ca2+-independent glutamate release, energy status and cytosolic free Ca2+ concentrations in isolated nerve terminals following metabolic inhibition: possible relevance to hypoglycaemia and anoxia. Neuroscience 27:175–182

Lipton P, Lobner D (1990) Mechanisms of intracellular calcium accumulation in the CA1 region of rat hippocampus during anoxia in vitro. Stroke 21:III-60–III-64

Lipton P, Whittingham TS (1984) Energy metabolism and brain slice function. In: Dingledine R (ed) Brain slices. Plenum, New York, pp 113–147

Lobner D, Lipton P (1990) Sigma-ligands and non-competitive NMDA antagonists inhibit glutamate release during cerebral ischemia. Neurosci Lett 117:169–174

Manev H, Favaron M, Guidotti A, Costa E (1989) Delayed increase of Ca2+ influx elicited by glutamate: role in neuronal death. Mol Pharmacol 36:106–112

Mayer ML, Miller RJ (1990) Excitatory amino acid receptors, second messengers and regulation of intracellular Ca^{2+} in mammalian neurons. Trends Pharmacol Sci 11:254–260

Nellgard B, Wieloch T (1992) Postischemic blockade of AMPA but not NMDA receptors mitigates neuronal damage in the rat brain following transient severe ischemia. J Cereb Blood Flow Metab 12:2–11

Okada Y, Michinori T, Yoneda K (1988) The protective effect of hypothermia on reversibility in the neuronal function of the hippocampal slice during long lasting anoxia. Neurosci Lett 84:277–282

Phillis JW, Walter GW, O'Regan MH, Stair RE (1987) Increases in cerebral cortical perforate adenosine and inosine concentrations during hypoxia and ischemia. J Cereb Blood Flow Metab 7:679–686

Raley-Susman KM, Lipton P (1990) In vitro ischemia and protein synthesis in the rat hippocampal slice: the role of calcium and NMDA receptor activation. Brain Res 515:27–38

Rothman SM, Olney JW (1987) Excitotoxicity and the NMDA receptor. Trends Neurosci 10:299–302

Rothman SM, Thurston JH, Hauhart RE, Clark GD, Solomon JS (1987) Ketamine protects hippocampal neurons from anoxia in vitro. Neuroscience 21:673–678

Sick TJ, Solow EL, Robert EL Jr (1987) Extracellular potassium ion activity and electrophysiology in the hippocampal slice: paradoxical recovery of synaptic transmission during anoxia. Brain Res 418:227–234

Siesjö BK, Bengtsson F (1989) Calcium fluxes, calcium antagonists, and calcium-related pathology in brain ischemia, hypoglycemia, and spreading depression: a unifying hypothesis. J Cereb Blood Flow Metab 9:127–140

Yoneda Y, Ogita K (1991) Neurochemical aspects of the N-methyl-D-aspartate receptor complex. Neurosci Res 10:1–33

Glial Swelling and Damage by Arachidonic Acid

F. Staub[1], A. Winkler[1], J. Peters[1], O. Kempski[2], and A. Baethmann[1]

Introduction

Arachidonic acid (AA) is a major constituent of membrane phospholipids in brain tissue. Normally, the free fatty acid is present only in small amounts, but it accumulates under adverse conditions such as ischemia (Kinouchi et al. 1990). The release of free fatty acids involves activation of phospholipases and breakdown of membrane phospholipids. AA in particular is considered to mediate pathological processes. It is a polyunsaturated compound and precursor of prostaglandins, leukotrienes, and oxygen-derived free radicals (Wolfe 1982). In cerebral ischemia concentrations of free AA of up to 0.5 m*M*/kg were found (Kinouchi et al. 1990). But also brain injury and seizures cause increased levels of free AA (Siesjö et al. 1982; Baethmann et al. 1989).

Noxious properties of AA have been observed in many investigations. For instance, it was demonstrated that lipid peroxidation or disturbance of the respiratory chain in mitochondria is induced by the fatty acid (Braughler and Hall 1989; Hillered and Chan 1988). Moreover, AA has a significant role in brain edema. The fatty acid was found to induce swelling of cerebral slices and to increase permeability of the blood–brain barrier (Chan and Fishman 1978; Unterberg et al. 1987). Since AA may lead to cytotoxic brain edema and damage of brain cells, we have currently investigated glial cell volume and viability during administration of the fatty acid. Additional experiments were performed to establish whether the respective effects can be inhibited by blocking of the cyclo- and lipoxygenase pathway or of lipid peroxidation. The study was carried out in vitro while controlling the relevant parameters to analyze the significance of a single pathophysiological factor in isolation (Kempski et al. 1983). For that purpose, suspensions of C6 glioma cells or astrocytes from primary culture were incubated under continuous recording of pH, pO_2, and temperature. Cell volume and viability were assessed by flow cytometry.

[1] Institute for Surgical Research, Ludwig Maximilians University, 81366 Munich, Germany
[2] Institute for Neurosurgical Pathophysiology, Johannes Gutenberg-University, 55101 Mainz, Germany

Cerebral Ischemia and Basic Mechanisms
Ed. by A. Hartmann, F. Yatsu, and W. Kuschinsky

Materials and Methods

C6 glioma cells were cultivated as monolayers in Petri dishes using Dulbecco's modified minimal essential medium (DMEM) with the addition of 25 m*M* bicarbonate. The medium was supplemented with 10% fetal calf serum (FCS), 100 IU/ml penicillin G, and 50 μg/ml streptomycin. The cells were grown in a humidified atmosphere of 5% CO_2 and room air at 37 °C. Glial cells from primary culture were prepared from 3-day-old rats according to a modified method of Frangakis and Kimelberg (1984). The culture conditions were identical with those described above. Only confluent cultures were used for the experiments. The cells were harvested with 0.05% trypsin–0.02% EDTA in phosphate-buffered saline and washed twice thereafter. After resuspension in serum-free medium the glial cells were transferred to a Plexiglas incubation chamber which was supplied with electrodes for control of pH, temperature, and pO_2. A gas-permeable silicon rubber tube in the chamber provided the cell suspension with a mixture of O_2, CO_2, and N_2. Details of the model have been published elsewere (Kempski et al. 1983).

The volume of the glial cells was determined by flow cytometry using an advanced coulter system with hydrodynamic focusing (Kachel et al. 1977). The technique was also employed for measurement of cell viability by exclusion of the fluorescence dye propidium iodide (final concentration: 40 μg/ml; Staub et al. 1990). Excitation was induced through a 500 nm short-pass filter. Maximum emission of propidium iodide at 630 nm was measured by using a 580 nm long pass filter. A window integration system was employed for discrimination of propidium-positive (dead) from nonfluoresceing (viable) cells.

The experiments with AA were performed after a 45-min control period used for measurements of cell volume, viability, and medium osmolality. Subsequently, the suspension was added with AA in a dose range of 0.01–1 m*M* (final concentration). Cell volume and viability were measured for 60 min during incubation with AA. Additional experiments at an AA concentration of 0.1 m*M* were conducted with inhibition of the cyclo- and lipoxygenase pathway by BW755C (0.2 m*M*), or during blocking of lipid peroxidation using the aminosteroid U-74389F (0.1 m*M*). The inhibitors were administered to the suspension 15 min prior to addition of AA. In further experiments cell swelling by 0.1 m*M* AA was studied in Na^+-free medium, where Na^+ ions were replaced by choline, and bicarbonate by 10 m*M* HEPES.

Results

Administration of AA to the suspension caused an immediate dose-dependent swelling of C6 glioma cells. Concentrations as low as 0.01 m*M* led to an

Table 1. Volume response of C6 glioma cells after administration of AA (0.1 mM) to the suspension medium

	Control			Incubation with arachidonic acid					
Time (min)	−15	−10	−5	1	5	10	20	40	60
AA 0.1 mM	100.40 0.26	99.93 0.17	99.67 0.25	104.45 1.16	110.04 1.49	112.17 1.90	110.99 2.22	112.39 2.83	109.58 2.20
AA 0.1 mM + BW755C	99.88 0.29	99.90 0.20	100.23 0.33	103.84 1.77	109.60 1.58	111.18 2.04	112.57 2.39	109.29 1.85	107.57 2.58
AA 0.1 mM + U-74389F	100.27 0.33	99.40 0.49	100.33 0.35	99.41* 0.70	100.18* 1.29	98.93* 1.35	100.66* 0.94	100.74* 1.01	100.97* 1.11
AA 0.1 mM Na^+-free medium	98.63 0.75	101.25 0.40	100.12 0.44	100.87* 0.46	101.65* 0.41	103.77* 0.86	101.11* 0.50	101.68* 0.59	102.00* 1.99

Cell swelling was significant ($p < 0.01$) compared to a control group without addition of AA (not shown) during the entire observation period. The cell volume response to AA is further shown in the presence of BW755C (0.2 mM), U-74389F (0.1 mM), or in Na^+-free incubation medium. The cell volume is given as a percentage of the cell size obtained during the last 15 min of the control period. The mean ± SEM of four to seven experiments in each group is shown.
* $p < 0.01$ vs. AA 0.1 mM.

increase of the cell volume to 103.0% ± 1.0% (mean ± standard error of mean, SEM) of control within 10 min ($p < 0.01$). A volume increase to 110.0% ± 1.5% (Table 1) or 118.8% ± 1.5% of control was obtained when the cells were administered with 0.1 or 1 mM AA, respectively ($p < 0.001$). After an initial rapid swelling the cell volume was constant for the remaining observation period (Table 1). Viability of the C6 glioma cells was 90.6% ± 0.8% under control conditions and it remained unchanged after addition of AA up to a level of 0.1 mM. At 0.5 mM, the number of viable cells started to significantly decrease after 10 min, reaching 76.8% ± 1.7% within 60 min ($p < 0.01$). Upon incubation with AA at 1 mM, cell viability immediately began to decline. Only 45.6% ± 4.8% of the C6 glioma cells were viable after 60 min ($p < 0.001$). The swelling and damage of C6 glioma cells by AA was confirmed in experiments using astrocytes from primary culture. Addition of AA at a dose level of 0.05 or 0.1 mM led to swelling of astrocytes ($p < 0.01$), which was comparable to that of C6 glioma cells. Contrary to the findings of C6 glioma cells, however, AA concentrations as low as 0.1 mM led to a decrease in cell viability ($p < 0.05$). Additional experiments were performed to analyze mechanisms of AA-induced glial swelling. Inhibition of metabolization of AA by cyclo- and lipoxygenase using the double pathway inhibitor BW755C did not affect glial swelling from AA (Table 1). However, preincubation with the aminosteroid U-4389F prevented cell swelling from AA almost completely ($p < 0.01$, Table 1). Similar results were obtained when the experiments were conducted with choline for replacement of Na^+ ions in the suspension medium and with

HEPES replacing bicarbonate. Under these conditions the largest volume increase seen after administration of AA was only 103.8% ± 0.9% of control after 10 min ($p < 0.01$, Table 1).

Discussion

The present results demonstrate a powerful potential of AA to induce glial swelling. The fatty acid led to a dose-dependent cell volume increase at concentrations which have been observed in brain tissue in vivo under pathophysiological conditions, such as focal injury or ischemia (Baethmann et al. 1989; Kinouchi et al. 1990). In addition, viability of the C6 glioma cells was affected by AA levels which may accumulate in the brain, for instance, in cerebral ischemia (Kinouchi et al. 1990).

Not only AA but also its metabolites such as prostaglandins or leukotrienes are considered to have a mediator function under pathophysiological conditions (Wolfe 1982). The present results on inhibition of the cyclo- and lipoxygenase pathway by BW755C suggest, however, that AA itself is the cell swelling-inducing agent and not its metabolites. On the other hand, administration of the aminosteroid U-74389F inhibited cell swelling from AA almost completely, indicating that lipid peroxidation is a major factor in AA-induced cell swelling. Lipid peroxidation may commence if oxygen-derived free radicals accumulate in the presence of free fatty acids (Chan et al. 1988; Braughler and Hall 1989). It is conceivable that superoxide radicals generated by conversion of AA were interacting with AA as substrate, thereby initiating formation of lipid- and lipidperoxide radicals. A chain reaction may ensue, mainly affecting fatty acids of cell membranes, consequently leading to damage of the double-lipid layer (Braughler and Hall 1989). As shown, glial swelling from AA was almost completely prevented when the experiments were conducted in a Na^+-free suspension medium making obvious the cellular uptake of Na^+ and thereby water as the ultimate mechanism of this phenomenon.

Taken together, AA causes swelling and damage of C6 glioma cells and astrocytes from primary culture most likely by activation of lipid peroxidation. By damaging the lipid bilayer of cell membranes lipid peroxides increase Na^+ permeability and thus influx of water into the cells. Ultimately, Na^+ together with Cl^- and water accumulate in the cell as a final step of glial swelling. It may be considered, in addition, that activation of lipid peroxidation raises intracellular Ca^{2+} as another factor of glial damage.

Acknowledgements. The excellent technical and secretarial assistance of Ingrid Kölbl, Ulrike Görke, Helga Kleylein, and Monika Stucky is gratefully acknowledged. We appreciate the generous gifts of BW755C by Dr. Cooke, Wellcome Research Laboratories, Beckenham, Kent, UK, and of U-74389F by Dr. McCall, Upjohn Laboratories, Kalamazoo, MI, USA. This work was supported by the Deutsche Forschungsgemeinschaft: Ba 452/6-7.

References

Baethmann A, Maier-Hauff K, Schürer L, Lange M, Guggenbichler C, Vogt W, Jacob K, Kempski O (1989) Release of glutamate and of free fatty acids in vasogenic brain edema. J Neurosurg 70:578–591

Braughler JM, Hall ED (1989) Central nervous system trauma and stroke. I. Biochemical considerations for oxygen radical formation and lipid peroxidation. Free Radic Biol Med 6:289–301

Chan PH, Fishman RA (1978) Brain edema: induction in cortical slices by polyunsaturated fatty acids. Science 201:358–360

Chan PH, Chen SF, Yu ACH (1988) Induction of intracellular superoxide radical formation by arachidonic acid and by polyunsaturated fatty acids in primary astrocytic culture. J Neurochem 50:1185–1193

Frangakis MV, Kimelberg HK (1984) Dissociation of neonatal rat brain by dispase for preparation of primary astrocyte cultures. Neurochem Res 9:1689–1698

Hillered L, Chan PH (1988) Role of arachidonic acid and other free fatty acids in mitochondrial dysfunction in brain ischemia. J Neurosci Res 20:451–456

Kachel V, Glossner E, Kordwig G, Ruhenstroth-Bauer G (1977) Fluvo-Metricell, a combined cell volume and cell fluorescence analyzer. J Histochem Cytochem 25:804–812

Kempski O, Chaussy L, Groß U, Zimmer M, Baethmann A (1983) Volume regulation and metabolism of suspended C6 glioma cells: an in vitro model to study cytotoxic brain edema. Brain Res 279:217–228

Kinouchi H, Imaizumi S, Yoshimoto T, Motomiya M (1990) Phenytoin affects metabolism of free fatty acids and nucleotides in rat cerebral ischemia. Stroke 21:1326–1332

Siesjö BK, Ingvar M, Westerberg E (1982) The influence of bicuculline-induced seizures on free fatty acid concentration in cerebral cortex, hippocampus, and cerebellum. J Neurochem 39:796–802

Staub F, Baethmann A, Peters J, Weigt H, Kempski O (1990) Effects of lactacidosis on glial cell volume and viability. J Cereb Blood Flow Metab 10:866–876

Unterberg A, Wahl M, Hammersen F, Baethmann A (1987) Permeability and vasomotor response of cerebral vessels during exposure to arachidonic acid. Acta Neuropathol (Berl) 73:209–219

Wolfe LS (1982) Eicosanoids: prostaglandins, thromboxanes, leukotrienes, and other derivates of carbon-20 unsaturated fatty acids. J Neurochem 38:1–14

Free Radicals, Protein Synthesis

Disturbances of Cerebral Protein Synthesis in Global and Focal Ischemia

K.A. HOSSMANN

Introduction

In the classical literature on cerebral ischemia, the high vulnerability of the brain has been attributed to the misrelationship between the high energy demands and the low energy reserves of this organ which results in the rapid depletion of energy metabolites when oxygen and glucose supply is interrupted. During the past decade, however, increasing evidence has accumulated which suggests that ischemic lesions may result from pathobiochemical changes which are independent of the energy state of the tissue. In particular, acidosis, free radical reactions, the massive release of neurotransmitters and the associated changes of calcium homeostasis seem to contribute to the manifestation of ischemic cell death (for reviews see Siesjö and Bengtsson 1989; Choi and Rothman 1990; Ginsberg 1990).

A factor which has received much less attention in the past, but which seems to be of equal importance for the development of ischemic lesions, is a disturbance of protein synthesis (Bodsch et al. 1985; Hossmann 1992). In fact, evidence has been provided that such disturbances may evolve despite preservation or recovery of energy metabolites in areas destined to die, and that therapeutical interventions that reverse the inhibition of protein synthesis also improve neuronal survival.

In the following, changes of the overall protein synthesis rate after global and focal ischemia will be reviewed. These changes reflect the mean behavior of the protein-synthetizing machinery but certain protein species such as heat shock proteins, the protein products of immediate early response genes or enzymes with short half-lives may deviate substantially from this pattern. A more comprehensive account which also includes these selective gene expressions will be given elsewhere (Hossmann 1992).

Max Planck Institute for Neurological Research, Department of Experimental Neurology, Gleueler-Str. 50, 50931 Cologne 41, Germany

Cerebral Ischemia and Basic Mechanisms
Ed. by A. Hartmann, F. Yatsu, and W. Kuschinsky

Global Ischemia

Reversible Alterations of Protein Synthesis

The first investigations on the effect of brain ischemia on protein synthesis were carried out after prolonged cerebrocirculatory arrest (Kleihues and Hossmann 1971, 1973). These investigations were prompted by the observation that energy metabolism and certain electrophysiological functions such as evoked potentials and EEG recovered after up to 1-h complete ischemia in normothermia if care was taken to prevent postischemic disturbances of recirculation (Hossmann and Sato 1970; Hossmann et al. 1973, 1977). Since these observations were in sharp contrast to the previous assumption that normothermic neurons die after a few minutes of circulatory arrest, measurements of protein synthesis were performed to find out if more complex biochemical functions were also restored. It turned out that this was the case, but the recovery of protein synthesis – measured as the incorporation of labeled amino acids into total brain proteins – progressed much more slowly than that of energy metabolism or the restoration of ion homeostasis. In fact, in normothermic cats and monkeys submitted to 1-h complete ischemia, recovery of protein synthesis to 50% required more than 6 h and recovery to normal about 24 h (Kleihues and Hossmann 1971; Bodsch et al. 1986). Energy state and ion homeostasis, in contrast, began to recover within a few minutes of recirculation (Hossmann et al. 1977; Behar et al. 1989) and returned to near normal in less than 1 h (Hossmann and Sato 1970). Electron micrographs and polysomal density gradients further showed that polyribosomes remained intact during ischemia but transiently disaggregated after the beginning of recirculation in parallel with the postischemic inhibition of protein synthesis (Kleihues and Hossmann 1971). The biochemical assays of protein synthesis were supplemented by autoradiographic studies in order to establish the anatomical pattern of postischemic recovery (Kleihues and Hossmann 1973; Bodsch et al. 1986; Hossmann et al. 1987). These studies revealed that after up to 1-h complete ischemia protein synthesis recovered within 24 h in the vast majority of neurons except the selectively vulnerable regions (see below). After 1-year survival the anatomical pattern of amino acid incorporation and of neuronal integrity did not change further (Hossmann et al. 1987) indicating that recovery of protein synthesis was permanent. Postischemic restoration of protein synthesis, in consequence, is a reliable predictor of postischemic cell survival.

These observations have been confirmed in various animal species and models of global ischemia. Wasterlain (1974) reduced cerebral blood flow in rats by increasing intracranial pressure to 1500 mm water for 30 min and observed a reversible disaggregation of polyribosomes and an inhibition of in vitro protein synthesis. Carai et al. (1979) occluded one carotid artery in gerbils for up to 3 h and also observed a reversible disaggregation of polyribosomes and a reduction of amino acid incorporation into brain proteins. Yanagihara (1976) incubated brain slices in vitro under anoxic conditions for 3–30 min and found a gradual decline of protein synthesis in parallel with the disaggregation of ribosomes. Dienel et al. (1980) submitted rats to 10- or 30-min four-vessel occlusion followed by recirculation for up to 2 days. Protein synthesis measured in vitro was severely suppressed after ischemia but returned to control after 4 h following 10-min and after 2–4 days after 30-min ischemia. Chavko et al. (1987) produced partial spinal cord ischemia by occluding the abdominal aorta for 40 min. They described disaggregation of polyribosomes and inhibition of in vitro protein synthesis after 10-min recirculation, followed by gradual recovery within 1 day. Many investigators also confirmed the dissociation between the rapid postischemic recovery of energy metabolism and the much slower restoration of protein synthesis (Dienel et al. 1980; Nowak et al. 1985; Raley and Lipton 1990). Moreover, it was pointed out that the maximal inhibitory effect on protein synthesis was produced already by 5-min ischemia, and that there was little further change when ischemia was prolonged to 20 min (Nowak et al. 1985).

Irreversible Alterations of Protein Synthesis

Although most of the earlier studies on protein synthesis were carried out to explore the resistance of the brain to prolonged circulatory arrest, more recent investigations have focussed on the potential role of protein synthesis for the manifestation of selective brain injury. This development was initiated by the observation of Ito et al. (1975) that a few minutes of global ischemia may cause selective neuronal death in some circumscribed regions such as the CA1 sector of hippocampus although the rest of the brain remained intact. Hemodynamic disturbances or disturbances of energy metabolism could be excluded as mechanisms of this injury (Munekata and Hossmann 1987; Mies et al. 1990); therefore, the question arose whether postischemic disturbances of protein synthesis may be involved.

The first to report such a relationship were Dienel et al. (1980). These authors submitted rats to four-vessel occlusion of 10–30-min duration, followed by postischemic recirculation for up to 48 h. Autoradiographic assays of protein synthesis revealed a gradual recovery to normal in all parts of the brain except in the vulnerable CA1 sector. In gerbils the same pattern of persisting inhibition of protein synthesis in the selectively vulnerable areas of the brain was observed already after an ischemia time of 5 min (Bodsch et al. 1985; Thilmann et al. 1986; Widmann et al. 1991) or even 2 min (Araki et al. 1990). Interestingly, these short ischemia times also induced a global inhibition of protein synthesis (Thilmann et al. 1986; Widmann et al. 1991) and a global disaggregation of ribosomes (Munekata et al. 1987) in the resistant parts of the brain (see above). However, in contrast to the selectively vulnerable areas, disturbances in these regions recovered within

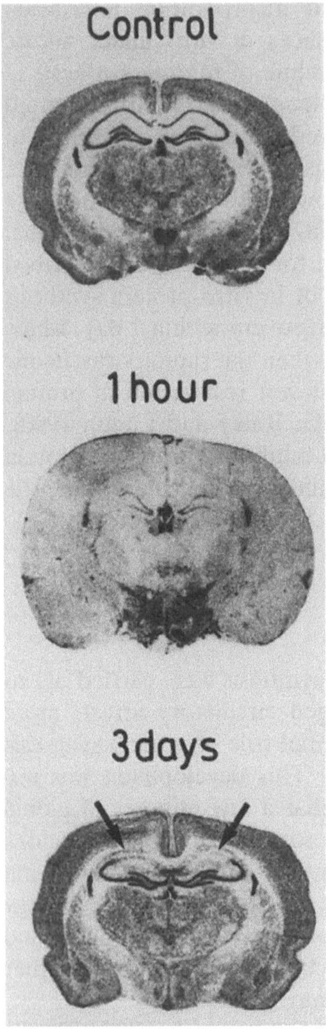

Fig. 1. Autoradiograms of ^{14}C-leucine incorporation into proteins of rat brain before (*Control*) and at two recirculation times (*1 hour* and *3 days*) after 30-min four-vessel occlusion. Note severe global depression of protein synthesis in all brain regions after 1-h recirculation, followed by recovery in all parts except the selectively vulnerable CA1 sector of hippocampus (*arrows*). Persistent inhibition of protein synthesis is thought to contribute to delayed neuronal death in this area

3–6h. Selective vulnerability, in consequence, is not the result of the selective inhibition of protein synthesis but of the selective failure of recovery (Fig. 1).

The close relationship between recovery of protein synthesis and cellular survival is also supported by histo-autoradiographic studies. Neurons exhibiting the typical morphological pattern of ischemic cell injury fail to incorporate amino acids into proteins (Kleihues and Hossmann 1973). Removal of the endogenous adenosine action by treatment with the receptor antagonist theophylline aggravates the disaggregation of ribosomes and shortens the ischemic period which can be tolerated without subsequent damage in gerbil (Dux et al. 1990). Conversely, neurons destined to survive, including the resistant interneurons in the selectively vulnerable CA1 sector of hippocampus, recover their protein-synthesizing capacity (Johansen and Diemer 1990).

The importance of protein synthesis for cell survival has also been documented after repetitive ischemia. Tomida et al. (1987) reported that repetition of brief episodes of ischemia at 1-h intervals results in a substantial aggravation of ischemic injury even in brain regions outside the vulnerable areas. Recent autoradiographic studies confirmed that in these regions protein synthesis failed to recover (Widmann et al. 1992). Conversely, prevention of neuronal death in the selectively vulnerable areas by barbiturates or hypothermia was associated with a reaggregation of ribosomes and a normalization of protein synthesis (Xie et al. 1989b; Miyazawa et al. 1991). Interestingly, neither these interventions nor the equally efficient treatment with the glutamate antagonist 2,3-dihydroxy-6-nitro-7-sulfamyl-benzo(F)trioxaline (NBQX) were able to prevent the initial postischemic inhibition of protein synthesis (Bergstedt and Wieloch 1991). This is in line with the contention that it is the failure of recovery and not the initial inhibition of protein synthesis that determines neuronal death after ischemia.

Focal Ischemia

Disturbances of protein synthesis have also been described after focal reduction of cerebral blood flow (Yanagihara 1976, 1978; Morimoto et al. 1978; Morimoto and Yanagihara 1981; Orunesu et al. 1980: Hossmann et al. 1985; Dwyer et al. 1987 Yoshimine et al. 1987; Abe et al. 1988; Xie et al. 1989a; Mies et al. 1991). The extent and severity of these disturbances are less predictable than after global ischemia because they depend not only on the duration of ischemia and recirculation but also on the severity of flow reduction.

After *permanent occlusion* of a major cerebral artery, the local density of ischemia depends on the collateral blood supply to the territory of the occluded vessel. In accordance with the threshold concept of ischemic injury (Astrup et al. 1981), protein synthesis is a function of blood flow. Two hours after combined middle cerebral and carotid artery occlusion in rat, reduction of blood flow to 50–80 ml/100 g per minute led to a modest cessation and the further reduction to 40–50 ml/100 g per minute to the complete cessation of the synthesis of most proteins (Jacewicz et al. 1986). Xie et al. (1989b) investigated this relationship in more detail, using double tracer autoradiography for the regional measurement of blood flow and protein synthesis. They noted a beginning decline of amino acid incorporation at flow values below 100 ml/100 g per minute, and a complete cessation at values below 40 ml/100 g per minute. These flow values are substantially higher than the threshold required to maintain normal brain energy state which in the gerbil

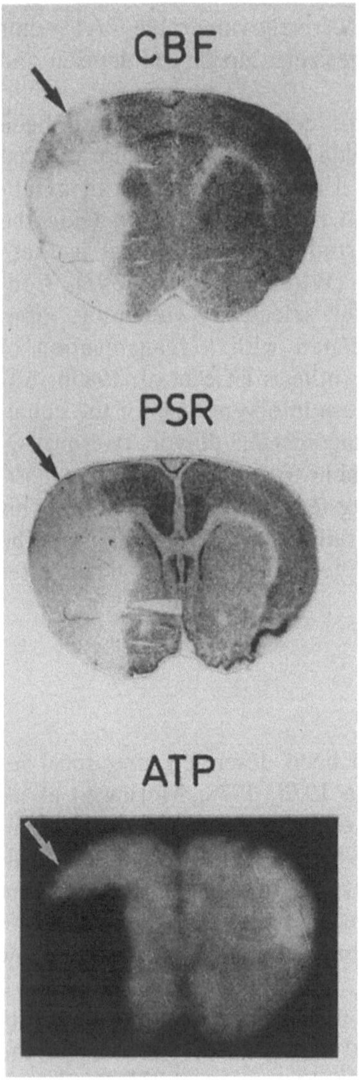

Fig. 2. Autoradiographic and bioluminescent imaging of cerebral blood flow (*CBF*), protein synthesis rate (*PSR*), and ATP of rat brain at 1 h following permanent middle cerebral artery occlusion. Blood flow was measured with ^{131}I-iodoantipyrin and protein synthesis with ^{14}C-leucine using double-tracer autoradiography. ATP was assessed on the adjacent tissue section by substrate-induced bioluminescence. Note the dissociation between preserved ATP and depressed protein-incorporated ^{14}C radioactivity in the low perfusion region of the left parietal cortex (*arrows*). (Modified from Mies et al. 1991)

amounts to about 20 ml/100 g per minute (Paschen et al. 1992). In the periphery of focal ischemia, inhibition of protein synthesis may, therefore, occur in the absence of energy failure (Fig. 2).

On the other hand, persisting inhibition of protein synthesis may cause a gradual deterioration of the energy state, as revealed by sequential measurements of the flow/metabolism threshold relationship (Mies et al. 1991). In rats submitted to permanent middle cerebral artery occlusion, the threshold for suppression of protein synthesis (defined as the flow value at which 50% of the investigated volume elements exhibit complete metabolic suppression) remained at a remarkably constant level of about 55 ml/100 g per minute over a period of 12 h. The threshold of energy state, in contrast, rose from 18 ml/100 g per minute after 1 h to 23 ml/100 g per minute after 6 h and further to 34 ml/100 g per minute after 12 h of vascular occlusion. In the stroke-prone spontaneously hypertensive rat in which measurements of blood flow and metabolism were carried out under conditions of chronic infarction, the region of ATP depletion and inhibition of protein synthesis was identical (Paschen et al. 1985). These observations demonstrate that, under chronic conditions of focal ischemia, the thresholds of protein synthesis and energy metabolism converge.

Disturbances of protein synthesis after *reversible focal ischemia* have only occasionally been described but the pattern observed reflects features of both focal and global ischemia. Morimoto and Yanagihara (1981) occluded the right common carotid artery in gerbils for 30 min or 3 h, followed by 3-h recirculation. During ischemia polyribosomes remained intact and the protein-synthesizing capacity tested in vitro was not impaired. Recirculation resulted in the disaggregation of polyribosomes and a severe suppression of protein synthesis which slowly recovered after 30-min but not after 3-h ischemia. Postischemic aggravation of the disturbance of the protein-synthesizing capacity of isolated polyribosomes was also observed by Orunesu et al. (1980) who used the same experimental model for induction of reversible ischemia. Xie et al. (1988) embolized monkeys with fragments of autologous blood clots which produce transient episodes of multifocal brain ischemia. They demonstrated sharply demarcated areas of complete inhibition of protein synthesis which clearly outlasted the ischemic episodes because they were still present after 2 h following the embolization although the electrophysiological alterations had already recovered.

An interesting observation was reported by Abe et al. (1988) who occluded the middle cerebral artery of rat for 1 h followed by 2 h and 7 days of reperfusion. These authors demonstrated that protein synthesis recovered in the territory of the middle cerebral artery but they also found that it remained irreversibly suppressed in neurons of the CA1 sector of hippocampus although this area was outside the territory of the occluded artery. This finding demonstrates that the inhibition of protein synthesis is not restricted to the ischemic territory, and – perhaps more importantly – that inhibition of protein synthesis in the selectively vulnerable sector is irreversible, even if this inhibition is not directly caused by the ischemic event.

Mechanisms of Ischemic Disturbances of Protein Synthesis

Up to now few experiments have been carried to elucidate the mechanism of the disturbances of protein synthesis after global or focal ischemia. The following discussion is highly speculative and the concept proposed requires experimental validation. However, the comparison of the pattern and the time course of the metabolic disturbance with the known pathophysiological events associated with either focal or global ischemia converge in a rather simple concept that is able to explain most of the observed phenomena.

Complete cessation of blood flow in the normothermic brain results in the rapid cessation of the energy-producing metabolism and, in consequence, the breakdown of all endergonic processes including mRNA translation. It does not cause a breakdown of the protein-synthesizing machinery, as evidenced by the undisturbed synthesis rate of ribosomes incubated in vitro after a period of ischemia (Cooper et al. 1977). The breakdown of this machinery occurs either during the recirculation after reversible ischemia or in the periphery of the territory of a permanently occluded artery, i.e., in the presence of energy metabolites: during postischemic recirculation as soon as energy metabolism has recovered or in the periphery of an infarct where blood flow remains above the threshold of energy depletion. The failure of protein synthesis is associated with a disaggregation of ribosomes, indicating that the disturbance is due to an inhibition of peptide chain initiation (Kleihues et al. 1975, Cooper et al. 1977).

A critical step in polypeptide chain initiation is the formation of a stable ternary complex consisting of met-tRNA, GTP, and the eukaryotic initiation factor eIF-2. This complex is inactivated by phosphorylation of eIF-2 which, in turn, is mediated by at least four different protein kinases (for review see Jagus et al. 1981). One of these is the calcium/phospholipid-dependent protein kinase C which phosphorylates the β subunit of eIF-2 (Schatzman et al. 1983; Alcazar et al. 1988). In fact, addition of calcium ions to heme-stabilized reticulocyte lysates leads to a dose-dependent translational inhibition by as much as 92% (de Haro et al. 1983).

If a similar inactivation by calcium ions of peptide chain initiation occurs in the brain, the phenomena described here can be explained. During complete ischemia intracellular calcium activity increases due to the opening of agonist- and voltage-gated calcium channels and the release from intracellular calcium stores (for review see Siesjö and Bengtsson 1989). Peptide chain initiation factors are not immediately phosphorylated because ATP declines before calcium increases. This explains that ribosomes sampled after the onset of ischemia maintain their protein-synthesizing capacity. Inactivation of polypeptide chain initiation occurs during postischemic recirculation because ATP is quickly restored. Intracellular calcium activity, in contrast, remains elevated for some time because the restoration of the intra/extracellular calcium homeostasis requires extensive ion pumping and – probably more importantly – because the elevated levels of extracellular excitatory amino acids prevent the closure of receptor-operated ion channels (Benveniste et al. 1984; Hagberg et al. 1985). As a result, elongation and termination – but not initiation – of peptide chains recover and ribosomes disaggregate.

This concept is confirmed by a recent study of Raley-Susman and Lipton (1990) who investigated protein synthesis in the rat hippocampal slice following in vitro ischemia. They observed that exposure of the slices to buffer lacking calcium and containing the *N*-methyl-D-aspartate (NMDA) receptor blocker ketamine prevented both the inhibition of protein synthesis and the morphological changes in the neurons. Conversely, the rapid flooding of the cytosol with calcium after the onset of ischemia (Harris et al. 1981) could explain that maximal inhibition of protein synthesis is produced already after 5-min circulatory arrest (Nowak et al. 1985).

The postischemic recovery process depends on the normalization of transmitter and calcium homeostasis. In the resistant parts of the brain intracellular calcium normalizes and protein synthesis recovers within a few hours after brief episodes of ischemia (Dux et al. 1987; Widmann et al. 1991), and within 1 day if ischemia is prolonged to 1 h (Bodsch et al. 1986; Hossmann et al. 1983). In the selectively vulnerable parts of the brain, in contrast, the disturbance of calcium homeostasis persists, and peptide chain initiation is not reversed. Persisting disturbances of ion homeostasis may also be responsible for the irreversible inhibition of protein synthesis after repetitive ischemia (Widmann et al. 1992). This would be in line with the observation that repetitive ischemia enhances disturbances of electrolyte homeostasis and, in consequence, postischemic brain edema (Vass et al. 1988). Finally, this concept could explain that protein synthesis of selectively vulnerable neurons may transiently begin to recover before it is irreversibly suppressed (Bodsch et al. 1986; Thilmann et al. 1986; Johansen and Diemer 1990; Widmann et al. 1991). In fact, electron-microscopic evaluation of mitochondrial calcium sequestration – which parallels the alterations of cytosolic calcium activity (Nicholls 1985) – revealed a triphasic pattern in CA1 neurons: the massive calcium deposits visible 15 min after 5-min ischemia transiently disappeared between 30 min and 2 h but started to accumulate again after 6 h in parallel with the secondary suppression of protein synthesis (Dux et al. 1987). The temporal and topical pattern of the postischemic impairment of protein synthesis can, therefore, be readily associated with the known alterations of postischemic ion and transmitter homeostasis.

A similar scenario can be drawn for the alterations during and after focal ischemia. In the center of the ischemic focus, protein synthesis is inhibited due to energy failure. In the peripheral parts of the infarct or in regions in which spontaneous recirculation occurs, energy metabolism is preserved but intracellular calcium homeostasis is disturbed because ischemia generates spreading depression-like depolarizations (Nedergaard and Astrup 1986; Iijima et al. 1992) and because excitatory amino acids – which are released in the center of the infarct (Shimada et al. 1989) – diffuse into the surrounding tissue. This explains that protein synthesis begins to decline at much higher flow values than energy metabolism. Cell survival in the periphery of an infarct will, therefore, depend on the prevention of normalization of ion and transmitter homeostasis. Recently, evidence has been provided that the NMDA antagonist MK-801 is able to completely suppress the peri-infarct depolarizations (Iijima et al. 1992). This may be the reason that this or other glutamate and calcium antagonists are able to reduce the size of an infarct although blood flow does not improve (Park et al. 1988, 1989; Nakayama et al. 1988; Dirnagl et al. 1990; Iijima et al. 1992).

In conclusion, inhibition of protein synthesis seems to be an important pathogenetic factor for both irreversible neuronal death in the vulnerable brain regions after global ischemia, and in the periphery of an infarct after focal ischemia. There are indications that this inhibition is a functional disturbance related to the inactivation of polypeptide chain initiation factors which can be ameliorated by pharmacological interventions. Reversal of the inhibition of protein synthesis is, therefore, a promising therapeutical challenge which deserves further attention.

References

Abe K, Araki T, Kogure K (1988) Recovery from edema and of protein synthesis differs between the cortex and caudate following transient focal cerebral ischemia in rats. J Neurochem 51:1470–1476

Alcazar A, Mendez E, L-Fando J, Salinas M (1988) Specific phosphorylation of the beta subunit of eIF-2 factor from brain by three different protein kinases. Biochem Biophys Res Commun 153:313–320

Araki T, Kato H, Inoue T, Kogure K (1990) Regional impairment of protein synthesis following brief cerebral ischemia in the gerbil. Acta Neuropathol (berl) 79:501–505

Astrup J, Siesjö BK, Symon L (1981) Thresholds in cerebral ischemia. The ischemic penumbra. Stroke 12:723–725

Behar K, Rothman DL, Hossmann KA (1989) NMR-spectroscopic investigation of the recovery of energy and acid-base homeostasis in cat brain after prolonged ischemia. J Cereb Blood Flow Metab 9:655–665

Benveniste H, Drejer J, Schousboe A, Diemer NH (1984) Elevation of the extracellular concentrations of glutamate and aspartate in rat hippocampus during transient cerebral ischemia monitored by intracerebral microdialysis. J Neurochem 43:1369–1374

Bergstedt K, Wieloch T (1991) Effect of hypothermia and postischemic NBQX administration of 14C-leucine incorporation into proteins in the rat brain following 15-minute transient 2-VO ischemia. J Cereb Blood Flow Metab 11 Suppl 2:118

Bodsch W, Takahashi K, Barbier A, Grosse Ophoff B, Hossmann K-A (1985) Cerebral protein synthesis and ischemia. Prog Brain Res 63:197–210

Bodsch W, Barbier A, Oehmichen M, Grosse Ophoff B, Hossmann K-A (1986) Recovery of monkey brain after prolonged ischemia. II. Protein synthesis and morphological alterations. J Cereb Blood Flow Metab 6:22–33

Carai AM, Orunesu G, Pau AM, Sehrbundt Viale E, Turtas S, Viale GL (1979) Untersuchungen über die Ribosomen in dem ischämischen und postischämischen Gehirn des Gerbils. Zentralbl Neurochir 40:151–158

Chavko M, Danielisova V, Marsala J (1987) Molecular mechanisms of ischemic damage of spinal cord. Gerontology 33:220–226

Choi DW, Rothman SM (1990) The role of glutamate neurotoxicity in hypoxic-ischemic neuronal death. Annu Rev Neurosci 13:171–182

Cooper HK, Zalewska T, Kawakami S, Hossmann K-A, Kleihues P (1977) The effect of ischaemia and recirculation on protein synthesis in the rat brain. J Neurochem 28:929–934

De Haro C, de Herreros AG, Ochoa S (1983) Activation of the hemestabilized translational inhibitor of reticulocyte lysates by calcium ions and phospholipid. Proc Natl Acad Sci USA 80:6843–6847

Dienel GA, Pulsinelli WA, Duffy TE (1980) Regional protein synthesis in rat brain following acute hemispheric ischemia. J Neurochem 35:1216–1226

Dirnagl U, Jacewicz M, Pulsinelli W (1990) Nimodipine posttreatment does not increase blood flow in rats with focal cortical ischemia. Stroke 21:1357–1361

Dux E, Mies G, Hossmann K-A, Siklos L (1987) Calcium in the mitochondria following brief ischemia of gerbil brain. Neurosci Lett 78:295–300

Dux E, Fastbom J, Ungerstedt U, Rudolphi K, Fredholm BB (1990) Protective effect of adenosine and a novel xanthine derivative propentofylline on the cell damage after bilateral carotid occlusion in the gerbil hippocampus. Brain Res 516:248–256

Dwyer BE, Nishimura RN, Powell CL, Mailheau SL (1987) Focal protein synthesis inhibition in a model of neonatal hypoxic-ischemic brain injury. Exp Neurol 95:277–289

Ginsberg MD (1990) Local metabolic responses to cerebral ischemia. Cerebrovasc Brain Metab Rev 2:58–93

Hagberg H, Lehmann A, Sandberg M, Nyström B, Jacobson I, Hamberger A (1985) Ischemia-induced shift of inhibitory and excitatory amino acids from intra- to extracellular compartments. J Cereb Blood Flow Metab 5:413–429

Harris RJ, Symon L, Branston NM, Bayhan M (1981) Changes in extracellular calcium activity in cerebral ischaemia. J Cereb Blood Flow Metab 1:203–209

Hossmann K-A (1992) Disturbances of cerebral protein synthesis and ischemic cell death. Prog Brain Res 96:161–177

Hossmann K-A, Sato K (1970) Recovery of neuronal function after prolonged cerebral ischemia. Science 168:375–376

Hossmann K-A, Lechtape-Grüter H, Hossmann V (1973) The role of cerebral blood flow for the recovery of the brain after prolonged ischemia. Z Neurol 204:281–299

Hossmann K-A, Sakaki S, Zimmermann V (1977) Cation activities in reversible ischemia of the cat brain. Stroke 8:77–81

Hossmann K-A, Paschen W, Csiba L (1983) Relationship between calcium accumulation and recovery of cat brain after prolonged cerebral ischemia. J Cereb Blood Flow Metab 3:346–353

Hossmann K-A, Mies G, Paschen W, Csiba L, Bodsch W, Rapin JR, Le Poncin-Lafitte M, Takahashi K (1985) Multiparametric imaging of blood flow and metabolism after middle cerebral artery occlusion in cats. J Cereb. Blood Flow Metab 5:97–107

Hossmann K-A, Schmidt-Kastner R, Grosse Ophoff B (1987) Recovery of integrative central nervous function after one hour global cerebro-circulatory arrest in normothermic cat. J Neurol Sci 77:305–320

Iijima T, Mies G, Hossmann K-A (1992) Repeated negative DC deflections in rat cortex following middle cerebral artery occlusion are abolished by MK-801: effect on infarct volume. J Cereb Blood Flow Metab 12:727–733

Ito U, Spatz M, Walker JT Jr, Klatzo I (1975) Experimental cerebral ischemia in mongolian gerbils. I. Light microscopic observations. Acta Neuropathol (Berl) 32:209–223

Jacewicz M, Kiessling M, Pulsinelli WA (1986) Selective gene expression in focal cerebral ischemia. J Cereb Blood Flow Metab 6:263–272

Jagus J, Anderson WF, Safer B (1981) The regulation of initiation of mammalian protein synthesis. Prog Nucleic Acid Res 25:127–185

Johansen FF, Diemer NH (1990) Temporal profile of interneuron and pyramidal cell protein synthesis in rat hippocampus following cerebral ischemia. Acta Neuropathol (Berl) 81:14–19

Kleihues P, Hossmann K-A (1971) Protein synthesis in the cat brain after prolonged cerebral ischemia. Brain Res 35:409–418

Kleihues P, Hossmann K-A (1973) Regional incorporation of L-(3-^{3}H)tyrosine into cat brain proteins after 1 hour of complete ischemia. Acta Neuropathol (Berl) 25:313–324

Kleihues P, Hossmann K-A, Pegg AE, Kobayashi K, Zimmermann V (1975) Resuscitation of the monkey brain after one hour complete ischemia. III. Indications of metabolic recovery. Brain Res 95:61–73

Mies G, Paschen W, Hossmann K-A (1990) Cerebral blood flow, glucose utilization, regional glucose, and ATP content during the maturation period of delayed ischemic injury in gerbil brain. J Cereb Blood Flow Metab 10:638–645

Mies G, Ishimaru S, Xie Y, Seo K, Hossmann K-A (1991) Ischemic thresholds of cerebral protein synthesis and energy state. J Cereb Blood Flow Metab 11:753–761

Miyazawa T, Widmann R, Hossmann K-A (1991) Effect of brain temperature on selective vulnerability and protein synthesis after 30 min forebrain ischemia in rats. J Cereb Blood Flow Metab 2 Suppl: 844–844

Morimoto K, Yanagihara T (1981) Cerebral ischemia in gerbils: polyribosomal function during progression and recovery. Stroke 12:105–110

Morimoto K, Brengman J, Yanagihara T (1978) Further evaluation of polypeptide synthesis in cerebral anoxia, hypoxia and ischemia. J Neurochem 31:1277–1282

Munekata K, Hossmann K-A (1987) Effect of 5-minute ischemia on regional pH and energy state of the gerbil brain: relation to selective vulnerability of the hippocampus. Stroke 18:412–417

Munekata K, Hossmann K-A, Xie Y, Seo K, Oschlies U (1987) Selective vulnerability of hippocampus: ribosomal aggregation, protein synthesis, and tissue pH. In: Powers WJ, Raichle ME (eds) Cerebrovascular diseases. Raven, New York, pp 107–117

Nakayama H, Ginsberg MD, Dietrich WD (1988) (S)-emopamil, a novel calcium channel blocker and serotonin S_2 antagonist, markedly reduces infarct size following middle cerebral artery occlusion in the rat. Neurology 38:1667–1673

Nedergaard M, Astrup J (1986) Infarct rim: effect of hyperglycemia on direct current potential and [^{14}C]2-deoxyglucose phosphorylation. J Cereb Blood Flow Metab 6:607–615

Nicholls DG (1985) A role for the mitochondrion in the protection of cells against calcium overload? Prog Brain Res 63:97–106

Nowak TS Jr, Fried RL, Lust WD, Passonneau JV (1985) Changes in brain energy metabolism and protein synthesis following transient bilateral ischemia in the gerbil. J Neurochem 44:487–494

Orunesu G, Pau A, Sehrbundt Viale E, Turtas S, Viale GL (1980) Aminoacid incorporation into polyribosomes of ischemic and reperfused gerbil brain. Acta Neurochir (Wien) 51:-247–252

Park CK, Nehls DG, Graham DI, Teasdale GM, McCulloch J (1988) Focal cerebral ischaemia in the rat: treatment with the glutamate antagonist MK-801 after induction of ischaemia. J Cereb Blood Flow Metab 8:757–762

Park CK, Nehls DG, Teasdale GM, McCulloch J (1989) Effect of the NMDA antagonist MK-801 on local cerebral blood flow in focal cerebral ischemia in the rat. J Cereb Blood Flow Metab 9:617–622

Paschen W, Mies G, Bodsch W, Yamori Y, Hossmann K-A (1985) Regional cerebral blood flow, glucose metabolism, protein synthesis, serum protein extravasation, and content of biochemical substrates in stroke-prone spontaneously hypertensive rats. Stroke 16:841–845

Paschen W, Mies G, Hossmann K-A (1992) Threshold relationship between cerebral blood flow, glucose utilization and energy metabolites during development of stroke in gerbils. Exp Neurol 117:325–333

Raley KM, Lipton P (1990) NMDA receptor activation accelerates ischemic energy depletion in the hippocampal slice and the demonstration of a threshold for ischemic damage to protein synthesis. Neurosci Lett 110:118–123

Raley-Susman KM, Lipton P (1990) In vitro ischemia and protein synthesis in the rat hippocampal slice: the role of calcium and NMDA receptor activation. Brain Res 515:27–38

Schatzman RC, Grifo JA, Merrick WC, Kuo JF (1983) Phospholipid-sensitive Ca^{2+}-dependent protein kinase phosphorylates the beta subunit of eukaryotic initiation factor 2 (eIF-2). FEBS Lett 159:167–170

Shimada N, Graf R, Rosner G, Wakayama A, George CP, Heiss W-D (1989) Ischemic flow threshold for extracellular glutamate increase in cat cortex. J Cereb Blood Flow Metab 9:603-606

Siesjö BK, Bengtsson F (1989) Calcium fluxes, calcium antagonists, and calcium-related pathology in brain ischemia, hypoglycemia, and spreading depression: a unifying hypothesis. J Cereb Blood Flow Metab 9:127–140

Thilmann R, Xie Y, Kleihues P, Kiessling M (1986) Persistent inhibition of protein synthesis precedes delayed neuronal death in postischemic gerbil hippocampus. Acta Neuropathol (Berl) 71:88–93

Tomida S, Nowak TS Jr, Vass K, Lohr JM, Klatzo I (1987) Experimental model for repetitive ischemic attacks in gerbil: the cumulative effect of repeated ischemic insults. J Cereb Blood Flow Metab 7:773–782

Vass K, Tomida S, Hossmann K-A, Nowak TS Jr, Klatzo I (1988) Microvascular disturbances and edema formation after repetitive ischemia of gerbil brain. Acta Neuropathol (Berl) 75:288–294

Wasterlain CG (1974) Brain ribosomes in intracranial hypertension. J Neurochem 23:253–259

Widmann R, Kuroiwa T, Bonnekoh P, Hossmann K-A (1991) (14C)leucine incorporation into brain proteins in gerbils after transient ischemia: relationship to selective vulnerability of hippocampus. J Neurochem 56:789–796

Widmann R, Weber C, Bonnekoh P, Schlenker M, Hossmann K-A (1992) Neuronal damage after repeated 5 min ischemia in the gerbil is preceded by prolonged impairment of protein metabolism. J Cereb Blood Flow Metab 12:425–433

Xie Y, Munekata K, Seo K, Hossmann K-A (1988) Effect of autologous clot embolism on regional protein biosynthesis of monkey brain. Stroke 19:750–757

Xie Y, Mies G, Hossmann K-A (1989a) Ischemic threshold of brain protein synthesis after unilateral carotid artery occlusion in gerbils. Stroke 20:620–626

Xie Y, Seo K, Hossmann K-A (1989b) Effect of barbiturate treatment on post-ischemic protein-biosynthesis in gerbil brain. J Neurol Sci 92:317–328

Yanagihara T (1976) Cerebral anoxia: effect on neuron-glia fractions and polysomal protein synthesis. J Neurochem 27:539–543

Yanagihara T (1978) Experimental stroke in gerbils: effect on translation and transcription. Brain Res 158:435–444

Yoshimine T, Hayakawa T, Kato A, Yamada K, Matsumoto K, Ushio Y, Matsumoto K, Mogami H (1987) Autoradiographic study of regional protein synthesis in focal cerebral ischemia with TCA wash and image subtraction techniques. J Cereb Blood Flow Metab 7:387–393

Neurotrophin Expression and Gene Regulation in a Focal Cerebral Ischemia Model in the Rat

C.Y. Hsu[1], G. An[1], J.S. Liu[1], G. Sun[2], and T.N. Lin[1]

Introdution

Signal Transduction After Injury to the Central Nervous System

Injury to the central nervous system (CNS) is accompanied by increased release of neurotransmitter into extracellular space. Neurotransmitter actions may invoke two sequential responses. The first is an immediate and short-lived event mediated by second messengers such as calcium, IP_3, diacylglycerol, cAMP, cGMP, etc. The second is a long-lasting process mediated by the so-called third messengers affecting the phenotypes of neuronal function through alteration of gene expression. Calcium appears to play an important role in linking the second-and third-messenger signaling processes. Protein kinase C (PKC), an enzyme that is activated by diacylglycerol resulting from the neurotransmitter-mediated hydrolysis of inositol phospholipids, relays information of a variety of extracellular signals across the cell membrane to regulate many intracellular process [2]. PKC is also involved in gene regulation [3]. Figure 1 shows signaling processes leading to gene regulation.

Immediate Early Genes and Gene Regulation

In eukaryote the common pathway to regulate gene expression is via changes in activity of sequence-specific transcription factors. Recent studies of gene regulation have focused on the rapid activation of transcription factor genes. The proto-oncogene products Fos and Jun are potential transcription factors which can initiate the genetic response to CNS stimulation or injury. Fos and Jun interact via the leucine-zipper to form heterodimers which bind specifically to the AP-1 binding sites or 12-*o*-tetradecanoylphorbol-13-acetate (TPA)-responsive elements of many genes to enhance their expression (Fig. 2). Proto-oncogenes including the c-*fox* and c-*jun* families are

[1] Division of Restorative Neurology and Human Neurobiology, Baylor College of Medicine, Houston, TX 77030, USA
[2] Department of Biochemistry, University of Missouri, Columbia, MO 65212, USA

Cerebral Ischemia and Basic Mechanisms
Ed. by A. Hartmann, F. Yatsu, and W. Kuschinsky

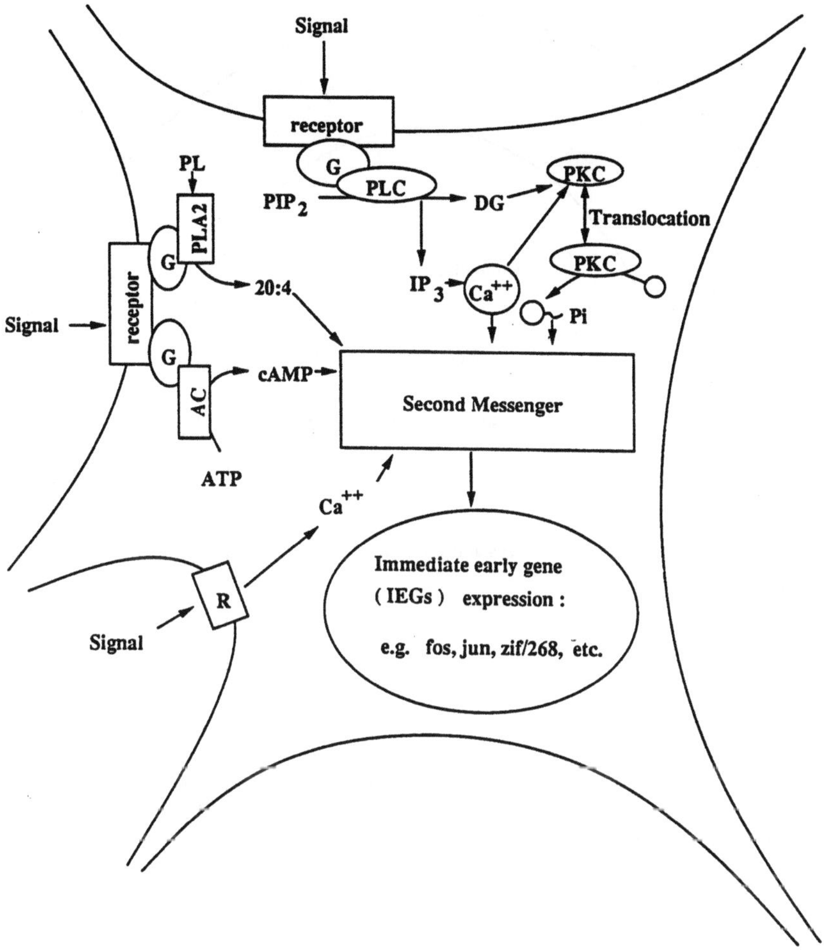

Fig. 1. Simplified scheme of signal transduction and initial steps of gene regulation

members of the immediately early genes (IEGs) that are rapidly induced by stimulation or injury [4]. In the spinal cord, c-*fos* mRNA levels rise after peripheral sensory stimulation [5] and in the brain, after pentylenetetrazol (PTZ)-induced seizures [6], kindling stimulation [7] or lesion-induced seizures [8]. Cerebral ischemia has also been shown to be a potent stimulus to alter expression of IEGs [9–14].

Neurotrophin Genes

Activation of proto-oncogene expression (*fos* and *jun* families) enhances expression of nerve growth factor (NGF) gene via AP-1 transcription factor.

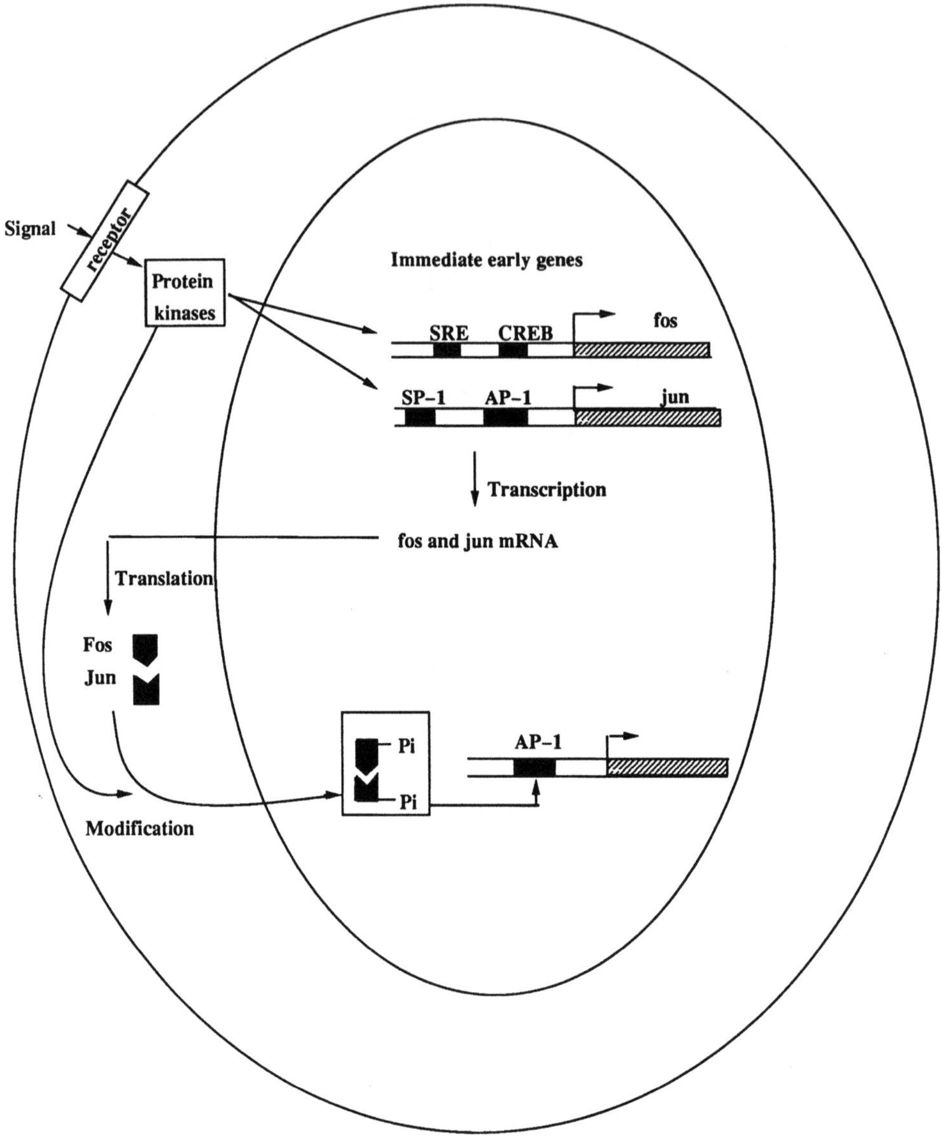

Fig. 2. Simplified scheme of regulatory elements that control transcription

NGF is the first neurotrophic factor discovered. It is now categorized in a neurotrophin superfamily which was established following recent molecular cloning of brain-derived neurotrophic factor (BDNF) [14], neurotrophin-3 (NT-3) [15], neurotrophin-4 (NT-4) [16], and others. Neurotrophins are expressed in different regions of brain with variable intensity. The neocortex and hippocampus, well known as target regions, are most prominent in the expression of these neurotrophin genes. The roles of these neurotrophins in neuronal viability and development have now been recognized [17, 18]. They have been shown to promote differentiation and survival of specific types of neurons [19]. During development, neuronal death may be caused by a limited availability of neurotrophic factors [20]. It is also likely that the expression of neurotrophins may determine the viability of diseased neurons.

In this chapter, we summarize selected aspects of the postischemic biochemical events relevant to a delayed expression of neurotrophin genes after ischemic insult. These studies in a focal cerebral ischemia–reperfusion model in the rat constitute our initial attempts to understand the biochemical mechanism of CNS adaptation to injury via signal transduction and gene regulation. The ultimate goal of these studies is to characterize the regulatory processes of neurotrophin expression which may underlie the molecular mechanism of functional recovery after ischemic brain injury. Other investigators have shown that the induction of IEGs and neurotrophin genes following both global and focal cerebral ischemia [9–15].

Materials and Methods

Stroke Model

The focal cerebral ischemia–reperfusion model in the rat has been detailed in previous publications [21–23]. Housing and anesthesia concurred with guidelines established by the institutional animal welfare committee, in accordance with the *Public Health Service Guide for the Care and Use of Laboratory Animals*, the USDA Regulations, and the AVMA Panel on Euthanasia guidelines. Surgery leading to severe focal ischemia confined to the cerebral cortex of the right middle cerebral artery (MCA) territory entailed microsurgical ligation of the MCA and both common carotid arteries (CCAs). Blood flow in all three arteries after ischemia for 30 or 90 min was restored by release of arterial occlusion. The rectal temperature was monitored and maintained at 37.0 ± 0.5°C. Physiological parameters including plasma blood glucose, arterial blood pressure, and pulse rate were monitored and maintained within normal ranges during surgery and ischemia as described previously [22, 23]. Ischemia for 30 min resulted in minimal ischemic brain injury in this model. If ischemia is maintained for 90 min, a consistent large infarct develops in the right MCA cortex (Fig. 3) [23]. After ischemic insult the rats were kept in air-ventilated incubators at 24 ± 0.5°C for different reperfusion periods (0, 30, 60, 90 min; 4 h; 1, 3 days; 1, 2, 3, 4 weeks). At the end of experiments rats were decapitated; the brains were dissected and the cortex was quickly isolated and frozen in liquid nitrogen until further processes. In some experiments in vivo microdialysis was conducted.

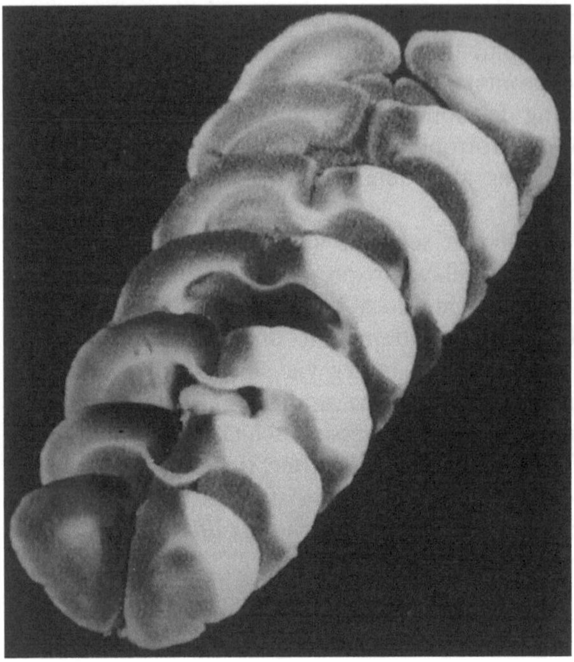

Fig. 3. Brain slices showing a cortical infarction in the right MCA cortex (from An et al., Ann Neurol 1993;33:457–464, reproduced with permission)

Microdialysis

Amino acids were recovered from the cortical extracellular space employing custom-made microdialysis probes. After having been pre-equilibrated overnight in artificial cerebrospinal fluid (CSF), a microdialysis probe was guided stereotaxically through a burr hole into the cerebral cortex of anesthetized animals according to the following coordinates: AP 0, ML ± 4.5, angled 5° laterally from the sagittal plane. This positioning centered the probe in the cortical ischemic zone to be generated by temporary occlusion of the right MCA and both CCAs. Another probe was placed in the left counterpart which sustained only mild reduction of blood flow (reduction by 8%–12%). The probes were stabilized by bracing the PE10 collar against the margins of the burr hole and then immobilized by dental acrylate. Dialysate was collected at a rate of 5 μl/min into chilled microfuge tubes at 15-min intervals starting 60 min after insertion of the probe, a period sufficient to attain stable glutamate levels in our model. Following collection of two 15-min baseline samples, ischemia was induced by temporary occlusion of the right MCA and both CCAs for 90 min. Microdialysis was continued over 15-min intervals throughout the ischemic period (90 min) and for an additional 60 min during reperfusion. Dialysate samples were maintained at

−80°C until amino acid separation and quantitation. The recovery rates for the microdialysis probes (mean ± SD) were 7.4% ± 1.5% (n = 20) for glutamate. Data shown are dialysate glutamate concentrations without correction for recovery rate.

Separation and Quantitation of Amino Acids

The concentration of amino acids in the dialysate was determined by pre-column phenylisothiocyanate derivatization followed by a Waters HPLC Pico-Tag column separation and detection at absorbance 254 nm.

Separation and Quantitation of Polyphosphoinositides

Separation of ^{32}P-labeled phospholipids using high-performance thin-layer chromatography and quantitation of each phospholipid fraction have been described in details elsewhere [24].

RNA Isolation and Northern Blot Analysis

Brains taken for mRNA isolation were frozen immediately in liquid nitrogen after removal and then stored at −80°C. Total RNA was isolated by a single-step acid guanidinium thiocyanate-phenol-chloroform extraction method [25]. For RNA Northern blot analysis, equal amounts of total RNAs (10 μg/lane) from each experimental group were electrophoresed on 1.2% agarose gel in the presence of 2.2 *M* fromaldehyde and transblotted onto Nytran membranes or nitrocellulose filters (Schleicher and Schuell, Keene, NH) [26]. Filters were prehybridized at 65°C in a solution containing 6× SSC, 10 m*M* EDTA, 5× Denhardt's solution, 0.5% SDS, and 100 μg/ml of sheared salmon sperm DNA. The ^{32}P-labeled c-*fox*, c-*jun*, *junB*, *junD*, nur77, zif/268, NGF, BDNF, and NT-3 cDNA inserts and b-actin probes were added directly to the prehybridization solution at a radioactivity of 1×10^6 cpm/ml. Radioactive probes were prepared by the random-primer labeling method suggested by the supplier (Amersham, Arlington Heights, IL). After hybridizing for 24–48 h at 68°C, the filters were washed twice in 2× SSC/0.1% SDS at room temperature for 15 min, followed by two 30-min washes at 68°C in 1× SSC/0.1% SDS and one 30-min wash at 68°C in 0.1× SSC/0.1% SDS. The filters were then dried and exposed to Kodak XAR-5 film. The autoradiographic bands in the film were measured using a laser densitometry.

Nuclear Run-on Transcription Assay

Nuclei isolation and in vitro transcription reactions were performed as described previously [27]. The brain tissues were rinsed with 5 ml ice cold PBS, then homogenized in the lysis buffer containing 10 m*M* Tris-HCl (pH 7.4), 10 m*M* NaCl, 3 m*M* $MgCl_2$, and 0.5% Nonidet P-40. The lysates were centrifuged at 500× g for 5 min at 4°C. The pellets, which contained most of the nuclei, were resuspended in a storage buffer of 50 m*M* Tris-HCl (pH 8.3), 5 m*M* $MgCl_2$, 0.1 m*M* EDTA, and 40% glycerol. For transcription reactions, 200 μl nuclei ($1-5 \times 10^7$) were mixed with 200 μl reaction mixture consisting of 10 m*M* Tris-HCl (pH 0.8), 5 m*M* $MgCl_2$, 300 m*M* KCl, 5 m*M* DTT, 1 m*M* each of ATP, CTP, and GTP, and 200 μCi [a-^{32}P]UTP(3000 Ci/mmol). The reaction mixture was then incubated at 30°C for 30 min. After the incubation, the RNAs were purified by a single-step acid guanidinium thiocyanate-phenol-chloroform extraction method [28]. 0.5 μg each of the c-*fos*, c-*jun*, *junB* and *junD* CDNA inserts, the b-actin plasmid, and pUC 19 plasmid DNA were denatured in 0.4 N NaOH at 65°C for 10 min, then neutralized and blotted onto Nytran membranes using a Slot-blotter apparatus (Bio-Rad, Richmond, CA). Hybridization was performed as described for Northern blot analysis.

In Situ Hybridization

Rats were perfused with 4% paraformaldehyde, and cryoprotected in 10% sucrose overnight; 20-μm brain slices were prepared and mounted on polylysine-coated slides. Brain slices were subjected to 0.001% proteinase K digestion at 37°C for 20 min, then immersed in 0.1 *M* TEA containing 0.25% acidic acid anhydride for 10 min. The brain slices were subsequently dehydrated in 50%, 70%, 95%, 100% ethanol for 3 min each and then hybridized with random-primed ^{32}P-labeled CDNA probes (10^7 cpm/ml) overnight at 43°C. The brain slices were then washed sequentially in 2×, 1×, 0.2×, 0.1× SSC at 43°C for 60 min each. Dehydration in 50%, 70%, 95%, 100% ethanol for 3 min each was repeated. Brain slices were then exposed to Kodak XAR-5 film.

Nuclear Extract Preparation and Gel Mobility Shift Assays

Cerebral cortex (control and 90-min ischemia followed by 30 min, 4 h, and 24 h of reperfusion) were immersed in 3 vol buffer containing 10 m*M* HEPES (pH 7.9), 1.5 m*M* $MgCl_2$, 10 m*M* KCl, 0.2 m*M* PMSF and 0.5 m*M* DDT, and homogenized immediately for 10 s, then incubated for 10 min on ice. The cell lysates were centrifuged at 3300 × *g* for 15 min. The nuclei pellets were resuspended to 3 ml in a low-salt buffer containing 20 m*M*

HEPES (pH 7.9), 25% glycerol, 1.5 m*M* $MgCl_2$, 20 m*M* KCl, 0.2 m*M* EDTA, 0.2 m*M* PMSF, and 0.5 m*M* DTT; 1 ml of high-salt buffer (same as low-salt buffer except with 1.2 *M* KCl instead of 20 m*M*) was added to the nuclei suspension and incubated on ice for 30 min. The samples were centrifuged in a Beckman JA-18 rotor at 13000 rpm (25000 × *g*) for 30 min at 4°C. The supernatant was dialyzed against a buffer containing 20 m*M* HEPES (pH 7.9), 29% glycerol, 100 m*M* KCl, 0.2 m*M* EDTA, 0.2 m*M* PMSF, and 0.5 m*M* DTT. The nuclear extracts were stored in aliqouts under −80°C.

Mobility shift DNA protein-binding assays were performed as described by Ausubel et al. [28] with some modifications. The AP-1 consensus oligonucleotide (Promega, Madison, WI) was end-labeled with ^{32}P-r-ATP by T4 polynucleotide kinase. The assays were performed in a volume of 20 μl containing 10 μg protein extract, 100000 cpm of ^{32}P-labeled AP-1 sequence, 5% glycerol, 1 m*M* $MgCl_2$, 0.5 m*M* EDTA, 0.5 m*M* DTT, 50 m*M* NaCl, 10 m*M* Tris-HCl (pH 7.5), and 4 μg salmon sperm DNA. The assays were examined by electrophoresis in a 4% polyacrylamide gel and 50 m*M* Tris, 0.4 *M* glycine, 2 m*M* EDTA. In the control and competition experiments, the reaction mixtures were preincubated with a 100-fold excess of cold AP-1 oligonucleotide, or with either anti-Fos or anti-JunB antibody prior to probe addition.

Results

In the right (ischemic) MCA cortex, ischemia induced a drastic increase in dialysate concentrations of several amino acids, including glutamate (35-fold) (Fig.4), aspartate (26-fold), gamma-amino-butyric acid (18-fold), taurine (tenfold), phosphoethanolamine (sixfold), and glycine (twofold). Upon reperfusion, the dialysate concentrations of these amino acids rapidly returned toward baseline values. More than 30 amino acids that were also assayed showed no significant changes. In the left MCA cortex (control), dialysate concentrations of these amino acids were not significantly affected (Fig. 4). Focal cerebral ischemia induced a significant decrease in labeled phosphatidyl 4,5-bisphosphates (PIP_2) content by 66%−78% in the ischemic right MCA cortex. PIP_2 level was not affected in the left MCA cortex or sham-operated control brain (Fig. 5).

Northern blot analysis showed that focal cerebral ischemia caused an intense but transient expression of c-*fos* mRNA which peaked at 30−90 min during reperfusion. The mRNA signals were no longer distinguishable from baseline after 4 h of reperfusion (Fig. 6). In situ hybridization showed that focal ischemia induced two temporal and spatial patterns of c-*fos* expression. Ischemia lasting 30 min causing no evident structural damage resulted in c-*fos* expression exclusively in the ischemic right MCA cortex (Fig. 7). Ischemic lasting 90 min leading to cortical infarction was associated with an

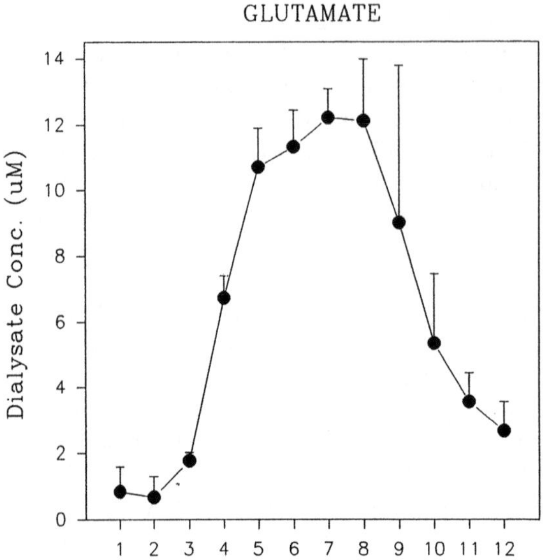

Fig. 4. Dialysate glutamate concentrations before (fractions 1–2), during (fractions 3–8), and after (fractions 9–12) focal cerebral ischemia. The period of each fraction was 15 min. *Solid circles*, samples from the right MCA cortex (ischemic side)

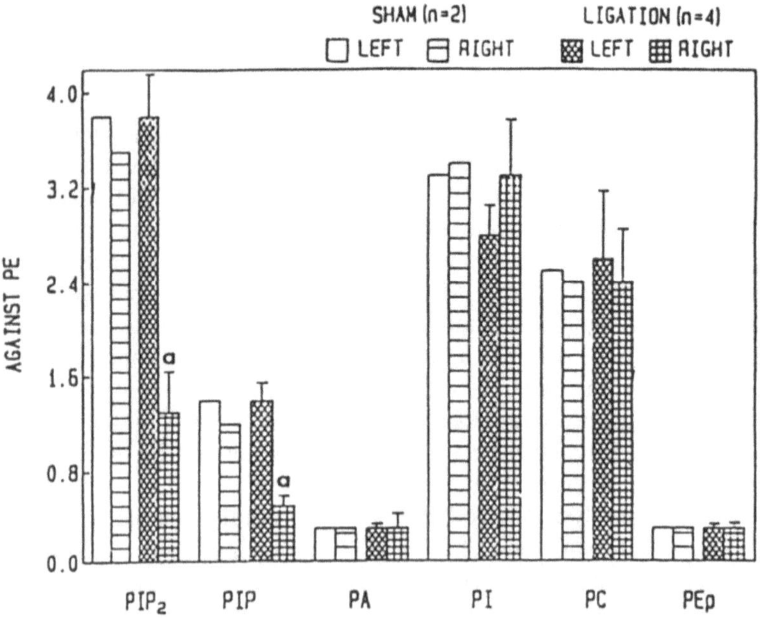

Fig. 5. Degradation of ^{32}P-labeled inositol phospholipids following ischemic insult. *RIGHT*, right MCA sample (cortex sustained severe ischemia); *LEFT*, left MCA sample (control, mild ischemia). Controls from sham-operated brains with no ischemia to MCA cortex are also shown (from Lin et al., Stroke 1991;22:495–498, reproduced with permission)

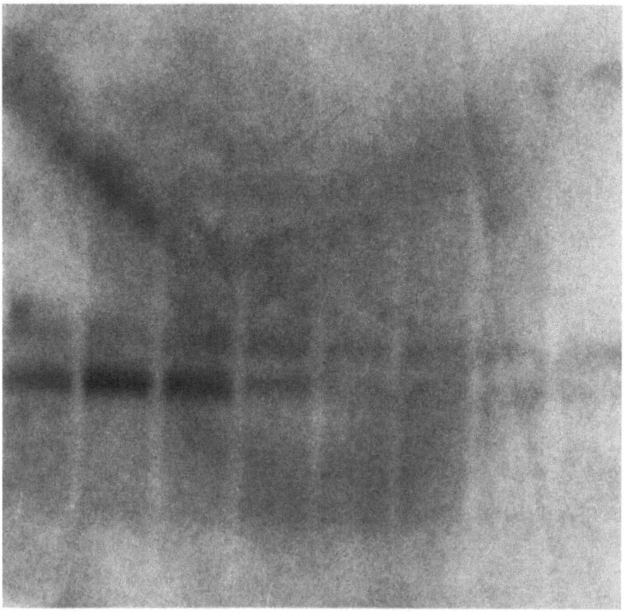

Fig. 6. Northern blot showing an increased expression of c-*fos* mRNA signals immediately after ischemia (lane 1) and during reperfusion for 30 (lane 2), 60 (lane 3), and 90 min (lane 4). No evident increase in signals was noted at 4 h (lane 5) and beyond

expression of c-*fos* mRNA in regions adjacent to the ischemic right MCA cortex especially ipsilateral hippocampus which sustained only very mild ischemia. There was no detectable increase in c-*fos* mRNA signals in the left MCA cortex which sustained only mild ischemia (Fig. 6). A similar transient increase in the mRNA level of other immediate early genes (including, c-*jun*, *junB*, *zif*/268, and *nur-77*) were also observed. On the other hand, preliminary studies also indicated that there was no detectable changes in the expression of c-*myc*, which also belongs to the IEG family. An *hsp*-like mRNA which was constitutively expressed in the rat brain was not affected by ischemic insult except that ipsilateral hippocampus also showed slight increase in mRNA signals after ischemia.

Nuclear run-on assays indicated that the increase in mRNA levels of c-*fos*, *junB* and c-*jun* are due to the increase in transcription of these genes. Mobility shift assays were used to determine the DNA binding activity of transcriptional factor AP1 (Fos-Jun heterodimer). Nuclear extract from control cortex showed a basal binding activity to the AP1 DNA sequence. Ischemia for 90 min followed by reperfusion resulted resulted in a four- to sixfold increase of AP1 binding activity that persisted for at least 24 h. Binding activity to AP1 DNA could be reduced by cold AP1, anti-Fos or anti-JunB antibodies.

A transient increase in the expression of NGF and BDNF mRNA was seen after reperfusion of 30 min or 4 h in the ischemic cortex. There was an

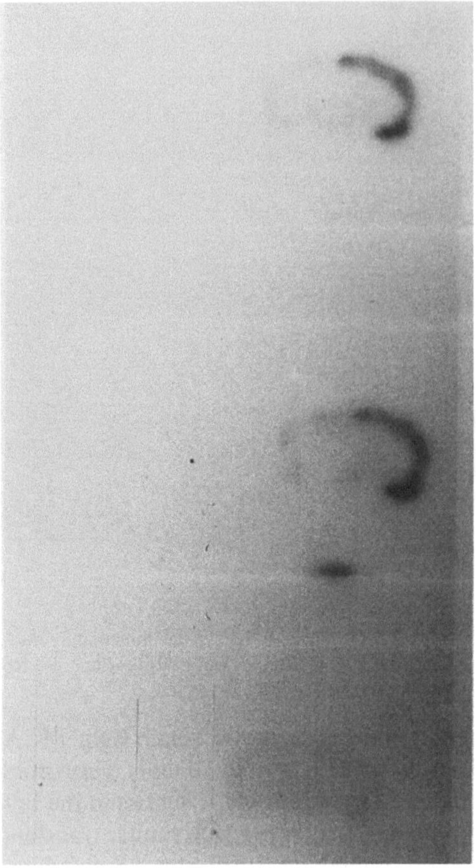

Fig. 7. In situ hybridyzation showing increased c-*fos* mRNA signals in the right MCA cortex 30 min after ischemia for 90 min. The *upper two panels* are from ischemic brains and the *bottom panel* from a sham-operated rat

obvious suppression of gene expression 1–3 days after reperfusion. A second peak of increased gene expression appeared 2–4 weeks after ischemic insult. The biphasic pattern of expression was also observed in the contralateral hemisphere, but to a lesser extent. In situ hybridyzation study using ^{32}P-NGF and BDNF probes showed increased signal in the ischemic core as well as the adjacent cortex and hippocampus.

Discussion

Focal cerebral ischemia in the right MCA cortex followed by reperfusion led to reversible or irreversible ischemic brain injury dependent on the duration of ischemic insult [23]. Microdialysis study showed a drastic increase in extracellular glutamate and other amino acid neurotransmitter in the right MCA cortex which sustained severe ischemia but not in the left MCA cortex which sustained only mild ischemia and no evident morphological injury. The increase in extracellular concentrations of several amino acid neurotransmitter suggests that receptor-mediated biochemical events including signal transduction mediated by second messengers are likely to occur during ischemia. The observation of a substantial decrease in PIP_2 in the right MCA cortex with severe ischemia but not the left MCA cortex with mild ischemia is consistent with the contention of a postischemic activation of phospholipase C cascade leading to diacylglycerol release and IP_3 formation. The former activates PKC and the latter mobilizes intracellular calcium. An obvious consequence of these signaling processes is an alteration of gene expression.

Increased expression of several IEGs after focal cerebral ischemia appears to underlie the pathophysiological significance of a postischemic activation of neurotransmitter receptors and related second messenger (IP_3, diacylglycerol, PKC calcium, etc.) events. It is interesting to note the co-induction of both c-*fos* and *junB*. Results from the nuclear fun-on experiment indicate that the increased expression of these two genes is controlled at the transcriptional stage. Thus, an increase in IEG mRNA levels is due to an increase in the transcription rate of these genes, and not due to the alteration of mRNA stability or other post-transcriptional processed. A cAMP-responsive element (CRE) and a serum-responsive element (SRE) located in the 5′-flanking region of c-*fos* gene may play a role in controlling the induction of c-*fos* expression [4]. Although there is no CRE or SRE in the promoter region of c-*jun* or *junB* genes, another regulatory element, the Sp1 site, is found in the promoter region and may regulate the expression of these two genes [29] (Fig. 2). Our preliminary results showed that nuclear extracts from ischemic cortex do contain higher binding activity to both CRE and Sp1 than control cortex (G. An and C.Y. Hsu, unpublished data), indicating that these two elements may be involved in regulating the expression of c-*fos* and c-*jun* families.

The induction of c-*fos*, *junB*, and c-*jun* expression following ischemia–reperfusion implies that the activity of transcription factor AP-1 may also increase. This was demonstrated directly by mobility shift assays. The AP-1 factor has been shown to transcriptionally activate a number of genes [3, 4]. The increased AP-1 binding activity may also induce the expression of the so-called late effector genes which are important for basic cellular functions and cell survival. Among these late genes, the neurotrophin superfamily is a possible target for AP-1 activity. An AP-1 binding site has been found at the junction of first exon and first intron of the NGF gene [30]. Recent experimental evidence suggests that the expression of NGF is regulated by Fos [31] and is induced by TPA [32]. Our finding of increased expression of these late neurotrophin genes including NGF is consistent with a probable pathophysiological significance of proto-oncogene expression [9–12] leading to increased AP1 binding activity following focal cerebral ischemia–reperfusion. Further studies are needed to characterize the mechanism of postischemic gene regulation and expression of neurotrophins [13].

Acknowledgements. We thank Dr. Y.Y. He and Dr. J.J. Xue for technical support. Experiments described in this review are supported by NIH-NINDS grants NS25545 and NS28996.

References

1. Choi DW, Rothman SM (1990) The role of glutamate neurotoxicity in hypoxic-ischemic neuronal death. Annu Rev Neurosci 13:171–182
2. Nishizuka Y (1986) Studies and perspectives of protein kinase C. Science 233:305–312
3. Nishizuka Y (1989) The family of protein kinase C for signal transduction. JAMA 262:1826–1833
4. Morgan JI, Curran T (1991) Stimulus-transcription coupling in the nervous system: involvement of the inducible proto-oncogenes fos and jun. Annu Rev Neurosci 14:421–451
5. Hunt SP, Pini A, Evan G (1987) Induction of c-fos like protein in spinal cord neurons following sensory stimulation. Nature 328:632–634
6. Morgan JI, Cohen DR, Hempstead JL, Curran T (1987) Mapping pattern of c-fos expression in the central nervous system after seizure. Science 237:192–197
7. Dragunow M, Robertson HA (1987) Kindling stimulation induces c-fos protein in granule cells of the rat dentate gyrus. Nature 329:441–442
8. White JD, Gall CM (1987) Differential regulation of neuropeptide and protooncogene MRNA content in the hippocampus following recurrent seizure. Mol Brain Res 3:21–29
9. Nowak, TS Jr, Ikeda J, Nakajma T (1990) 70-kDa heat stock protein and c-fos gene expression after transient ischemia. Stroke 21 (Suppl):III107–III111
10. Sharp FR, Lowenstein D, Simon R, Hisanaga K (1991) Heat shock protein hsp72 induction in cortical and striatal astrocytes and neurons following infarction. J Cereb Blood Flow Metab 11:621–626
11. Welsh FA, Moyer DJ, Harris VA (1992) Regional expression of heat shock protein-70 mRNA and c-fos mRNA following focal ischemia in rat brain. J Cereb Blood Flow Metab 12:204–210
12. Kindy MS, Carney JP, Dempsey RJ, Carney JM (1991) Ischemic induction of proto-oncogene expression in gerbil brain. J Mol Neurosci 2:217–228
13. Lindvall O, Ernfors P, Bengzon J, Kokaia Z, Smith ML, Siesjo BK, Persson H (1992) Differential regulation of mRNAs for nerve growth factor, brain-derived neurotrophic factor, and neurotrophin 3 in adult rat brain following cerebral ischemia and hypoglycemic coma. Proc Natl Acad Sci 89:648–653
14. Leibrock J, Lottspeich F, Hohn A, Hofer M, Hengener B, Masiakowski P, Thoenen H, Barde YA (1989) Molecular cloning and expression of brain-derived neurotrophic factor. Nature 341:149–152
15. Maisonpierre PC, Bellascio L, Squinoto S, Ip NY, Furth ME, Lindsay RM, Yancopoulos GD (1990) Neurotrophin-3 a neurotrophic factor related to NGF and BDNF. Science 247:1446–1451
16. Hallbook F, Ibanez CF, Persson H (1991) Evolutionary studies of nerve growth factor family reveals a novel member abundantly expressed in xenopus ovary. Neuron 6:845–858
17. Kromer LF (1986) Nerve growth factor treatment after brain injury prevent neuronal death. Science 235:214–216
18. Sharp FR, Gonzale MF, Hisanaqa K, Mobley WC, Saqar SM (1989) Induction of the c-fos gene product in rat fore brain following cortical lesions and NGF injections. Neurosci Lett 100:117–122

19. Hohn A, Leibrock J, Bailey K, Barde YA (1990) Identification and characterization of a novel member of the nerve growth factor/brain derived neurotrophic factor family. Nature 344:339–341
20. Barde YA (1989) Trophic factors and neuronal survival. Neuron 2:1525–1534
21. Chen ST, Hsu CY, Hogan EL, Macriq H, Balentine JD (1986) A model of focal ischemic stroke in the rat: reproducible extensive cortical infarction. Stroke 17:738–743
22. Liu TH, Beckman JS, Freeman BA, Hogan EL, Hsu CY (1989) Poly-ethylene glycol-conjugated superoxide dismutase and catalase reduce ischemic brain injury. Am J Physiol 256:H589–593
23. Yip PK, He YY, Hsu CY, Garg N, Marangos P, Hogan EL (1991) Effect of plasma glucose on infarct size in focal cerebral ischemia-reperfusion. Neurology 41:899–905
24. Lin TN, Liu TH, Xu J, Hsu CY, Sun GY (1991) Brain polyphoshoinositide metabolism during focal ischemia in rat cortex. Stroke 22:495–498
25. Chomczynski P, Sacchi N (1987) Single-step of RNA isolation by acid guanidinium thiocyanate-phenol-chloroform extraction. Anal Biochem 162:156–159
26. Thomas PS (1980) Hybridization of denatured RNA and small DNA fragments transferred to nitrocellulose. Proc Natl Acad Sci USA 77:5201–5205
27. Celano, P, Berchtold C, Casero RA Jr (1989) A simplification of the nuclear run-off transcription assay. Biotechniques 7:942–944
28. Ausubel F, Brent R, Kingston RE, Moore DD, Seidman JG, Smith JA, Struhl K (1989) Current protocols in molecular biology. John Wiley and Sons, New York, pp 12.2.1–12.2.7
29. De Groot RP, Auwerx J, Karperien M, Staels B, Kruijer W (1991) Activation of jun B by PKC and PKA signal transduction through a novel cis-acting element. Nucleic Acids Res 19:775–780
30. Zheng M, Heinrich G (1988) Structural and functional analysis of the promoter region of the nerve growth factor gene. Mol Brain Res 3:133–140
31. Hengerer B, Lindholm D, Heumann R, Ruther U, Wagner EF, Thoenen H (1990) Lesion-induced increase in nerve growth factor mRNA is mediated by c-fos. Proc Natl Acad Sci USA 87:3899–3903
32. D'Mello SR, Heinrich G (1990) Induction of nerve growth factor gene expression by 12-O-tetradecanoyl phorbol 13-acetate. J Neurochem 55:718–721

Nitroarginine Reduces Infarction After Middle Cerebral Artery Occlusion in Rats

J. CHEN[1], K.A. CONGER[2], M.-J. TAN[2], and J.S. BECKMAN[1]

Introduction

Neuronal production of nitric oxide (·NO) provides a common link between two seemingly independent mechanisms of brain injury—excitatory neurotransmitters and oxygen radicals. The connection results from the recent demonstration that glutamate stimulates neurons to produce nitric oxide (Garthwaite et al. 1988). Neurons produce nitric oxide by oxidizing arginine with a calcium-activated enzyme that is physiologically activated by the *N*-methyl-D-asparate (NMDA) subclass of receptors (Garthwaite 1991). Neuronal nitric oxide helps regulate local cerebral blood flow, contributes in synaptic plasticity (Nowak 1992), and may have a role in the normal development of brain (Gally et al. 1990). Ischemia and hypoxia initiate events that mimic the normal physiological regulation of the NMDA subclass of glutamate receptors, which may overproduce nitric oxide when tissue is reperfused. Oxidative metabolism disrupted by ischemia will produce the oxygen radical, superoxide (O_2^-), that reacts with the elevated concentrations of nitric oxide to form the destructive species peroxynitrite ($ONOO^-$) and thereby exacerbating cerebral injury.

Nitric Oxide and Superoxide-Mediated Tissue Injury

Superoxide dismutase can reduce injury from ischemia/reperfusion, inflammation, hyperoxic exposure, and other diseases, which indirectly implicates superoxide anion (O_2^-) in many pathological processes (Fridovich 1986). While superoxide can be directly toxic, its limited reactivity with many biological molecules raises questions about its toxicity per se (Baum 1984;

[1] Department of Anesthesiology The University of Alabama at Birmingham, Birmingham, AL 35294, USA
[2] Department of Neurology, The University of Alabama at Birmingham, Birmingham, AL 35294, USA

Cerebral Ischemia and Basic Mechanisms
Ed. by A. Hartmann, F. Yatsu, and W. Kuschinsky

Sawyer and Valentine 1981). Because nitric oxide contains an unpaired electron and is paramagnetic, it rapidly reacts with superoxide to form peroxynitrite anion in high yield (Blough and Zafiriou 1985; Saran 1990).

$$\cdot NO + O_2^- \Rightarrow ONOO^- \qquad \text{Eq. 1}$$

The rate constant for this reaction has recently been reported to be $6.7 \times 10^9 M^{-1} s^{-1}$ (Huie and Padmaja 1993). Superoxide dismutase stabilizes the endothelium-derived relaxing factor (EDRF) (Rubanyi and Vanhoutte 1986), suggesting that peroxynitrite is produced in vivo. In alkaline solutions, peroxynitrite anion is rather stable and can be stored in the freezer for months in concentrations as high as 1 M. However, it has a pK_a of 6.8 at 37°C and the protonated form peroxynitrous acid (ONOOH) is a reactive oxidant.

$$ONOO^- + H^+ \Leftrightarrow ONOOH \qquad \text{Eq. 2}$$

Peroxynitrous acid can form an activated intermediate, which acts as an oxidant resembling hydroxyl radical (.OH) plus nitrogen dioxide (NO_2).

$$ONOOH \Leftrightarrow O{=}N\diagdown O^{\cdot} \cdots \cdot OH + R \Rightarrow ROH\cdot + NO_2 \qquad \text{Eq. 4}$$

The activated intermediate will also rearrange to form nitrate directly. Thus, only 25%–30% of peroxynitrite reacts as hydroxyl radical and nitrogen dioxide under physiological conditions with the remainder recombining to form nitrate (Beckman et al. 1990). At physiological pH, peroxynitrite anion is sufficiently stable to diffuse over several cell diameters to critical cellular targets before becoming protonated and decomposing. Peroxynitrite can diffuse at least 1000 times further than hydroxyl radical (Hutchinson 1957). The formation of hydroxyl radical-like oxidant is only one of several cytotoxic reactions carried out by peroxynitrite. Peroxynitrite directly oxidizes sulfhydryl groups at a 1000-fold greater rate than hydrogen peroxide at pH 7.4 (Radi et al. 1991b) and initiates lipid peroxidation (Radi et al. 1991a). It is a powerful bactericidal agent (Zhu et al. 1992) produced by alveolar macrophages (Ischiropoulos et al. 1992). In addition, peroxynitrite reacts with transition metals to form a powerful nitrating agent with reactivity suggestive of the nitronium ion (NO_2^+) (Ischiropoulos et al. 1990). Thus, peroxynitrite is a reactive species with a sufficiently long lifetime to be highly toxic (Ischiropoulos et al. 1992).

Previously, we have shown that polyethylene-glycol superoxide dismutase can reduce cerebral infarct volume in an MCA-carotid occlusion model (Liu et al. 1989). Recently, inhibition of nitric oxide synthesis with nitroarginine has been shown to reduce infarct volume in mice (Nowicki et al. 1991) and to protect cultured cortical neurons from glutamate toxicity (Dawson et al. 1991b). In the present study, we tested whether inhibition of nitric oxide is cerebral protective in the same model used for our previous superoxide dismutase studies.

Materials and Methods

Nitroarginine Treatment

The hydrochloride salt of nitroarginine (Aldrich) was prepared by dissolving nitroarginine in a minimal volume of 3 N HCl and evaporating under vacuum to dryness. A 6 mg/ml solution in saline was adjusted to between pH 6 and 7 and infused through the femoral vein into the rats.

Male Long-Evans rats weighing 200–260 g were employed in the experiment. To prevent hyperglycemia, rats were starved overnight and 0.15 ml 2% lidocaine was given around all incisions. Body temperature was maintained at 36.5–37°C of a rectal temperature probe controlling a heating lamp until surgery was completed and then in a neonatal incubator maintained at 36°C until the rats recovered from anesthesia. Rats were anesthetized with 100 mg/kg ketamine (i.p.) and 6 mg/kg of Rompun (i.m.). The operations for permanent occlusion of the middle cerebral artery and common carotid arteries were performed as described before (Liu et al. 1989). *N*-Nitroarginine (5 mg/kg) or the same volume of saline was given through the right femoral venous line 80 min after occlusion of the carotid arteries. Rats were sacrificed 24 h after reperfusion under anesthesia with the brain remove intact and placed into ice-cold saline for 5 min. The brain was sliced into eight 2-mm thick pieces and then stained with 2%2,3,5-triphenyltetrazolium chloride in phosphate buffered saline (pH 7.4) for 30 min. Stained brain slices were stored with 10% formaldehyde in the dark until infarct areas were outlined and digitized. Investigators were blinded both in the administration of treatment regimes and in the subsequent infarct volume determination.

Results

A total of 28 rats were randomized to receive either *N*-nitroarginine or saline. There was no mortality, but four rats (three with nitroarginine and one with saline) were excluded as a result of technical problems with the surgery. Infarct areas in the anterior sections were significantly smaller in the nitroarginine group (n = 11) compared to the saline group (n = 13), using a two-way analysis of variance (Fig. 1). Brain infarct volume in the nitroarginine group was 28% smaller than saline. However, the differences in total infarct volume were not significantly different by the less powerful Student's t test. A sample size of 18–20 appears to be necessary to distinguish a 20% reduction of infarct volume in this model (Liu et al. 1989). Attempts to pre-administer nitroarginine at the same dosage resulted in substantial mortality during the surgical procedure or during recovery (>50%).

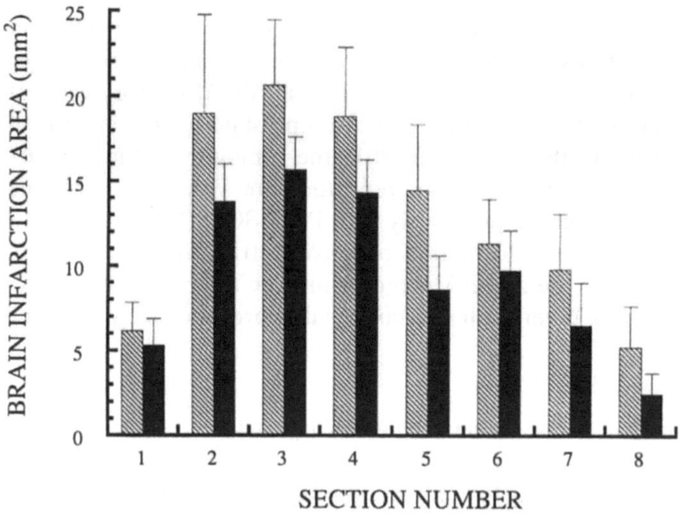

Fig. 1. Infarct area from the right cortex of rats subjected to permanent MCA occlusion with 90 min of bilateral common carotid artery occlusion. The sections are 2 mm thick and presented from anterior to posterior of the brain. *Solid columns*, received 5 mg/kg nitroarginine; *hatched columns*, received normal saline 10 min before the release of the common carotid artery occlusion

Discussion

Nitric oxide may be an important mediator of brain injury (Beckman 1991). Dawson et al. (1991a) found that both methyl- and nitro-L-arginine protect cultured cortical neurons from glutamate toxicity. Negative reports have also appeared concerning the protective effect of nitric oxide inhibition for neurons treated with glutamate (Demerlè-Pallardy et al. 1991), though some of these discrepancies might be due to subtle differences in media culture such as arginine concentration or the differing contamination with other cells types. Relatively low dosages of nitroarginine administered after initiation of ischemia were shown to reduce cortical infarct volume in a mouse middle cerebral artery occlusion model of stroke (Nowicki et al. 1991). Nitric oxide and peroxynitrite has also been implicated in acute immune-stimulated lung injury (Mulligan et al. 1991) and in myocardial ischemia (Matheis et al. 1992).

Activation of the NMDA receptor is an important determinant of selective neuronal loss from brief focal ischemic insults (Simon et al. 1984). Remarkably, NMDA antagonists also reduce infarct volume following focal cerebral ischemia (Park et al. 1988). What determines whether only a few neurons slowly die versus the rapid and total collapse of an ischemic region into infarction? Glutamate antagonists, inhibitors of nitric oxide synthesis and free radical scavengers can reduce infarct volume by approximately the same extent in middle cerebral artery occlusion models of stroke. Oh and Betz (1991) found that NMDA antagonists and the free radical scavenger dimethylthiourea both reduced infarct volumes in rat MCA territory without a synergistic interaction. Nowicki et al. (1991) found slightly greater protection with *N*-nitroarginine than with MK-801, the inhibitor of the NMDA receptor. The reduction of infarct volume reported here with *N*-nitroarginine was about the same as observed previously with PEG-SOD (Liu et al. 1989). We suggest that the potential formation of peroxynitrite from nitric oxide produced by activation of the NMDA receptor may be the common connection between these three diverse pharmacological approaches for protecting brain.

Comparative Toxicity of Nitric Oxide Versus Peroxynitrite

Nitric oxide is produced by activated macrophages and neutrophils as a microbicidal mechanism (Hibbs et al. 1987, 1988). However, the toxicity of nitric oxide per se at the low concentrations produced in vivo has been overestimated due to extrapolation from its reactivity at high concentrations. At low concentrations, similar to those which can be produced in vivo, nitric oxide can be safely administered for weeks in the gas phase to treat pulmonary hypertension (Frostell et al. 1991). Nitric oxide is toxic at high concentrations, principally due to reaction with oxygen to form more reactive species, NO_2, N_2O_3, and N_2O_4, so the toxicity of nitric oxide falls rapidly with the *square* of nitric oxide concentration. While the orange-brown color of nitrogen dioxide is formed immediately when gaseous nitric oxide contacts air, its rate is at least 10 million times slower at the maximum physiological nitric oxide concentrations (about $0.1\,\mu M$) (Shibuki 1990; Shibuki and Okada 1991). Over 1 h is required for physiological concentrations of nitric oxide to decompose to nitrite and nitrate (Zafiriou and McFarland 1980). However, the reaction of nitric oxide with superoxide greatly increases the oxidizing power of both species. The name superoxide is a misnomer from the 1930s and in fact superoxide is a weak reducing agent (Sawyer and Valentine 1981). There is now abundant evidence that superoxide is involved in cerebral injury (Beckman et al. 1988; Kontos 1985).

Superoxide Dismutase and Cerebral Injury

Liposome-entrapped superoxide dismutase reduces infarct volume in a similar MCA occlusion model in the rat (Imaizumi et al. 1990). Superoxide has been implicated in cerebral injury resulting from trauma (Kontos and Wei 1986), hypertension (Kontos 1985), cold edema (Ando et al. 1989; Chan et al. 1987), intraventricular hemorrhage (Ment et al. 1985), and ischemia (Araki et al. 1992; Cao et al. 1988). Transgenic mice overexpressing superoxide dismutase are also protected against cold edema (Chan et al. 1991). Superoxide dismutase also reduces neuronal death in culture resulting from brief glucose deprivation (Saez et al. 1987) and from hypoxia (Kinoshita et al. 1991). Cultured neurons derived from trans-genic mice overexpressing superoxide dismutase are protected from glutamate toxicity (Chan et al. 1990). Low molecular weight antioxidants also protect cultured neurons from glutamate toxicity (Miyamoto et al. 1989).

Although PEG-SOD injected into the systemic circulation can reduce cerebral infarct volume (Liu et al. 1989), it does not readily cross the blood–brain barrier (Haun et al. 1991; Yoshida et al. 1992). There may be a parallel with intestinal ischemia/reperfusion, the first organ system where superoxide was implicated in free radical injury. A major effect of superoxide dismutase in the intestine is to inhibit neutrophil adherence to ischemic endothelium (Granger et al. 1989). More recently, Kubes et al. (1991) have shown that inhibition of nitric oxide synthesis greatly enhances neutrophil adherence. Nitric oxide is rapidly destroyed by superoxide, which may contribute to the efficacy of superoxide dismutase in protecting ischemic gut. On the other hand, once inflammatory cells have adhered and become activated, they produce large quantities of nitric oxide, superoxide, and peroxynitrite as a cytotoxic mechanism (Hibbs et al. 1987). Nitric oxide production by activated inflammatory cells is a significant source of injury in immune-stimulated pulmonary injury (Mulligan et al. 1991).

Early Activation of Neutrophils and Microglia by Ischemia and Brain Injury

Activation of inflammatory cells may be a key for propagating injury in the ischemic zone and expansion of the infarct into penumbra. Growing evidence suggests that neutrophils are adhering to ischemic brain endothelium at an earlier time that previously thought. Microglia-derived macrophages are important for the clearing of cellular debris following injury and are observed in large numbers after several days following injury. However, recent histological observations indicate that microglia may be activated in the first hours of ischemia and brain trauma (Gehrmann et al. 1992). Activated microglia produce superoxide (Colton and Gilbert 1987) and also may be able to express the inducible form of the nitric oxide synthase (Zielasek et al. 1992). We have established that rat alveolar macrophages produce substantial amounts of peroxynitrite (Ischiropoulos et al. 1992). Thus, peroxynitrite might be produced by activated microglia.

The Two Faces of Nitric Oxide in Disease

In the present study, we found that pre-administration of 5 mg/kg nitroarginine before MCA–carotid occlusion caused substantial mortality. Nitroarginine at this dosage reduces cerebral blood flow by 40% without affecting the systemic blood pressure. Thus, nitroarginine may be severely disrupting collateral circulation, which is an important component of the MCA-carotid ischemia model (Chen et al. 1986).

A further complication of inhibiting nitric oxide synthesis by nitroarginine and methylarginine may be the enhanced adherence of neutrophils to ischemic brain. A similar controversy has arisen in heart where nitric oxide and nitrovasodilators improve myocardial function (Johnson et al. 1991), while inhibition of endogenous nitric oxide synthesis with nitroarginine or treatment with free radical scavengers can protect against hypoxic myocardial injury (Matheis et al. 1992). Neutrophils have a wellestablished role in propagating myocardial ischemic injury and nitrovasodilators may be reducing the adhesion of neutrophils and platelets to ischemic endothelium. A similar phenomenon may be occurring in ischemic brain, where preadministration of nitroarginine enhanced the binding of neutrophils to ischemic endothelium and enhancing injury. Delayed administration until the time of reperfusion may be protective by reducing the excessive of nitric oxide. Thus, deciphering the role of nitric oxide in cerebral ischemia will be complex, but a better understanding will help illuminate the interrelationships from several diverse mechanisms of cerebral injury.

Acknowledgments. This work was supported by NIH Grants NS24338 and HL-46407 as well as a Grant-in-Aid from the American Heart Association. J.S. Beckman is an Established Investigator of the American Heart Association.

References

Ando Y, Inoue M, Hirota M, Morino Y, Araki S (1989) Effect of a superoxide dismutase derivative on cold-induced brain edema. Brain Res 477:286–291

Araki N, Greenberg JH, Uematsu D, Sladky JT, Reivich M (1992) Effect of superoxide dismutase on intracellular calcium in stroke. J Cereb Blood Flow Metab 12:43–52

Baum RM (1984) Superoxide theory of oxygen toxicity is center of heated debate. Chem Eng News April 9:20–28

Beckman JS (1991) The double edged role of nitric oxide in brain function and superoxide-mediated pathology. J Dev Physiol 15:53–59

Beckman JS, Liu TH, Hogan EL, Lindsay SL, Freeman BA, Hsu CY (1988) Evidence for a role of oxygen radicals in cerebral ichemic injury. In: Cerebrovascular diseases. Raven, New York, pp 373–380

Beckman JS, Beckman TW, Chen J, Marshall PM, Freeman BA (1990) Apparent hydroxyl radical production from peroxynitrite: implications for endothelial injury by nitric oxide and superoxide. Proc Natl Acad Sci USA 87:1620–1624

Blough NV, Zafiriou OC (1985) Reaction of superoxide with nitric oxide to form peroxonitrite in alkaline aqueous solution. Inorg Chem 24:3504–3505

Cao W, Carney JM, Duchon A, Floyd RA, Chevion M (1988) Oxygen free radical involvement in ischemia and reperfusion injury to brain. Neurosci Lett 88:233–238

Chan PH, Longar S, Fishman RA (1987) Protective effects of liposome-entrapped superoxide dismutase on posttraumatic brain edema. Ann Neurol 21:540–547

Chan PH, Chu L, Chen SF, Carlson EJ, Epstein CJ (1990) Reduced neurotoxicity in transgenic mice overexpressing human copper-zinc-superoxide dismutase. Stroke 21 Suppl 3:III-80–III-82

Chan PH, Yang GY, Chen SF, Carlson E, Epstein CJ (1991) Cold-induced brain edema and infarction are reduced in transgenic mice overexpressing CuZn-superoxide dismutase. Ann Neurol 29:482–486

Chen ST, Hsu CY, Hogan EL, Maricq H, Balentine JD (1986) A model of focal ischemic stroke in the rat: reproducible extensive cortical infarction. Stroke 17:738–743

Colton AC, Gilbert DL (1987) Production of superoxide by a CNS macrophage, the microglia. FEBS Lett 223:284–288

Dawson TM, Bredt DS, Fotuhi M, Hwang PM, Snyder SH (1991a) Nitric oxide synthase and neuronal NADPH diaphorase are identical in brain and peripheral tissues. Proc Natl Acad Sci USA 88:7797–7801

Dawson VL, Dawson TM, London ED, Brent DS, Snyder SH (1991b) Nitric oxide mediates glutamate neurotoxicity in primary coritical cultures. Proc Natl Acad Sci USA 88:6368–6371

Demerlè-Pallardy C, Lonchampt M-O, Chabrier P-E, Braquet P (1991) Absence of implication of L-arginine/nitric oxide pathway in neuronal cell injury induced by L-glutamate or hypoxia. Biochem Biophys Res Commun 181:456–464

Fridovich I (1986) Biological effects of the superoxide radical. Arch Biochem Biophys 247:1–11

Frostell D, Fratacci MD, Wain JC, Jones R, Zapol WM (1991) Inhaled nitric oxide. A selective pulmonary. vasodilator reversing hypoxic pulmonary vasoconstriction. Circulation 83:2038–2047

Gally JA, Montague PR, Reeke GN Jr (1990) The NO hypothesis: possible effects of a short-lived, rapidly diffusible signal in the development and function of the nervous system. Proc Natl Acad Sci USA 87:3547–3551

Garthwaite J (1991) Glutamate, nitric oxide and cell-cell signalling in the nervous system. Trends Neurosci 14:75–82

Garthwaite J, Charles SL, Chess-Williams R (1988) Endothelium-derived relaxing factor release on activation of NMDA receptors suggests role as intercellular messenger in the brain. Nature 336:385–388

Gehrmann J, Bonnekoh P, Miyazawa T, Hossman KA, Kreutzberg GW (1992) Immunocytochemical study of an early microglial activation in ischemia. J Cereb Blood Flow Metab 12:257–269

Granger DN, Beniot JN, Suzuki M, Grisham MB (1989) Leukocyte adherence to venular endothelium during ischemia-reperfusion. Am J Physiol 257:G683–G688

Haun SE, Kirsch JR, Helfaer MA, Kubos KL, Traystman RJ (1991) Polyethylene glycol-conjugated superoxide dismutase fails to augment brain superoxide dismutase activity in piglets. Stroke 22:655–659

Hibbs JB Jr, Taintor RR, Vavrin Z (1987) Macrophage cytotoxicity: role of l-arginine deminiase and imino nitrogen oxidation to nitrite. Science 235:473–235

Hibbs JB Jr, Taintor RR, Vavrin Z, Rachlin EM (1988) Nitric oxide: a cytotoxic activated macrophage effector molecule. Biochem Biophys Res Commun 157:87–94

Huie RE, Padmaja S (1993) The reaction rate of nitric oxide with superoxide. Free Rad Res. Commun 18:195–199

Hutchinson F (1957) The distance that a redical formed by ionizing radiation and diffuse in a yeast cell. Radiat Res 7:473–483

Imaizumi S, Wollworth V, Fishman RA, Chan PH (1990) Liposome-entrapped superoxide dismutase reduce cerebral infarction in cerebral ischemia in rats. Stroke 21:1312–1317

Ischiropoulos H, Chen J, Tsai JHM, Martin JC, Smith CD, Beckman JS (1990) Peroxynitrite (ONOO−) reacts with superoxide dismutase to give the reactive nitonium ion (Abstr). Free Radic Biol Med 9(S1):131

Ischiropoulos H, Zhu L, Beckman JS (1992a) Peroxynitrite formation from activated rat alveolar macrophages. Arch Biochem Biophys 298:446–451

Ischiropoulos H, Zhu L, Chen J, Tsai HM, Martin JC, Smith CD, Beckman JS (1992b) Peroxynitrite-mediated tyrosine nitration catalyzed by superoxide dismutase. Arch Biochem Biophys 298:431–437

Johnson G III, Tsao P, Lefer A (1991) Cardioprotective effects of authentic nitric oxide in myocardial ischemia with reperfusion. Crit Care Med 19:244–252

Kinoshita A, Yamada K, Kohmura E, Hayakawa T (1991) Human recombinant superoxide dismutase protects primary cultured neurons against hypoxic injury. Pathobiology 59: 340–344

Kontos HA (1985) Oxygen radicals in cerebral vascular injury. Circ Res 57:508–516

Kontos HA, Wei EP (1986) Superoxide production in experimental brain injury. J Neurosurg 64:803–807

Kubes P, Suzuki M, Granger DN (1991) Nitric oxide: an endogenous modulator of leukocyte adhesion. Proc Natl Acad Sci USA 88:4651–4655

Liu TH, Beckman JS, Freeman BA, Hogan EL, Hsu CY (1989) Polyethylene glycolconjugated superoxide dismutase and catalase reduce ischemic brain injury. Am J Physiol 256:H589–H593

Matheis G, Sherman MP, Buckberg GD, Haybron DM, Young HH, Ignarro LJ (1992) Role of L-arginine-nitric oxide pathway in myocardial reoxygenation injury. Am J Physiol 262: H616–H620

Ment LA, Steward WB, Duncan CC (1985) Superoxide dismutase protects in a beagle puppy model of neonatal intraventricular hemorrhage. J Neurosurg 62:563–567

Miyamoto M, Murphy TH, Schnaar RL, Coyle JT (1989) Antioxidants protect against glutamate-induced cytotoxicity in a neuronal cell line. J Pharmacol Exp Ther 250:1132–1140

Mulligan MS, Hevel JM, Marletta MA, Ward PA (1991) Tissue injury caused by deposition of immune complexes is L-arginine dependent. Proc Natl Acad Sci USA 88:6338–6342

Nowak R (1992) Corners of the mind: the cellular basis of memory and learning. J NIH Res 4:49–55

Nowicki JP, Duval D, Poignet H, Scatton B (1991) Nitric oxide mediates neuronal death after focal cerebral ischemia in the mouse. Eur J Pharmacol 204:339–340

Oh SM, Betz AL (1991) Interaction between free radicals and excitatory amino acids in the formation of ischemic brain edema in rats. Stroke 22:915–921

Park CK, Nehls DG, Graham DI, Teasdale GM, McCulloch J (1988) Focal cerebral ischaemia in the cat: treatment with the glutamate antagonist MK-801 after induction of ischaemia. J Cereb Bloool Flow Metab 8:757–762

Radi R, Beckman JS, Bush KM, Freeman BA (1991a) Peroxynitrite-induced membrane lipid peroxidation. The cytotoxic potential of superoxide and nitric oxide. Arch Biochem Biophys 288:481–487

Radi R, Beckman JS, Bush KM, Freeman BA (1991b) Sulfhydryl oxidation by peroxynitrite: the cytotoxic potential of superoxide and nitric oxide. J Biol Chem 266:4244–4250

Rubanyi GM, Vanhoutte PM (1986) Superoxide anions and hyperoxida inactivate endothelium-derived relaxing factor. Am J Physiol 250:H822–H827

Saez JS, Kessler JA, Bennett MVL, Spray DC (1987) Superoxide dismutase protects cultured neurons against death by starvation. Proc Natl Acad Sci USA 84:3056–3059

Saran M (1990) Reaction of NO with O_2^-. Implications for the action of endothelium-derived relaxing factor (EDRF). Free Radic Res Commun 10:221–226

Sawyer DT, Valentine J (1981) How super is superoxide? Acct Chem Res 14:393–400

Shibuki K (1990) An electrochemical microprobe for detecting nitric oxide release in brain tissue. Neurosci Res 9:69–76

Shibuki K, Okada D (1991) Endogenous nitric oxide release required for long-term synaptic depression in the cerebellum. Nature 349:326–329

Simon RP, Swan JH, Griffins T, Meldrum BS (1984) Blocakde of N-methyl-D asparate receptors may protect against ischemic damage in the brain. Science 226:850–885

Yoshida K, Buton GF, McKinney JS, Young H, Ellis EF (1992) Brain and tissue distribution of polyethylene glycol-conjugated superoxide dismutase in rats. Stroke 23:865–869

Zafiriou O, McFarland M (1980) Determination of trace levels of nitric oxide in aqueous solution. Anal Chem 52:1662–1667

Zielasek J, Tausch M, Toyka KV, Hartung HP (1992) Production of nitrite by neonatal rat microglial cells/brain macrophages. Cell Immunol 141:111–120

Zhu L, Gunn C, Beckman JS (1992) Bactericidal activity of peroxynitrite. Arch Biochem Biophys 298:452–457

No Appreciable Swelling of Cultured Neurons After Oxygen Deprivation, and Cell Damage Occasionally Aggravated by Oxygen Resupply

M. Tomita[1], Y. Fukuuchi[1], and S. Terakawa[2]

Introduction

Cerebral ischemia has been extensively studied by biochemical and histological techniques. However, its dynamic aspects remain very unclear because of a lack of satisfactory means for observing the process of neuronal damage in real time. Recent advances in technology employing video microscopy and digital image processing have made it possible to visualize cells with a substantial improvement in spatiotemporal resolution. Such techniques enables us to investigate the individual responses as well as the temporal profile of cultured brain cells during deprivation of oxygen and its resupply.

Methods

As a representative of neurons, we used cultured neuronal cells (N_{18}; neuroblastoma cells provided by Prof. Yutaka Nagata, Department of Physiology, Fujita-Gakuen University School of Medicine). The cells were inoculated into a flask containing 10% Dulbecco's modified Eagle's medium plus 10% Gibco newborn calf serum (DMEM), and then incubated for 1 week at 37°C under 5% CO_2 plus oxygen. Two days before the experiment, cells were harvested from the flask, re-inoculated and cultured further on a collagen-coated thin cover glass slip placed in a small dish containing the DMEM medium. Observations of the cells were made employing a microscopic system developed by one of the authors (S.T.) as follows. The cover glass slip on which the cells had grown was taken out of the dish and placed airtightly using petroleum jelly on a slide glass having a square window so as to form a microscopic observation chambe with the cells at the bottom. The slide glass with the observation chamber was mounted on the stage of an inverted microscope and continuously superfused with

[1] Department of Neurology, School of Medicine, Keio University, 35 Shinanomachi, Shinjuku-ku, Tokyo 160, Japan
[2] Department of Cell Physiology, National Institute for Physiological Sciences, 38 Nishigounaka, Myodaiji-cho, Okazaki 444, Japan

Cerebral Ischemia and Basic Mechanisms
Ed. by A. Hartmann, F. Yatsu, and W. Kuschinsky

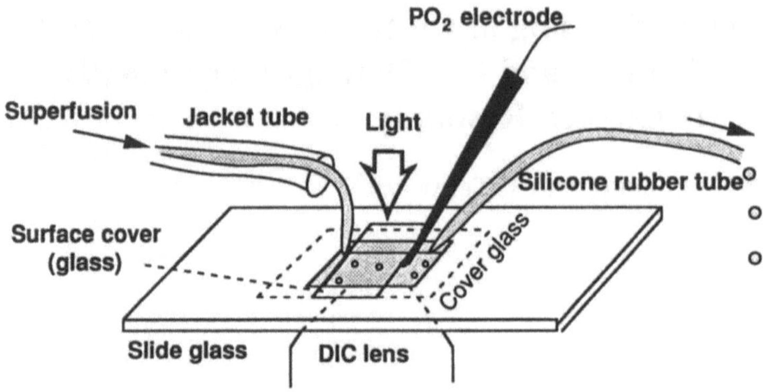

Fig. 1. Observation chamber consisting of a slide glass with a square window and a cover glass slip on which cells have been cultured. The superfusing fluid of DMEM was monitored for its oxygen tension (PO_2) with a platinum electrode. The observation chamber was covered with another cover glass slip to avoid air contamination

deoxygenated or oxygenated DMEM at 37°C pumped through a silicone rubber tube (Fig. 1). The silicone rubber tube was jacketed with a larger plastic tube, through which either oxygen gas or nitrogen gas was supplied to the running DMEM by siliconemediated diffusion. The cells were observed under an inverted Nomarski microscope equipped with a ×100 DIC objective lens and a ×2.5 insertion lens. Optical images were obtained with a charge-coupled device (CCD) camera (SSC-M350, Sony, Tokyo). The small detection area (4.7 × 6.3 mm) of this camera increased the electrical magnification of the video image. The video image was contrast enhanced digitally in real time and recorded on videotape. Following control observation of cells superfused with DMEM saturated with 100% oxygen gas, the superfusion tube was switched to nitrogen gas. After being subject to 20-min anoxia, the cells were resupplied with oxygen. The changes in cell volume, bleb formation at the cell membrane, and movements of mitochondria, granules and/or intracellular organelles were observed and videotaped continuously before, during anoxia, and after exposure to the oxygenated DMEM. Six serial experiments with anoxia and restoration of oxygen were scrutinized in detail by play-back of the videotape.

To measure the PO_2 of the superfusing fluid, a hand-made platinum electrode was employed (Fig. 2). A piece of platinum wire of 100 μm in diameter was sealed in a thin glass tube (1.0 mm in outer diameter). One end of the glass tube was beaded and the platinum wire was cut to expose its surface at the center of the bead. The sensitive surface of the electrode was covered with a thin polyethylene membrane, and saline was filled between the membrane and the platinum so as to provide electrical connection to an Ag-AgCl reference electrode. The electrical circuit is illustrated in Fig. 2. The output of the electrode was fed into a DC recorder with a parallel

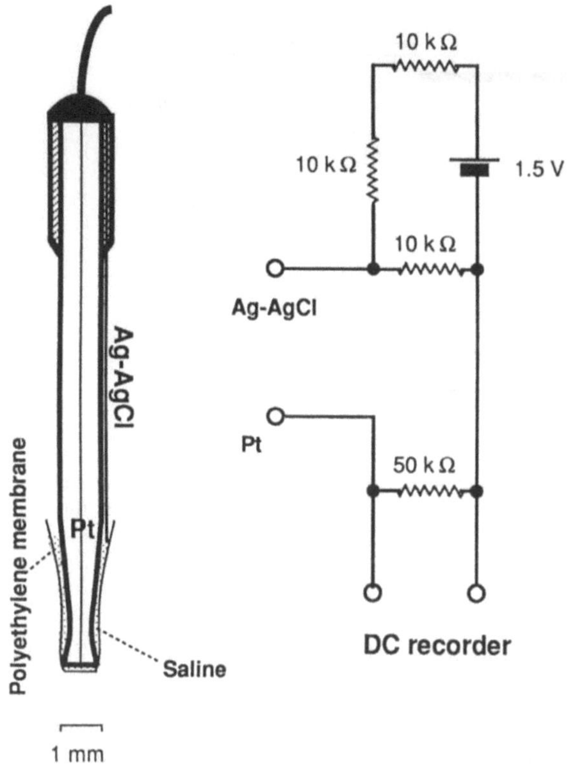

Fig. 2A,B. Platinum electrode (**A**) and electrical circuit (**B**) for measuring PO_2

resistance of 50 kΩ. The assembled electrode was then inserted into the edge of the observation chamber, as shown in Fig. 1. Continuous recording was performed in the control state, during complete anoxia, and during resupply of oxygen. Calibration of the PO_2 electrode was carried out with DMEM solutions saturated with nitrogen gas and with oxygen gas, respectively. The PO_2 was expressed as a percentage of an oxygen-saturated level (taken as 100%).

Results

Figure 3 shows an actual record of the PO_2 in the medium near the cells. With superfusion of oxygen-saturated DMEM, it rose to 100% within 5 min. When the superfusing solution was switched to deoxygenated DMEM, the value rapidly decreased to reach a level of less than 4%. When the oxygenated DMEM was again superfused, the PO_2 of the solution in the observation chamber returned to 100%. Simultaneous observation of cells in

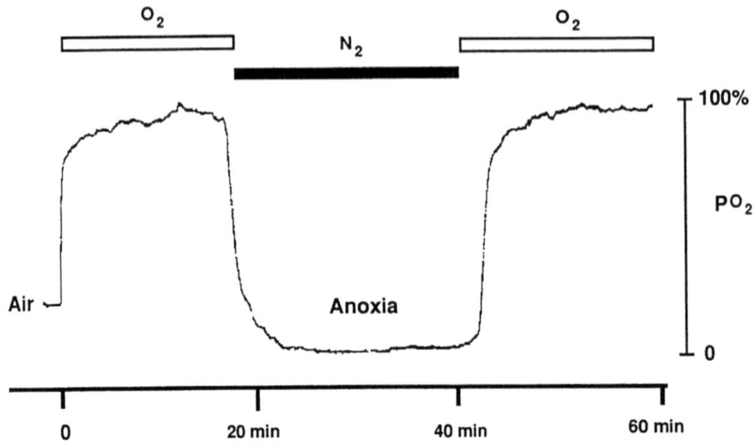

Fig. 3. Continuous recording of PO_2 of the superfusing medium in the measuring chamber where cells had been placed. Note that the PO_2 in the medium decreased to less than 4% during anoxia

six respective experiments with the above PO_2 changes yielded the following findings:

1. With anoxia, the cells began to protrude blebs or grape-like hydrops which often coalesced together. The processes were retracted in two cases. The cell surface became rough and rocky in two cases (Fig. 4). However, no appreciable cell swelling was observed.
2. Mitochondria were seen to be actively moving near the cell membrane in the control state. They usually displayed rhythmic changes of dispersion–congregation at almost two cycles per second. With anoxia, the mitochondria tended to clump together, losing their moving activities for translocation and fluctuation. No generalized swelling of the individual mitochondrial particles was noted.
3. No swelling of the nucleus was noted in all six cases.
4. The majority of granules in the control cells were quite busy, displaying up-and-down, to-and-fro, gliding, or travelling movements all the time. Such movements of the granules became less active with decrease in the PO_2 of the medium (Fig. 5). Marked decreases in the moving activities of the mitochondria and granules were observed in two cases (nos. 2 and 4, Table 1).

After 20-min anoxia, the cells were again exposed to oxygen-saturated DMEM. Following restoration of the PO_2 to 95%–100%, the cells appeared to become resuscitated in terms of their granule movements and vivid appearance in four cases. However, in the remaining two cases (nos. 2 and 4, Table 1), the resupply of oxygen led to full stoppage of all movements of

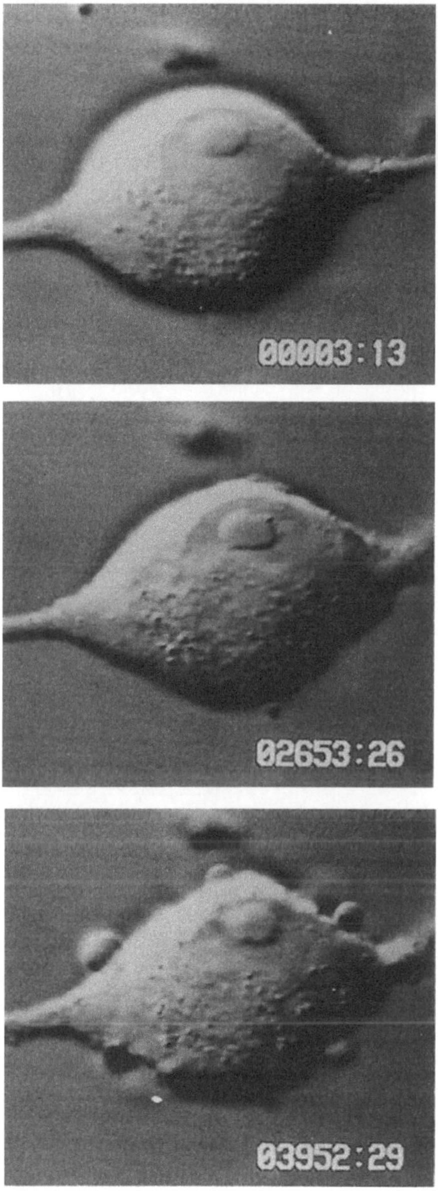

A

B

C

Fig. 4A–C. Sequential images of a cell in the control period (**A**), during anoxia (**B**), and during resupply of oxygen-saturated DMEM (**C**). In this particular case, the resupply of oxygen appeared to aggravate degeneration of the cell, causing it to exhibit a rocky appearance resembling the moon surface

the mitochondria, intracellular organelles and/or particles. The cytosol appeared to become coagulated (coagulation degeneration).

A summary of the changes in the cells induced by anoxia is given in Table 1.

Table 1. Summary of anoxic changes and recovery

No.	Cell body	Appearance N_2	Membrane Blebs C	N_2	O_2	Nucleus N_2	Mitochondria Activity C	N_2	O_2	Granules Movement C	N_2	O_2	Outcome with O_2
1	Swelling (−)	Process retracted	(−)	(+)	(−)	Blurring	(++)	(+)	(++)	(++)	(+)	(++)	Recovered
2	Swelling (−)	Rocky	(−)	(+)	(++)	–	(++)	(+)	(−)	(++)	(+)	(−)	Coagulated
3	Swelling (−)	Unchanged	(−)	(+)	(−)	Blurring	(++)	(+)	(++)	(++)	(+)	(++)	Recovered
4	Swelling (±)	Process retracted, rocky	(−)	(+)	(+)	–	(+)	(+)	(−)	(+)	(+)	(−)	Coagulated
5	Swelling (−)	Unchanged	(−)	(+)	(−)	Unchanged	(++)	(+)	(++)	(+++)	(++)	(+++)	Recovered
6	Swelling (±)	Unchanged	(−)	(±)	(−)	Unchanged	(++)	(+)	(++)	(++)	(+)	(++)	Recovered

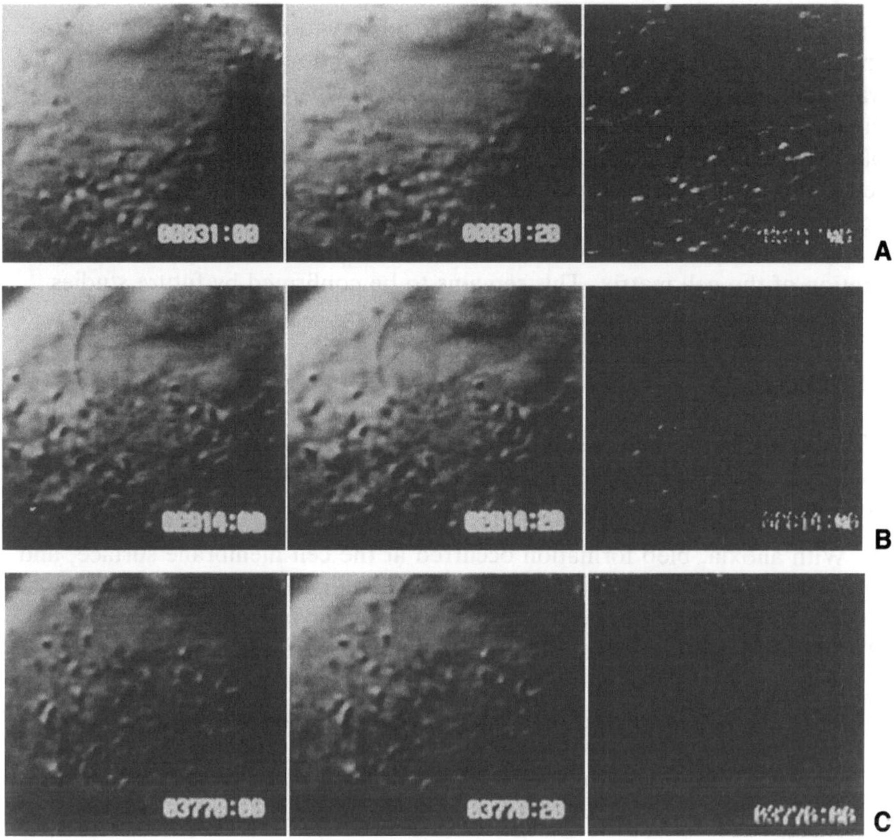

Fig. 5A–C. To examine the activity of granular movements, differential images were obtained by subtracting two images from video frames at a 20-frame (0.67 s) interval. **A** In the control state; **B** during anoxia; **C** after oxygen resupply. Note that the granules in differential images became fewer in number, indicating that the activity of the cell was greatly reduced

Discussion

Webster and Ames [2] observed that retinas deprived of oxygen and glucose for only 3 min showed generalized swelling of the mitochondria in the bipolar and ganglion cells. Extension of the deprivation caused further mitochondrial swelling and led to progressive swelling of the Golgi membranes and granular endoplasmic reticulum. These changes were reported to be almost completely reversible for up to 20 min but became irreversible by 30 min, at which time multiple discontinuities had appeared in the cells and organelles. Our findings indicated that oxygen deprivation for 20 min caused swelling of neither the cell body nor mitochondria of isolated cultured neurons. However, it did suppress the moving activities of the mitochondria and granules. The above discrepancy in observations could reflect differences in the cell population of glucose concentration. Resupply of nutrients was reported by Webster and Ames [2] to reverse the subcellular alterations. Resupply of oxygen also led to recovered activities of the organelles in our experiments in four out of six cases. However, unexpectedly, in the remaining two cases, the cell injury caused by the anoxia appeared to be aggravated by oxygen resupply. It is speculated that generation of oxygen free radicals and calcium ions could be involved in such coagulation degeneration of the cell matrix. This remains to be confirmed by future studies.

Summary

Video-microscopic observations aided by high-speed image processing were make on cultured neuroblastoma cells (N_{18}), which were subjected to anoxia for 20 min and then reoxygenated. We observed the following changes:

1. With anoxia, bleb formation occurred at the cell membrane surface, and there was clumping of mitochondria, reduction of granular movements, but on swelling of the cell body or intracellular organelles.
2. Dual effects were induced by resupplied oxygen: recovery or further deterioration of the above changes.
3. Deterioration occurred in two out of six cases, demonstrating coagulative degeneration with complete stoppage of both the mitochondria and granules. The cells exhibited a moon surface-like rocky appearance.

References

1. Tomita M, Gotoh F (1992) Cascade of cell swelling-thermodynamic potential discharge of brain cells after membrane injury. Am J Physiol 262:H603–H610
2. Webster HF, Ames A (1965) Reversible and irreversible changes in the fine structure of nervous tissue during oxygen and glucose deprivation. J Cell Biol 26:885–909

Gene Expression

Changes in Gene Expression After Brief Ischemic Insults as Potential Mediators of Induced Tolerance: Postischemic Temperature as a Variable Affecting the Stress Response

S. Suga[1,2] and T.S. Nowak, Jr.[1,3]

Introduction

Recent studies have described the induction of tolerance to otherwise damaging ischemic insults following short priming challenges that do not themselves cause severe injury. This effect has been most convincingly demonstrated following brief 2-min ischemic insults in the gerbil [7, 8] after which vulnerability of CA1 neurons to a standard 5-min insult was reduced. Increased expression of the inducible 70-kDa heat shock/stress protein hsp72 in the protected neuron population correlated reasonably well with the time course of observed tolerance during 1–4 days of recirculation after the initial insult [7]. Parallel results have been obtained in vitro demonstrating reduced sensitivity of cultured neurons to glutamate toxicity following prior heat shock sufficient to induce hsp72 expression [11, 21]. Induced ischemic tolerance in vivo following hyperthermic stress has also been described [4, 9], although hsp72 expression in CA1 neurons after such insults has not been convincingly demonstrated [12]. Taken together the above studies suggest that the induced expression of hsp72 in otherwise vulnerable neurons may contribute to induced ishemic tolerance.

In our own hands the expression of hsp72 immunoreactivity in CA1 neurons after short periods of ischemia has not been a consistent finding. Similarly, while protective effects of 2-min ischemia against later 5-min insults have been evident in many of our experiments, these observations lack sufficient reproducibility to be considered of predictive value. We have therefore begun to examine potential sources of variability in the experimental model. The effect of temperature on the impact of an ischemic insult has been well appreciated [2, 3], and some reports have indicated that temperature changes occurring during early recirculation may also have an important influence on the course of subsequent pathology [10]. In the

[1] Stroke Branch, National Institute of Neurological Disorders and Stroke, National Institutes of Health, Bethesda, Maryland, USA

[2] Department of Neurosurgery, Keio University School of Medicine, 35 Shinanomachi, Shinjuku-ku, Tokyo 160, Japan (permanent address)

[3] Department of Neurology, University of Tennessee College of Medicine, 855 Monroe Ave., Room 415, Memphis, TN 38163, USA (current address)

Cerebral Ischemia and Basic Mechanisms
Ed. by A. Hartmann, F. Yatsu, and W. Kuschinsky

present experiments we provide an initial characterization of the effect of postischemic temperature change on the stress response following brief ischemia in the gerbil. Because of potential complications in the detection of the immunoreactive hsp72 protein we have used in situ hybridization detection of an inducible mRNA of the hsp70 family to focus on transcriptional aspects of the stress response.

Materials and Methods

Ischemia was produced in adult (50–70 g) female Mongolian gerbils (Tumblebrook Farm, West Brookefield, MA, USA) by bilateral carotid artery occlusion under anesthesia with 2% halothane in 30% O_2/70% N_2O, according to a protocol approved by the Animal Care and Use Committee, NINDS. The short 2-min periods of ischemia employed in the present study were produced by manual application of tension to a loop of suture placed behind the arteries.

Rectal temperatures were maintained at 37°–38°C at the time of occlusion by means of a heating pad. Under routine conditions the anesthesia was discontinued upon recirculation. Rectal temperatures were monitored at 15-min intervals during the initial 3 h of recirculation in order to define the time course of spontaneous postischemic temperature changes. In another set of experiments the anesthesia was continued throughout this 3-h interval, with rectal temperatures adjusted by means of a heating lamp either to maintain normothermia or to match the temperature profile established for the awake animals.

For in situ hybridization studies, animals were killed by decapitation and brains were removed and frozen in methylbutane chilled to −40°C. Frozen sections (15 μm) were collected on polylysine-coated slides, air dried, and stored at −70°C. Hybridization was accomplished as previously described [15]. In brief, sections were fixed in 4% paraformaldehyde, 100 m*M* sodium phosphate (pH 7.4) for 5 min, and rinsed successively – twice in 70% ethanol and twice in 2× SSC (1 min each). Slides were transferred to 100 m*M* triethanolamine HCl (pH 8.0) and acetic anhydride was added dropwise with stirring to a final concentration of 0.25%. After 10 min the slides were transferred to 200 m*M* Tris, 100 m*M* glycine (pH 6.5) for 30 min. Sections were dehydrated through graded ethanol and finally extracted for 5 min in chloroform. Hybridization was at 37°C overnight in a mixture consisting of 50% formamide, 2× SSC, 10% dextran sulfate, 0.5 mg/ml sonicated salmon DNA, 0.25 mg/ml yeast tRNA, 1× Denhardt's solution (0.02% each Ficoll, polyvinylpyrrolidone, and bovine serum albumin) and 100 m*M* dithiothreitol. Slides were then washed in 50% formamide, 2× SSC at 42°C with four changes during 1 h, rinsed in 2× SSC at room temperature, and dehydrated through graded ethanol. Autoradiographic images were obtained on Kodak SB5 film. Regional hybridization was quantitated in some experiments by video camera-based densitometry (Image 1.32) calibrated with simultaneously exposed ^{14}C standards (Amersham Life Sciences, Arlington Heights, IL, USA).

The probe used in these studies was an oligonucleotide that selectively detects an inducible member of the rodent hsp70 family, as previously characterized [13, 15, 17]. Because an identity of the detected RNA as the sequence encoding hsp72 has not been rigorously established we will use "hsp70" as a general term to designate the hybridizing mRNA. The probe was 3′ end labeled with $\alpha[^{35}S]$thio-dATP using terminal deoxynucleotide transferase [22] in a kit from Boehringer Mannheim, Indianapolis, IN, USA.

Results

In an initial study the relationship between postischemic temperature and hsp70 mRNA expression was examined retrospectively in animals that were monitored for spontaneous changes in temperature during early recirculation. Gerbils were subjected to 2-min ischemia, with rectal temperatures (T_R) monitored just before occlusion and again at 30-min recirculation. Animals were killed at 3 h or 24 h after the insult, and hsp70 mRNA expression was graded as either positive or negative for hsp70 hybridization. Shown in Fig. 1 are autoradiograms of hybridized sections obtained from the brains of gerbils subjected to 2-min ischemia and 3-h recirculation, illustrating examples of positive and negative responses. In positive animals the induced mRNA was detected in all of the major hippocampal neuron populations, although the signal intensity in CA3 was relatively lower than that seen in dentate and CA1. In animals killed at 24-h recirculation hsp70 hybridization was restricted to CA1 neurons (data not shown). Based on the hybridization results gerbils were assigned to positive or negative response groups and the T_R data were statistically evaluated. Comparable results were obtained at the 3-h and 24-h recirculation times so the data were pooled for analysis. T_R at the onset of ischemia did not differ between the response groups and averaged 37.6 ± 0.5°C. However, as shown in Table 1, a significant hyperthermia was evident at 30-min recirculation in the hsp70-positive animals that was not seen in the hsp70-negative group.

When hsp70 hybridization was examined as a function of temperature in the above animals, it was evident that T_R above 39°C was generally associated with hsp70 mRNA expression, while gerbils with T_R of 38°–39°C were positive or negative in approximately equal proportions. This is illustrated in data from another series of animals in which quantitative estimates of hsp70 hybridization were obtained. Results are shown in Fig. 2 for dentate granule cells as a representative cell population at 3-h recirculation and for CA1 at 24 h. Although only a few animals were evaluated in these experiments, it is apparent that $T_R \geqslant 39$°C is a reliable predictor of hsp70

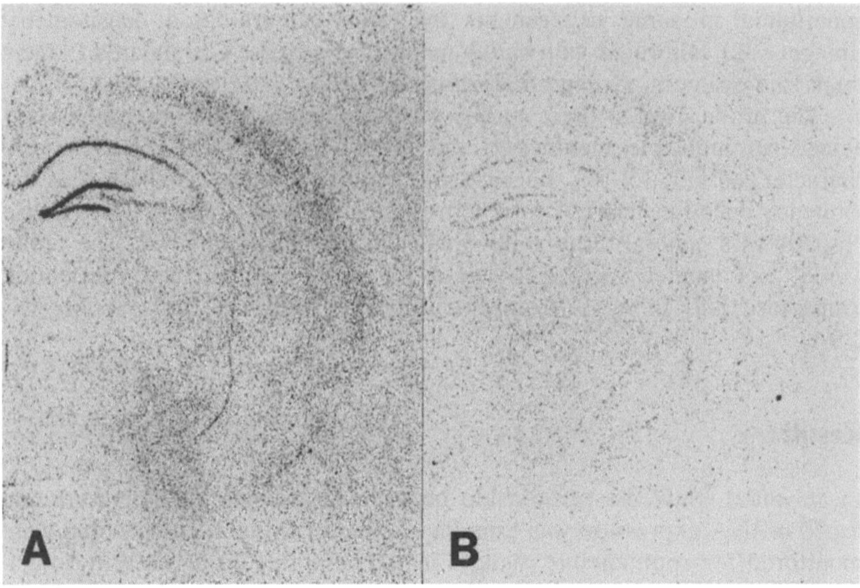

Fig. 1A,B. Variability of hsp70 mRNA expression after 2-min ischemia in the gerbil. **A** Positive hybridization at 3-h recirculation is evident throughout hippocampus, but most prominently in dentate and CA1, as well as in outer cortical layers. **B** Similarly treated animal that failed to show significant hsp70 expression

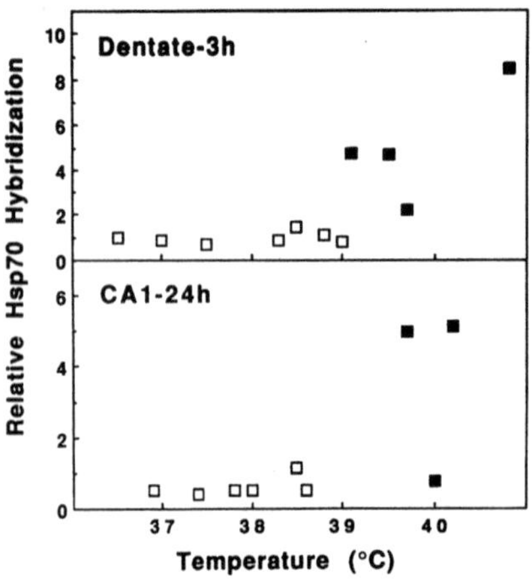

Fig. 2. Dependence of hsp70 mRNA expression on postischemic recirculation temperature. Gerbils were subjected to 2-min ischemia and immediately released from anesthesia, after which rectal temperatures were determined at 30-min recirculation. Brain sections were prepared and hybridized, and relative hsp70 mRNA expression was evaluated in the indicated cell populations by densitometry as described in the text

Table 1. Postischemic recirculation temperature as a correlate of the stress response

Response Group	Postischemic T_R (°C)
Hsp70 negative (n = 11)	37.8 ± 0.8
Hsp70 positive (n = 7)	39.7 ± 0.7

Animals were scored as positive or negative for hsp70 expression as indicated in Fig. 1 and in the text. Temperatures were obtained at 30-min recirculation and represent the mean ± SD for determinations on the indicated number of animals (n).

Table 2. Hsp70 hybridization in gerbil hippocampus after controlled-temperature recirculation

	3 h Recirculation		24 h Recirculation	
	Dentate (nCi/g)	CA1 (nCi/g)	Dentate (nCi/g)	CA1 (nCi/g)
Normothermic	49.5 ± 13.6	23.5 ± 5.6	9.9 ± 4.3	46.1 ± 15.6
Hyperthermic	216.3 ± 34.3*	172.9 ± 33.3*	18.9 ± 2.7	317.9 ± 90.1*

Tabulated values indicate mean ± SEM for data from five to six animals. Blank values for each region obtained from hybridized sections of control animals were subtracted from the initial measurements to obtain the results shown. The two hippocampi from each animal were considered independently for statistical purposes.
* $p \leq 0.05$ vs. normothermic group.

expression in hippocampus at early recirculation times. The impact of temperature is less clear with regard to the prolonged expression in CA1 at 24 h, for which hyperthermia appears to be necessary but not sufficient to determine positive expression.

The significance of temperature as a variable influencing hsp70 expression was strikingly evident in comparisons of groups of animals in which temperature was specifically regulated during early recirculation. Based on the above observations, a temperature profile was established for positively responding animals that included a rise in T_R to 39.5°C by 30 min that was sustained through approximately 90 min of recirculation, after which it gradually returned to normothermia by 3 h (not shown). For these experiments gerbils were therefore subjected to 2-min ischemia followed by recirculation under sustained anesthesia with T_R controlled either at a constant 37°C or to match the hyperthermic profile. Results shown in Table 2 demonstrate that animals subjected to hyperthermic recirculation showed significantly greater hsp70 hybridization in both dentate and CA1 at 3 h, and in CA1 at 24 h, than did animals maintained under normothermic conditions.

Discussion

The above results demonstrate that postischemic temperature increases are of critical importance in determining the expression of hsp70 mRNA following short ischemic insults, with temperatures in excess of 39°C required for a robust response. This is in good agreement with those studies that have indicated a correlation between postischemic hyperthermia and neuronal damage after ischemia [10], although there have also been reports that postischemic temperature is not a significant variable [26]. The differences in experimental detail that account for these observed variations in thermal sensitivity remain to be understood. In this regard it should be noted that ischemia of 2-min duration is a threshold insult for inducing the stress response [16, 18] (Ikeda et al., submitted), as well as for triggering postischemic hypoperfusion [6] and presumably other postischemic sequelae, and it would be expected that insults of such short duration would be extremely sensitive to any source of variability affecting the impact of the injury.

We have not yet determined whether there is any significant increase in histologically evident neuronal damage under conditions of hyperthermic vs. normothermic recirculation after 2 min ischemia. We have seen apparently complete normalization of amino acid incorporation in CA1 within 24 h following 2-min ischemia regardless of temperature changes during recirculation (unpublished observations). Since lasting protein synthesis deficits have been consistently observed in vulnerable neurons populations after ischemia [5, 24, 27], this result offers some support for the suggestion that CA1 neurons in the gerbil are not likely to be injured after 2 min ischemia even under conditions of postischemic hyperthermia. On the other hand, we have also noted that hsp70 mRNA, once induced, is persistently expressed in CA1 neurons during 24–48 h after 2-min ischemia [20] (Ikeda et al., submitted). Since the duration of hsp70 mRNA expression in a given cell population is well correlated with its relative vulnerability to ischemic injury [15], it may be suggested that 2-min ischemia under conditions that result in a vigorous stress response is already a reasonably severe insult in the gerbil. It should be noted that postischemic hyperthermia is not the primary determinant of prolonged hsp70 mRNA expression associated with neuronal vulnerability, since temperature is not an absolute predictor of hybridization intensity in CA1 at 24 h (Fig. 2).

The above observations have established an influence of postischemic temperature on transcriptional expression of hsp70 after brief ischemia. Possible effects of hyperthermia on translation of the newly synthesized mRNA remain to be evaluated. As noted above, overall amino acid incorporation activity has certainly recovered within 24-h recirculation after 2-min insults, but the early time course of recovery could be affected by temperature and other variables. The modest increases in temperature observed after ischemia are below the threshold for acute interference with translational mechanisms in brain and other tissues [14]. We have recently described other changes in gene expression that could contribute to induced tolerance, demonstrating increases in an immunoreactive protein of the Jun family of transcription factors in CA1 neurons during 6–24-h recirculation after 2-min but not after 5-min ischemia in the gerbil [19, 20] (Suga and Nowak, submitted). In contrast, protein synthesis in CA1 apparently does not recover sufficiently rapidly even after 2-min ischemia to allow very short-lived mRNAs such as c-*fos* to be translated. These observations indicate that translational activity is significant in CA1 between approximately 3 h and 6 h after 2-min insults, allowing the accumulation of Jun immunoreactivity within this interval, apparently without a striking dependence on temperature. Presumably hsp70 mRNA, when present, is also actively translated during this time.

It is of interest that in the studies just mentioned there was no striking variability in Jun expression comparable to that described here for hsp70, suggesting that regulatory pathways mediating these responses may differ in their intrinsic thermal sensitivity. The mechanisms by which mild hyperthermia interacts with other postischemic events to allow hsp70 expression remain to be elucidated. Neuronal hsp70 mRNA expression is observed in hippocampal neurons under some conditions in response to hyperthermic stress [1], although as noted earlier for translation effects the threshold for thermal induction of hsp70 is higher than temperatures reached after ischemia and are typically in excess of 40°C [13, 17]. In addition, hyperthermic induction of hsp70 includes a dominant glial and vascular component [12] not seen after short global ischemia [23, 25]. It is apparent that ischemia and postischemic temperature interact to generate the signals responsible for hsp70 expression under the present conditions.

Conclusion

These results demonstrate the importance of postischemic hyperthermia as a factor affecting the stress response after brief ischemia in the gerbil. Its significance in other ischemia models or under other experimental conditions remains to be established. At least we have identified a source of considerable variability in the impact of short priming insults that will allow the selection of animals reproducibly expressing hsp70. If the stress response is in fact an important contributor to the induction of ischemic tolerance it would be predicted that animals exhibiting postischemic hyperthermia would also show more reproducible protection against subsequent insults, and experiments are in progress to test this hypothesis. It would be expected that expression of no single protein could fully account for a phenomenon as complex as ischemic tolerance, and the task for the future will be to determine how the growing number of changes in gene expression recognized to occur after short insults interact to account for the observed effect.

Since preliminary observations would suggest that not all changes in gene expression after ischemia show a temperature dependence comparable to that of hsp70, it may in any case be possible to establish the association between a given transcriptional response and ischemic tolerance based on their respective thermal sensitivities. Such studies should be a useful step toward further molecular definition of the tolerant state.

References

1. Blake MJ, Nowak TS Jr, Holbrook NJ (1990) In vivo hyperthermia induces expression of HSP70 mRNA in brain regions controlling the neuroendocrine response to stress. Mol Brain Res 8:89–92
2. Buchan A, Pulsinelli WA (1990) Hypothermia but not the N-methyl-D-aspartate antagonist, MK-801, attenuates neuronal damage in gerbils subjected to transient global ischemia. J Neurosci 10:311–316
3. Busto R, Dietrich WD, Globus MY-T, Valses I, Scheinberg P, Ginsberg MD (1987) Small differences in intrahemispheric brain temperature critically determine the extent of ischemic neuronal injury. J Cereb Blood Flow Metab 7:729–738
4. Chopp M, Chen H, Ho K-L Dereski MO, Brown E, Hetzel FW, Welch KMA (1989) Transient hyperthermia protects against subsequent forebrain ischemic cell damage in the rat. Neurology 39:1396–1398
5. Dienel GA, Pulsinelli WA, Duffy TE (1980) Regional protein synthesis in rat brain following acute hemispheric ischemia. J Neurochem 35:1216–1226
6. Ikeda J, Hirakawa K, Nowak TS Jr, Klatzo I (1990) Changes in rCBF and edema after transient cerebral ischemia. Ischemic threshold of postischemic hypoperfusion. No To Shinkei 42:951–957
7. Kirino T, Tsujita Y, Tamura A (1991) Induced tolerance to ischemia in gerbil hippocampal neurons. J Cereb Blood Flow Metab 11:299–307
8. Kitagawa K, Matsumoto M, Tagaya M, Hata R, Ueda H, Niinobe M, Handa N, Fukunaga R, Kimura K, Mikoshiba K, Kamada T (1990) "Ischemic tolerance" phenomenon found in brain. Brain Res 528:21–24
9. Kitagawa K, Matsumoto M, Tagays M, Kuwabara K, Hata R, Handa N, Fukunaga R, Kimura K, Kamada T (1991) Hyperthermia-induced neuronal protection against ischemic injury in gerbils. J Cereb Blook Flow Metab 11:449–452
10. Kuroiwa T, Bonnekoh P, Hossmann K-A (1990) Prevention of postischemic hyperthermia prevents ischemic injury of CA1 neurons in gerbils. J Cereb Blood Flow Metab 10:550–556
11. Lowenstein DH, Chan PH, Miles MF (1991) The stress protein response in cultured neurons: characterization and evidence for a protective role in excitotoxicity. Neuron 7:1053–1060
12. Marini AM, Kozuka M, Lipsky RL, Nowak TS Jr (1990) 70-Kiladalton heat shock protein induction in cerebellar astrocytes and cerebellar granule cells in vitro: comparison with immunocytochemical localization after hyperthermia in vivo. J Neurochem 54:1509–1516
13. Miller EK, Raese JD, Morrison-Bogorad M (1991) Expression of heat shock protein 70 and heat shock cognate 70 messenger RNAs in rat cortex and cerebellum after heat shock or amphetamine treatment. J Neurochem 56:2060–2071
14. Nowak TS Jr (1988) Effects of amphetamine on protein synthesis and energy metabolism in mouse brain: role of drug-induced hyperthermia. J Neurochem 50:285–294
15. Nowak TS Jr (1991) Localization of 70 kDa stress protein mRNA induction in gerbil brain after ischemia. J Cereb Blood Flow Metab 11:432–439
16. Nowak TS Jr, Osborne OC (1991) Threshold ischemic duration for stess protein induction in gerbil brain (Abstr). Stroke 22:131

17. Nowak TS Jr, Bond U, Schlesinger MJ (1990) Heat shock RNA levels in brain and other tissues after hyperthermia and transient ischemia. J Neurochem 54:451–458
18. Nowak TS Jr, Osborne OC, Ikeda J (1992) Role of altered gene expression in development of neuronal changes after ischemia. In: Ito U, Kirino T, Kuroiwa T, Klatzo I (eds) Maturation phenomena in cerebral ischemia. Springer, Berlin, Heidelberg New York, pp 121–128
19. Nowak TS Jr, Osborne OC, Suga S (1992) Changes in gene expression after transient ischemia as potential markers for excitotoxic pathology. In: Globus MY-T, Dietrich WD (eds) The role of neurotransmitters in brain injury. Plenum, New York, pp 227–232
20. Nowak TS Jr, Osborne OC, Suga S (1993) Stress protein and proto-oncogene expression as indicators of neuronal pathophysiology after ischemia. Prog Brain Res 96:195–208
21. Rordorf G, Koroshetz WJ, Bonventre JV (1991) Heat shock protects cultured neurons from glutamate toxicity. Neuron 7:1043–1051
22. Roychoudhury R, Wu R (1980) Terminal transferase-catalyzed addition of nucleotides to the 3′ termini of DNA. Methods Enzymol 65:43–62
23. Simon RP, Cho H, Gwinn R, Lowenstein DH (1991) The temporal profile of 72-kDa heatshock protein expression following global ischemia. J Neurosci 11:881–889
24. Thilmann R, Xie Y, Kleihues P, Kiessling M (1986) Persistent inhibition of protein synthesis precedes delayed neuronal death in postischemic gerbil hippocampus. Acta Neuropathol (Berl) 71:88–93
25. Vass K, Welch WJ, Nowak TS Jr (1988) Localization of 70 kDa stress protein induction in gerbil brain after ischemia. Acta Neuropathol (Berl) 77:128–135
26. Welsh FA, Harris VA (1991) Postischemic hypothermia fails to reduce ischemic injury in gerbil hippocampus. J Cereb Blood Flow Metab 11:617–620
27. Widmann R, Kuroiwa T, Bonnekoh P, Hossmann K-A (1991) [^{14}C]Leucine incorporation into brain proteins in gerbils after transient ischemia: relationship to selective vulnerability of hippocampus. J Neurochem 56:789–796

Heat Shock Gene Expression as a Marker of Ischemic and Other Types of Injury

F.R. Sharp[1,2], H. Kinouchi[3], J. Koistinaho[1,2], P.H. Chan[3], and S.M. Sagar[1,2]

Introduction

While searching for markers of physiologically activated cells we discovered that the c-*fos* protooncogene could be induced by certain physiological stimuli including cortical stimulation, seizures, and osmotic stimulation [28]. It has since been shown that a variety of protooncogenes can be induced by physiological stimuli as well as by ischemia and axotomy [14, 15, 22]. We also found (concurrently with Dr. T. Nowak, Jr.) that though physiologcal stimuli would not induce the heat shock genes, a variety of injuries did induce these genes in brain. The following discussion will briefly summarize the biology of the heat shock gene response, the history of previous studies of heat shock protein expression in brain, and our work on the effect of seizures and ischemia on the induction of the *hsp70* gene in specific cell types and the relevance of these observations to understanding the pathogenesis of ischemic brain injury.

Heat Shock or Stress Genes

Heating or stressing cells depresses synthesis of almost all proteins. The synthesis of one class of proteins, however, is maintained or even increased during the stress. These stress or heat shock proteins comprise a large family of genes in plants, bacteria, and all animals that are highly conserved between species. These genes have been defined by their induction during periods of stress or heating and by the presence of heat shock elements in their promoters [6, 19]. Specific genes appear to be induced by certain stresses and not others. For example, the *grp78* gene is induced by hypo-

[1] Department of Neurology (V127), Department of Veterans Affairs Medical Center, 4150 Clement Street, San Francisco, CA 94121, USA
[2] Department of Neurology, University of California at San Francisco, San Francisco, CA 94143, USA
[3] Department of Neurosurgery, University of California at San Francisco, San Francisco, CA 94143, USA

Cerebral Ischemia and Basic Mechanisms
Ed. by A. Hartmann, F. Yatsu, and W. Kuschinsky

glycemia and calcium but not to a large extent by other stresses like heat. The *hsp70* gene is induced by a variety of stimuli, including heat, that cause proteins to become denatured [27].

There are three major classes of stress genes: the low molecular weight 20–30 kDa proteins; the middle molecular weight 60–70 kDa proteins; and the high molecular weight 90–110 kDa proteins [6, 19]. In each of these classes some of the genes are expressed in normal as well as stressed cells (constitutively expressed), and some genes are only induced when the cells are stressed (inducible). Though the functions of many of these genes are still unknown, they are generally involved in a variety of important processes in the cells that require protein–protein interactions.

One member of the high molecular weight class (HSP90) normally binds to and regulates availability of the progesterone receptor. Though functions of the lower molecular weight proteins are less clear, one serves as an aggregate in lens proteins in the eye, and others (the ubiquitin proteins) regulate proteolysis in normal and stressed cells [6, 19].

The 60–70 kDa class of heat shock proteins appear to serve many "chaperone" functions [8, 11, 27]. HSP60 plays a role in transporting proteins across mitochondrial membranes and perhaps regulating their folding once inside the target organelle. One of the HSP70 family appear to serve a similar role for ribosomes. HSC73, which is constitutively expressed in most cells, binds to the exposed loop of the clathrin light chain to promote uncoating of clathrin-coated vesicles. HSC73 may also regulate proteolysis in liposomes. One or several members of the HSP70 family appear to bind to virtually all newly forming proteins on ribosomes, presumably to prevent abnormal folding of peptide chains as they are being formed [2].

The highly inducible *hsp70* gene product HSP70, which is the subject of most of our work, is not present in normal rat brain. Following a variety of injuries HSP70 is rapidly induced. The function of HSP70 remains speculative. Pelham hypothesized that denatured proteins activate the gene and that HSP70 then binds to partially denatured proteins, via an ATP-dependent mechanism, to prevent further loss of tertiary structure of these proteins [27].

The current mechanism of *hsp70* activation is believed to be that heat stress induces oligomerization of proteins called heat shock factors (HSFs) [34]. HSFs bind to heat shock elements in the promoter of the *hsp70* gene and initiate transcription. *hsp70* mRNA is transported out of the nucleus to ribosomes, where HSP70 protein is synthesized [24]. HSP70 protein then binds to denatured proteins in cell cytoplasm, and HSP70 enters the nucleus to bind to and prevent denaturation of preribosomes and nuclear proteins. It is possible that HSFs also bind to the promoter regions of non-heat shock genes to downregulate these genes during periods of stress [34].

Some of the recent excitement about the study of stress genes is the observation that induction of the stress response appears to protect cells from subsequent lethal injuries. This has been shown in the retina for light injury, in cultured neurons for glutamate injury, and in whole brain for ischemic injury [1, 5, 21].

Currie and White first showed that trauma induced stress proteins in rat brain [7]. They also reported induction of the 70-kDa stress protein in one rat brain following carotid ligation. Shortly after this, Brown and colleagues reported induction of stress proteins in brain following hyperthermia and drug treatments [3]. Nowak first demonstrated the systematic induction of HSP70 following transient ischemia in the gerbil brain [23]. Similar studies from the Pulsinelli laboratory by Dienel et al. [9] and others [16] confirmed that HSP70 protein, as well as several other stress proteins, was induced following global ischemia and that this induction followed a specific time course. The study of stress proteins in ischemia was slowed because of the lack of methods to study the response at the cellular level. This was overcome when a specific monoclonal antibody to HSP70 was developed by Welch and Suhan [33].

HSP70 Protein Induction by Seizures

Using this antibody to HSP70, the Nowak and Sharp laboratories reported HSP70 induction following global ischemia [10, 12, 13, 29, 31], focal ischemia [23], and status epilepticus produced by systemic kainic acid [12, 32] and fluothyl [20]. Gonzalez et al. and Brown et al. emphasized the utility of using the heat shock response to map and study injured cells in brain [4, 12].

Generalized status epilepticus produced by kainic acid induced HSP70 protein almost exclusively in neurons in all brain regions known to sustain damage from kainate [12, 32]. HSP70 was rarely induced in glia. This is consistent with the concept that kainic acid is a neurotoxin and does not injure glial cells. We suggested, furthermore, that a subpopulation of the neurons in which HSP70 was induced must be the neurons within the CNS which have kainic acid receptors that mediate the injury. However, we also showed that kainic acid injections into the hippocampus induced HSP70 in neurons in hippocampus and some distance away within the amygdala [12]. Presumably, directly activated hippocampal neurons, which are known to project to the amygdala, led to overactivation and transynaptic injury and HSP70 induction in the amygdala neurons. These results are important for showing that both local and transynaptic activation of receptors can injure cells and induce HSP70 in those cells. Studies by Lowenstein et al. in our laboratory further showed that the duration of status epilepticus correlated with degree of HSP70 induction, with no stress protein being induced in neurons in subjects with very brief durations of sustained electrical activity, whereas increasing durations of seizures correlated with induction of HSP70 in increasing numbers of neurons [20].

Global Ischemia Induces the HSP70 Protein

Vass, Welch, and Nowak first demonstrated the cellular distribution of HSP70 induction in the gerbil global ischemia model [31]. The greatest induction occurred in CA3 and dentate gyrus, regions known to survive the ischemic injury. Little HSP70 was induced in CA1 neurons that were known to die. These authors hypothesized that HSP70 was primarily induced in neurons that survived the ischemic injury [31].

Using a different model of global ischemia, bilateral carotid occlusion combined with systemic hypotension, our laboratory obtained similar results in some circumstances and quite different results in others [13, 29]. Short durations of ischemia (5 min) induced HSP70 in either small groups of CA1 pyramidal neurons or in all CA1 pyramidal neurons combined with dentate hilar neurons. Longer durations of ischemia (10 min) led to little HSP70 immunostaining in CA1, whereas there was marked immunostaining in CA3 and dentate gyrus neurons (similar to Vass et al.). In a few animals there was no neuronal staining at all in the hippocampal formation, with HSP70 staining only of endothelial cells. This data was interpreted to be compatible with the known selective vulnerability of different populations of neurons and other cell types to ischemic injury in this model. HSP70 was induced first in the cells most sensitive to ischemia, the CA1 pyramidal neurons, and then the hilar interneurons; next in CA3 pyramidal neurons; and last in dentate gyrus granule cell neurons. As the severity of ischemia increased, HSP70 was first induced in glial cells and lastly in vascular endothelial cells, the cells most resistant to ischemic damage [13].

These results also suggest that a transcriptional or translational block occurred as the degree of ischemia worsened. Though mildly ischemic CA1 neurons synthesize HSP70, longer durations of ischemia resulted in little HSP70 expression in these neurons, showing that they were no longer capable of synthesizing HSP70 protein. In addition, severe ischemic insults resulted in the inability of any neurons in the hippocampal formation to synthesize HSP70, though the endothelial cells were still capable of HSP70 synthesis. These data suggest the following order of senstivity to ischemia and subsequent *hsp70* transcriptional and translational failure following ischemia: (1) CA1 pyramidal neurons, (2) CA4 hilar interneurons, (3) CA3 pyramidal neurons, (4) dentate granule cell neurons, (5) glia, (6) endothelial cells.

The same global ischemia model produced different patterns of HSP70 induction in cortex and striatum [29]. Though HSP70 was induced primarily in neurons in the entire distribution of the carotids, unilaterally or bilaterally, in some animals, HSP70 was often induced in discrete patches or islands in both cortex and striatum. The islands were composed of HSP70-immunostained glia on the perimeter and occasional stained neurons in the interior. Since the islands formed cylinders in three dimensions, it was proposed that HSP70 was induced around ischemic vessels in cortex and striatum. Since the factors that contribute to this unique pattern were unclear, we embarked upon focal ischemia studies.

Focal Ischemia Induces HSP70

Recent studies from our group by Kinouchi et al. [17] have shown that very brief periods of focal middle cerebral artery ischemia induced HSP70 primarily in neurons in cortex and striatum. Longer durations of temporary middle cerebral artery (MCA) ischemia resulted in infarction in the core of the ischemic areas in cortex and striatum. HSP70 was induced almost exclusively within endothelial cells in areas of infarction in both cortex and striatum. At the borders of infarction, in the penumbral areas, HSP70 was inconsistently induced in neurons. In most though not all cases HSP70 was induced in glial cells on the perimeters of the penumbral regions.

Our data show that, following ischemia of sufficient duration to produce focal infarction, transcription and/or translation of the *hsp70* gene is disrupted in nearly all neurons and glia in the region that will infarct. However, transcription and translation of the *hsp70* gene can persist in endothelial cells of blood vessels in brain regions that are clearly infarcted by histological criteria. Unfortunately, our studies do not answer the question of whether the endothelial cells in which the HSP70 protein is induced in the area of infarction will die or not.

hsp70 In Situ Hybridization Studies in Focal Ischemia

To determine whether disruption of *hsp70* transcription or translation occurs in neurons in areas of infarction, *hsp70* mRNA in situ hybridization studies have been carried out [18]. The results show that brief durations of MCA ischemia that induce HSP70 primarily in neurons also induces *hsp70* mRNA primarily in neurons. Long durations of MCA ischemia that produce infarction do not induce *hsp70* mRNA or HSP70 protein in neurons in areas of infarction; whereas *hsp70* mRNA and HSP70 protein are induced in endothelial cells in areas of infarction. At the borders of the infarction *hsp70* mRNA and HSP70 protein appear to be induced in both neurons and glial cells. Therefore, in our focal ischemia studies both transcription and translation of the *hsp70* gene are disrupted in neurons and glia in regions of infarction. This constrasts to results in gerbils where global ischemia induces *hsp70* mRNA in CA1 pyramidal neurons, whereas there is little or no HSP70 protein induced [25]. The differences could be due to fundamental differences in global ischemic injury versus focal infarction, differences in species, and possibly to the degree of ischemic injury. That is, mild degrees of ischemic injury might induce transcription and translation of the *hsp70* gene, moderate ischemic injury might induce transcription without translation, and severe ischemic injury would prevent both transcription and translation of the *hsp70* genes. It is possible that blockade of translation but not transcription of the *hsp70* gene is a marker of delayed neuronal death that occurs in CA1 neurons in the global ischemia model.

MK801, Ketamine, and Phencyclidine Induce HSP70

Many laboratories have shown that the noncompetitive *N*-methyl-D-aspartate (NMDA) antagonists MK801, ketamine, and phencyclidine (PCP) can decrease the volume of infarction following focal ischemic insults. All of these drugs produce side effects in animals and can produce psychosis in humans. Olney and coworkers first reported that all NMDA antagonists produced abnormal vacuoles in posterior cingulate and retrosplenial cortical neurons in rats [26].

Our laboratory has now shown that MK801, ketamine, and PCP induce HSP70 protein and *hsp70* mRNA in the vacuolated neurons in posterior cingulate and retrosplenial cortical neurons of rats [30] (see Fig. 1). The induction of HSP70 is maximal at 24–48 h following MK801, ketamine, and PCP and can persist for 1–2 weeks. Moreover, these drugs do not induce neuronal injury in newborn and juvenile rats, but only in teenage (30 day) and adult rats. Lastly, we have shown that antipsychotic medications, including haloperidol and rimcazole, can prevent the brain injury produced by these compounds (MK801, PCP, and ketamine).

These data are important for showing that certain toxic drugs used to treat stroke in experimental animals can themselves induce heat shock protein. Moreover, the neuronal injury produced by MK801, ketamine, and PCP can be prevented by antipsychotic medications. These results are of general importance for showing that heat shock protein induction can be a sensitive indicator of injury produced by toxins, seizures, and stroke. Lastly, since it is believed that most of the neurons injured by MK801, ketamine, and PCP survive, it is likely that HSP70 can demonstrate injured cells in brain and other regions that can survive. This is of great potential importance for therapies aimed at preventing all types of neural injury.

Lastly, it is proposed that brain injury produced by PCP, ketamine, and MK801 is mediated by sigma receptors. PCP induces HSP70 in neurons in posterior cingulate, retrosplenial cortex, neocortex layers 2, 3, 5, and 6, piriform cortex, amygdala, and possibly hippocampus. PCP, MK801, and ketamine are all noncompetitive blockers of the NMDA receptor, and appear to block calcium entry through the channel of the NMDA receptor. Since haloperidol and rimcazole, drugs that block the sigma receptor but have no effect on the NMDA receptor, prevent the injury produced by PCP, MK801, and ketamine, this has led us to the following hypothesis. PCP, MK801, and ketamine block NMDA receptors on GABAergic post-

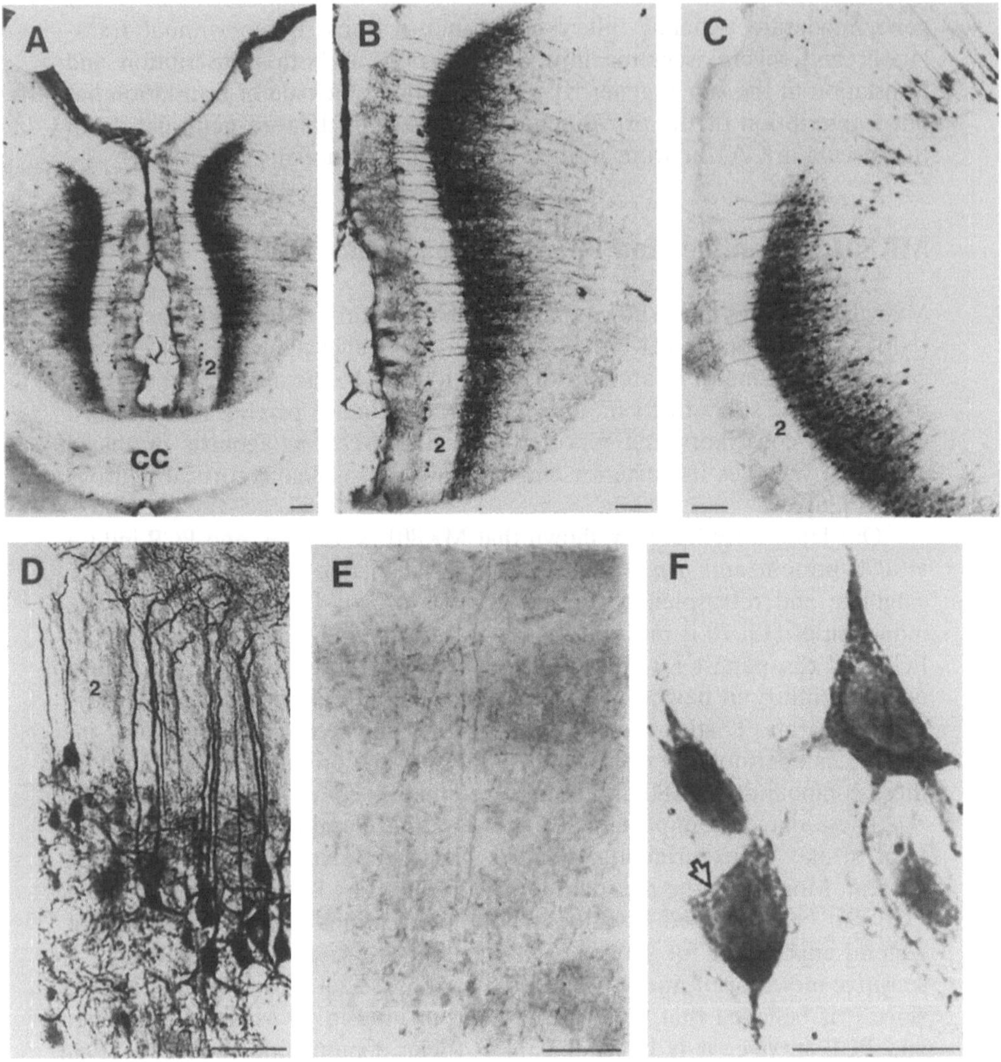

Fig. 1A–F. MK801 (1 mg/kg) given intraperitoneally induces the HSP70 heat shock protein in posterior cingulate (**A,B,D**) and retrosplenial (**C,F**) cortical neurons. There are a few HSP70 immunoreactive neurons in layer 2 (**A,B**), most of the neurons are in layer 3 (**A–D,F**), and there are many neurons in layer 5 of agranular, retrosplenial cortex (**C**). Note the vacuoles in HSP70 immunostained neurons (*arrow*, **F**). Prior administration of haloperidol (1 mg/kg) completely blocked the formation of vacuoles and induction of HSP70 protein (**E**)

synaptic neurons. This decreases inhibitory activity which is presynaptic onto incoming sigma afferents onto the original glutaminergic neuron. This results in increased activity of the sigma afferents and produces an excitotoxic injury due to sigma receptor overactivation. Such a hypothesis can be tested, and is extremely important for ischemia research since it suggests that sigma receptor overactivation can also cause injury and could mediate ischemia-induced injury, just as the NMDA and kainic acid receptors appear to do.

Acknowledgments. We wish to thank Matt Morton, Michael Butman, Pat Jasper, and Katy Hicks for excellent technical assistance on many portions of these studies. This work was supported by the Merit Review program of the VA Research Service, and by NIH grants NS 28167 to FRS, and NS 14543 to FRS, SMS and PHC.

References

1. Barbe MF, Tytell M, Gower DJ, Welch WJ (1988) Hyperthermia protects against light damage in the rat retina. Science 241:1817–1820
2. Beckmann RP, Mizzen LA, Welch WJ (1990) Interaction of HSP70 with newly synthesized proteins: implications for protein folding and assembly. Science 248:850–854
3. Brown IR (1983) Hyperthermia induces the synthesis of a heat shock protein by polysomes isolated from the fetal and neonatal mammalian brain. J Neurochem 40:1490–1493
4. Brown IR, Rush S, Ivy GO (1989) Induction of a heat shock gene at the site of tissue injury in the rat brain. Neuron 2:1559–1564
5. Chopp M, Chen H, Ho KL, Dereski MO, Brown E, Hetzel FW, Welch KMA (1989) Transient hyperthermia protects against subsequent forebrain ischemic cell damage in the rat. Neurology 39:1396–1398
6. Craig EA (1989) Essential roles of 70 kDa heat inducible proteins. BioEssays 11:48–52
7. Currie RW, White FP (1981) Trauma-induced protein in rat tissues: a physiological role for a "heat shock" protein? Science 214:72–73
8. Deshaies RV, Koch BD, Schekman R (1989) The role of stress proteins in membrane biogenesis. TIBS 13:384–388
9. Dienel GA, Kiessling M, Jacewicz M, Pulsinelli WA (1986) Synthesis of heat shock proteins in rat brain cortex after transient ischemia. J Cereb Blood Flow Metab 6:505–510
10. Ferriero DM, Soberano HQ, Simon RP, Sharp FR (1990) Hypoxia-ischemia induces heat shock protein-like (HSP72) immunoreactivity in neonatal rat brain. Dev Brain Res 53:145–150
11. Gething MJ, Sambrook J (1992) Protein folding in the cell. Nature 355:33–45
12. Gonzalez MF, Shiraishi K, Hisanga K, Sagar SM, Mandabach M, Sharp FR (1989) Heat shock proteins as markers of neural injury. Mol Brain Res 6:93–100
13. Gonzalez MF, Lowenstein DH, Hisanaga K, Simon RP, Sagar SM, Sharp FR (1991) Induction of heat shock protein 72-like immunoreactivity in the hippocampal formation following global ischemia. Brain Res Bull 26:241–250
14. Herdegen T, Kummer W, Fiallos CE, Leah J, Bravo R (1991) Expression of c-Jun, Jun B, and Jun D proteins in rat nervous system following transection of vagus nerve and cervical sympathetic trunk. Neurosci 45:413–422
15. Hunt SP, Pini A, Evan G (1987) Induction of c-fos like protein in spinal cord neurons following sensory stimulation. Nature 328:632–634
16. Kiessling M, Dienel GA, Jacewicz M, Pulsinelli WA (1986) Protein synthesis in post-ischemic rat brain: a two-dimensional electrophoretic analysis. J Cereb Blood Flow Metab 6:642–649
17. Kinouchi H, Sharp FR, Hill MP, Koistinaho J, Sagar SM, Chan PH (1993) Induction of 70-kDa heat shock protein and hsp70 mRNA following transient focal cerebral ischemia in the rat. J Cereb Blood Flow Metab 13:105–115

18. Kinouchi H, Sharp FR, Hill MP, Koistanaho J, Sagar SM, Chan PH (1992) Induction of hsp70 mRNA in the penumbra following transient and permanent focal cerebral ischemia in the rat. Brain Res, in press
19. Lindquist S (1986) The heat shock response. Annu Rev Biochem 55:1151–1191
20. Lowenstein DH, Simon RP, Sharp FR (1990) The pattern of 72-kDa heat shock protein-like immunoreactivity in the rat brain following fluothyl-induced status epiletpticus. Brain Res 531:173–182
21. Lowenstein DH, Chan PH, Miles MF (1991) The stress protein response in cultured neurons: characterization and evidence for a protective role in excitotoxicity. Neuron 7:1053–1060
22. Morgan JI, Curran T (1989) Stimulus-transcription coupling in neurons: role of cellular immediate-early genes. Trends Neurosci 12:459–462
23. Nowak TS Jr (1985) Synthesis of a stress protein following transient ischemia in the gerbil. J Neurochem 45:1635–1641
24. Nowak TS (1990) Protein synthesis and the heat shock/stress response after ischemia. Cereb Brain Metab Rev 2:345–366
25. Nowak TS Jr, Bond U, Schlesinger MJ (1990) Heat shock RNA levels in brain and other tissues after hyperthermia and transient ischemia. J Neurochem 54:451–458
26. Olney JW, Labruyere J, Price MT (1989) Pathological changes induced in cerebrocortical neurons by phencyclidine and related drugs. Science 244:1360–1362
27. Pelham HRB (1986) Speculations on the functions of the major heat shock and glucose-regulated proteins. Cell 46:959–961
28. Sagar SM, Sharp FR, Curran T (1988) Expression of c-fos protein in brain: metabolic mapping at the cellular level. Science 240:1328–1331
29. Sharp FR, Lowenstein D, Simon RP, Hisanaga K (1991) Heat shock protein hsp70 induction in cortical and striatal astrocytes and neurons following infarction. J Cereb Blood Flow Metab 11:621–627
30. Sharp FR, Jasper P, Hall J, Noble L, Sagar SM (1991) MK801 and ketamine induce heat shock protein HSP72 in injured neurons in posterior cingulate and retrosplenial cortex. Ann Neurol 30:801–809
31. Vass K, Welch WJ, Nowak TS Jr (1988) Localization of 70kDa stress protein induction in gerbil brain after ischemia. Acta Neuropathol 77:128–135
32. Vass K, Berger ML, Nowak TS Jr, Lassmann H (1989) Induction of stress protein HSP70 in nerve cells after status epilepticus in the rat. Neurosci Lett 100:259–264
33. Welch WJ, Suhan JP (1986) Cellular and biochemical events in mammalian cells during and after recovery from physiological stress. J Cell Biol 103:2035–2052
34. Westwood JT, Clos J, Wu C (1991) Stress-induced oligomerization and chromosomal relocalization of heat-shock factor. Nature 353:822–827

Gene Expression and Functional Changes After Acute Ischemia: Age-Related Differences in Outcome and Mechanisms

J.M. Carney, M.S. Kindy, C.D. Smith, K. Wood, T. Tatsuno, J.F. Wu, W.R. Landrum, and R.A. Floyd

Introduction

Ischemia reperfusion injury to brain is a common process in a number of clinical conditions. Among these conditions are stroke, concussion, and subarachnoid hemorrhage. The ultimate process of postischemic injury remains to be described. Several possible mechanisms have been proposed. Early in the process of reperfusion injury it has been demonstrated tat there is a significant increase in the amount of free radicals generated in the brain [2, 3, 11]. One of the consequences of radical damage is lipid and protein oxidation [4, 12, 14]. Recent studies have demonstrated that this ligand-gated channel can be oxidatively modified. One of the most likely ligand-gated channel candidate systems for its involvement in postischemic brain damage is the *N*-methyl-D-aspartate (NMDA) receptor subtype of the glutamate receptor activated channel. Stimulation of this receptor has been demonstrated to increase the influx of calcium through the NMDA-regulated channel and via other calcium channels [1, 8, 17]. One of the consequences of the sudden increase in intracellular free calcium is the activation of lipases and proteases [1, 2, 36]. Blockade of ligand-regulated and voltage-sensitive calcium channels by different pharmacologic agents prevents ischemia/reperfusion injury [1, 2, 17]. Early in the reperfusion process there is a significant increase in the expression of immediate early genes (IEGs). Among the IEGs expressed, *hsp*, c-*fos* and c-*jun* have been demonstrated to increase following ischemia reperfusion injury.

Aging has been associated with increases in oxidative events and a reduction in calcium sequestration [13]. In addition, it is generally recognized that ischemia/reperfusion injury of the brain is more damaging in older patients. This chapter presents data that demonstrates that there is a significant change in sensitivity to ischemia as a function of age. Moreover, we present results that demonstrate that the changes in gene expression following ischemia are correlated with age-related changes in edema and behavior following ischemia.

Department of Pharmacology, University of Kentucky Medical School, Lexington, KY 40536, USA

Cerebral Ischemia and Basic Mechanisms
Ed. by A. Hartmann, F. Yatsu, and W. Kuschinsky

Cerebral ischemia is a condition that can result in devastating consequences. There are a number of biochemical and physiological consequences of cerebral ischemia. Among these are free radical-mediated membrane damage [2, 3, 14, 20, 21]; production of prostaglandins, leukotrikenes, and other arachidonic acid metabolites; generation of cytotoxic edema [1, 2]; activation of α-amino-3-hydroxy-5-methyl-4-isoxozolepropionic acid (AMPA) and *N*-methyl-D-aspartate receptors (NMDA) [8, 17, 23]; cellular loading with calcium, activation of cellular lipases and proteases [1, 2]; and activation of nitric oxide synthetase. Temporally associated with these events are changes in specific gene expression. Following ischemia reperfusion injury, a number of genes are expressed. Among them are heat shock protein-70 (*hsp*), c-*fos*, c-*jun*, BNDF and ornithine decarboxylase (ODC) [10, 19, 22, 26, 27, 29]. The progressive intracellular changes that lead to the eventual death of the cell are associated with the early changes that result in the expression of these genes and possibly other genes. Age-related differences in sensitivity to these changes may determine inherent differences in susceptibility to stroke. Pharmacologic manipulation of this expression or the initiating conditions could prevent the reperfusion-related neuropathologic changes. The following sections will review the temporal changes in gene expression following ischemia/reperfusion injury and the age-related differences in susceptibility to these changes.

Materials and Methods

Male gerbils were purchased from Tumblebrook Farm (West Brookfield, MA). Both young adult (3 months of age) and retired breeder gerbils (18–20 months of age) were used. Prior to experimentation animals were housed in group cages. All gerbils were housed in the University of Kentucky Medical Center animal facility under a 12-h light/dark cycle. All experiments were conducted during the light phase of the cycle.

Animals were surgically prepared with bilateral carotid artery occluders under surgical anesthesia (40 mg/kg pentobarbital) 2 days prior to experimentation. Following surgery gerbils were individually housed to prevent accidental carotid occlusion by cage mates. The 48-h post-surgical period also allowed for the metabolic clearance of the anesthetic prior to ischemia testing. We have previously demonstrated that the surgical placement of the carotid occluders using a double lumen catheter makes it possible to produce reliable occlusion and reperfusion of the carotid arteries in an unanesthetized, unmedicated preparation [5]. Because the gerbils were not anesthetized, it is possible to test the gerbils for behavioral sequelae immediately following ischemia/reperfusion injury.

Animals were sacrificed by decapitation at specified times following ischemia for RNA and free radical analysis. Brains were rapidly removed and frozen in liquid nitrogen. Brains were stored at −80°C until analyzed.

Measurement of Changes in Locomotor Activity

Spontaneous locomotor activity of individual gerbils was measured using photocell activity chambers as previously described [5, 31]. Gerbils were individually monitored in 60-cm diameter circular chambers that were fitted out with two orthogonally placed photocell circuits. Interruption of a single beam defined a count. Counts were cumulated over time using computer control systems [31]. Both successive behavioral changes and total cumulated counts were accumulated.

Measurement of Changes in Brain Phosphate Metabolism Using Nuclear Magnetic Resonance Spectroscopy

In vivo ^{31}P (80 MHz) spectra were obtained using a 4.7-Tesela horizontal 25.4-cm free-bore magnetic resonance Imager/Spectrometer (SISCO, Fremont, CA). The conscious gerbils were placed in a specially constructed restrainer that held the animal's head in a fixed position within the magnetic field and the surface coil. The head was positioned in the center of the magnetic field. A four-turn, 1.8-cm diameter copper wire transmit/receive surface coil was placed on the dorsal surface of the head. Proton imaging demonstrated that the placement was directly over the cerebral cortex. The specific peaks identified and analyzed using this procedure were inorganic phosphate (Pi), phosphocreatine (PCr) and the three peaks (alpha, beta, gamma) of ATP (the beta phosphate of ATP will be used in quantitation). Intracellular pH was calculated using the chemical shift of Pi, relative to PCr [9].

Brain Edema Measurements

Brain water content was measured using the gravimetric methods [1]. Brains were removed at specified times post-reperfusion and a small portion of neocortex removed and placed in a calibrated bromobenzene/kerosine column. The specific gravity of the tissue was determined by comparing the sedimentation of the tissue to liquid density standards.

Isolation of RNA and Northern Blot Analysis

Total cellular RNA was isolated from gerbil brains according to the method of Chirgwin et al. [6]. The neocortex was homogenized in guanidine thiocyanate and the nucleic acids precipitated with ethanol. RNA was analyzed for specific gene expression by electrophoresis through 1.0% agarose, 2.2 *M* formaldehyde gels [24]. The RNA was transferred to nitrocellulose and hybridized in 50% formamide, 100 μg/ml denatured salmon sperm DNA, 5 × Denhardt's, 0.1% SDS, 20 m*M* sodium phosphate (pH 6.8) and 1 *M* NaCl with ^{32}P-labeled cDNA probes for 12–16 h at 42°C. Blots were visualized by autoradiography and quantitated using an LKB ultrascan densitometry.

Results

Exposure to different periods of ischemia with reperfusion demonstrated substantial, significant, and age-related differences in sensitivity (Fig. 1). Adult gerbils (3–4 months) were significantly more resistant to the lethal effects of ischemia than the 18–20-month-old retired breeders. These differences in sensitivity are consistent with previous reports [12–14, 16]. In addition to the differences in lethality, older gerbils were more susceptible to ischemia-induced neuronal loss. The inherent differences in susceptibility to postischemic neurotoxicity did not appear to be the result of a difference in the level of hydroxyl–free radical production, based upon the salicylate trapping assay. Both the young and retired breeder gerbils demonstrated the same relative level of hydroxyl radical production (Fig. 2).

However, there were significant differences in gene expression following ischemia reperfusion (Fig. 3). There was a significant difference in the baseline level of gene expression in the two age groups . The levels of *Hsp*-70 and c-*fos* were elevated in the older gerbils compared to the adult gerbils. In addition, the degree of postischemic induction of the early inducible genes and later genes was markedly reduced compared to adult gerbils. While the degree of expression was different in the two age groups, the temporal pattern of gene expression appeared to be quite similar. *hsp*-70 was the earliest to be expressed of the genes monitored. This was followed by the later expression of c-*fos* and ODC (Fig.4).

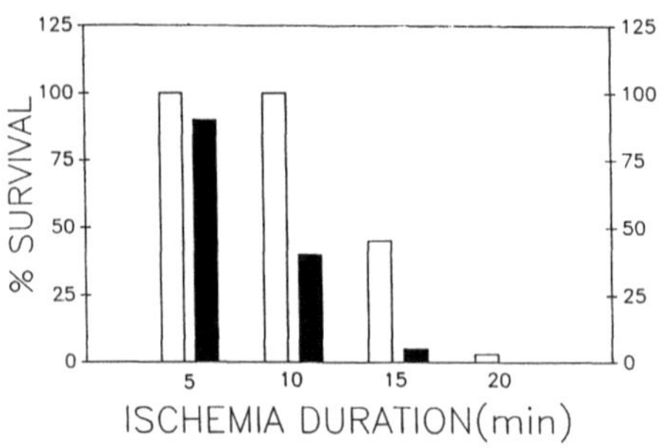

Fig. 1. Inherent differences in ischemia-related lethality in adult (3–4 months old) and retired breeder (18+ months old) male gerbils. Each histogram represents the mean survival 24 h after the onset of reperfusion for groups of 12 gerbils. *Open columns*, adult gerbils; *Solid columns*, retired breeders

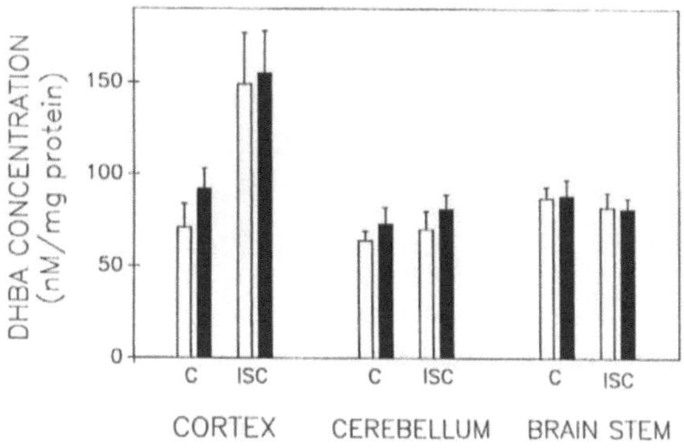

Fig. 2. Regional differences in hydroxyl radical production in adult (*open bars*) and retired breeder (*solid bars*) gerbils. Each histogram is the mean (±SEM) of six subjects. The same animals were used in the determination of DHBA in the brain regions indicated

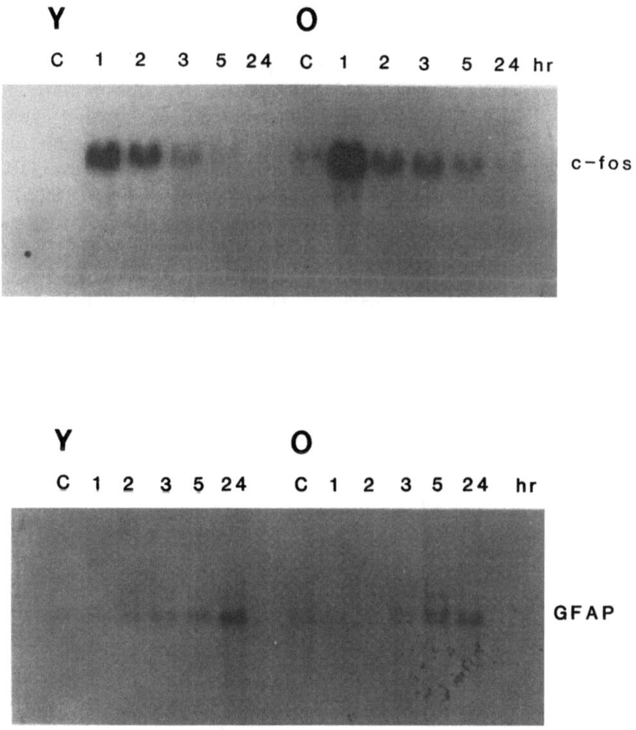

Fig. 3A,B. Gene expression in the ischemic brain. RNA isolated from the cortex of young (*Y*) and old (*O*) gerbils following 5 min of ischemia and various times of reperfusion and from sham-operated animals (*C*). **A** Total RNA (20 μg) from each sample was examined by Northern blot analysis for c-*fos* mRNA. **B** The blot in **A** was stripped and rehybridized with the mouse HSP70 cDNA

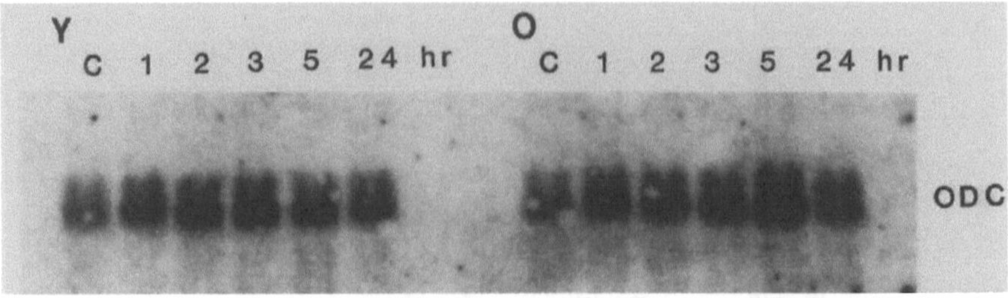

Fig. 4. Ornithine decarboxylase (ODC) gene expression following ischemia in young and old gerbils. RNA isolated from the cortex of young (*Y*) and old (*O*) gerbils following 5 min of ischemia and various times of reperfusion, and from sham-operated animals (*C*); 20 μg RNA was hybridized with a cDNA to the mouse ODC gene

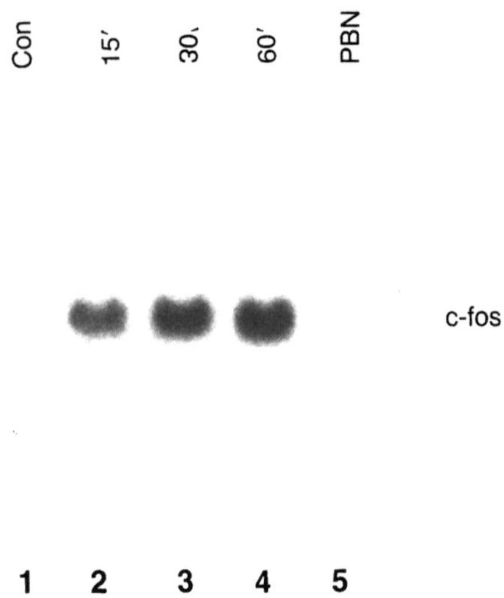

a

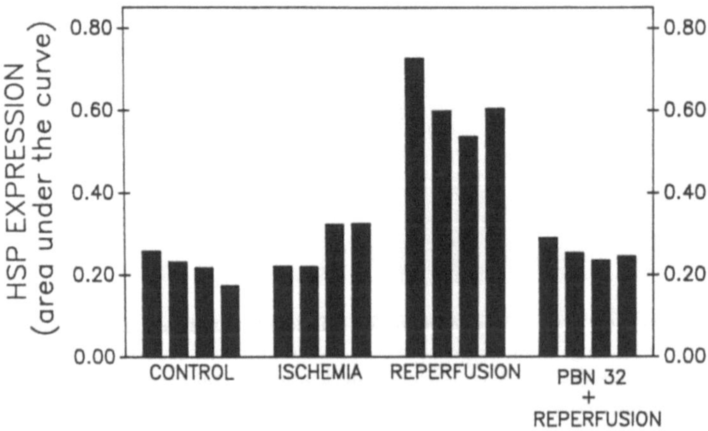

b

Fig. 5a,b. Effects of PBN pretreatment of gerbils prior to ischemic injury on gene expression. **a** Gerbils were pretreated with saline (*lanes 1*, *2*, *3*, and *4*) or PBN (300 mg/kg, 30 min before ischemia, *lane 5*) then subjected to 5 min of ischemia and various times of reperfusion. The indicated times (min) refer to the reperfusion time after 5 min ischemia. PBN-pretreated gerbils were allowed a 60-min reperfusion time. RNA was hybridized with the c-fos cDNA. **b** Gerbils were pretreated with saline or PBN (30 min prior to ischemia) then subjected to 10 min of ischemia and various times of reperfusion. RNA was hybridized with the HSP-70 cDNA. Each histogram represents the neocortical HSP expression of a single gerbil

Previous work has demonstrated the correlation of reperfusion production of oxygen free radical species and the expression of early genes. Under conditions of no reperfusion there was no production of oxygen free radicals and no increase in the levels of early genes [3, 19]. Pretreatment with drugs that reduce the level of oxygen free radical production also reduces the level of gene expression (Fig. 5).

Comparison of the physiologic, metabolic, and behavioral effects of ischemia/reperfusion injury demonstrated a consistent and significant age-related difference in sensitivity. Aged gerbils demonstrated a greater degree of cortical edema following 10 min of cerebral ischemia, followed by 1-h reperfusion (Fig. 6). Similarly, recovery of both intracellular pH and cerebral energetics (based upon the PCr/Pi for neocortex) was protracted in the aged gerbils (Fig. 7). These exaggerated recoveries were mirrored in the

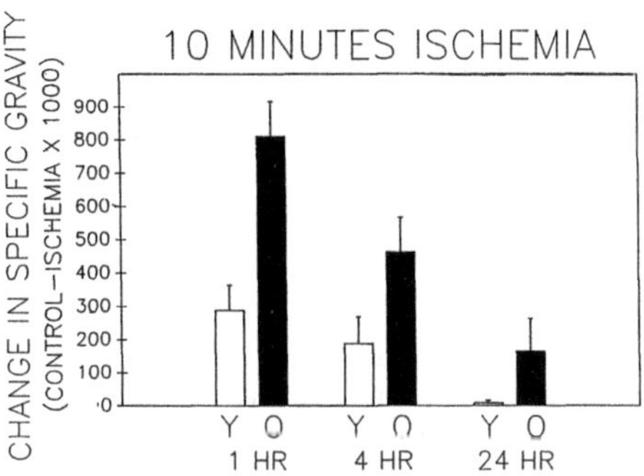

Fig. 6. Age-related differences in post-ischemic edema following ischemia/reperfusion injury. Groups of six gerbils were used for each of the age and reperfusion time groups. Data are presented as the mean (± SEM) for six subjects per group. *Y*, young gerbils; *O*, old gerbils

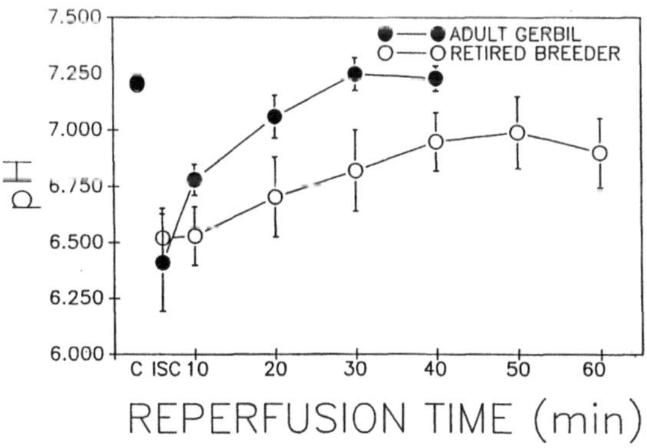

Fig. 7. Age-related differences in postischemic recovery of intracellular pH in adult (*solid circles*) and retired breeder (*open circles*) gerbils following 10 min of bilateral carotid occlusion. Intracellular pH was determined in the unanesthetized gerbils using NMR spectroscopy and the chemical shift of the inorganic phosphate. The data are plotted as the mean (±SEM) for a group of six gerbils in each of the two age groups. Control (*C*) indicates the within subject pH prior to ischemia. The young and aged gerbils had the same intracellular pH at the start of the experiment

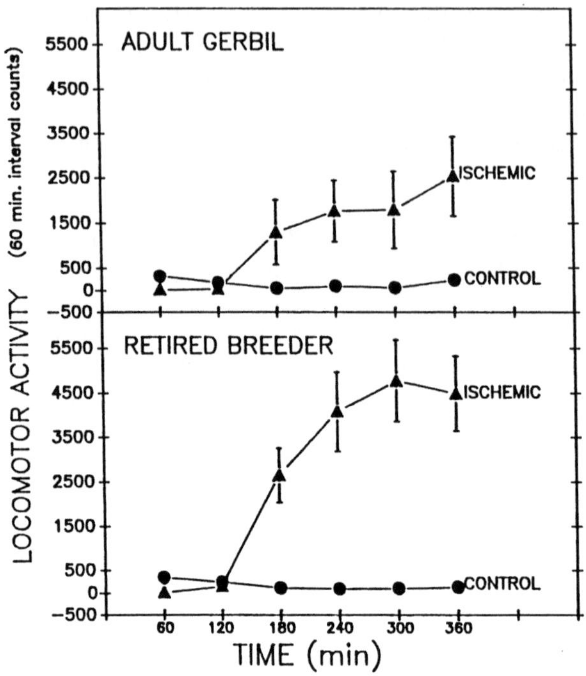

Fig. 8. Age-related differences in postischemic hyperactivity following 10 min of transient ischemia. Animals were immediately placed into the locomotor activity chambers following the onset of reperfusion and monitored for 6 h. Data are presented as the mean (±SEM) for groups of six gerbils in each age group. *Solid circles*, control level of activity of a yoked group of gerbils from each age group that were not exposed to ischemia/reperfusion injury

marked increase in the postischemic hyperactivity seen in the aged gerbils compared to adult gerbils (Fig. 8).

Discussion

The induction of immediate, early genes occurs rapidly after the onset of reperfusion [19, 25–27, 29]. These changes are similar to what has been demonstrated in vitro [33, 34]. It is important to note that these changes in gene expression do not occur if there is no reperfusion of the tissue. A number of changes occur during the ischemic period that may be involved in the promotion of these genes. First, there is a rapid change in the intracellular pH (from 7.2 to 6.4). This can result in changes in the binding of inhibitory or regulatory proteins to their substrates and changes in the structure of these and other proteins. Second, due to the loss of ATP, there may be changes in the sequestration of ions and alterations in osmotic gradients. Third, the significant increase in intracellular inorganic phosphate may result in the loss of antioxidants and calcium from the mitochondria [35]. Upon reperfusion, this alteration in function of the mitochondria can result in an increase in intracellular calcium and an activation of systems that generate free radicals and activate proteases. The elevation of intracellular calcium may be the signal for the promotion of gene expression.

The early expression of c-*fos* and c-*jun* allows for the formation of a transcriptional complex that allows for the specific association with the TPA-responsive element (TRE) found in the promotor regions of a number of genes [30]. The transient nature of the expression of these genes suggests that their function is to activate a system that is responsive to the transient stress associated with ischemia/reperfusion injury [19]. This is supported by the observation that neuronal elements tend to express these genes more than glial elements [19]. A similar pattern of selective activation in neurons has been demonstrated for heat shock proteins following ischemia [26, 27].

The induction of these immediate early genes is followed later by the induction of ODC mRNA. The ODC gene contains a cAMP-responsive element which binds the CREB and results in the activation of transcription [7, 9, 10]. Both elevation of intracellular cAMP and elevation of intracellular free calcium levels (which occurs during reperfusion) can result in increased expression of ODC [10]. Pretreatment with pentobarbital significantly reduces the postreperfusion production of hydroxyl free radicals [3] and prevents the increases in c-*fos*, c-*jun* [19], and ODC [10]. Thus, the delayed expression of ODC may indeed be initiated by the early intracellular changes that occur during the first few minutes of reperfusion.

The demonstration of age-related differences in postischemic gene expression and function indicates that there are probably several different factors that can serve to modify the pattern of postischemic gene expression. Since the level of hydroxyl free radicals produced following ischemia reperfusion was not substantially different, some other factors must be involved in determining the differences. One difference is that aged brain is known to be significantly more oxidized [4]. During the initial reperfusion period there is a marked increase in oxidized protein [28] and hydroxyl radical production [3]. In addition, as present here, the aged brain has a higher baseline level of immediate early genes prior to ischemia/reperfusion injury. All of these observations would support the possibility that the aged brain is already under a higher level of oxidative stress [2, 12, 13]. The failure of the aged brain to demonstrate the same level of gene expression may reflect the refractory nature of the system [19]. If there is a higher constant level of expression of c-*fos*, then following ischemia reperfusion certain of the neuronal elements may not express normally. If the induction of the early immediate genes is a protective response, then dysfunction in this programmed expression may increase the likelihood that the oxidative stress of reperfusion will be lethal to the cell. In addition, previous studies have demonstrated that the mitochondria from aged brains are less able to sequester calcium compared to young adult brains. This progressive decline in calcium regulation could significantly affect the timing and regulation of gene expression.

Taken together, age is a significant factor in determining the molecular and physiologic responses to ischemia/reperfusion injury. Given the epidemiology of stroke and the increase in the mean age of the population, the role of aging in determining outcome following ischemia reperfusion injury is an important area for further research.

References

1. Baethmann A, Maier-Hauff K, Kempski O, Uterberg A, Wahl M, Schurer L (1988) Mediators of brain edema and secondary brain damage. Crit Care Med 16:972–978
2. Braughler JM, Hall EE (1989) Central nervous system trauma and stroke. Free Radic Biol Med 6:289–301
3. Cao W, Carney JM, Duchon A, Floyd RA, Chevion M (1988) Oxygen free radical involvement in ischemia and reperfusion injury to brain. Neurosci Lett 88:233–238
4. Carney JM, Starke-Reed PE, Oliver CN, Landrum WR, Cheng MS, Wu JF, Floyd RA (1991) Reversal of age-related increase in brain protein oxidation, decrease in enzyme activity and loss in temporal and spatial memory by chronic administration of the spin-trapping compound N-tert-butyl-alpha-phenylnitrone. Proc Natl Acad Sci USA 88:3633–3636
5. Chandler MJ, DeLeo JA, Carney JM (1985) An unanesthetized gerbil model of cerebral ischemia-induced behavioral changes. J Pharmacol Methods 14:137–146
6. Chirgwin JM, Pryzbala AE, MacDonald RY, Rutter WJ (1979) Isolation of biologically active ribonucleic acid from sources enriched in ribonuclease. Biochemistry 18:5294–5299
7. Coffino P, Chen EL (1988) Nucleotide sequence of the mouse ornithine decarboxylase gene. Nucleic Acids Res 16:2731–2732
8. Cole AJ, Saffen DW, Barban JM, Worley PF (1989) Rapid increase of an immediate early gene messenger RNA in hippocampal neurons by synaptic NMDA receptor activation. Nature 340:474–476
9. Dempsey RJ, Combs DJ, Donaldson DL, Thomas G, Smith C (1990) In vivo 31-phosphorous spectroscopy during transient cerebral ischemia in the gerbil. Neurol Res 12:106–110
10. Dempsey RJ, Carney JM, Kindy MS (1991) Modulation of ornithine decarboxylase mRNA following transient ischemia in the gerbil brain. J Cereb Blood Flow Metab 11:979–985
11. Feix JB, Kalyanaraman B (1991) Production of singlet oxygen-derived hydroxyl radical adducts during merocyanine-540-mediated photosensitization: analysis by ESR-spin-trapping and HPLC with electrochemical detection. Arch Biochem Biophys 291:43–51
12. Floyd RA (1990) Role of oxygen free radicals in carcinogenesis and brain ischemia. FASEB J 4:2587–2597
13. Floyd RA (1991) Oxidative damage to behavior during aging. Science 254:1597
14. Floyd RA, Carney JM (1991) Age influence on oxidative events during brain ischemia/reperfusion. Arch Gerontol Geriatr 12:155–177
15. Floyd RA, Henderson R, Watson JJ (1986) Use of salicylate with high pressure liquid chromatography and electrochemical detection (LCED) as a sensitive measure of hydroxyl free radicals in adriamycin treated rats. J Free Radic Biol Med 2:13–18
16. Floyd RA, West MS, Eneff KL (1990) Conditions influencing yield and analysis of 8-hydroxy-2′-deoxyguanosine in oxidatively damaged DNA. Anal Biochem 188:155–158
17. Hahn JS, Aizenman E, Lipton SA (1988) Central mammalian neurons normally resistant to glutamate toxicity are made sensitive by elevated extracellular calcium: toxicity is blocked by N-methyl-D-aspartate antagonist MK-801. Proc Natl Acad Sci USA 85:6560–6565

18. Jarrot DM, Domer FR (1980) A gerbil model of cerebral ischemia suitable for drug evaluation. Stroke 11:203–209
19. Kindy MS, Carney JP, Dempsey RJ, Carney JM (1991) Ischemic induction of protonocogene expression in gerbil brain. J Mol Neurosci 2:217–228
20. Kirino T (1982) Delayed neuronal death in the gerbil hippocampus following ischemia. Brain Res 239:57–69
21. Kontos HA, Wei EP (1986) Superoxide production in experimental brain injury. J Neurosurg 64:803–807
22. Lindvall O, Ernfors P, Bengzon J, Kokia Z, Smith ML, Siesjö B, Persson H (1992) Differential regulation of mRNAs for nerve growth factor, brain-derived neurotrophic factor and neurotrophic 3 in the adult rat following cerebral ischemia and hypoglycemic coma. Proc Natl Acad Sci USA 89:648–652
23. Maner H, Favaron M, Guidotti A, Costa E (1989) Delayed increase of calcium influx elicited by glutamate: role in neuronal death. Mol Pharmacol 36:106–112
24. Maniatis T, Fritsch, EF. Sambrook J (1982) Molecular cloning: a laboratory manual. Cold Spring Harbor Laboratory, Cold Spring Harbor
25. Morgan JI, Curran T (1986) Role of ion flux in the control of c-fos expression. Nature 322:552–555
26. Nowak TS Jr (1985) Synthesis of a stress protein following transient ischemia in the gerbil. J Neurochem 5:1635–1641
27. Nowak TS Jr (1989) Heat shock RNA levels in brain and other tissues after hyperthermia and transient ischemia. J Neurochem 49:451–560
28. Oliver CN, Starke-Reed PE, Stadtman ER, Liu GJ, Carney JM, Floyd RA (1990) Oxidative damage to brain proteins, loss of glutamine synthetase activity and production of free radical during ischemia/reperfusion injury to gerbil brain. Proc Natl Acad Sci USA 87:5144–5147
29. Onodera H, Kogure K, Ono Y, Igarashi K, Kiyota Y, Nagaoka A (1989) Protooncogene c-fos is transiently induced in the rat cerebral cortex after forebrain ischemia. Neurosci Lett 98:101–104
30. Rauscher FJ III, Cohen DR, Curran T, Bos TJ, Vogt PK, Bohman D, Tijan R, Franza BR (1988) fos-Associated protein p39 is the product of the jun proto-oncogene. Science 240:1010–1016
31. Rayfield F, Carney JM (1981) Controlling behavioral experiments with BASIC on 6502 based microcomputer. Behav Res Methods Instrum 11:585–588
32. Rothman SM, Olney J (1986) Glutamate and the pathophysiology of hypoxic-ischemia brain damage. Ann Neurol 19:105–111
33. Sassone-Corsi P (1988) cAMP induction of early adenovirus promoters involves sequences required for E1A trans-activation. Proc Natl Acad Sci USA 85:7192–7196
34. Sassone-Corsi P, Verma IM (1987) Modulation of c-fos gene transcription by negative and positive cellular factors. Nature 326:507–510
35. Savage MK, Reed DJ (1992) Respiratory inhibitors influence glutathione status during the permeability transition in mitochondria (Abstr 1612). Annual Meeting of Society for Toxicology
36. Sisjö BK (1988) Mechanisms of ischemic brain damage. Crit Care Med 16:954–963
37. Smith CD, Carney JM, Starke-Reed PE, Oliver CN, Stadtman ER, Floyd RA, Markesbery WR (1991) Excess brain protein oxidation and enzyme dysfunction in normal aging and in Alzheimer disease. Proc Natl Acad Sci USA 88:10540–10543
38. Stadtman ER, Oliver CN (1991) Metal-catalyzed oxidations of proteins. J Biol Chem 266:2005–2008
39. Zaleska MM, Nagy K, Floyd RA (1989) Iron-induced peroxidation and inhibition of dopamine synthesis in striatum synaptosomes. Neurochem Res 14:597–605

Models of Neural Circuit Reorganization After Injury

F.J. SEIL

Introduction

While regeneration in the peripheral nervous system is a well-recognized phenomenon, it had long been believed that central nervous system (CNS) regeneration either did not occur, or was extremely limited. It remains true that regenerative sprouting, defined as the regrowth of a damaged axon, is commonly initiated, but is usually not successful in the adult mammalian CNS. Several factors contribute to this regenerative failure, including (a) a generally regressive reaction and reduced survival of adult CNS neurons after axonal injury [2]; (b) a nonsupportive environment for axonal growth [43]; and (c) the presence in the CNS of oligodendrocyte- and myelin-associated factors that are inhibitory to axonal growth [29]. Other contributory elements may include a down-regulation in the mature state of the production of molecules that promote neuron survival or neurite extension during development [1].

It has been shown that when injured CNS axons are placed in a supportive environment, such as a peripheral nerve bridge, they are capable of considerable regenerative growth, even exceeding their normal lengths [9, 27]. If such axons succeed in contacting target neurons upon reentry into the CNS, they can form functional synapses [50]. Unfortunately, axonal growth generally stops within 1–2 mm after reentry into the CNS, as substrate conditions in the mature CNS are nonpermissive for regenerative growth.

However, regenerative sprouting is not the only strategy available for the reassembly of damaged neural circuitry. Another possible mechanism is collateral sprouting, in which uninjured axons innervating the same target territory as axons that have sustained damage may sprout collateral axons that innervate the target sites vacated by the damaged fibers. Collateral sprouting represents a regenerative response in which there is a territorial expansion of innervation by surviving intact axons. Studies beginning in the late 1950s have shown that collateral sprouting is a mechanism that is available for circuit reorganization in the injured CNS (reviewed in [8, 32]).

Office of Regeneration Research Programs (151N), VA Medical Center, Portland, OR 97201, USA

Cerebral Ischemia and Basic Mechanisms
Ed. by A. Hartmann, F. Yatsu, and W. Kuschinsky

In Vivo Studies of Collateral Sprouting and CNS Reorganization

In a classic study by Liu and Chambers [18], all lumbar and sacral dorsal roots but one (L6) were sectioned unilaterally in cats. The result was an increased intraspinal projection of the spared root, compared with the same root on the unlesioned side. The increase was found in all of the spinal cord segments to which the spared root normally projected (dorsal horn, zona intermedia, Clarke's nucleus) and in the nucleus gracilis. The increased projection was concluded to be a consequence of injury-induced collateral sprouting from the spared dorsal root.

The first morphological demonstration of axonal sprouting in the CNS came from studies of the partially deafferented adult rat septal nucleus by Raisman [24] and Raisman and Field [26]. Neurons in the lateral septal nucleus receive hippocampal afferents via the fimbria, and these afferents terminate exclusively on dendrites (Fig. 1A). A second projection from the lateral hypothalamus and ventral tegmental area by way of the median forebrain bundle (MFB) terminates on both dendrites and somata of lateral septal nucleus neurons. The MFB was lesioned (Fig. 1B) and several weeks were allowed for clearance of all of the degenerated synapses. The fimbria was then lesioned and the septal nucleus was examined with the electron microscope 2 days later. Considerable numbers of degenerating axosomatic as well as axodendritic synapses were found, indicative of collateral sprouting from fimbrial axons after the MFB lesion. Thus the intact fimbrial axons had increased their projection to include somata as well as dendrites after loss of the MFB axons, and a constant number of synapses in the septal nucleus was maintained. Further extension of these studies [12, 25] showed that if one fimbria was lesioned, its projection territory was taken over by

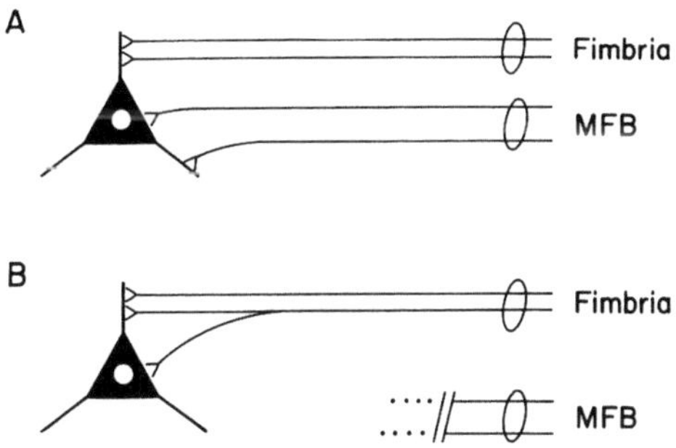

Fig. 1A,B. Collateral axonal sprouting in the septal nucleus. **A** Afferent fibers from the hippocampus project via the fimbria to the dendrites of lateral septal nucleus neurons. A second group of afferents originates in the lateral hypothalamus and ventral tegmental area and projects via the medial forebrain bundle (*MFB*) to lateral septal nucleus neuron dendrites and somata. **B** After lesioning the MFB, fimbrial axons expand their target territory by projecting collateral sprouts to somata, as well as maintaining their axonal projection to the dendrites of lateral septal nucleus neurons. (From [32])

axons from the contralateral fimbria (normally the fimbrial projection to parts of the septal nucleus is bilateral). Only after bilateral fimbrial lesions did nonfimbrial axons sprout and innervate sites occupied by fimbrial axons, thus establishing an hierarchical order for synapse replacement, with like (homotypical) terminals having precedence over unlike (heterotypical) terminals.

Collateral sprouting in the CNS was further demonstrated in functional and ultrastructural studies in the red nucleus of adult cats by Murakami et al. [22], Nakamura et al. [23], Tsukahara [46], and Tsukahara et al. [47, 48]. The red nucleus is another area in which there is a convergence of inputs from two different sources (Fig. 2A). Axons from the contralateral cerebellar interpositus nucleus project to the somata of red nucleus neurons, while afferents from ipsilateral sensorimotor cortex synapse on distal portions of dendrites. With intracellular recording electrodes placed in the somata of large red nucleus neurons, interpositus nucleus stimulation evoked fast-rising excitatory postsynaptic potentials (EPSPs), indicative of axosomatic synapses, while stimulation of sensorimotor cortex provoked slow-rising EPSPs, consistent with synapses on dendrites at a distance from the record-

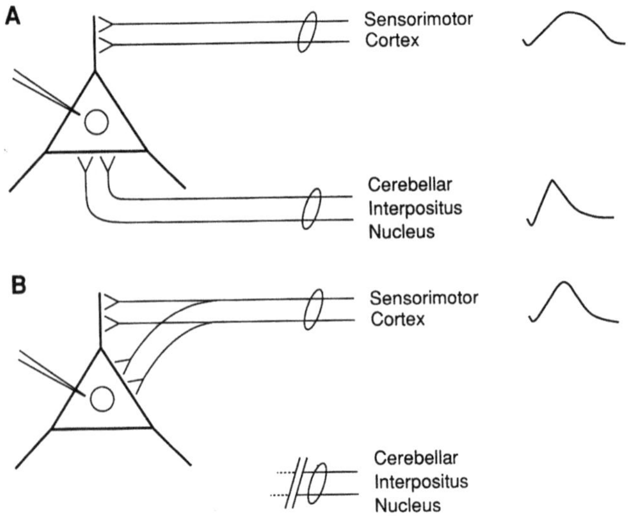

Fig. 2A,B. Collateral innervation in the red nucleus. **A** Red nucleus neurons receive projections to distal dendrites from the ipsilateral sensorimotor cortex and somatic afferents from the contralateral cerebellar interpositus nucleus. With a recording electrode placed in the soma, electrical stimulation of the sensorimotor cortex induces a slow-rising excitatory postsynaptic potential (EPSP) in the red nucleus neuron, while stimulation of the interpositus nucleus produces a fast-rising EPSP. **B** After lesioning the interpositus nucleus, sensorimotor cortex axons undergo collateral sprouting and project to somata and proximal as well as distal dendrites. Electrical stimulation of the sensorimotor cortex results in an EPSP with an intermediate rise time ing electrode. The interpositus nucleus was lesioned (Fig. 2B), and 10 days or more thereafter, recording from red nucleus neurons revealed a decrease in the rise time of EPSPs evoked by sensorimotor cortex stimulation, suggesting that cortical axons had sprouted to increase their projection sites to neuronal somata and proximal dendrites, an interpretation supported by electron-microscopic examination. These studies not only demonstrated collateral sprouting and synaptic replacement in response to injury, but showed that the replaced synapses, although heterotypical, were functional.

Collateral sprouting and laminar reorganization in the lesioned adult rat hippocampus were studied in great detail by Cotman et al. [7], Lynch et al. [20, 21] (reviewed in [28]), and similar results were obtained. It is now well established that collateral axonal sprouting and circuit reorganization after injury do occur in the mature mammalian CNS.

An In Vitro Model of Neural Circuit Reorganization

Organotypic CNS tissue cultures provide a model system in which highly organized groups of neurons and glia can be manipulated to allow study of some of the principles of circuit reorganization in response to injury [33, 37]. Cerebellar explants derived from newborn mice develop the complex interneuronal and neuron–glia relationships characteristic of the cerebellum in vivo [30]. Cortical and deep nucleus areas are defined, and the five major cortical neuronal groups are represented in culture, including granule cells, the only excitatory cortical neurons, and Purkinje, basket, stellate, and Golgi cells. The granule cells project their axons, the parallel fibers, to all other cortical neurons. The presumptive granule cell neurotransmitter is glutamic acid [15]. The only axons to project from the cerebellar cortex are Purkinje cell axons, which innervate deep nucleus neurons. GABA is the Purkinje cell neurotransmitter [15]. Purkinje cell axons give off recurrent axon collaterals that project to basket, stellate, and Golgi cells, as well as to other Purkinje cells. Both Purkinje cell axons and axon collaterals become myelinated in vitro, and the myelinated axons projecting from cortex to deep nuclei form a white matter tract, as in vivo. Basket, stellate, and Golgi cells are inhibitory interneurons, with basket and stellate cells projecting to Purkinje cell somata and dendrites, respectively, while Golgi cells give rise to complex axons that terminate on granule cell dendrites. The major difference between cerebellar cultures and cerebellum in vivo is the absence of extracerebellar afferents, including mossy and climbing fibers, in the isolated culture system. The cortical organization of an organotypic cerebellar culture is schematized in Fig. 3A.

Granule cells are dividing neuroblasts in the mouse at birth. Exposure of cerebellar cultures to the DNA synthesis inhibitor, cytosine arabinoside (ara-C), for the first 5 days in vitro destroyed oligodendrocytes, functionally compromised astrocytes, and drastically reduced the granule cell population

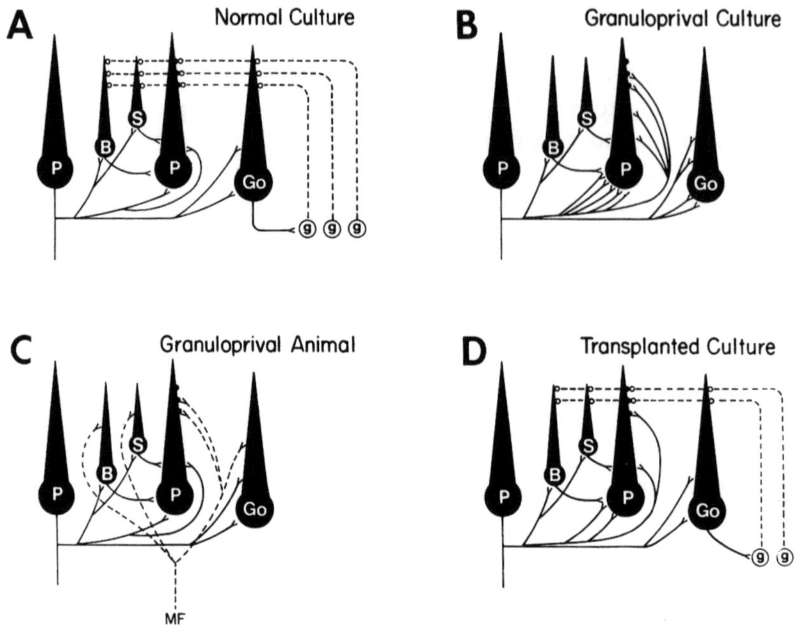

Fig. 3A–D. Circuit reorganization in cerebellar cultures. **A** Cortical region of a normal organotypic cerebellar culture, showing the main projections of Purkinje (*P*), basket (*B*), stellate (*S*), Golgi (*Go*), and granule (*g*) cells. Inhibitory fibers are represented as *solid lines* and excitatory fibers as *dashed lines*. Extracerebellar afferents (mossy and climbing fibers) are absent or severely reduced. **B** After elimination of the granule cells by exposure to cytosine arabinoside (granuloprival culture), there is a marked sprouting of inhibitory Purkinje cell recurrent axon collaterals, which increase their projection to the somata of other Purkinje cells and also project to the Purkinje cell dendritic spines normally occupied by excitatory parallel fibers. **C** After a similar destruction of granule cells in vivo (granuloprival animal), there is a sprouting of excitatory extracellular afferents (*MF*), which project to sites normally innervated by parallel fibers. There is no recurrent axon collateral sprouting in this case. **D** Transplantation of granuloprival cerebellar cultures with granule cells results in a replacement of most of the heterotypical recurrent collateral–dendritic spine synapses with homotypical parallel fiber–dendritic spine synapses and in a reduction of most of the sprouted recurrent axon collaterals. (From [37])

[6, 36, 41]. The consequence of the effect of ara-C exposure on glia was a lack of myelination and a failure of formation of the normal astrocytic sheaths around Purkinje cell somata and dendrites.

Virtual elimination of the granule cells produced, in effect, a major denervation of the remaining cortical neurons. In response to the absence of parallel fibers, there was a remarkable sprouting of Purkinje cell recurrent axon collaterals. Sprouted recurrent axon collaterals projected to Golgi cells, hyperinnervated unensheathed Purkinje cell somata, and formed heterotypical synapses with Purkinje cell dendritic spines, as schematized in Fig. 3B. The stimulus for Purkinje cell recurrent axon collateral sprouting in the absence of parallel fibers is unknown, but postulates include the attraction of postsynaptic sites (however mediated) [44] or target cell production of a sprouting factor that is normally held in check by a presynaptic inhibitory factor [10]. The sprouting was not mediated by the neurotransmitters, glutamic acid, or GABA, as incorporation of large concentrations of these neurotransmitters in the culture nutrient medium did not prevent recurrent axon collateral sprouting [34].

The presence of recurrent axon collateral terminals on Purkinje cell dendritic spines indicated that sites normally contacted by excitatory parallel fibers were now occupied by inhibitory terminals. That such heterotypical synapses were both functional and inhibitory was shown by extracellular electrophysiologic recording. Antidromic stimulation of Purkinje cell axons provokes a transitory increase (disinhibition) of spontaneous cortical spike discharges in cerebellum in vivo and in normal cerebellar cultures [11, 36]. In reorganized ara-C-treated cerebellar cultures, antidromic stimulation of Purkinje cell axons induced a profound inhibition of spontaneous cortical activity, an atypical response due to the collateral hyperinnervation [36]. This form of reorganization is different from that which occurred in the in vivo cerebellum with granule cell destruction (granuloprival cerebellum). In this case, as diagrammed in Fig. 3C, there was a sprouting of excitatory extracerebellar afferents that projected to the Purkinje cell dendritic spines, without an appreciable sprouting of Purkinje cell recurrent axon collaterals [19]. Such synapses were also heterotypical, but they were excitatory, indicating that excitatory terminals have preference over inhibitory terminals as replacements for absent excitatory axons, even though the excitatory neurotransmitters may be different. It is only in the extreme circumstance of isolated cultures where no source of excitatory axons was available that inhibitory fibers sprouted to replace an excitatory projection.

Purkinje cell survival was considerably increased (three- to four-fold) in ara-C-treated explants [31]. A possible reason for this is the greatly expanded projection field for Purkinje cell axons in the form of synaptic sites available because of the absence of parallel fibers, even if the available sites are the dendritic spines of other Purkinje cells. The increased survival of Purkinje cells in culture conditions may be analogous to the rescue of neurons from naturally occurring cell death during development in vivo by enlarging the area of the target to which the neurons project [14]. Support for this concept comes from the reduction of Purkinje cells that occurs after transplantation of ara-C-treated cultures with granule cells, as will be discussed in the following section.

Circuit Remodeling in Vitro After Transplantation

Could replacement of the missing elements in reorganized ara-C-treated cerebellar cultures induce a second round of circuit remodeling? To address this question, cerebellar cultures exposed to kainic acid were superimposed upon "host" ara-C-treated cultures [5, 38]. Kainic acid destroys all cerebellar cortical neurons except granule cells, while glia remain functionally intact [35]. Cultures exposed to kainic acid are therefore complementary to ara-C-treated cultures and can serve as a "transplant" source for granule cells and glia.

Transplantation of ara-C-treated cerebellar cultures with granule cells and glia resulted in myelination of Purkinje cell axons and collaterals and in astrocytic ensheathment of Purkinje cell somata and dendrites. Purkinje cell numbers were reduced to normal. Sprouted recurrent axon collaterals were markedly reduced and heterotypical recurrent collateral–Purkinje cell dendritic spine synapses were replaced with homotypical parallel fiber–Purkinje cell dendritic spine synapses (Fig. 3D). This indicated not only the priority of the originally programmed axons for their synaptic sites, but also demonstrated that in the CNS, as in the peripheral nervous system [16, 52], sprouted axon collaterals may retreat upon the reappearance of primary axons. The synapses that had been formed by the sprouted collaterals were eliminated, even though they were functional, and were replaced by the appropriate synapses. Antidromic stimulation of Purkinje cell axons no longer inhibited spontaneous cortical activity.

Axosomatic hyperinnervation of Purkinje cells was reduced after transplantation with granule cells and glia, concurrent with the reduction of sprouted recurrent axon collaterals (Fig. 3D). In order to determine whether the reduction of axosomatic synapses was related to astrocytic ensheathment of Purkinje cells or was merely coincident with the reduction of excess sprouted recurrent axon collaterals, ara-C-treated cerebellar cultures were transplanted with fragments of 7-day-old mouse optic nerve as a source of glia without granule cells [39]. Neither the number of Purkinje cells nor the sprouted recurrent axon collaterals were significantly reduced in such cultures, indicating that the formation of parallel fiber–Purkinje cell dendritic spine synapses was critical for these changes to occur. However, astrocytic ensheathment of Purkinje cells did occur after optic nerve transplantation, and axosomatic synapses were reduced (Fig. 4), suggesting that astrocytes played a role in the regulation of synaptic density around Purkinje cells. These findings are consistent with in vivo studies in which double synapses formed on rat hypothalamic magnocellular neurons after dehydration-induced retraction of glial sheaths, which was reversed with rehydration [13, 49].

An unexpected finding in ara-C-treated cerebellar cultures transplanted with optic nerve fragments was the occurrence of clusters of Purkinje cell dendritic spines without presynaptic elements, with a doubling of the total number of free dendritic spines [39]. Similar spine-rich clusters had not been observed in nontransplanted ara-C-exposed cultures, nor in ara-C-treated cerebellar cultures transplanted with granule cells and glia or with purified dissociated oligodendrocytes [5, 6, 40]. While astrocytic processes some-

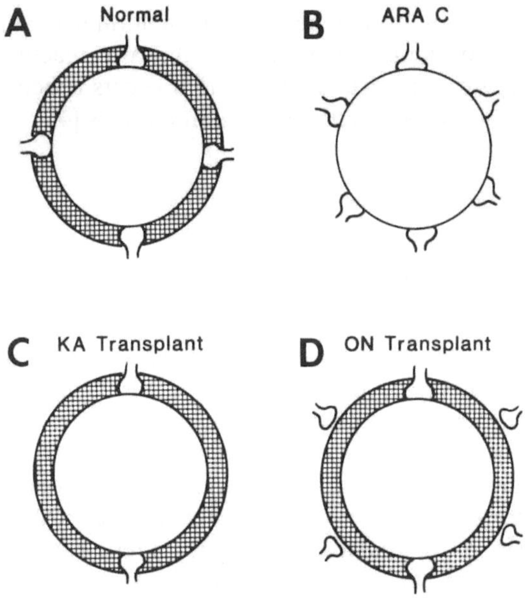

Fig. 4A–D. Synapses on Purkinje cell somata (*circles*) with or without astrocytic sheaths (*cross-hatched areas*). **A** Purkinje cell soma in a normal cerebellar culture. An astrocytic sheath is present, which is penetrated by terminals of basket cell axons and Purkinje cell recurrent axon collaterals (in about equal numbers) to form axosomatic synapses. **B** In an ara-C-treated culture there is an absence of astrocytic ensheathment of Purkinje cells and a hyperinnervation of Purkinje cell somata by sprouted recurrent axon collaterals. **C** After transplantation with kainic acid (*KA*) treated cultures as a source of granule cells and glia, there is astrocytic ensheathment of Purkinje cells in host ara-C-treated cultures, and a reduction of sprouted recurrent axon collaterals and of Purkinje cell axosomatic synapses. The number of axosomatic synapses is even less than in normal cultures because there is also some reduction of the basket cell population by exposure to ara-C [6]. **D** In ara-C-treated cerebellar cultures transplanted with fragments of optic nerve (*ON*) as a source of glia, there is no reduction of sprouted recurrent axon collaterals, but the number of Purkinje cell axosomatic synapses is reduced, concurrent with ensheathment of Purkinje cells by astrocytes, suggestive of an astrocytic role in the regulation of synaptic density on Purkinje cell somata. (From [39])

times appeared near the clusters, there was no astrocytic apposition of the free spines. This suggested that the Purkinje cell dendritic spine proliferation may have been induced by an astrocyte derived diffusible factor. Cytosine arabinoside – treated cerebellar cultures were therefore exposed to astrocyte conditioned medium, with a resultant 2.5-fold increase in the number of Purkinje cell dendritic spines [42]. This suggested a possible mechanism for the elaboration of target postsynaptic membrane prior to the arrival of innervating axons during development or during postdevelopmental plastic or regenerative events. In the postdevelopmental state, astrocytes might elaborate a dendritic spine proliferation factor as part of an astrocytic response to denervation, with resultant expansion of neuronal postsynaptic sites in preparation for reinnervation by regenerating axon terminals. In vivo studies with denervated hippocampus have shown an early reactive gliosis, followed by a later increase in protein synthesis that occurs concurrently with an increase in polyribosomes underlying dendritic spines [45].

Discussion

The following inferences can be made on the basis of studies with the above described tissue culture models:

1. Cerebellar cultures undergo structural and functional circuit reorganization in response to a form of injury consisting of loss or functional compromise of some of their neuronal and glial elements.
2. During reorganization there is an hierarchical order of synapse formation, with similar axons having priority over dissimilar axons as replacements for presynaptic components. In the extreme case of isolated cerebellar cultures, excitatory terminals can be replaced by inhibitory terminals.
3. Further reorganizational changes can occur if the missing elements are restored.
4. When missing neuronal elements are introduced into a reorganized system, appropriate synapses are formed, replacing heterotypical synapses even if the latter have become functional.
5. In the CNS, as in the peripheral nervous system, sprouted axon collaterals may withdraw or degenerate upon the reappearance of primary axons.
6. Replacement of missing or functionally compromised glia can result in restoration of the normal glial milieu.
7. Astrocytes may have a role in the regulation of synaptic density around certain neurons, especially those with astrocytic sheaths.
8. Astrocytes may produce a factor that induces elaboration of target neuronal postsynaptic membrane prior to innervation.

Whether or not similar principles of circuit reorganization that apply to experimental animal or tissue culture models of injury are operative in human subjects after injuries such as stroke had not been established. A functional reorganization of the brain after capsular infarction in humans was shown in a recent study using positron emission tomography [51]. The results suggested to the authors that the human brain has a considerable capacity for reorganization and functional plastic changes after injury. It is possible that cellular mechanisms such as collateral sprouting and synaptic reorganization underlie some of the changes that occur during recovery from stroke.

Given this possibility, it would be reasonable to ask if the consequences of collateral sprouting and synaptic reorganization are beneficial for the injured subject. This question appears to have more than one answer, depending upon which system is examined. There is some evidence that heterotypical synapses formed as a consequence of collateral sprouting in the red nucleus induced partial recovery of normal movement [45]. On the other hand, collateral sprouting of extracerebellar efferents and subsequent innervation of Purkinje cell dendritic spines in animals without cerebellar granule cells did not prevent severe ataxia [19]. It has been postulated that collateral sprouting may interfere with regenerative sprouting in the CNS, as local growth after lesioning may be achieved at the expense of regenerative growth in the mammalian spinal cord [3]. It is thus possible that the results of collateral sprouting and either homotypical or heterotypical synapse formation vary from one CNS area to another, and even from projection to projection, as to whether function is improved, unaffected, or made worse. It therefore behooves us to learn more about the consequences of circuit reorganization in various parts of the CNS, and about the principles that govern such reorganization, so that we may learn how to manipulate the system to maximize recovery.

An optimistic note for the possibility of reconstructing damaged CNS circuitry by replacement of missing elements from exogenous sources is sounded by the described "transplantation" experiments with cerebellar culture models and by numerous animal studies of fetal tissue grafting [4]. The fact that there appear to be hierarchical orders for synapse formation is encouraging for the prospect of restoring circuits like those originally present. Transplantation of CNS tissue may be a future method of therapeutic intervention for improvement of functional recovery after stroke. There has been considerable progress in the field of CNS transplantation, most notable as possible therapy for some of the neurodegenerative disorders [17]. Clinical applications of such technology based on sound scientific principles remain a future consideration, especially for stroke, but there is increasing hope that this approach may be of benefit in human disease.

Acknowledgments. Studies from the author's laboratory were supported by the US Department of Veterans Affairs and by National Institutes of Health grant NS 17493.

References

1. Ard MD, Bunge RP (1988) Heparan sulfate and laminin immunoreactivity in cultured astrocytes: relationship to differentiation and neurite growth. J Neurosci 8:2844–2858
2. Barron KD (1989) Neuronal responses to axotomy: consequences and possibilities for rescue from permanent atrophy or cell death. In: Seil FJ (ed) Neural regeneration and transplantation. Liss, New York, pp 79–99 (Frontiers of clinical neuroscience, vol 6)
3. Bernstein JJ, Wells MR, Bernstein ME (1978) Spinal cord regeneration: synaptic renewal and neurochemistry. In: Cotman CW (ed) Neuronal plasticity. Raven, New York, pp 49–71

4. Björklund A (1991) Neural transplantation – an experimental tool with clinical possibilities. Trends Neurosci 14:319–322
5. Blank NK, Seil FJ (1983) Reorganization in granuloprival cerebellar cultures after transplantation of granule cells and glia. II. Ultrastructural studies. J Comp Neurol 214:267–278
6. Blank NK, Seil FJ, Herndon RM (1982) An ultrastructural study of cortical remodeling in cytosine arabinoside induced granuloprival cerebellum in tissue culture. Neuroscience 7:1509–1531
7. Cotman CW, Matthews DA, Taylor D, Lynch G (1973) Synaptic rearrangement in the dentate gyrus: histochemical evidence of adjustments after lesions in immature and adult rats. Proc Natl Acad Sci USA 70:3473–3477
8. Cotman CW, Nieto-Sampedro M, Harris EW (1981) Synapse replacement in the nervous system of adult vertebrates. Physiol Rev 61:684–784
9. David S, Aguayo A (1981) Axonal elongation into PNS "bridges" after CNS injury in adult rats. Science 214:931–933
10. Diamond J, Cooper E, Turner C, MacIntyre L (1976) Trophic regulation of nerve sprouting. Science 193:371–377
11. Eccles JC, Ito M, Szentagothia J (1967) The cerebellum as a neuronal machine. Springer, Berlin Heidelberg New York
12. Field PM, Coldham DE, Raisman G (1980) Synapse formation in the adult rat brain: preferential reinnervation of denervated fimbrial sites by axons of the contralateral fimbria. Brain Res 189:103–113
13. Hatton GI (1985) Reversible synapse formation and modulation of cellular relationships in the adult hypothalamus under physiological conditions, In: Cotman CW (ed) Synaptic plasticity. Guilford, New York, pp 373–404
14. Hollyday M, Hamburger V (1976) Reduction of the naturally occurring motor neuron loss by enlargement of the periphery. J Comp Neurol 170:311–320
15. Ito M (1984) The cerebellum and neural control. Raven, New York
16. Jackson PC, Diamond J (1981) Regenerating axons reclaim sensory targets from collateral nerve sprouts. Science 214:926–928
17. Lindvall O (1991) Prospects of transplantation in human neurodegenerative diseases. Trends Neurosci 14:376–384
18. Liu C-N, Chambers WW (1958) Intraspinal sprouting of dorsal root axons. Arch Neurol Psychiatry 79:46–61
19. Llinas R, Hillman DE, Precht W (1973) Neuronal circuit reorganization in mammalian agranular cerebellar cortex. J Neurobiol 4:69–94
20. Lynch GS, Matthews DA, Mosko S, Parks T, Cotman CW (1972) Induced acetylcholine-rich layer in rat dentate gyrus following entorhinal lesions. Brain Res 42:311–318
21. Lynch GS, Gall C, Rose G, Cotman CW (1976) Changes in the distribution of the dentate gyrus associational system following unilateral or bilateral entorhinal lesion in the adult rat. Brain Res 110:57–71
22. Murakami F, Katsumaru H, Saito K, Tsukahara N (1982) A quantitative study of synaptic reorganization in red nucleus neurons after lesion of the nucleus interpositus in the cat: an electrom microscopic study involving intracellular injection of horseradish peroxidase. Brain Res 242:41–53
23. Nakamura Y, Mizuno N, Konishi A, Sato M (1974) Synaptic reorganization of the red nucleus after chronic deafferentation from cerebellorubral fibers: an electron microscopic study in the cat. Brain Res 82:298–301
24. Raisman G (1969) Neuronal plasticity in the septal nuclei of the adult rat. Brain Res 14:25–48
25. Raisman G (1985) Synapse formation in the septal nuclei of adult rats. In: Cotman CW (ed) Synaptic plasticity. Guilford, New York, pp 13–38
26. Raisman G, Field PM (1973) A quantitative investigation of the development of collateral reinnervation after partial deafferentation of the septal nuclei. Brain Res 71:1–16

27. Richardson PM, McQuinness UM, Aguayo AJ (1980) Axons from CNS neurons regenerate into PNS grafts. Nature 284:264–265
28. Scheff SW (1989) Synaptic reorganization after injury: the hippocampus as a model system. In: Seil FJ (ed) Neural regeneration and transplantation. Liss, New York, pp 137–156 (Frontiers of clinical neuroscience, vol 6)
29. Schwab M, Caroni P (1988) Rat CNS myelin and a subtype of oligodendrocytes in culture represent a non-permissive substrate for neurite growth and fibroblast spreading. J Neurosci 8:2381–2393
30. Seil FJ (1979) Cerebellum in tissue culture. Rev Neurosci 4:105–177
31. Seil FJ (1987) Enhanced Purkinje cell survival in granuloprival cerebellar cultures. Dev Brain Res 35:312–316
32. Seil FJ (1989) Axonal sprouting in response to injury. In: Seil FJ (ed) Neural regeneration and transplantation. Liss, New York, pp 123–135 (Frontiers of clinical neuroscience, vol 6)
33. Seil FJ (1989) Tissue culture studies of neural plasticity. Restor Neurol Neurosci 1:1–11
34. Seil FJ, Leiman AL (1985) Evidence against neurotransmitter mediation of sprouting in granuloprival coeruleocerebellar cultures. Exp Neurol 87:270–277
35. Seil FJ, Blank NK, Leiman AL (1979) Toxic effects of kainic acid on mouse cerebellum in tissue culture. Brain Res 161:253–265
36. Seil FJ, Leiman AL, Woodward WR (1980) Cytosine arabinoside effects on developing cerebellum in tissue culture. Brain Res 186:393–408
37. Seil FJ, Blank NK, Leiman AL (1983) Circuit reorganization in granuloprival and transplanted cerebellar cultures. In: Seil FJ (ed) Nerve, organ and tissue regeneration: research perspectives. Academic, New York, pp 283–300
38. Seil FJ, Leiman AL, Blank NK (1983) Reorganization in granuloprival cerebellar cultures after transplantation of granule cells and glia. I. Light microscopic and electrophysiological studies. J Comp Neurol 214:258–266
39. Seil FJ, Meshul CK, Herndon RM (1988) Synapse regulation by transplanted astrocytes: a tissue culture study. In: Gash DM, Sladek JR Jr (eds) Prog Brain Res 78:395–399
40. Seil FJ, Johnson ML, Sancto RP, Herndon RM, Mass MK (1989) Myelination of axons within cytosine arabinoside treated mouse cerebellar explants by cultured rat oligodendrocytes. Brain Res 503:111–117
41. Seil FJ, Herndon RM, Tiekotter KL, Blank NK (1991) Reorganization of organotypic cultures of mouse cerebellum exposed to cytosine arabinoside: a timed ultrastructural study. J Comp Neurol 313:193–212
42. Seil FJ, Eckenstein FP, Reier PJ (1992) Induction of dendritic spine proliferetion by an astrocyte secreted factor. Exp Neurol 117:85–89
43. Smith GM, Rutishauser U, Silver J, Miller RH (1990) Maturation of astrocytes in vitro alter the extent and molecular basis of neurite outgrowth. Dev Biol 138:377–390
44. Sotelo C, Privat A (1978) Synaptic remodeling of the cerebellar circuitry in mutant mice and experimental malformations. Acta Neuropathol (Berl) 43:19–34
45. Steward O, Davis L, Dolti C, Phillips LL, Rao A, Banker G (1988) Protein synthesis and processing in cytoplasmic microdomains beneath postsynaptic sites on CNS neurons. Mol Neurobiol 2:227–261
46. Tsukahara N (1985) Synaptic plasticity in the red nucleus and its possible behavioral correlates, In: Cotman CW (ed) Synaptic plasticity. Guilford, New York, pp 201–229
47. Tsukahara N, Hultborn H, Murakami F (1974) Sprouting of cortico-rubral synapses in red nucleus neurones after destruction of the nucleus interpositus of the cerebellum. Experientia 30:57–58
48. Tsukahara N, Hultborn H, Murakami F, Fujito Y (1975) Electrophysiological study of formation of new synapses and collateral sprouting in red nucleus neurons after partial denervation. J Neurophysiol 38:1359–1372
49. Tweedle CD, Hatton GI (1984) Synapse formation and disappearance in adult rat supraoptic nucleus during different hydration states. Brain Res 309:373–376

50. Vidal-Sanz M, Bray GM, Aguayo AJ (1991) Regenerated synapses persist in the superior colliculus after the regrowth of retinal ganglion cell axons. J Neurocytol 20:940–952
51. Weiller C, Chollet F, Friston KJ, Wise RJS, Frackowiak RSJ (1992) Functional reorganization of the brain in recovery from stratocapsular infarction in man. Ann Neurol 31:463–472
52. Zander E, Weddell G (1951) Reaction of corneal nerve fibers to cornea injury. Br J Ophthalmol 35:61–87

Clinical Relevance of Gene Expression in Cerebral Ischemia

F.M. Yatsu

Introduction

The monumental and Herculean task of averting irreversible ischemic brain damage with neuroprotective agents to preserve the complex brain/mind relationship, the highest expression of evolutionary development, was substantially advanced by the discovery that excitatory amino acid neurotransmitters aggravated ischemic injury while their blockade was protective. Galvanized by this discovery, a broad assault has taken place on identifying metabolic and cellular factors accounting for ischemic vulnerability and developing derivative interventional strategies. This volume is testimony to the exciting search for vulnerable molecular factors conveying ischemic vulnerability, but, likely, successful intervention for neuroprotection will require multiple therapies to defend against the fast-moving cascade of disintegrating cellular and molecular events. The role of gene expression in cerebral ischemia has caught the imagination of investigators because of the powerful role genes play in ultimately regulating cellular physiology and function, and the awesome domination genes have in being the final authority for cellular and species functioning.

The approach to determining the clinical relevance of gene expression in cerebral ischemia can be divided into experimental and clinical, or theoretical and applicable approaches. For example, regulation and activity of gene expression in ischemia can be investigated in both in vivo and in vitro conditions for the purposes of exploring the changes and regulation of genes which may play a role in either protecting against cellular injury, such as heat shock proteins, or in insuring cellular regrowth or neuroplasticity. While such studies may detail subtle regulatory factors of gene expression for short- and long-term recovery processes, the clinical relevance of such studies may be uncertain since these changes take place while the complex and subtle neuronal activity underlying the neurophysiology of cognition may be impaired.

Studies on gene expression in cerebral ischemia, as well as derivative investigations on transcriptional, transducing, and translational factors

---

Department of Neurology, University of Texas Medical School, Houston, Texas, USA

Cerebral Ischemia and Basic Mechanisms
Ed. by A. Hartmann, F. Yatsu, and W. Kuschinsky

as well as on functional important proteins, have shown promise in identifying critically important foci of gene regulation which may bear upon both ischemic vulnerability and on therapeutic manipulations to optimize neuronal stability and recovery. The early expression of growth factor genes, so-called proto-oncogenes and related stress-related proteins, was recognized as an early accompaniment of cerebral ischemia (Jacewicz et al. 1986; Jorgensen et al. 1989). However, the subtleties and complexities of their expression and their relationship to early neuronal survival and to long-term recovery with neuroplasticity remains largely to be explored, although substantial insights into the potential influence of early expression of proto-oncogenes and of heat shock proteins is emerging. For example, Jorgensen et al. (1991) find delayed postischemia synthesis of the gene regulatory protein c-*fos* preceding neuronal death for the vulnerable CA1 hippocampal neurons, which suggests that c-*fos* expression may be either a futile attempt at restoring neuronal function, a paraphenomenon, or even a detrimental factor. However, experiments by Combs et al. (1992) on the effects of hyperglycemia on c-*fos* induction in ischemic brain suggests a protective role for c-*fos*. Since hyperglycemia enhances ischemic injury to brain, the added suppression of c-*fos* expression accompanying hyperglycemia indicates a potentially protective role. The notion of a protective role of c-*fos* and its companion c-*jun* is also suggested by their parallel increase with the degree and duration of ischemia (Kindy et al. 1991). The ability to augment or suppress proto-oncogene expression by manipulation of receptor activity promises to provide an investigative tool to identify more precisely the role of these proto-oncogenes. For example, in their study of various neruonal growth factors in ischemia and hypoglycemia, Lindvall et al. (1992) found increased expression of brain-derived neurotrophic factor (BDNF) mRNA and of nerve growth factor (NGF) in granule cells of the dentate gyrus while neurotrophin 3 was reduced in this region as well as in CA2 and the medial CA1 of the hippocampus. Of interest is their finding that increased synthesis of BDNF after ischemia can be blocked by the alpha-amino-3-hydroxy-5-methyl-isoxazole-4-propionic acid (AMPA) receptor inhibitor NBQX, while the *N*-methyl-D-aspartate (NMDA) receptor inhibitor MK801 was without effect. These findings suggest that receptor manipulation can either augment or suppress growth factor gene responses, thereby providing an ability to charaterize their role in ischemic neuronal vulnerability.

The role of specific proteins, such as heat shock or stress proteins and polyamines, believed to be protective for ischemic neurons, requires, as with proto-oncogenes, further studies. For example, the late ischemically induced increases in polyamine synthesis from ornithine by an induction of ornithine decarboxylase (OCD) has been inferred to restore normal function (Dempsey et al. 1991). However, polyamine synthesis is suppressed by phenobarbital, a γ-aminobutyric acid (GABA) agonist, while it is unaffected by MK801, the NMDA receptor antagonist. Since both agents are neuroprotective, the results suggest that polyamines may influence long-term neuronal recovery but not acute damage control. Despite this, the potential influence of OCD and polyamines as a late consequence in aggravating cerebral edema requires further assessment (Dempsey et al. 1990). A similar reasoning can be applied to the intensively studied heat shock protein (*hsp*-70). With cerebral ischemia, both the stress-induced and constitutively synthesized form of heat shock protein from the multigene family are augmented in their expression, as is the antecedent and presumably linked event of expressing the transcriptional zinc finger motif (Brown 1990; Abe et al. 1991). While protein synthesis with ischemia is delayed or unchanged (Nowak 1990; Xie et al. 1989; Nowak et al. 1990), the selective expression of *hsp*-70 suggests a functional role. Because the vulnerable CA1 neurons depress protein synthesis generally, while overexpressing *hsp*-70, Nowak (1990) concludes that these stress-induced proteins have a role, perhaps a detrimental one.

The issue of clinical relevance has two aspects of importance. The first is the ability to translate precisely and accurately the experimental data to the human condition, namely, for the status of the ischemically impaired tissue and its narrow window of vulnerability, and whether advanced available technology can precisely match animal studies with the patients' ischemic brain tissue. The close matching of vulnerability requires use of advance technology such as positron emission tomography (PET) and magnetic resonance spectroscopy (MRS) to evaluate ischemically impaired tissue in a dynamic fashion with precise identification of critically vulnerable processes.

The second issue is the need to resist temptations to rush headlong into the use of genetic engineering techniques in humans without sufficient experimental data. In our collective quest to find the ultimate prevention and "cure" of cerebral ischemia, the appeal and lure of quickly and hurriedly applying genetic engineering techniques may seduce our intellect to finding the quick cure at the expense of carefully planned and substantially documented studies justifying human experimentation. The fallout or consequence of falsely raising expectations in presumably correcting fundamental mechanisms, such as genes, based upon insufficient information is frustration and skepticism on the part of investigators, and cynicism and hostility from the public, the ultimate judge of human interventions. In considering experimentation on the brain and all that it implies for humanity and civilization, neuroscientists must be mindful of the adage *primum non nocere* as well as standards of conduct established by the Nuremberg Code and the Declaration of Helsinski in carrying out ethically and experimentally sound human research.

In summary, gene expression in cerebral ischemia holds great promise in providing clues for both minimizing ischemic injury and in maximizing recovery through plasticity, but efforts to apply genetic engineering techniques to the human condition of brain ischemia must be based upon faultless experimental data simulating the human condition.

References

Abe K, Kawagoe J, Sato S, Sahara M, Kogure K (1991) Induction of the "zinc finger" gene after transient focal ischemia in rat cerebral cortex. Neurosci Lett 123:248–250

Brown IR (1990) Induction of heat shock (stress) genes in the mammalian brain by hyperthermia and other traumatic events: a current prospective. J Neurosci Rev 27:247–255

Combs DJ, Dempsey RJ, Donaldson D, Kindy MS (1992) Hyperglycemia suppresses c-fos mRNA expression following transient cerebral ischemia in gerbils. J Cereb Blood Flow Metab 12:169–172

Dempsey RJ, Carney JM, Kindy MS (1990) Inducible ornithine decarboxylase expression in brain subject to vasogenic oedema after transient ischaemia: relationship to c-fos gene expression. Acta Neurochir Suppl (Wien) 51:189–191

Dempsey RJ, Carney JM, Kindy MS (1991) Modulation of ornithine decarboxylase mRNA following transient ischemia in the gerbil brain. J Cereb Blood Flow Metab 11:979–985

Jacewicz M, Kiessling M, Pulsinelli WA (1986) Selective gene expression in focal cerebral ischemia. J Cereb Blood Flow Metab 6:263–272

Jorgensen MB, Deckert J, Wright DC, Gehlert DR (1989) Delayed c-fos proto-oncogene expression in the rat hippocampus induced by transient global ischemia: an in situ hybridization study. Brain Res 484:393–398

Jorgensen MB, Johansen FF, Diemer NH (1991) Post-ischemic and kainic acid-induced c-fos protein expression in the rat hippocampus. Acta Neurol Scand 84:352–356

Kindy MS, Carney JP, Dempsey RJ, Carney JM (1991) Ischemic induction of protooncogene expression in gerbil brain. J Mol Neurosci 2:217–228

Lindvall O, Ernfors P, Bengzon J, Kokaia Z, Smith ML, Siesjö BK, Persson H (1992) Differential regulation of mRNAs for nerve growth factor, brain-derived neurotrophic factor, and neurotrophin 3 in the adult rat brain following cerebral ischemia and hypoglycemic coma. Proc Natl Acad Sci USA 89:648–652

Nowak TS Jr (1990) Protein synthesis and the heat shock/stress response after ischemia. Cerebrovasc Brain Metab Rev 2:345–366

Nowak TS Jr, Ikeda J, Nakajima T (1990) 70-kDa heat shock protein and c-fos gene expression after transient ischemia. Stroke 21 Suppl 11:III107–III111

Xie YX, Herget T, Hallmayer J, Starzinski-Powitz A, Hossmann KA (1989) Determination of RNA content in postischemic gerbil brain by in situ hybridization. Metab Brain Dis 4:239–251

Leukocytes, Microcirculation and the Coagulation System

Mechanisms of Leukocyte-Mediated No-Reflow and Tissue Injury in the Microcirculation

G.W. Schmid-Schönbein[1] and G.J. del Zoppo[2]

Introduction

Even though the number of leukocytes in the circulation, is small compared to red cells, evidence from molecular, biochemical, biophysical, cell biological, microcirculatory, physiological, pharmacological, and epidemiological studies suggests that they may be involved in a range of cardiovascular diseases. This adds a new dimension to the usual view of an *inflammatory response* by leukocytes, i.e., that leukocytes react to tissue injury like a fireman reacts to a fire after it has been started. Instead, there is the possibility that leukocytes may *initiate* tissue injury even to innocent bystanders, in such a way that they may in fact be the arsonists. In the following, we will summarize some of the underlying mechanisms that lead to leukocyte-mediated cytotoxicity with special emphasis to cerebral ischemia. In light of the rapidly increasing body of evidence to support this hypothesis, it is necessary to limit the discussion to the underlying mechanisms and the number of citations. The focus here will be on neutrophils and monocytes at the microvascular level since their role in ischemic conditions is currently best defined. We will discuss first basic biophysical and biochemical properties of leukocytes, their kinetics in the microcirculation, and then explore the current evidence for a leukocyte involvement in acute and chronic cardiovascular complications.

Neutrophil and Monocyte Biophysical Properties

Leukocytes have to deform when passing through capillaries. This deformation consists of a compression of the cells into a cylindrical shape at the entry to such narrow blood vessels, and it requires a driving pressure. Two issues, the cell morphology and deformability, come into focus in this re-

[1] Institute for Bioengineering Engineering, University of California, San Diego, La Jolla, CA 92093-0412, USA
[2] Department of Molecular and Experimental Medicine and Division of Hematology/Medical Oncology, The Scripps Research Institute, La Jolla, CA 92037, USA

Cerebral Ischemia and Basic Mechanisms
Ed. by A. Hartmann, F. Yatsu, and W. Kuschinsky

gard. Neutrophils and monocytes are quite different in this respect from red cells or platelets. Passive leukocytes are spherical in shape with numerous membrane folds on their surface providing the cell with an excess membrane, although the membrane itself is relatively nondistensible (Schmid-Schönbein et al. 1980a; Sung et al. 1982). In peripheral blood the leukocyte diameter is generally smaller than the dimension observed on blood smears (Schmid-Schönbein et al. 1984). But leukocyte diameters exceed the average capillary diameter, so that deformation is required for passage through the capillary network. The nucleus of circulating neutrophils makes up only about 20% of the cytoplasms; but the primary, secondary, and tertiary granules are present in large numbers (Schmid-Schönbein and Chien 1989). The neutrophilic granules are carriers of the lysosomal enzymes (Bainton et al. 1971; Dewald et al. 1983) and adhesion glycoproteins (Miller et al. 1987). The cell cytoplasms is made up largely of actin (Senda et al. 1969; Hartwig and Stossel 1975), actin-binding protein (Stossel and Hartwig 1976), myosin (Stossel and Pollard 1973; Tatsumi et al. 1973; Korn and Hammer 1988), and regulatory proteins for actin polymerization (Southwick and Stossel 1981; Yin et al. 1981; Southwick and Hartwig 1982). In the active state, leukocytes project spontaneously pseudopods, which are initially depleted of cytoplasmic organelles and enriched in the polymerized, fibrous form of actin (Valerius et al. 1981; Sullivan and Mandell 1983).

Whereas the red cell is made up of a soft and highly deformable membrane surrounding a low viscous interior of saturated hemoglobin suspension (viscosity of about 0.06 poise), the leukocyte cytoplasm is viscoelastic and substantially more rigid. Several biophysical models have been proposed (Schmid-Schönbein et al. 1981; LaCelle et al. 1982; Evans and Yeung 1989; Needham and Hochmuth 1990), and current estimates of the cytoplasmic viscous properties are three to four orders of magnitude higher than that of the red cells. The different types of leukocytes are similar in this respect. Pseudopods exhibit even more rigid viscoelastic properties. In the pseudopod, viscous creep is greatly reduced and stiff elastic properties dominate (Schmid-Schönbein et al. 1982) in spite of the fact that the pseudopod is the region which shows large deformation inside the cell (due to the actin polymerization). Thus, as we will discuss later, leukocytes with pseudopods have considerably greater difficulties to pass through capillaries (Sutton and Schmid-Schönbein 1992).

Leukocyte-Derived Superoxide Formation

Assembly of both a cytoplasmic and plasma membrane component leads to the activation of the nicotinamide-adenine-dinucleotide phosphate (reduced form; NADPH) oxidase and reduction of oxygen to superoxide (O_2^-) (Babior 1978). Superoxide is rapidly converted to hydrogen peroxide by superoxide dismutase (McCord and Fridovich 1969). These radicals are further converted to hydroxyl radicals or, in the presence of myeloperoxidase, to *N*-chloroamines, and both radicals may cause membrane and cell damage (McCord and Fridovich 1969). The NADPH oxidase can be stimulated by a range of receptor agonists and receptor-independent mechanisms [e.g., f-Met-Leu-Phe (fMLP) and analogs, complement, leukotrienes, cytokines, platelet-activating factor, endotoxins, various tissue breakdown products, and possibly by contact formation with other cells, e.g., mediated by the integrins], and it can be inhibited by adenosine, methyl xanthines, and albumin. The activation of the NADPH oxidase can be coupled to pseudopod formation (e.g., after application of fMLP), but in vivo these two activities may be uncoupled and individual leukocytes may exhibit different degrees of superoxide formation, pseudopod projection, and degranulation.

There are other sources of superoxide radical generation, such as xanthine oxidase, in mitochondria and endothelium (McCord 1987; Weiss 1989). Depending on the nature of the tissue stimulation, each source may contribute at different levels to the free radical-mediated tissue injury. In the following we will focus on situations which are associated with leukocyte accumulation in the tissue, so that the leukocyte-derived oxygen radicals predominate.

Leukocyte Adhesion

Adhesion requires formation of membrane contact area and therefore involves in many, but not all, cases unfolding of the plasma membrane surface folds. The exact mechanism by which this is achieved is not certain at the moment. Adhesive contact may just involve in part passive deformation by the attractive membrane forces between leukocyte and its substrate, whereas in other situations it may be facilitated by actin polymerization with an active spreading of the cell cytoplasm. In recent years a class of leukocyte membrane glycoproteins has been identified which appear to modulate adhesion to the endothelium and leukocyte activation. The relevance of these proteins has been emphasized in patients who are genetically deficient in these proteins (Harlan 1985; Ross et al. 1985). In neutrophils and monocytes, a family of heterodimer glycoproteins has been identified which mediate adhesion known as integrins (Sanchez-Madrid et al. 1983). The leukocyte integrins CD11a,b,c/CD18 share a common β chain (CD18) that is monocovalently associated with each α-chain (Hynes 1987). There is a corresponding family of glycoproteins, denoted as selectins, which are either constitutively expressed in ednothelial cells, or may be induced following stimulation with endotoxin or cytokines (McEver 1991). In nonactivated cells the integrins and certain selectins are usually present at a basal level, but stimulation leads to transfer of the adhesion glycoproteins from an intracellular pool to the plasma membrane where they are expressed (e.g., tertiary and secondary granules in neutrophils or Weibel-Palade bodies in endothelial cells) (Todd et al. 1984; Miller et al. 1987; Jones et al. 1988; Lacal et al. 1988). Stimulation of neutrophils (e.g., with fMLP) leads to expression of CD18, while at the same time the leukocyte adhesion molecule 1 (LAM-1) selectin appears to be shed from the membrane (von Andrian et al. 1991). Such *upregulation* results in elevation of membrane adhesion energy with enhanced attachment of leukocytes to the endothelium, the initial step in the transendothelial diapedesis. There are several counter-receptors for the integrins on the endothelial cells, for example, ICAM-1 (CD54) for the integrin CD11b/CD18 (Diamond et al. 1990) or an inducible cell ligand GMP-140 and ELAM-1 for LAM-1 (Picker et al. 1991; Spertini et al. 1991). ICAM-1 can also be found on parenchymal cells that are subject to a lethal attack by neutrophils (Smith et al. 1991a), an inducible endothelial cell ligand GMP-140 and ELAM-1 for LAM-1 (Picker et al. 1991; Spertini et al. 1991). Antibodies developed against surface epitopes of these adhesion glycoproteins have provided an important tool to explore the role of leukocyte adhesion in different immunological and inflammatory reactions. This is currently an active research area.

Microvascular Kinetics

Transport of leukocytes in the microcirculation is specific for the microanatomy of the host organ, the hemodynamics, and the rheological and adhesive properties of the leukocytes and the endothelium. Leukocytes behave differently than red cells and platelets. We will review this briefly based on results derived predominantly from peripheral organs, suitable for intravital microscopy. The situation in the brain is largely unexplored in this respect.

As leukocytes enter the arteriolar side of the microcirculation, they tend to be shunted along microvascular pathways with the highest flowrates (Ley et al. 1989). This tendency is due to the fact that at microvascular bifurcations leukocytes are preferentially entering the channel with the fastest flow (Schmid-Schönbein et al. 1980b; Fenton et al. 1985a,b; Gaehtgens et al. 1985). This effect exists for both red cell and leukocytes, but is more pronounced in the case of the leukocytes.

One of the implications for the tendency to shunt is that some regions of the microcirculation may not be perfused with leukocytes for prolonged intervals of time. An example is skeletal muscle. In its resting state, leukocytes are shunted into adjacent connective tissue, where capillaries are wider and it is easier for leukocytes to pass (Lindbom and Arfors 1985; Ley et al. 1989). Shunting is achieved by closure of terminal arterioles supplying a local capillary network, while the majority of the blood flow and the leukocyte flux continue along the arcade arterioles. The terminal arterioles are maintained to a large degree under myogenic autoregulation. In contrast, in the low-flow state, such as during ischemia or shock, the shunting mechanism is upset. Under low flow, the transmural pressure at the level of the terminal arterioles is reduced, they dilate and thereby permit entry of an increased number of leukocytes into the capillary network, but in the presence of a reduced capillary perfusion pressure. This is an undesirable condition which leads to microvascular trapping of leukocytes.

At the entry into a capillary, the transit time for a leukocyte tends to be longer than that for a red cell (Bagge et al. 1972; Fenton et al. 1985a,b; Yamakawa et al. 1987; Warnke and Skalak 1990). This is due to their large size and stiffness, and leads to temporary plugging of capillaries. The entry time depends critically on the capillary diameter and becomes more significant when the vessel diameters fall below 7–6 μm. Temporary plugging occurs typically at the capillary side-branches of the arterioles, a region of the microcirculation that is close to the precapillary sphincters. Leukocytes are passively deformed at the capillary entries, a process which is driven by the prevailing pressure drop.

Once inside a capillary, the leukocytes move with a lower velocity than the red cells (Mayrovitz 1982). This difference in velocity leads to the characteristic leukocyte–erythrocyte train formation inside capillaries and consequently systematic attachment of leukocytes to the endothelium in postcapillary venules. This is brought about by the following mechanism. The cell train formation is hydrodynamically unstable. The faster-flowing red cells tend to pass the slower-moving leukocytes and in the course displace the leukocytes from a center position towards the wall in the postcapillary venules so that attachment between leukocyte and endothelium under the relatively short-range forces (less than 0.1 μm) of membrane adhesion becomes possible (Schmid-Schönbein et al. 1980a–c). Systematic attachment of leukocytes to the endothelium is also possible at postcapillary convergent bifurcations where two blood streams join together and the red cells of one stream displace leukocytes in the neighboring stream towards the endothelium (Bagge and Karlsson 1980). Each leukocyte which passes through the microcirculation is pushed in close contact to the endothelium and thereby initiates the adhesion process. The effect is due to the red cells since, under experimental conditions when red cells are removed from the circulation and only leukocytes are present, few leukocytes become attached to the endothelium and most leukocytes that enter the microcirculation on the arteriolar side exit on the venular side (Bagge et al. 1983). In larger postcapillary venules, red cell aggregation causes radial displacement of leukocytes towards the endothelial wall (Goldsmith and Spain 1984; Nobis et al. 1985).

The red cell–leukocyte train formation in capillaries leads also to a substantial elevation of the hemodynamic resistance (Thompson et al. 1989). This observation raises a fundamental point about the influence of leukocytes in the microcirculation. The presence of leukocytes in whole organ resistance has in the past been neglected, mostly due to the relatively small volume ratio that is contributed to whole blood by the leukocytes (less than 1%). While this is a good approximation in large blood vessels, the situation may be different in the microcirculation. There are few quantitative data in the brain, but in skeletal muscle preparations even a modest number of leukocytes have been reported to make a significant contribution to whole organ resistance (Braide et al. 1984; Warnke and Skalak 1990). Direct measurements by means of whole organ perfusion studies show that the resistance imposed by a single leukocyte in the rat gracilis muscle is on average equal to that of about 750 red cells. If the leukocytes are activated and project pseudopods, this ratio increases (Sutton and Schmid-Schönbein 1992). These results indicate that leukocytes, inspite of their negligible cell count compared with red cells, cannot be neglected during normal physiological perfusion, and even less so in pathological situations.

Leukocyte Capillary Obstruction

As mentioned above, the pressure drop along the capillaries is the only driving force for passage of leukocytes. If the arterial pressure is reduced, either due to an atherosclerotic lesion in arteries, central hypotension (e.g., in shock) or due to other reasons, two complications arise. First, the capillaries will passively relax and thereby reduce their intraluminal dimensions (Fung and Sobin 1972; Skalak and Schmid-Schönbein 1986). Consequently, the leukocytes must enter even narrower channels than under normal physiological pressures. Second, since the pressure drop across the length of the capillaries is reduced, the leukocytes are insufficiently deformed and become trapped. This occurs preferentially in the vicinity of protruding endothelial nuclei and frequently involves neutrophils and monocytes (Fig. 1). No swelling of the endothelium is required, but adhesion between the endothelial and leukocyte membrane is required to prevent the leukocyte from dislodging out of the plugged position. The smaller lymphocytes are less prone to capillary entrapment.

Granulocyte capillary obstruction has been observed in ischemic heart muscle (Engler et al. 1983), brain (Yamakawa et al. 1987; Del Zoppo et al. 1991), skeletal muscle (Bagge et al. 1980), the lung (Wilson 1972; Worthen et al. 1989), and several other organs (Barroso-Aranda et al. 1988). Under conditions of reduced perfusion pressure, the entrapment of leukocytes in capillaries is predominantly a hemodynamic event, so restoration of the perfusion pressure serves as the main intervention. In acute ischemia, the majority of leukocytes involved are neutrophils (Fig. 2). During the initial phase of the reaction, most neutrophils remain intravascular. Obstructed capillaries contain also high numbers of red cells and also an enrichment of platelets, but both cell types are passively trapped due to the leukocytes. At the instant of time when the leukocyte moves out of an obstructed capillary,

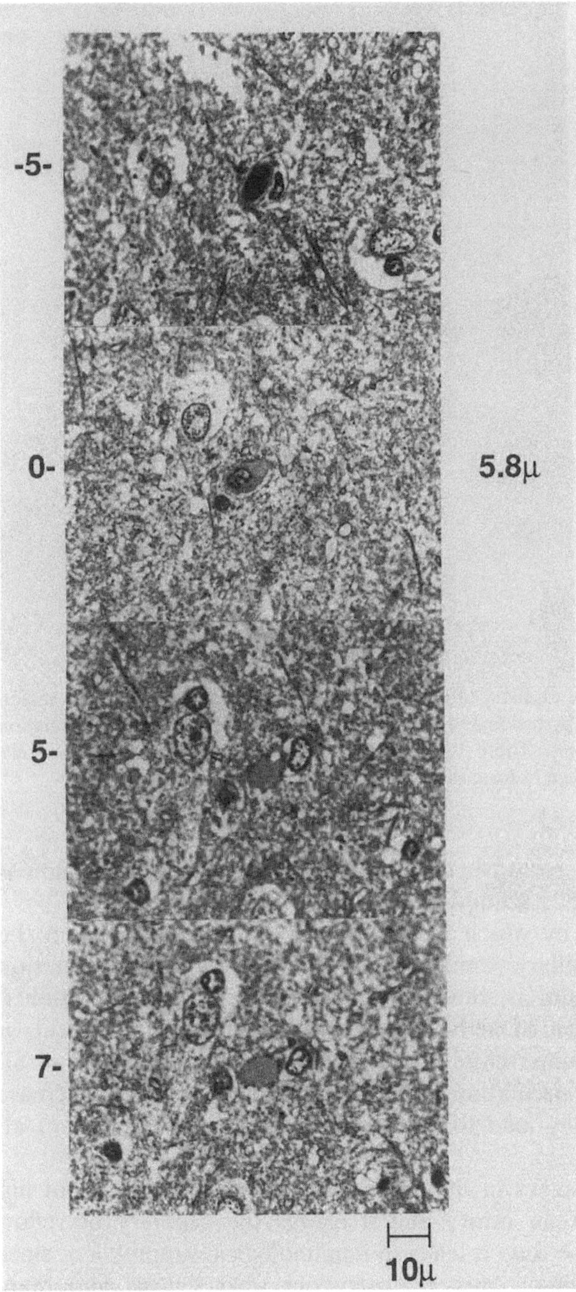

Fig. 1. Serial reconstruction of occluded microvessel at 5–7-μm intervals from basal ganglia following ischemia and reperfusion. A granulocyte can be seen obstructing a capillary with 5-μm diameter after an ischemia and reperfusion protocol. (From Del Zoppo et al. 1991)

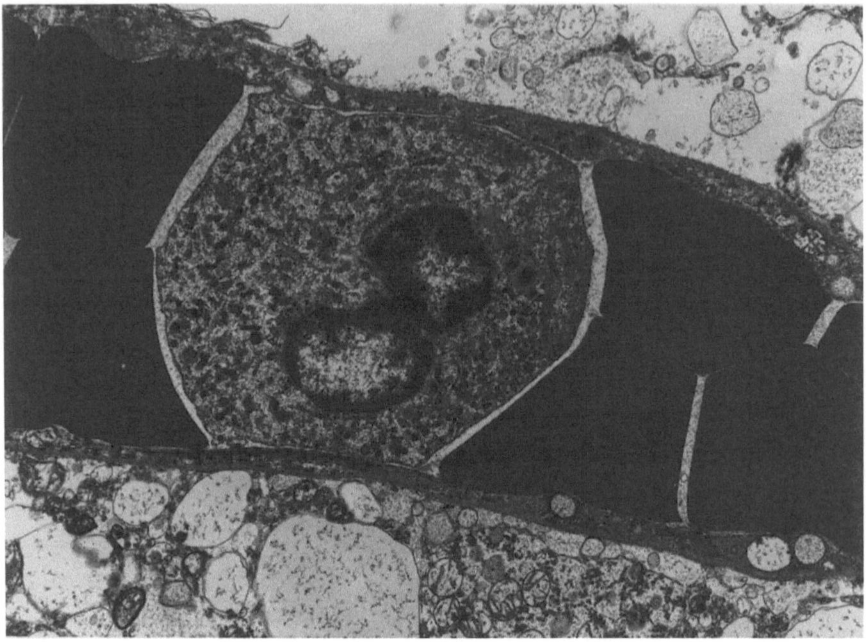

Fig. 2. Neutrophil obstructing a capillary (about 6 μm in diameter) in the basal ganglia region after 3-h middle cerebral artery occlusion and 1-h reperfusion. Note the close contact formation with the endothelium (<120 nm). There is no significant swelling of the endothelial cells although signs of ischemic damage in form of cytoplasmic vacuoles are present

all red cells and platelets are washed out of the capillary as well and flow is restored (Bagge et al. 1980; Schmid-Schönbein 1987).

Another mechanism by which leukocytes may become trapped in the capillaries and/or postcapillary venules occurs when the membrane adhesion energy to the endothelium is elevated, a situation that can be studied experimentally by infusion of activating agent, i.e., after complement or endotoxin infusion (Hammerschmidt et al. 1981; Barroso-Aranda et al. 1991). In the pulmonary circulation it has also been shown that an increase in neutrophil stiffness may lead to enhanced entrapment (Worthen et al. 1989).

Entrapment of leukocytes in the microcirculation has a number of implications. It initiates organ injury and it causes the capillary no-reflow phenomenon. At the same time it leads to immunological suppression, since entrapped leukocytes cannot properly carry out phagocytosis and there is reduced likelihood of contact formation between microorganisms and leukocytes. Attachment to a single site on the endothelium may initiate transformation of the leukocytes (mostly neutrophils and monocytes) into an active state. The cells then project pseudopods and eventually migrate across the endothelium into the interstitium. Attachment of the leukocytes to the postcapillary endothelium is frequently associated with oxygen free radical formation, especially in the gap region between leukocytes and endothelial membranes (Suematsu et al. 1989). This free radical formation involves the integins (CD11b/CD18) (Suematsu et al. 1993). Under experimental conditions of inflammation or ischemia with reperfusion, attachment of leukocytes to microvascular endothelium is largely limited to the venules. Attachment to the arterioles occurs less often, even when flowrates in the arterioles are reduced and stimulators are applied (House and Lipowsky 1987). This suggests that the membrane adhesion energy to the arteriolar endothelium is lower than to the venular endothelium, possibly by a reduced expression of adhesion glycoprotein.

An interesting exception to this situation has been observed after infusion of oxidized lipids into the circulation. This leads to activation and attachment of monocytes to the arterioles, a situation not observed otherwise (Lehr et al. 1991b). Mildly oxidized lipids are known to activate monocytes (Drake et al. 1991), which leads to the surface expression of the procoagulant tissue factor (TF) on monocytes. TF is the natural cofactor for factor VII/VIIa, which is required for thrombin generation and fibrin formation (Bach 1988; Davie et al. 1991). The expression of the glycoproteins in the capillary network per se is largely unexplored at the moment.

Leukocyte Interaction with Venular Endothelium

Leukocytes adhering to the endothelium of venules are exposed to a shear force and moment exerted by the passing blood stream, which deform the cell and make it roll on the endothelium or wash it away altogether (Lipowsky et al. 1991). These hemodynamic forces are opposed in the region of membrane contact to the endothelium by the adhesion energy, mediated in part by the integrins and selectins discussed above (Zimmerman and McIntyre 1988). In the membrane contact region there act both shear stress and normal stress. The type of motion which leukocytes exhibit on the postcapillary venular endothelium depends on the balance between these forces (Schmid-Schönbein et al. 1975; Lawrence et al. 1990). If membrane adhesion is weak (i.e., few glycoproteins are active or expressed) then under normal blood flow the leukocyte will be quickly washed away. If blood flow is reduced, the leukocytes may still be retarded in spite of weak adhesion. In contrast, if the membrane adhesion energy is high, then even strong shear forces by the blood may be insufficient to remove the cells. Leukocytes frequently roll along the endothelium, a situation in which the moment exerted by the blood stream exceeds the moment supported in the contact region, but not the shear force.

Recent evidence suggests that neutrophil rolling may be mediated by the selectins (e.g., LECAM-1) while transendothelial migration may require integrins (e.g., CD11b/CD18) (Lawrence and Springer 1991; Smith et al.

1991b; von Andrian et al. 1991). Neutrophils which establish adhesion with the endothelium and stop rolling may enter into an active stage. They exhibit amoeboid migration on the endothelium either against or in the direction of the blood stream. They crawl across the endothelium and may rest for prolonged periods between endothelial cells and the basement membrane, suggesting that the latter may be a mechanical barrier for neutrophils. Penetration may require proteolytic cleavage of basement membrane proteins, such as collagen (Weiss et al. 1985). Neutrophils may migrate in both directions across the endothelium. In the bone marrow, neutrophils migrate across pores through the endothelial cells rather than through interendothelial junctions. In the peripheral circulation a similar process may occur (Hammersen and Hammersen 1987).

Neutrophil Microvascular Kinetics in Acute Ischemia and Shock

Ischemia and reperfusion lead to accumulation of inflammatory cells in the microcirculation of the heart (Romson et al. 1983; Engler et al. 1986; Mullane et al. 1987; Breitbart et al. 1990), liver (Holman and Saba 1988; Komatsu et al. 1990), skeletal muscle (Kurthius et al. 1988; Smith et al. 1989), lung (Hogg 1987), intestine (Smith et al. 1987), and possibly many other organs.

In the brain similar evidence exists (Pozzilli et al. 1985; Hallenbeck et al. 1986). In fact, in the brain, the first direct evidence was obtained that ischemia leads to capillary occlusions and a no-reflow phenomenon (Ames et al. 1968). Direct observation of occluded vessels with histological techniques, both at the light- and electron-microscopic levels, shows that the no-reflow phenomenon is limited to the true capillaries (less than about $8\,\mu$m lumen diameter), whereas arterioles and venules are reperfused (Del Zoppo et al. 1991). Occluded capillaries are intact; there is mild, if any, swelling of the endothelium after short reperfusion periods; and packed red cells and occasional platelets are observed. Individual neutrophils are encountered which are deformed inside the capillaries into cylindrical shapes with close membrane abutment to the endothelium, similar to the situation in other organs (Bagge 1980; Engler et al. 1983). Attachment of neutrophils to postcapillary venules can also be detected, although this may not lead to microcirculatory occlusion while such neutrophils are a factor in tissue injury. Neutropenia has been shown to be beneficial in several models of brain ischemia (Dutka et al. 1989; Vasthare et al. 1990).

Similar accumulations of leukocytes in the microcirculation can be observed in hemorrhagic or endotoxic shock, although under these conditions virtually all organs are subject to leukocyte accumulation. A shift of the leukocyte population manifests itself by a dramatic drop of the circulating leukocyte count at the beginning of the shock period. The reasons for this initial leukocyte trapping in hemorrhagic versus endotoxic shock are, however, different: in hemorrhagic shock it is due to the pressure reduction, in endotoxic shock it is due to activation of the leukocytes and endothelium with enhancement of the membrane adhesion and possibly cell stiffening. Thus, in the presence of a stimulator such as endotoxin, neutrophils and monocytes adhere to microvessel walls in spite of normal perfusion. Accumulation of neutrophils leads to progressive organ injury and, since no organ is excluded from the entrapment, there is a tendency for multi-organ failure, one of the hallmarks of shock. Under experimental conditions a carefully executed neutropenia, in which widespread trapping of neutrophils in the vital organs is avoided, serves to protect against the cytotoxic effects of neutrophils in hemorrhagic shock (Barroso-Aranda et al. 1988).

Spontaneous Neutrophil Activation

One of the interesting features of neutrophils in the hemorrhagic shock model is that these cells may serve to predict survival. Under conditions of relatively short periods of pressure reduction (for example, at a mean arterial pressure of 40 mmHg for 90 min) a well-known feature is observed. In spite of presumably normal animals, some survive, whereas the remainder succumb at various stages to the shock insult. To gain further insight into the mechanism by which neutrophils influence the course of the syndrome, a distinction between active and passive cells was made. One specific form of neutrophil activation is the capacity to reduce nitro-blue tetrazolium (NBT) dye, which is associated with an enhanced generation of superoxide anion radicals. There is a distribution of such NBT-positive neutrophils such that some animals have higher counts, others may have values as low as zero. Animals with an initial count of activated circulating neutrophils *before* hemorrhagic shock above about 225 cells/mm^3 will not survive. In contrast, when the number of activated circulating neutrophils before pressure reduction is lower, a high percentage of the animals survive (Barroso-Aranda and Schmid-Schönbein 1989). There is thus a positive correlation between the number of activated circulating neutrophils and the lethality after entrapment in the microcirculation following the pressure reduction. The exact origin of the spontaneously activated circulating leukocytes is currently uncertain; however infections, intestinal endotoxin translocation, environmental factors like smoking (Drost et al. 1992), and hypertension (Schmid-Schönbein et al. 1991) may play a role.

Interventions

Many different pharmacological agents influence leukocyte function. The problem, however, is to interfere with the cytotoxic activity of these cells, while leaving the protective activity intact. Among potentially useful strategies are anti-inflammatory agents, agents directed against the complement system, against cytokine release, anti-oxygen free radical therapies, use of carbohydrate moieties or immuntherapeutic protection against adhesion receptors, or reduction of spontaneous activation. One of the ways by which we have succeeded in reducing the pool of spontaneously activated neutrophils is by chronic treatment with a methyl-xanthine derivative. This leads to improvement of leukocyte deformability (Schmalzer and Chien 1984; Needham and Hochmuth 1990); reduced adhesion to endothelium in vivo; and suppression of tumor necrosis factor formation, one of the powerful stimulators for neutrophils (Hakim and Mandell 1990). Consequently there is improvement in several clinical outcomes, including survival after shock (Barroso-Aranda and Schmid-Schöbein 1990a,b). Treatment of experimental animals with methyl-xanthine serves to minimize spontaneous neutrophil activation and organ injury, especially if pretreatment can be carried out. Pretreatment may also be beneficial in other situations, such as organ injury mediated during general surgery, as well as in chronic vascular complications (Colgan et al. 1990).

Since the discovery of adenosine as an endogenous downregulator of neutrophils (Cronstein et al. 1986, 1992), a protective action of this metabolite in the ischemic heart under experimental conditions could be demonstrated (Gruber et al. 1989; Kaminski and Proctor 1989). This protection is independent of the vasodilatory action of adenosine. In acute situations, application of monoclonal antibodies against the leukocyte CD11/CD18 has been reported to protect against neutrophil- and monocyte-mediated organ injury in different ischemic conditions (Arfors et al. 1987; Hernandez et al. 1987; Simpson et al. 1988; Vedder et al. 1988, 1990; Ma et al. 1991; Walsh et al. 1991). An improvement in microvascular reflow has been documented in experimental and focal central ischemia in the brain of primates (Mori et al. 1992), while the effect of the same intervention has also improved the clinical outcome in other forms of central nervous system ischemia (Clark et al. 1991). There are other mechanisms to interfere with leukocyte cytotoxicity, such as nitric oxide (Kubes et al. 1991), dietary fish oils (Lehr et al. 1991a), oxygen free radical scavengers, arachidonic acid inhibitors, and others. It is interesting to note that apparently different biochemical pathways which protect against tissue injury in ischemia have one factor in common: they prevent the accumulation of leukocytes in the microcirculation (Mullane et al. 1987). The key issue in any therapeutic intervention may be to prevent leukocyte accumulation and activation in local regions of the microcirculation. This may not be achieved solely by interventions against leukocyte adhesion in some situations, since reduction of adhesion leads inevitably to recruitment of leukocytes from the marginating pool back into the active circulation. Consequently elevated numbers of leukocytes are transported by the blood stream into the microcirculation, which can be counterproductive in some organs if local activation is still present via alternative biochemical pathways.

Microvascular Accumulation of Leukocytes in Chronic Conditions

The entrapment of leukocytes in chronic vascular disease is still less explored, but there are suggestions that complications arising from circulating activated leukocytes may play an important role in chronic vascular disease. An important problem in this regard is diabetic retinopathy. Microangiographic observations have indicated that prior to neuro-optical dysfunction there exists an interruption of capillary flow in the retinal microcirculation. The flow interruptions originate almost exclusively in vessels of the capillary network, while involvement of arterioles and venules in vascular lesions usually occurs at a later time. In the alloxane-treated rat diabetic model, monocytes are predominantly responsible for the capillary occlusion (Schröder et al. 1991). Vascular lesions occur exclusively in regions of monocyte accumulation, even after a prolonged diabetic state. Accumulation of the monocytes is accompanied by injury to the endothelium and interstitial cells and their phagocytosis by monocytes/macrophages. In addition, neovascularization of characteristically enlarged and convoluted venular microvessels occurs around local interstitial aggregates of monocytes/macrophages (Schröder et al. 1991). The situation may not be unlike lesions in atherosclerosis or in living injury, both exhibit a major involvement of monocytes (Jerome and Lewis 1984). Entrapment and activation of neutrophils has also been reported in claudication (Neumann et al. 1990).

As we discussed above, the cause for the abnormal trapping of monocytes in the microcirculation of diabetics appears to be an induction of spontaneous monocyte activation, since in such diabetic animals there is no apparent compromise of the central blood pressure. The spontaneous activation of the circulating neutrophils and monocytes correlates closely with elevation of the blood glucose level, even at early stages after initiation of the diabetic state. Upregulation of the adhesion receptors was seen on human diabetic monocytes (Setiadi et al. 1987) and constitutes a risk for microvascular trapping.

Technical Limitations and Epidemiology

There are two important technical problems that surface in numerous clinical and experimental studies which deserve special attention. First, although leukopenia serves as an important test case in experimental situations designed to investigate the role of leukocytes, it is plagued with difficulties. Among several procedures to induce leukopenia (e.g., radiation, cytotoxins, monoclonal antibodies directed against neutrophil epitopes), few actually serve to remove completely the neutrophils from the circulation, but instead merely lead to a shift of the circulating pool into the marginating pool. There is usually little control over the location of leukocyte trapping or the degree of activation involved. Consequently, it is insufficient to record after leukopenia a reduced leukocyte count in the central circulation, and instead it is necessary to demonstrate before experimentation that the target organ under investigation is free of leukocytes and unaffected by the steps necessary to achieve leukopenia.

Second, there is a problem with the detection of leukocytes and their spontaneous physiological activity in venous or arterial blood samples and the interpretation of results derived from venous blood. This is fundamental to clinical research. The uncertainty arises since fully activated leukocytes in the circulation do not remain there, but instead become trapped in the microcirculation. The circulating leukocytes represent only a subpopulation of either nonactivated cells or only partially activated or primed cells. Arterial or venous central blood is constantly filtered by the microcirculation with respect to circulating activated cells. The truly activated leukocytes, which may represent in many situations the true culprit for the ongoing organ injury, may not be detectable in venous blood samples. This is unfortunate but unavoidable. It may be necessary that alternative approaches be developed to detect leukocyte activation in blood samples, e.g., the plasma of the patients be applied to *control leukocytes* from volunteers without overt disease.

In spite of these limitations, it has been possible to document elevated leukocyte counts as risk factors in stroke or early following stroke (Prentice et al. 1982; Ernst et al. 1987). Blood samples of patients after stroke exhibit abnormal rheological properties (Ciuffetti et al. 1989), including upregulation of CD18 on neutrophils (Grau et al. 1992). We should consider the possibility that the cause of receptor upregulation may also underly infiltration of leukocytes into the cerebral microcirculation and subsequent tissue injury, rather than being the result of the cerebral vascular accident itself.

Conclusion

In light of the basic biophysical, biochemical, and microcirculatory conditions that are associated with the high propensity of leukocytes to accumulate in various tissue regions, it may be suggested that such complications are not limited to the cases discussed above. In spite of the extensive evidence that has come forward in the last decade, we may have still just seen the "tip of the iceberg." Leukocyte-mediated cytotoxicity may not be limited to the obvious and severe clinical complications, such as stroke or myocardial ischemia, but instead may also occur chronically in many tissues at levels not necessarily associated with complications which warrant clinical attention. More subtle microvascular and tissue restructuring may also be caused by similar events on a limited scale. Reduction in leukocyte-mediated injury may be most effective if treatment interferes early with the activation and local accumulation of inflammatory cells. Such strategies need to be designed without compromising the immunological and antimicrobial activity of leukocytes.

Acknowledgement. This research was supported by USPHS Grant HL-10881, by Program Project Grant 43026 (Dr. Shu Chien, Director), and by Grant NS 26945.

References

Ames A, Wright RL, Kowada M, Thurston JM, Majno G (1968) Cerebral ishemia. II. The no-reflow phenomenon. Am J Pathol 52:437–453

Arfors K, Lundberg C, Lindbom L, Lundberg K, Beatty P, Harlan J (1987) A monoclonal antibody to the membrane glycoprotein complex CD18 inhibits PMN accumulation and plasma leakage in vivo. Blood 69:338–340

Babior BM (1978) Oxygen dependent microbial killing by phagocytes. N Engl J Med 288:659–668

Bach RR (1988) Initiation of coagulation by tissue factor CRC Crit Rev Biochem 23:339–368

Bagge U (1980) White blood cell deformability and plugging of skeletal muscle capillaries in hemorrhagic shock. Acta Physiol Scand 108:159–163

Bagge U, Karlsson R (1980) Maintenance of white cell margination at the passage through small venular junctions. Microvasc Res 20:92–95

Bagge U, Skalak R, Attefors R (1972) Granulocyte rheology. Experimental studies in an in-vitro microflow system. Adv Microcirc 7:29–48

Bagge U, Amundson B, Lauritsen C (1980) White blood cell deformability and plugging of skeletal muscle capillaries in hemorrhagic shock. Acta Physiol Scand 108:159–163

Bagge U, Blixt A, Strid K-G (1983) The initiation of post-capillary margination of leukocyets: studies in vitro on the influence of erythocyte concentration and flow velocity. Int J Microcirc Clin Exp 2:215–227

Bainton DF, Ullyot JL, Fahrquar MG (1971) The development of neutrophilic polymorphonuclear leukocytes in human bone marrow. Origin and content of azurophil and specific granules. J Exp Med 134:907–934

Barroso-Aranda J, Schmid-Schönbein GW (1989) Transformation of neutrophils as measured by NBT-reduction as indicator of irreversibility in hemorrhagic shock. Am J Physiol 257:H846–H852

Barroso-Aranda J, Schmid-Schönbein GW (1990a) Pentoxifylline pretreatment decreases the pool of circulating activated neutrophils, in-vivo adhesion to endothelium, and improves survival from hemorrhagic shock. Biorheology 27:401–418

Barroso-Aranda J, Schmid-Schönbein GW (1990b) Pentoxifylline pretreament decreases neutrophil activation during endotoxic shock and improves survival. In: Hakim J, Mandell GL (eds) Pentoxifylline and analogues: effects on leukocyte function. Karger, Basel, pp 97–103

Barroso-Aranda J, Schmid-Schönbein GW, Zweifach BW, Engler RL (1988) Granulocytes and no-reflow phenomenon in irreversible hemorrhagic shock. Circ Res 63:437–447

Barroso-Aranda J, Schmid-Schönbein GW, Zweifach BW, Mathison JC (1991) Polymorphonuclear neutrophil contribution to induced tolerance by bacterial lipopolysaccaride. Circ Res 69:1196–1206

Braide M, Amundson B, Chien S, Bagge U (1984) Quantitative studies on the influence of leukocytes on the vascular resistance in a skeletal muscle preparation. Microvasc Res 27:331–352

Breitbart GB, Dillon PK, Suval WD, Pasberg FT, FitzPatrick M, Durán WN (1990) Leukopenia reduces microvascular clearance of macromolecules in ischemiareperfusion injury. Curr Surg 47:8–12

Ciuffetti G, Balendra R, Lennie SE, Anderson J, Lowe GD (1989) Impaired filterability of white cells in acute cerebral infarction. Br Med J 298:930–931

Clark WM, Madden KP, Rothlein R, Zivin JA (1991) Reduction of central nervous system ischemic injury in rabbits using leukocyte adhesion antibody treatment Stroke 22:877–883

Colgan M-P, Dormandy JA, Jones PW, Schraibman IG, Shanik DG, Young RAL (1990) Oxpentifylline treatment of venous ulcers of the leg. Br Med J 300:972–974

Cronstein BN, Levin RI, Belanoff J, Weissmann G, Hirschhorn R (1986) Adenosine: an endogenous inhibitor of neutrophil-mediated injury to endothelial cells. J Clin Invest 78:760–770

Cronstein BN, Levin RI, Philips M, Hirschhorn R, Abramson SB, Weissman G (1992) Neutrophil adherence to endothelium is enhanced via adenosine A1 receptors and inhibited via adenosine A2 receptors. J Immunol 148:2201–2206

Davie EW, Fujikawa K, Kisiel W (1991) The coagulation cascade: initiation, maintenance, and regulation. Biochemistry 30:10363–10370

Del Zoppo GJ, Schmid-Schönbein GW, Mori E, Copeland BR, Chang C-M (1991) Polymorphonuclear leukocytes occlude capillaries following middle cerebral artery occlusion and reperfusion. Stroke 22:1276–1283

Dewald B, Bretz U, Baggiolini M (1983) Exocytosis induced in neutrophils by chemotactic agents and other stimuli. Agents Actions Suppl:371–381

Diamond MS, Staunton DE, de Fougerolles AR, Stacker SA, Garcia-Aguilar J, Springer TA (1990) ICAM-1 (CD54): a counter receptor for Mac-1 (CD11b/CD18). J Cell Biol 111:3129–3139

Drake TA, Hannani K, Fei H, Lavi S, Berliner JA (1991) Minimally oxidized lowdensity lipoprotein induces tissue factor expression in cultured human endothelial cells. Am J Pathol 138:601–607

Drost EM, Selby C, Lannan S, Lowe GD, MacNee W (1992) Changes in neutrophil deformability following in vitro smoke exposure: mechanism and protection. Am J Respir Cell Mol Biol 6:287–295

Dutka AJ, Kochanek PM, Hallenbeck JM (1989) Influence of granulocytopenia on canine cerebral ischemia induced by air embolism. Stroke 20:390–395

Engler RL, Schmid-Schönbein GW, Pavelec RS (1983) Leukocyte capillary plugging in myocardial ischemia and reperfusion in the dog. Am J Pathol 111:98–111

Engler RL, Dahlgren MD, Peterson MA, Dobbs A, Schmid-Schönbein GW (1986) Early accumulation of polymorphonuclear leukocytes during three hour myocardial ischemia. Am J Physiol 251:H93–H100

Ernst E, Hammerschmidt E, Bagge U, Matrai A, Dormandy JA (1987) Leukocytes and the risk of ischemic disease. JAMA 257:2318–2324

Evans E, Yeung A (1989) Apparent viscosity and cortical tension of blood granulocytes determined by micropipet aspiration. Biophys J 56:151–160

Fenton BM, Carr RT, Cokelet GR (1985a) Nonuniform red cell distribution in 20 to 100 μm bifurcations. Microvasc Res 29:103–126

Fenton BM, Wilson DW, Cokelet GR (1985b) Analysis of the effects of measured white blood cell entrance times on hemodynamics in a computer model of a microvascular bed. Pflügers Arch. 403:395–401

Fung YC, Sobin SS (1972) Elasticity of the pulmonary alveolar sheet. Circ Res 30:451–468

Gaehtgens P, Ley K, Pries AR, Müller R (1985) Mutual interactions between leukocytes and microvascular blood flow. Prog Appl Microcirc 7:15–28

Goldsmith HL, Spain S (1984) Margination of leukocytes in blood flow through small tubes. Microvasc Res 27:204–222

Grau AJ, Berger E, Sung K-L P, Schmid-Schönbein GW (1992) Granulocyte adhesion, deformability, and superoxide formation in acute stroke. Stroke 22:33–39

Gruber HE, Hoffer ME, McAllister DR, Laikind PK, Lane TA, Schmid-Schönbein GW, Engler RL (1989) Increased adenosine concentration in blood from ischenic myocardium by AICA riboside: effects on flow, granulocytes, and injury. Circulation 80:1400–1411
Hakim J, Mandell GL (eds) (1990) Pentoxifylline and analogues: effect on leukocyte function. Karger, Basel
Hallenbeck J, Dutka A, Tanishima T, Kochanek P, Kumaroo K, Thompson C, Obreovitch T, Contreras T (1986) Polymorphonuclear leukocyte accumulation in brain regions with low blood flow during the early postischemic period. Stroke 17:246–253
Hammerschmidt DE, Harris PD, Wayland JH, Craddock PR, Jacob HS (1981) Complement-induced granulocyte aggregation in-vivo. Am J Pathol 102:146–150
Hammersen F, Hammersen E (1987) The ultrastructure of endothelial gap-formation and leukocyte emigration. Prog Appl Microcirc 12:1–34
Harlan JM (1985) Leukocyte-endothelial interactions. Blood 65:513–525
Hartwig JH, Stossel TP (1975) Isolation and properties of actin, myosin and a new actin-binding protein in rabbit alveolar macrophages. J Biol Chem 250:5696–5705
Hernandez L, Grisham M, Twohig B, Arfors K, Harlan J, Granger N (1987) Role of neutrophils in ischemia reperfusion induced microvascular injury. Am J Physiol 253:H699–H703
Hogg JC (1987) Neutrophil kinetics and lung injury. Physiol Rev 67:1249–1295
Holman JM, Saba TM (1988) Hepatocyte injury during post-operative sepsis: activated neutrophils as potential mediators. J Leuk Biol 43:193–203
House SD, Lipowsky HH (1987) Leukocyte-endothelium adhesion: microhemodynamics in mesentery of the cat. Microvasc Res 34:363–379
Hynes RO (1987) Integrins: a family of cell surface receptors. Cell 48:549–554
Jerome WG, Lewis JC (1984) Early atherogenesis in white Carneau pigeons Am J Pathol 116:56–68
Jones DH, Anderson DC, Burr BL, Rudloff HE, Smith CW, Krater SS, Schmalstieg FC (1988) Quantitation of intracellular Mac-1 (CD11b/Cd18) pools in human neutrophils. J Leuk Biol 44:535–544
Kaminski PM, Proctor KG (1989) Attenuation of no-reflow phenomenon, neutrophil activation, and reperfusion injury in intestinal microcirculation by topical adenosine. 65:426–435
Komatsu RA, Koo A, Guth PH (1990) Leukocyte flow dynamics in the rat liver microcirculation. Microvasc Res 40:1–13
Korn ED, Hammer JAI (1988) Myosins of nonmuscle cells. Annu Rev Biophys Chem 17:23–45
Kubes P, Suzuki M, Granger DN (1991) Nitric oxide: an endogenous modulation of leukocyte adhesion. Proc Natl Acad Sci USA 88:4651–6455
Kurthius RJ, Grisham MB, Granger CN (1988) Leukocyte depletion attenuates vascular injury in postischemic skeletal muscle. Am J Physiol 254:H823–H827
Lacal P, Pulido R, Sánchez-Madrid F, Cabañas C, Mollinedo F (1988) Intracellular localization of a leukocyte adhesion glycoprotein family in the tertiary granules of human neutrophils. Biochem Biophys Res Comm 154:641–647
LaCelle PL, Bush RW, Smith BD (1982) Viscoelastic properties of normal and pathologic human granulocytes and lymphocytes. In: Bagge U, Born GVR, Gaehtagens P (eds) White blood cells morphology and rheology as related to function. Nijhoff, The Hague
Lawrence MB, Springer TA (1991) Leukocytes roll on a selectin at physiologic flow rates: distinction from and prerequisite for adhesion through integrins. Cell 65:859–873
Lawrence MB, Smith CW, Eskin SG, McIntire LV (1990) Effect of venous shear stress on CD18-mediated neutrophil adhesion to cultured endothelium. Blood 75:227–237
Lehr HA, Hubner C, Finckh BDN, Beisiegel U, Kohlschutter A, Messmer K (1991a) Dietary fish oil reduces leukocyte/endothelium interaction following systemic administration of oxidatively modified low density lipoprotein. Circulation 84:1725–1731
Lehr HA, Hubner C, Nolte D, Finckh B, Beisiegel U, Kohlschutter A, Messmer K (1991b) Oxidatively modified human low-density lipoprotein stimulates leukocyte adherence to the microvascular endothelium in vivo. Res Exp Med 19:85–90

Ley K, Meyer J-U, Intaglietta M, Arfors K-E (1989) Shunting of leukoçytes in rabbit tenuissimus muscle. Am J Phhysiol 256:H85–H93
Lindbom L, Arfors K-E (1985) Mechanism and site of control for variation in the number of perfused capillaries in skeletal muscle. Int J Microcirc Exp Clin 4:19–30
Lipowsky HH, Riedel D, Shi GS (1991) In vivo mechanical properties of leukocytes during adhesion to venular endothelium. Biorheology 28:53–64
Ma X-L, Tsao PS, Lefer AM (1991) Antibody to CD-18 exerts endothelial and cardiac protective effects in myocardial ischemia and reperfusion. J Clin Invest 88:1237–1243
Mayrovitz HN (1982) The relationship between leukocytes and erythrocytes velocity in arterioles. In: Bagge U, Born GVR, Gaehtgens P (eds) White blood cells, morphology and rheology as related to function. Nijhoff, The Hague
McCord JM (1987) Oxygen-derived radicals: a link between reperfusion injury and inflammation.Fed Proc 46:2402–2406
McCord JM, Fridovich I (1969) Superoxide dismutase: an enzymatic function for erthryocuprein (hemocuprein). J Biol Chem 244:6049–6055
McEver RP (1991) Selectins: novel receptors that mediate leukocyte adhesion during inflammation. Thromb Haemost 65:223–228
Miller LJ, Bainton DF, Borregaard N, Springer TA (1987) Stimulated mobilization of monocyte Mac-1 and p150, 95 adhesion proteins from an intracellular vesicular compartment to the cell surface. J Clin Invest 80:535–544
Mori E, Del Zoppo GJ, Chambers JD, Copeland BR, Arfors K-E (1992) Inhibition of polymorphonuclear leukocyte adherence suppresses no-reflow after focal cerebral ischemia in baboons. Stroke 23:712–718
Mullane KM, Salmon JA, Kraemer R (1987) Leukocyte derived metabolites of arachidonic acid in ischemia-induced myocardial injury. Fed Proc 46:2422–2433
Needham D, Hochmuth RM (1990) The behavior of human neutrophils into a 4 μm pipet and measurement of cytoplasmic viscosity. J Biomech Eng 112:269–276
Neumann F-J, Waas W, Diehm C, Weiss T, Haupt HM, Zimmermann R, Tillmanns H (1990) Activation and decreased deformability of neutrophilis after intermittent claudication. Circulation 82:922–929
Nobis U, Pries AR, Cokelet GR, Gaehtgens P (1985) Radial distribution of white cells during blood flow in small tubes. Microvasc Res 29:295–304
Picker LJ, Warnock RA, Burns AR, Doerschuk CM, Berg EL, Butcher EC (1991) The neutrophil selectin LECAM-1 presents carbohydrate ligands to the vascular selectins ELAM-1 and GMP-140. Cell 66:921–933
Pozzilli C, Lenzi G, Argentino C, Carolei A, Rasura M, Signore A, Bozzao L, Pozzilli P (1985) Imaging of leukocyte infiltration in human cerebral infarcts. Stroke 16:251–256
Prentice RL, Szatrowski TP, Kato H, Mason MW (1982) Leukocyte counts and cerebrovascular disease. J Chronic Dis 35:703–714
Romson JL, Hook BG, Kunkel SL, Abrams GD, Schork MA, Lucchesi BR (1983) Reduction of the extent of ischemic myocardial injury by neutrophil depletion in the dog. Circulation 67:1016–1023
Ross GD, Thompson RA, Walport MJ, Springer TA, Watson JV, Ward RHR, Lida J, Newman SL, Harrison RA, Lachmann PJ (1985) Characterization of patients with an increased susceptibility to bacterial infections and a genetic deficiency of leukocyte membrane complement receptor type 3 and the related membrane antigen LFA-1. Blood 66:882–890
Sanchez-Madrid F, Nagy J, Robbins E, Simon P, Springer TA (1983) A human leukocyte differentiation antigen family with distinct alpha subcrits and a common beta subunit: the lymphocyte function-associated antigen (LFA-1), the C3b; complement receptor (OkM1/Mac-1) and the p150, 95 molecule. J Exp Med 158:1785–1803
Schmalzer EA, Chien S (1984) Filterability of leukocytes: effects of pentoxifylline. Blood 64:542–546

Schmid-Schönbein GW (1987) Mechanism of granulocyte-capillary-plugging. Prog Appl Microcirc 12:223–230
Schmid-Schönbein GW, Chien S (1989) Morphomtry of human leukocyte granules. Biorheology 26:331–343
Schmid-Schönbein GW, Fung YC, Zweifach BW (1975) Vascular endothelium-leukocyte interaction: sticking shear force in venules. Circ Res 36:173–184
Schmid-Schönbein GW, Shih YY, Chien S (1980a) Morphometry of human leukocytes. 56:-866–875
Schmid-Schönbein GW, Skalak R, Usami S, Chien S (1980b) Cell distribution in capillary networks. Microvasc Res 19:18–44
Schmid-Schönbein GW, Usami S, Skalak R, Chien S (1980c) The interaction of leukocytes and erthrocytes in capillary and postcapillary. Microvasc Res 19:45–70
Schmid-Schönbein GW, Sung K-LP, Tözeren H, Skalak R, Chien S (1981) Passive mechamical properties of human leukocytes. Biophys J 36:243–256
Schmid-Schönbein GW, Skalak R, Sung K-LP, Chien S (1982) Human leukocytes in the active state. In: Bagge U, Born GVR, Gaehtgens P (eds) White blood cells, morphology and rheology as related to function. Nijhoff, The Hague
Schmid-Schönbein GW, Jan K-M, Skalak R, Chien S (1984) Deformation of leukocytes on a hematological blood film. Biorheology 21:767–681
Schmid-Schönbein GW, Seiffge D, DeLano FA, Zweifach BW (1991) Circulating leukocyte count and activation in spontaneously hypertensive rats and their normotensive controls. Hypertension 17:323–330
Schröder S, Palinski W, Schmid-Schönbein GW (1991) Activated monocytes and granulocytes, capillary non-perfusion and neovascularization in diabertic retinopathy. Am J Pathol 139:81–100
Senda N, Shibata N, Tatsumi N, Kondo K, Hamada K (1969) A contractile protein from leukocytes – Its extraction and some of its properties. Biochim Biophys Acta 181:191–200
Setiadi H, Wautier JL, Courillon-Mallet A, Passa P, Caen J (1987) Increased adhesion to fibronectin and MO-1 expression by diabetic monocytes. J Immunol 138:3230–3234
Simpson P, Todd R, Fantoen J, Mickelson J, Griffin J, Lucchesi B (1988) Reduction of experimental canine myocardial reperfusion unjury by a monoclonal antibody that inhibits leukocyte adhesion. J Clin Invest 81:624–629
Skalak TC, Schmid-Schönbein GW (1986) Viscoelastic properties of microvessels in rat spinotrapezius muscle. J Biomech Eng 108:193–200
Smith CW, Entman ML, Lane CL, Beaudet AL, Ty TI, Youker K, Hawkins HK, Anderson DC (1991a) Adherence of neutrophils to canine cardiac myocytes in vitro is dependent on intercellular adhesion molecule-1. J Clin Invest 88:1396–1406
Smith CW, Kishimoto TK, Abbassi O, Hughes B, Rothelin R, McIntire LV, Butcher E, Anderson DC (1991b) Chemotactic factors regulate lectin achesion molecule 1 (LECAM-1)-dependent neutrophil adhesion to cytokine-stimulated endothelial cells in vitro. J Clin Invest 87:609–618
Smith JK, Grisham MB, Granger DN, Korthius RJ (1989) Free radical defense mechanisms and neutrophil infiltration in post-ischemic skeletal muscle. Am J Physiol 256:H789–H793
Smith SM, Holm-Rutili L, Perry MA, Grisham MB, Arfors K-E, Granger DN, Kvietys PR (1987) Role of neutrophils in hemorrhagic shock-induced gastric mucosal injury in the rat. Gastroenterology 93:466–471
Southwick FS, Hartwig JH (1982) Acumentin, a protein in macrophages which caps the "pointed" end of actin filaments. Nature 297:303–307
Southwick FS, Stossel TP (1981) Isolation of an inhibitor of actin polymerization from human polymorphonuclear leukocytes. J Biol Chem 256:3030–3036
Spertini O, Luscinskas FW, Kansas GS, Munro JM, Griffin JD, Gimbrone MA, Tedder TF (1991) Leukocyte adhesion molecule-1 (LAM-1, L-selectin) interacts with an inducible endothelial cell ligand to support leukocyte adhesion. J Immunol 147:2565–2573

Stossel TP, Hartwig JH (1976) Interactions of actin, myosin, and a new actin-binding protein of rabbit pulmonary macrophages. J Cell Biol 68:602–619
Stossel TP, Pollard TD (1973) Myosin in polymorphonuclear leukocytes. J Biol Chem 248:-8288–8294
Suematsu M, Schmid-Schönbein GW, Chavez-Chavez RH, Yee TT, Tamatami T, Miyasaka M, DeLano FA, Zweifach BW (1993) In-vivo visualization of oxidative changes in microvessels during neutrophil activation. Am J Physiol 264:H881–H891
Suematsu M, Kurose I, Asako H, Miura S, Tsuchiya M (1989) In vivo visualization of vxyradical-dependent photoemission during endothelium-granulocyte interaction in microvascular beds treated with platelet-activating factor. J Biochem 106:355–360
Sullivan JA, Mandell GL (1983) Mobility of human polymorphonuclear neutrophils: microscopic analysis of substrate adhesion and distribution of F-actin. Cell Motil 3:31–46
Sung K-LP, Schmid-Schönbein GW, Skalak R, Schuessler GB, Usami S, Chien S (1982) Influence of physiochemical factors on rheology of human neutrophils. Biophys J 39:101–106
Sutton DW, Schmid-Schönbein GW (1992) Elevation of organ resistance due to leukocyte perfusion. Am J Physiol (in press)
Tatsumi N, Shibata N, Okamura Y, Takeuchi U, Senda N (1973) Actin and myosin A from leukocytes. Biochim Biophys Acta 305:433–444
Thompson TN, La Celle PL, Cokelet GR (1989) Perturbation of red cell flow in small tubes by white blood cells. Pflugers Arch 413:372–377
Todd RFI, Arnout MA, Rosin RE, Crowley CA, Peters WA, Babior BM (1984) Subcellular localization of the large subunit of Mol (Mol alpha; formerly gp110), a surface glycoprotein associated with neutrophil adhesion. J Clin Invest 74:1280–1290
Valerius NH, Stendahl O, Hartwig JH, Stossel TP (1981) Distribution of actin-binding protein and myosin in polymorphonuclear leukocytes during locomotion and phagocytosis. Cell 24:195–202
Vasthare US, Heinel LA, Rosenwasser RH, Tuma RF (1990) Leukocyte involvement in cerebral ishcemia and reperfusion injury. Surg Neurol 33:261–265
Vedder N, Winn C, Rice C, Chi E, Arfors K, Harlan J (1988) A monoclonal antibody to the adherence promoting leukocyte glycoprotein CD18, reduces organ injury and improves survival from hemorrhagic shock and resuscitation in rabbits. J Clin Invest 81:939–944
Vedder NB, Winn RK, Rice CL, Chi EY, Arfors KE, Harlan JM (1990) Inhibition of leukocyte adherence by anti-CD18 monoclonal antibody attenuates reperfusion injury in the rabbit ear. Proc Natl Acad Sci USA 87:2643–2646
Von Andrian UH, Chambers JD, McEvoy LM, Baryate RF, Arfors KE, Butcher EC (1991) Two step model of leukocyte-endothelial cell interaction in inflammation: distinct roles for LECAM-1 and the leukocyte b_2 integrins in vivo. Proc Natl Acad Sci USA 88:7538–7542
Walsh CJ, Carey PD, Cook DJ, Bechard DE, Fowler AA, Sugerman HJ (1991) Anti-CD18 antibody attenuates neutropenia and alveolar cappilary-membrane injury during gram-negative sepsis. Surgery 110:205–211
Warnke KC, Skalak TC (1990) The effects of leukocytes on blood flow in a model skeletal muscle capillary network. Microvasc Res 40:118–136
Weiss JS (1989) Tissue destructon by neutrophils. N Engl J Med 320:365–376
Weiss SJ, Peppin G, Ortiz X, Ragsdale C, Test ST (1985) Oxidative autoactivation of latent collagenase by human neutrophils. Science 227:747–749
Wilson JW (1972) Leukocyte sequestration and morphologic augmentation in the pulmonary network following hemorrhagic shock and related forms of stress. Adv Microcirc 4:197–232
Worthen GS, Schwab B, Elson EL, Downey GP (1989) Cellular mechanics of stimulated neutrophils: stiffening of cells induces retention in pores in vitro and lung capillaries in vivo. Science 245:183–186

Yamkawa T, Yamaguchi S, Sugiyama I (1987) White blood cell plugging and blood flow maldistribution in the capillary network of cat cerebral cortex in acute hemorrhagic hypotension: an intravital microscopic study. Circ Shock 22:323–332

Yin HL, Hartwig JH, Maruyama K (1981) Ca^{++} control of actin filament length. Effects of macrophage gelsolin on actin polymerization. J Biol Chem 256:9693–9697

Zimmerman G, McIntyre T (1988) Neutrophil adherence to human endothelium in vitro occurs by CDw18 glycoprotein dependent and independent mechanisms. J Clin Invest 81:531–537

Microcirculatory Changes in Cerebral Ischemia

J. Cervós-Navarro and M. Korves

Microcirculation

The term microcirculation, first used by Fulton in 1957, is applied to the stretch involving the exchange of metabolites. Capillaries and venules are included as well as arterioles that also have a nutritive function for surrounding tissue, free capillaries, and spaces. Structural peculiarities in which the vessel wall of the cerebral microcirculation differ from nonneuronal vessels are well known, even if comparative studies of topistic regional differences of cerebral vessels are rarely found. The density of capillaries varies between the region of the brain, while it is accepted that the total number of capillaries in a certain region is a reflection of the normal average energy demand of this region, thus reflecting the metabolic level.

In hypoxic states of the brain, the blood flow is frequently increased substantially. Following an ischemic insult, the cerebral blood flow (CBF) exceeds normal levels, as is well known, independently of the metabolic demand (luxury perfusion). The metabolic exchange crossing the vessel wall during this period is unknown. The total intracranial blood volume is increased and suggests capillary recruitment, thus giving us a hint that normally not simultaneously perfused capillaries are circulated or filled with blood (Tomita 1988). The concept of a functional reserve is controversial and the morphological evidence restricted (Collins et al. 1987; Kuschinsky et al. 1987). It could be shown as a result of measurements of CBF in very small compartments that there is no homogeneously increased CBF in hypoxemia but a heterogeneously distributed increase depending on local metabolic demand (Smith et al. 1983). A possible underlying phenomena could be a capillary perfusion heterogeneity, thus the transit time and the corresponding extraction rate differ in closely related areas. Ennis et al. (1990) noted a 54% decrease in the number of perfused blood vessels after 3-h incomplete ischemia, and the early reperfusion phase was characterized by a mismatch between capillary perfusion and CBF.

Institute of Neuropathology, Free University of Berlin, Hindenburgdamm 30, 12203 Berlin 45, Germany

Cerebral Ischemia and Basic Mechanisms
Ed. by A. Hartmann, F. Yatsu, and W. Kuschinsky

This could reflect as well the no-reflow phenomenon, a term triggering discussions in the last two decades (Ames et al. 1968; Matakas et al. 1975; Cervós-Navarro et al. 1980; Kagström et al. 1983). It is attributed to intravascular factors like microthrombi and platelet aggregation (Hossmann et al. 1980; Obrenovitch et al. 1985; Buchweitz and Weiss 1986; Turcani et al. 1988; Schürer et al. 1987; Heye et al. 1991) or to vascular wall changes due to smooth muscle contraction (Takagi et al. 1977), endothelial edema (Paljärvi et al. 1983), and microvilli formation (Dietrich et al. 1984).

Influence of the Macrocirculation on the Microcirculation

Resistance of large arteries appears to be greater in the cerebral circulation than in other vascular beds. Large arteries contribute to an important degree to total cerebral vascular resistance and are major determinants of local microvascular pressure. Resistance of large arteries and cerebral microvascular pressure are affected by several physiological stimuli, including changes in systemic blood pressure, increases in cerebral metabolism, activity of sympathetic nerves, and humoral stimuli such as circulating vasopressin and angiotensin. Stimuli such as sympathetic stimulation and vasopressin produce selective responses of large arteries and, thereby, regulate microvascular pressure without a significant change in CBF. These findings induce the theory that the brain may be sensitive to changes in cerebral microvascular pressure, resulting in activation of compensatory neurohumoral mechanisms. Important changes occur in large cerebral arteries under pathophysiological conditions; for example, atherosclerosis potentiates constrictor responses of large cerebral arteries to serotonin and thromboxane, which may contribute to vasospasm and transient ischemic attacks (Faraci and Heistad 1990).

Vasoparalysis and Vasospasm

Spielmeyer, influenced by the idea that blood vessels go into spasm in response to a variety of stimuli, considered a vasospasm as the main cause of circulatory disorders. Scholz (1939) stressed the role of the vasomotor innervation for the regulation of the topistic alterations as a manifestation of failure of the vasomotor function. As a result of experimental data supporting this concept, vascular spasm continued to be considered as a major factor in producing patterns of focal brain injury until the developing awareness that asphyxia dilates rather than constricts blood vessels. However, since evidence of arteriolar spasm (Fig. 1) has been demonstrated in different conditions (Cervós-Navarro et al. 1978), ischemically induced vasospasm in the microcirculation has been of interest and should be reconsidered under the light of very potent secreted vasoconstrictor agents such as

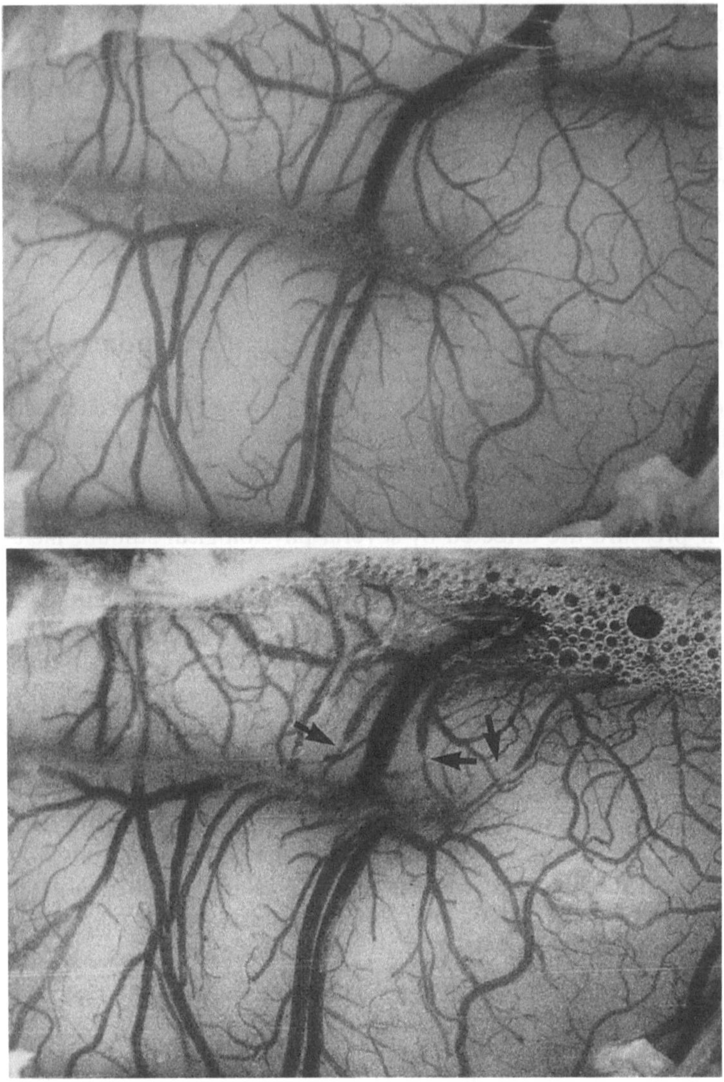
A
B

Fig. 1A,B. Leptomeningeal vessels before (**A**) and after irritation, showing intensive vasoconstriction (*arrows* **B**)

endothelin after endothelial irritation or after extremly pH changes (Fig. 2). Available data document a period of sensitivity of the cerebral vasculature for spasticity after ischemia (Hossmann et al. 1973; Ginsberg et al. 1978; Levy et al. 1979); this may contribute to the postischemic hypoperfusion especially during a period of longer-lasting reperfusion when the metabolic demand increases again. At this time a critical arteriolar supply of recovering areas could result in a secondary hypoxia.

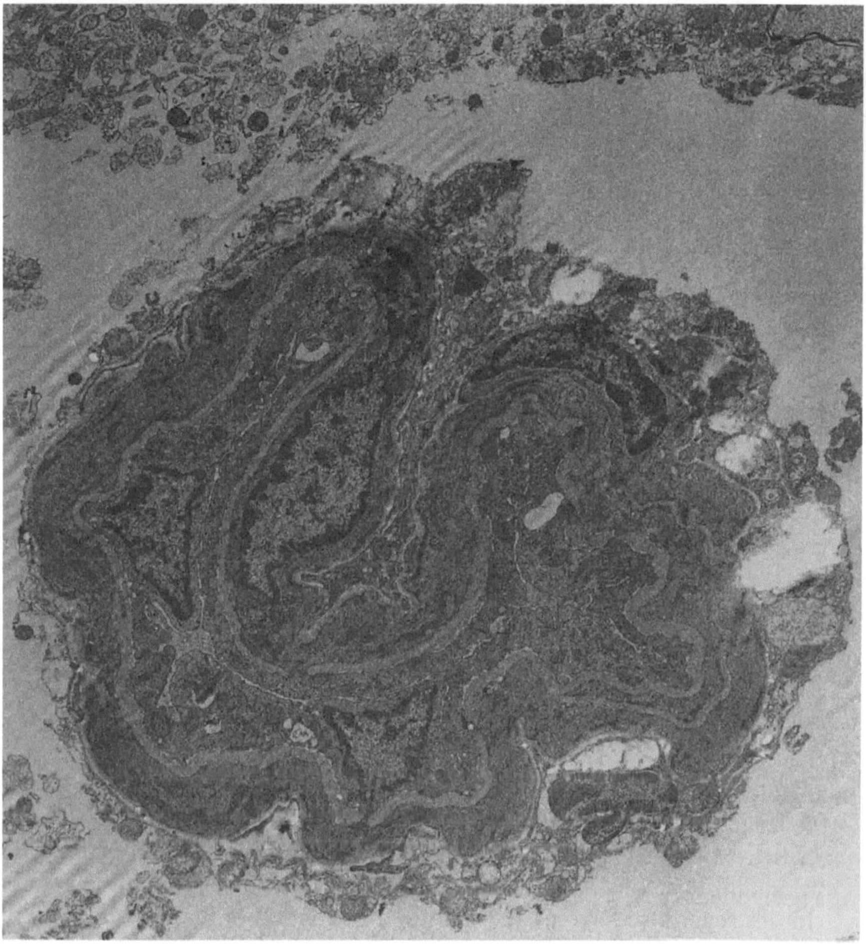

Fig. 2. Vasospasm of a intracerebral vessel induced by rapid changes in the pH of the circulating blood. ×8000

Owing to the fact that the periarteriolar tissue is free of capillaries (Cervós-Navarro and Iglesias Rozas 1978), we must assume a metabolic exchange at this site, thus a critical arteriolar blood supply may result in a lesion concentrated at this location. This is supported by periarteriolar necrosis in cases where arteriolar vasospasm can be postulated (Fig. 3). Evidence has accumulated on an active role of the endothelium on vasomotoric regulation. Humoral, neural, and mechanical stimuli are capable of inducing endothelium-dependent vasoconstriction. The most important vasoconstrictor synthesized by endothelial cells is endothelin (Yanagisawa et al. 1988). Furthermore, endothelial cells synthesize and release angiotensin-converting enzyme (ACE). Apart from the conversion of angiotensin I

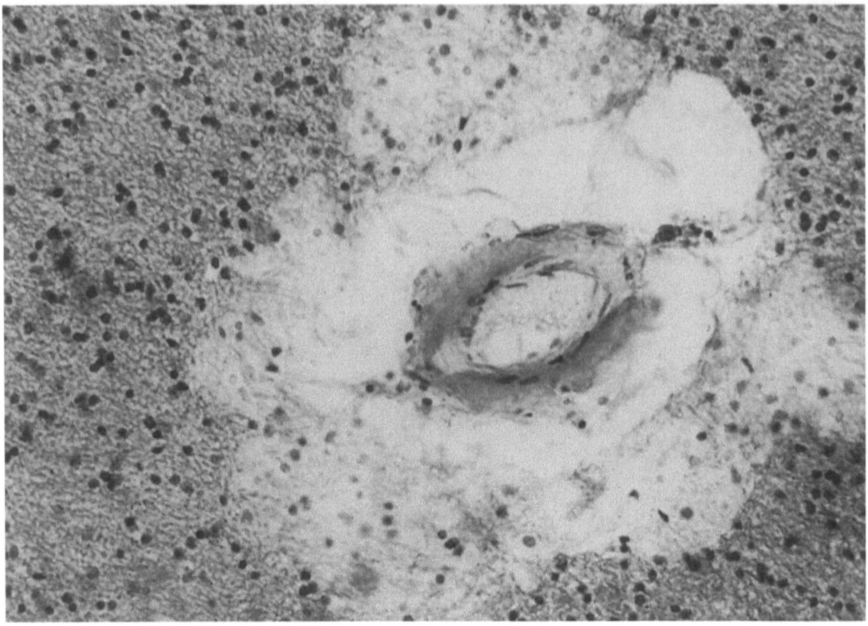

Fig. 3. Periarteriolar tissue necrosis. Hemotoxylin and eosin

to angiotensin II, ACE inactivates bradykinin generated from kininogen and it has vasodilatative abilities. In addition, endothelial cells metabolize vasoactive substances, e.g., adenosine, serotonin, and norepinephrine (Gordon and Pearson 1987).

In this regard we have to add that the distribution of the two subtypes of P2 purinoceptors in the brain circulation (Torregrosa et al. 1990) has not yet been investigated. The resulting effect of these substances, especially in regard to the cerebral microcirculation, could only be estimated with the knowledge of the receptor distribution in the brain vessels.

Endothelial cells produce different substances which influence vasoconstriction and vasodilatation, the endothelium-derived relaxing factor (EDRF) which is identical to nitric oxide (NO), a potent vasodilatator as well as an inhibitor of platelet aggregation.

Even if the available evidence strongly suggests that the EDRF generated by acetylcholine in the cerebral microcirculation is a nitrosothiol, that does not mean that all other capillary beds are regulated by the same pathway. In the microcirculation of skeletal muscle, the vasodilatation from bradykinin is mediated partly by prostacyclin and party by an EDRF similar to that generated by acetylcholine. The EDRF generated by bradykinin in the cerebral vascular bed is an oxygen radical generated in association with enhanced arachidonate metabolism via cyclooxygenase.

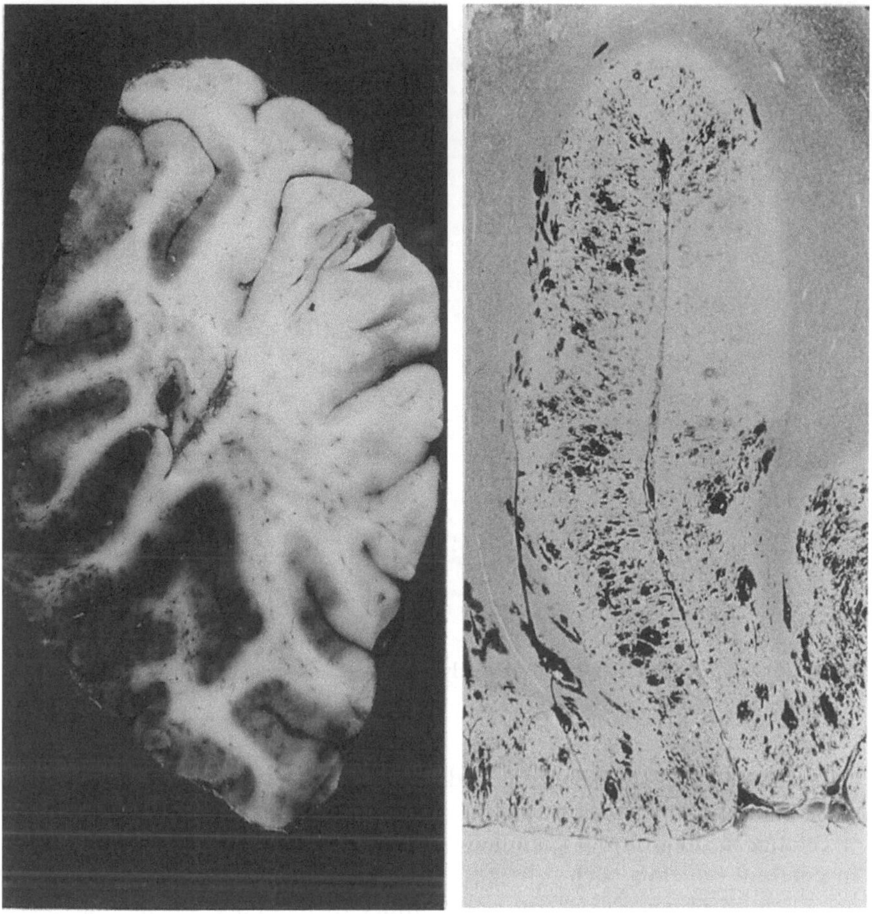

A B

Fig. 4A,B. Hemorhagic cortical infartion with intensive vascular stasis

Of significance is the effect of relaxing substances in the venous stasis. They can play a role in diapedesis and are an important factor in the evolution of ischemic to hemorhagic infarct (Fig. 4).

Leukocyte Activation in the Microcirculation

A leukocytic stage in recent infarcts (Fig. 5) lasting for some hours was a well-known finding for the neuropathologist (Cervós-Navarro et al. 1980). Evidence for a role of granulocytes in ischemic organ injury has been provided. Compared to red cells, granulocytes are large cells and have a stiff cytoplasm, making them prone to entrapment in the microcirculation. After

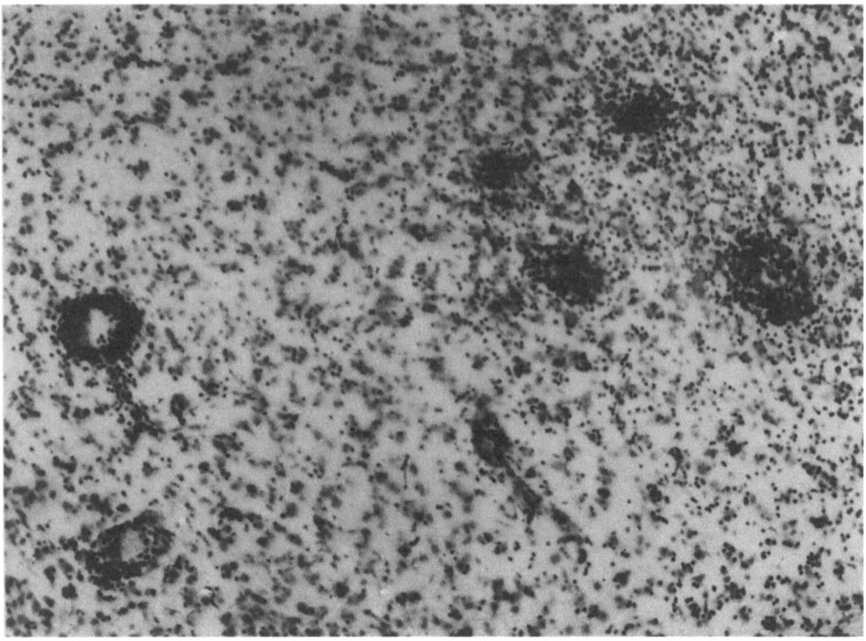

Fig. 5. Recent cerebral infarct with vascularly associated leukocyte/granulocyte adhesion

activation, granulocytes become adhesive (Zimmerman et al. 1992), they can elaborate superoxide radicals and release proteolytic enzymes. In the circulation a subgroup of granulocytes is in a spontaneously activated state. In cerebral ischemia such cells become trapped in the microcirculation and are likely to impose a risk for organ injury (Schmid-Schönbein and Engler 1990).

Hemostatic Disturbances in the Microcirculation

The main factor disturbing the normal blood flow structure, and therefore the normal rheological properties of blood, is the intravascular aggregation of blood cells. By labelling thrombi with intravenous H-serotonin, Dougherty et al. (1977) and Petito (1979) observed an increased platelet aggregation in the ischemic hemisphere, which was confirmed electronmicroscopically (Fig. 6). In ischemic brain damage in humans, microthrombi could be demonstrated in the ischemic area (Fig. 7) and the adjacent tissue (Cervós-Navarro et al. 1986; Sampaolo et al. 1987). Microthrombosis reduces regional CBF (Obrenovitch and Hallenbeck 1985), impairs the function of the blood–brain barrier, and leads to functional membrane failure and structural damage of nerve and glial cells (Heiss 1983).

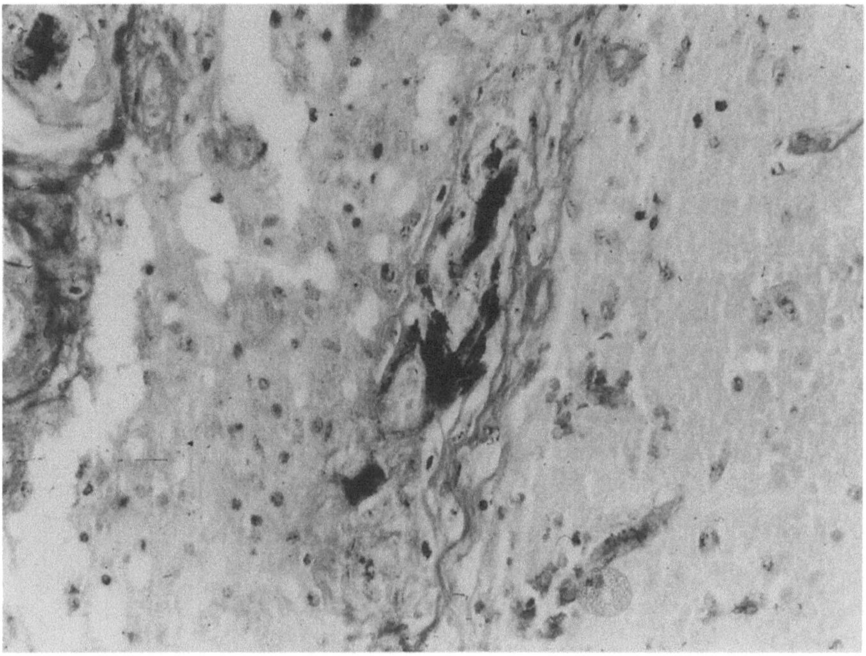

Fig. 6. Microthrombus in the cerebral circulation shown by light microscopy combined with a fibrin stain (PHE)

Brain ischemia leads to formation of microthrombi as a result of focal and systemic factors (Hossmann and Hossmann 1977). The majority of microthrombi can be found in the area of the ipsilateral hemisphere and occur in relation to local tissue alterations. Microthrombosis in the contralateral hemisphere, which could be demonstrated in animals, is probably due to systemic hypercoagulopathy.

Since one of the possible factors related to the infarction size is microthrombi and consecutive microcirculatory disturbance, searching for the underlying basic mechanisms of the endothelial cell has come under intensive review.

Role of the Endothelium in the Regulation of the Hemostatic System

The transformation of fluid blood to a gel-like mass when it escapes from the circulating blood is normally designated as clotting of blood, whereas hemostasis describes the process of induced or spontaneous stopping of bleeding.

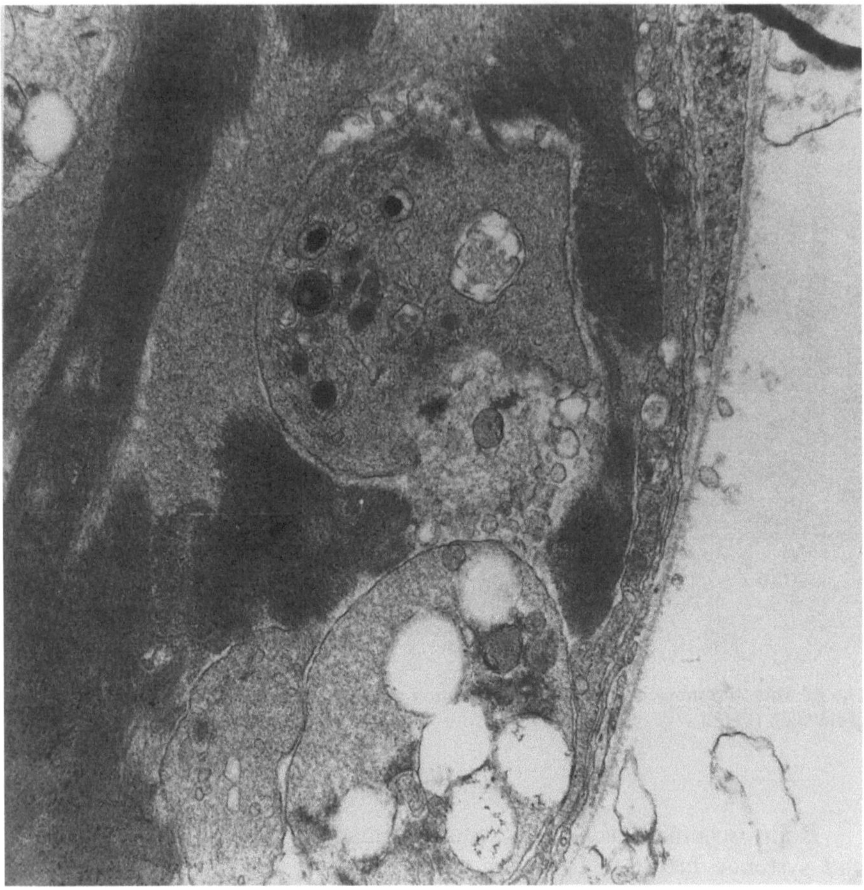

Fig. 7. Patelets attached to the vascular wall. ×18 000

The system of hemostasis is complex: several systems are positively or negatively influenced by each other. The constituents are: (a) cellular and humoral components of the circulating blood; (b) components of the vessel wall; and (c) vasomotoric reactions influencing the hemodynamics. The cellular components of the blood enclose platelets and leukocytes. The humoral components of the blood comprise plasmatic coagulation as well as fibrinolytic factors. Cellular as well as humoral components of the immune system interact with normal endothelial cells of the vessel wall and, equally importantly, with the subendothelial extracellular matrix.

Even a few years ago the endothelium was considered to be a passive barrier preventing the adhesion of blood cells and the formation of a thrombus. It has become possible to cultivate endothelial cells, thus allowing the investigation of the role of these cells in the hemostatic system. It must be emphasized that the interaction between circulating components and endothelial cells demonstrates a system of balance between prothrombotic and antithrombotic potentials.

Endothelial cells cover the luminal surface of the vascular wall, separating the circulating blood from the stationary tissue. In this function the endothelial cells actively influence the homeostasis of cellular and humoral components of blood coagulation and fibrinolysis (Gimbrone 1986; Müller-Berghaus 1987).

The endothelial cell is endowed with a great number of recognition sites to monitor the regulation of hemostasis on the endothelial cell surface, thus maintaining the surface nonthrombogenetic. The importance of the microcirculatory system is evidenced by the fact that more than 80% of the endothelial cells are found in the capillary bed, with an extremely high ratio of cell surface per unit blood volume; modulation of hemostasis occurs mainly in the peripheral vascular system (Gimbrone 1986).

The endothelial cell participates actively in the regulation of the hemostatic processes by its ability express and secrete proteins (coagulation factors, fibrinolytic as well as inhibitory factors), lipid components (e.g., eicosanoids), and glycosaminoglycans (e.g., heparan sulfate). Many functional, structural, as well as quantitative differences in cellular activities exist between endothelial cells originating from different types of vessel, tissue, and species, thus data must be interpreted carefully.

Endothelial Cell and Platelet Interaction

Endothelial cells influence the function of platelets as they synthesize and release activators as well as inhibitors of platelet aggregation (Moncada et al. 1987). The Willebrand factor is synthesized in endothelial cells, thus contributing to the adhesion of platelets to the subendothelium. The platelet-activating factor (PAF) can be released from endothelial cells and induce adhesion and aggregation of platelets and possibly also granulocytes (Prescott et al. 1990; Zimmerman et al. 1992). What is confusing is that PAF can be released by platelets and induces the segregation of EDRF, which is the most potent inhibitor of platelet aggregation (Nievelstein and de Groot 1988).

During the destruction of the membrane phospholipids, arachidonic acid is produced; during the further destruction oxygen-containing radicals are produced with consequent membrane damage. In addition, arachidonic acid is able to destroy the cell by another pathway, owing to its ability to produce eicosanoide. As a part of the group of eicosanoides, thromboxane A_2 is a potent activator of platelet aggregation with consequent reduction of the blood flow. PAF, a product of the phospholipid degradation after ischemia, increases the resulting platelet aggregation as a result of the same effect.

Endothelium-derived relaxing factor is released after the action of thrombin, ATP/ADP, bradykinin, and vasopressin on endothelial cells.

Prostacyclin is another powerful inhibitor of platelet aggregation and will be released when contact with thrombin, the complement factor 5a, bradykinin, histamine, or leukotriene has happened.

Hypoxia can induce vasodilatation, in part mediated by prostacyclin, whereas EDRF generation is antagonised by hypoxia. Prostacyclin as well as EDRF are secreted on the luminal side of the endothelial cells thus modulating platelet aggregation, whereas prostacyclin released on the abluminal site of the cells interacts with smooth muscle cells. In addition to the inhibitory effect of EDRF and prostacyclin on platelet aggregation, endothelial cells contribute to the inhibition of platelet aggregation by releasing ectoenzymes (nucleotidases). ATP is converted to ADP and consecutively to AMP and adenosine by this ectonucleotides. ADP is a inhibitor of platelet aggregation, and adenosine increases the concentration of cAMP which also interferes with the aggregation of platelets. Thus platelet aggregation and disaggregation are influenced by endothelial cells (Moncada et al. 1987; Vanhoutte 1988).

Vice versa, platelets influence the function of endothelial cells; platelets secrete vasopressin, ATP/ADP, PAF, and serotonin, all of them influencing the metabolism of endothelial cells and concurrently the production of prostacyclin and EDRF.

Adhesion Molecules and Hemostasis

The first step in the formation of microthrombi is the adhesion of platelets to the endothelial cell. Endothelial cells synthesize and secrete different adhesion proteins which are involved in the adhesion of cells to their substratum (Mosher 1989; Cheng and Kramer 1989; Giltay and Van Mourik 1988). The known counterparts for human arterial blood clotting on the platelet surface is the glycoprotein Ib binding part specifically coupling with the Willebrand factor as the ligand (Roth 1992) under the condition of shear forces. Collagen, fibronectin, vitronectin, thrombospondin, and the Willebrand factor are the important vascular or subendothelial counterparts of adhesion. Specific receptors for vitronectin, fibronectin, and collagen have been demonstrated on the surface of endothelial cells. Vitronectin contained in the extracellular matrix not only mediates the adhesion of endothelial cells to the matrix but also binds directly to plasminogen activator–inhibitor I (Declerck et al. 1988). Extracellular matrix-associated vitronectin stabilized PAI in the subendothelium, contributing to the inhibition of local fibrinolysis. The co-localization of urokinase receptors with integrins in focal adhesion suggests a regulatory role for extracellular vitronectin in cell surface-associated plasminogen activation (Pöllänen et al. 1991).

In summary, the endothelium synthesizes, secretes, and expresses prothrombotic as well as antithrombotic substances, thus contributing to the balance of the hemostatic system, it is also involved in all three essential compartments of hemostasis, the vasomotor system, the platelet–coagulation system and the fibrinolytic system.

References

Ames A, Wright R, Kowada M, Thurston JM, Majno G (1968) Cerebral ischemia. II. The no reflow phenomenon. Am J Pathol 52:437

Buchweitz E, Weiss H (1986) Alteration in perfused capillary morphometry in awake vs. anesthetized brain. Brain Res 337:105

Cervós-Navarro J, Iglesias Rozas JR (1978) The arteriole as a site of metabolic exchange. Adv Neurol 20:17–24

Cervós-Navarro J, Schneider H (1980) Pathologie des Nervensystems. I. Durchblutungsstörungen und Gefäßerkrankungen des Zentralnervensystems. Springer, Berlin Heidelberg New York

Cervós-Navarro J, Matakas F, Roggendorf W, Christmann U (1978) The morphology of spastic intracerebral arterioles. Neuropathol Appl Neurobiol 4:3679–379

Cervós-Navarro J, Figols J, Ebhardt G (1986) Microthrombosis: a contributing factor to the progression of cerebral infarction in man. Life Sci 115:109–119

Cheng YF, Kramer RH (1989) Human microvascular endothelial cells express integrin-related complexes that mediate to the extracellular matrix. J Cell Physiol 139:275–286

Collins RC, Wagman IL, Lymer L, Matter JM (1982) Distribution and recruitment of capilaries in rat brain (Abstr). J Cereb Blood Flow Metab 7:S336

Declerck PJ, De Mol M, Alessi MC, Baudner S, Paques EP, Preissner ET, Müller-Berghaus G, Collen D (1988) Purification and characterization of a plasminogen activator inhibitor 1 binding protein from human plasma. Identification as a multimeric form of S protein (vitronectin). J Biol Chem 263:15454–15461

Dietrich WD, Busto R, Ginsberg MD (1984) Cerebral endothelial microvilli: formation following global forebrain ischemia. J Neuropathol Exp Neurol 43:72

Dougherty JH, Levy DE, Weksler BB (1977) Platelet activation in acute cerebral ischemia. Lancet 16:821–824

Ennis SR, Keep RF, Schielke GP, Betz AL (1990) Decrease in perfusion of cerebral capillaries during incomplete ischemia and reperfusion. J Cereb Blood Flow Metab 10:213

Faraci FM, Heistad DD (1990) Regulation of large cerebral arteries and cerebral microvascular pressure. Univ Iowa Coll Med Circ Res 66(1):8–17

Fulton GF (1957) Microcirculation terminology. Angiology 8:102

Giltay JC, Van Mourik JA (1988) Structure and function of endothelial cell integrins. Haemostasis 18:376–389

Gimbrone MA (ed) (1986) Vascular endothelium in hemostasis and thrombosis. Churchill Livingstone, Edinburgh

Ginsberg MD, Budd WW, Welsh FA (1978) Diffuse cerebral ischemia in the cat. Local blood flow during severe ischemia and recirculation. Ann Neurol 3:482

Gordon JL, Pearson JD (1987) Biology of vascular endothelium. In: Bloom AL, Thomas DP (eds) Hemostasis and thrombosis. Basic principles and clinical practice. Churchill Livingstone, Edinburgh, pp 303–311

Heiss WD (1983) Flow thresholds of functional and morphological damage of brain tissue. Stroke 14:329–331.

Heye N, Paetzold C, Cervós-Navarro J (1991) The role of microthrombi and microcirculartory factors in localization and evolution of focal cerebral ischemia. Neurosurg Rev 14:7–16

Hossmann K-A, Hossmann V (1977) Coagulopathy following experimental cerebral ischemia. Stroke 8:249–254

Hossmann K-A, Schuier FJ (1980) Experimental brain infarction in rats. Stroke 11:583–592

Hossmann K-A, Lechtape-Grüter H, Hossmann V (1973) The role of the blood flow for the recovery of the brain after prolonged ischemia. Z Neurol 204:281

Hossmann K-A, Hossmann V, Tagaki S (1980) Effect of intravascular platelet aggregation on blood recirculation following prolonged ischemia on the cat brain. J Neurol 222:159

Kagström E, Smith ML, Siesjö BK (1983) Local cerebral blood flow in the recovery period following complete cerebral ischemia in the rat. J Cereb Blood Flow Metab 1:313

Kuschinsky W, Klein B, Schröck H, Vetterlein F (1987) The local density of perfused cappillaries in the rat brain. J Cereb Blood Flow Metab 7 Suppl 1:333

Levy DE, Van Uitert RL, Pike CL (1979) Delayed postischemic hypoperfusion, a potentially damaging consequence of stroke. Neurology (Minneap) 29:1245

Matakas F, Cuypers J, Cervós-Navarro J (1975) Disturbances of cerebral microcirculation after cerebral ischemia. Neuropathol Pol 13:3

Moncada S, Palmer RM, Higgs EA (1987) Prostacyclin and endothelium derived relaxing factor: biological interactions and significance. In: Thrombosis and haemostasis 1987. International Society on Thrombosis and Haemostasis and Leuven University Press, Leuven, pp 597–618

Mosher DF (1989) Fibronectin. Academic, San Diego

Müller-Berghaus GF (1987) Septicemia and the vessel wall. In: Thrombosis and haemostasis 1987. International Society on Thrombosis and Haemostasis and Leuven University Press, Leuven, pp 619–671

Nievelstein PFEM, de Groot PG (1988) Interactions of blood platelets with the vessel wall. Haemostasis 18:342–359

Obrenovitch PT, Hallenbeck JM (1985) Platelet accumulation in regions of low blood flow during postischemic period. Stroke 16:224–234

Paljärvi L, Rehncrona S, Döderfelt B, Olsson Y, Kalimo H (1983) Brain lactic acidosis and ischemic cell damage: quantitative ultrastructural changes in capillaries of rat cerebral cortex. Acta Neuropathol (Berl) 60:232

Petito C (1979) Platelet thrombi in experimental cerebral infarction. Stroke 10:192–196

Pöllänen J, Stephens RW, Vaheri A (1991) Directed plasminogen activation at the surface of normal and malignent cells. Adv Cancer Res 57:273–328

Prescott SM, McIntyre TM, Zimmerman GA (1990) The role of platelet activating factors in endothelial cells. Thromb Haemost 64:99–103

Roth GJ (1992) Platelets and blood vessels: the adhesion event. Immunol Today 13(3)

Sampaolo S, Cervos-Navarro J, Djouchadard, Figols J (1987) Clinical and experimental evidence of mikrothrombosis in cerebral ischemia. In: Hartmann A, Kuschinsky W (eds) Cerebral ischemia and hemorheology. Springer, Berlin Heidelberg New York, pp 386–393

Schmid-Schönbein GW, Engler RL (1990) Perspectives of leukocyte activation in the microcirculation. Biorheology 27(6):859–869

Scholz W (1939) Histologische Untersuchungen über Form, Dynamik und pathologisch anatomische Auswirkungen funktioneller Durchblutungsstörungen des Hirngewebes. Z. Gesamte Neurol Psychiatr 167:424

Schürer L, Grögaard B, Arfors K, Baethmann A (1987) Effects of antineutrophil serum (ANS) and superoxid (SOD) on postischemic hypoperfusion of the brain in rats. J Cereb Blood Flow 7:72

Shah AB, Beamer N, Coul BM (1985) Enhanced in vivo platelet activation in subtypes of ischemic stroke. Stroke 16:643–647

Smith ML, Kagström E, Siesjö BK (1983) Local cerebral blood flow in the rat brain during hypercapnia and hypoxia. Acta Physiol Scand 118:439

Takagi S, Cocito L, Hossmann K-A (1977) Blood recirculation and pharmacological responsiveness of the cerebral vasculature following prolonged ischemia of cat brain. Stroke 8:707

Tomita M (1988) Significance of cerebral blood volume. In: Tomita M, Sawada T, Naritomi H, Heiss WD (eds) Cerebral hyperemia and ischemia: from the standpoint of cerebral blood volume. Elesevier, New York, p 3

Torregrosa G, Miranda FJ, Salom JB, Alabadi JA, Alvarez C, Alborch E (1990) Heterogeneity of P2 purinoceptors in brain circulation. J Cereb Blood Flow Metab 10:572
Turcani P, Gotoh F, Ishihara N, Tanaka K, Gomi S, Takashima S, Mihara B (1988) Are blood platelets involved in the pathogenesis of ischemic brain edema in gerbils? Stroke 19:486
Vanhoutte PM (1988) Plateletes, endothelium and blood vessel wall. Experientia 44:105–109
Yanagisawa M, Kurhara H, Kimura S, Tomobe Y, Kobayashi M, Mitsui Y, Yazaki Y, Goto K, Masaki T (1988) A novel potent vasoconstrictor peptide produced by vascular endothelial cells. Nature 332:411–415
Zimmerman GA, Prescott M, McIntyre TM (1992) Endothelial cell interactions with granulocytes: tethering and signaling molecules. Immunol Today 13(3):93–99

Current Evidence on a Pathophysiological Function of Leukocyte/Endothelial Interactions in Cerebral Ischemia

C. ABELS[1], F. RÖHRICH[2], E. UHL[2], S. CORVIN[1], A. VILLRINGER[3], U. DIRNAGL[3], A. BAETHMANN[1], and L. SCHÜRER[2]

Introduction

Accumulation of leukocytes, particularly of polymorphonuclear neutrophils (PMNLs), is considered as an important response in tissue injury or infection. Recent investigations in a variety of organs, such as bowel (Granger et al. 1989), skeletal muscle (Belkin et al. 1989), and brain (Bednar et al. 1991; Grögaard et al. 1989; Mori et al. 1992), suggest an involvement of leukocytes in ischemia mechanisms. Although recirculation after ischemia is a most important factor of functional recovery, interactions between circulating blood elements and the tissue injured by ischemia could lead to further damage (Harlan 1987). Owing to specific rheologic and adhesive properties, leukocytes may plug capillaries in ischemic tissue (Engler et al. 1983; del Zoppo et al. 1991). An increase in vascular resistance leading to an impaired tissue reperfusion could result in additional damage from secondary ischemia. Further, a release of toxic material of PMNLs such as superoxide anion radicals or elastase might be involved (Weiss 1989).

Respective pathophysiologic mechanisms have been observed in extracerebral tissue in ischemia and shock. Factors which contribute to postischemic reperfusion injury in the central nervous system are not yet clearly defined. A basis of damage by PMNLs is that these cells accumulate in the tissue suffering from ischemia. Hallenbeck et al. (1986) were able to demonstrate significant postischemic accumulation of PMNLs in brain tissue in an air embolism model in dogs. This evidence has been considered sufficient to indicate a significant role of leukocyte/endothelial interactions in ischemia/reperfusion in the brain, although embolization by air is not comparable to ischemia from reduction or arrest of the cerebral blood flow. Intravascular air may cause excessive vascular injury (Garcia et al. 1981),

[1] Institute for Surgical Research, Klinikum Grosshadern, Ludwig Maximilians University, 81377 Munich, Germany
[2] Department of Neurosurgery, Klinikum Grosshadern, Ludwig Maximilians University, 81377 Munich, Germany
[3] Department of Neurology, Klinikum Grosshadern, Ludwig Maximilians University, 81377 Munich, Germany

Cerebral Ischemia and Basic Mechanisms
Ed. by A. Hartmann, F. Yatsu, and W. Kuschinsky

leeading to white blood cell accumulation, possibly without their activation. The purpose of the present experiments was to elucidate whether accumulation of leukocytes and leukocyte/endothelial interactions, respectively, occur in ischemia and reperfusion of rat brain.

Methods

Animal Preparation. For the experiments male albino rats of 250–300 g body weight were used. The animals were anesthetized, tracheotomized, and artificially ventilated after muscle paralysis. A closed cranial window (dura removed) was implanted over the parietal cortex (Kawamura et al. 1990) after insertion of arterial and venous catheters. The brain surface was superfused with artificial cerebrospinal fluid (CSF), and the intracranial pressure was maintained at 3–7 mmHg by adjustment of the outflow catheter. Endexpiratory pCO_2, arterial blood pressure, intracranial pressure, and airway pressure were monitored continuously. Arterial blood gases, hematocrit, and differential blood cell counts were controlled at intervals.

Induction of Ischemia. Either a modified four-vessel occlusion model according to Pulsinelli and Buchan (1988), or a modified two-vessel occlusion model (Nordström and Siesjö, 1978) was employed. Duration of ischemia was 10 or 20 min, respectively, recirculation 60 or 120 min.

Labeling and Imaging of Leukocytes. White blood cells were intravitally labeled by i.v. rhodamin 6G. The fluorescence dye selectively stains leukocytes and in addition serves as a vessel marker (Corvin et al. 1990). For visualization of the cerebral microcirculation, video-fluorescence microscopy (Kawamura et al. 1990) or confocal laser scanning microscopy (Dirnagl et al. 1991) was employed. Thereby, intraparenchymal brain vessels as well as the pial microcirculation could be studied as to the behavior of leukocytes prior to, during, and after cerebral ischemia.

Measurements. For evaluation of leukocyte behavior, the following parameters were determined off-line from video tapes: (a) centerline velocity of leukocytes in pial venules; (b) number of "stickers," that is, leukocytes firmly adhering to the vascular wall for at least 10 s or longer in a vascular segment of 100-μm length during an observation period of 1 min; (c) number of leukocytes slowly rolling along the venular endothelium for 1 min; (d) number of white blood cells emigrating from pial venules into the subarachnoid space.

Results

Video-Fluorescence Microscopy. The results are summarized in Table 1. The velocity of white blood cells through pial venules was utilized as an indicator of the quality of ischemia. Centerline velocity of leukocytes was close to zero during vessel occlusion. With onset of recirculation, centerline velocity increased markedly up to more than 200% of the preischemic value. Subsequently, it fell gradually towards the control value. The number of leukocytes rolling slowly along the venular wall at a low velocity (15–30 μm/s) was small in the control period, and it did not change during the experimental phase. The number of leukocytes firmly adhering to the venular endothelium, i.e., sticking, was low also in the control phase. It must be emphasized that induction of sticking was observed in none of the experiments (n = 10) with 20-min global cerebral ischemia.

Confocal Laser Scanning Microscopy. The results are summarized in Table 2. In capillaries of the cerebral parenchyma, 50 and 150 μm below the cortical surface, the number of leukocytes with an arrest of movement for more than 2 s during an observation period of 1 min was not significantly increased during the experiment, neither was there an enhancement of leukocyte sticking in pial venules. The number of leukocytes counted in the perivascular space around pial venules in an area of 268 × 153 μm amounted to 0.8 ± 1.8 under control conditions. It was increased to 11.8 ± 17.5 during reperfusion. The considerable standard error of the latter value is attributable to the fact that, for as yet unknown reasons, significant emigration (diapedesis) of white blood cells was found in two out of five experiments. Occasionally, migration of leukocytes through the vascular wall of pial venules was directly visible and, hence, taped. The migration process through the vessel wall was complete within 45 s in most cases.

Discussion

Whether or not leukocytes have a pathophysiologic function in the development of postischemic brain damage can be tested by the following criteria:

1. Accumulation and/or activation of leukocytes during cerebral ischemia and/or reperfusion
2. Induction of tissue damage in relation to the accumulation and/or activation of leukocytes
3. Attenuation of tissue damage from ischemia by elimination or inactivation of white blood cells

Most of the studies in the past have focussed on the third requirement. Evidence for the first requirement has rarely been provided during *early*

Table 1. Video-fluorescence microscopy: behavior of leukocytes in pial vessels of rats

	Before ischemia	During ischemia	Time after end of ischemia (min)						
			5	20	40	60	80	100	120
Centerline velocity (% control)	100	3.3 ± 1.8	222.7 ± 49.0	197.9 ± 46.2	113.4 ± 23.0	83.0 ± 8.2	104.0 ± 21.9	100.1 ± 14.9	98.9 ± 14.7
Roller/venule (min^{-1}* 100 μm^{-1})	2.2 ± 0.7	n.d.	3.4 ± 1.1	3.3 ± 1.3	2.2 ± 0.8	2.1 ± 0.8	2.1 ± 0.8	2.3 ± 0.7	1.5 ± 0.3
Sticker/venule (min^{-1}* 100 μm^{-1})	1.9 ± 0.4	0.7 ± 0.4	1.4 ± 0.5	1.9 ± 0.5	1.8 ± 0.4	2.0 ± 0.8	2.1 ± 1.0	2.1 ± 0.7	2.2 ± 0.7

Mean ± SEM; n.d., not determined.

Table 2. Confocal laser-scanning microscopy (five experiments)

	Before ischemia	60 min after the end of ischemia
Number of firmly adhering leukocytes		
Capillaries ($min^{-1} \times field^{-1}$)	1.0 ± 0.8	0.8 ± 1.8
Pial veins ($min^{-1} \times 100\,\mu m^{-1}$)	0.6 ± 0.2	1.6 ± 1.9
Number of leukocytes in extravascular space		
Field 268 × 153 μm	0.8 ± 1.8	11.8 ± 17.5

reperfusion after ischemia. Hallenbeck et al. (1986) reported accumulation of PMNLs in postischemic brain parenchyma following 1 h of air embolism. This study has frequently been considered as evidence of postischemic accumulation of leukocytes in the brain. It is obvious, however, that embolization by air is not comparable to ischemia from reduction or arrest of cerebral blood flow (Helps and Gorman 1991). It has been shown that injection of air causes excessive endothelial damage (Garcia et al. 1981), leading to white blood cell accumulation, possibly without activation of leukocytes.

According to the present experiments, cerebral ischemia in rats does not stimulate leukocyte/endothelium interactions in pial venules, or induce leukocyte plugging of parenchymal capillaries. During control conditions few or no white blood cells were found in postcapillary venules, indicative of a careful preparation technique. In the mesentery, for example, surgical manipulations and eversion of the gut cause a significant increase of rolling of leukocytes (Fiebig et al. 1991).

The pathophysiologic function of PMNLs in cerebral ischemia is controversially discussed. It is important to note that in the few existing studies various models of focal or global cerebral ischemia were employed with different duration of ischemia and postischemic observation periods. Previously, we found an improvement of postischemic cerebral reperfusion in rats made neutropenic by a polyclonal antiserum (Grögaard et al. 1989), although tissue damage was not reduced (Schürer et al. 1991). Bednar et al. (1991) using a hypotension/embolic stroke model reported improvement of postischemic flow plus reduction of infarct size by neutropenia in rabbits. Mori et al. (1992), after reversible middle cerebral artery (MCA) occlusion, succeeded in reducing the number of cerebral blood vessels with so-called no-reflow by blocking of leukocyte adhesion receptors. Takeshima et al. (1992) were administering monoclonal antibodies against leukocyte adhesion receptors 40 min after start of focal ischemia by MCA occlusion. Neither postischemic blood flow, evoked potentials, nor tissue damage was therapeutically influenced. Negative results were also published by Clark et al. (1991a,b) in an ischemia model based on infusion of microspheres. Neurologic scores of animals with blocking of leukocyte adhesion receptors were no different from the controls.

In the present studies, employing in vivo video-fluorescence microscopy of the brain surface, no changes of the number of slowly rolling or of sticking leukocytes were observed in pial venules during ischemia or early reperfusion. Diapedesis of leukocytes into the parenchyma was not found either. This is basically in agreement with in vivo experiments of this laboratory using laser-scanning video-microscopy of intraparenchymal brain capillaries and of pial blood vessels at the brain surface. Again, no significant increase in plugging of capillaries or adherence at the vascular endothelium by leukocytes was found. Leukocyte emigration, however, was observed in two out of five animals 1 h after ischemia. The reason for diapedesis occurring in a few experiments but not in others is currently not understood. Protocol differences, such as type and duration of ischemia, might have played a role. Further experiments with this novel method are required.

Conclusion

Current data are not sufficient to clearly identify leukocytes as a pathogenetic factor in ischemia/reperfusion. It remains to be seen whether occasionally occurring diapedesis of white blood cells is attributable to an ischemia-specific activation of the white blood cells or not. For that purpose, standardized experimental protocols with well-defined end points, such as infarct size and loss of nerve cells, are required.

References

Bednar MM, Raymond S, McAuliffe T, Lodge PA, Gross CE (1991) The role of neutrophils and platelets in a rabbit model of thromboembolic stroke. Stroke 22:44–50

Belkin M, LaMorte WL, Wright GJ, Hobson RW (1989) The role of leukocytes in the pathophysiology of skeletal muscle ischemic injury. J Vasc Surg 10:14–19

Clark WM, Madden KP, Rothlein R, Zivin JA (1991a) Reduction of central nervous system ischemic injury by monoclonal antibody to intracellular adhesion molecule. J Neurosurg 75:623–627

Clark WM, Madden KP, Rothlein R, Zivin JA (1991b) Reduction of central nervous system ischemic injury in rabbits using leukocyte adhesion antibody treatment. Stroke 22:877–883

Corvin S, Schürer L, Abels C, Kempski O, Baethmann A (1990) Effect of stimulation of leukocyte chemotaxis by fMLP on white blood cell behavior in the microcirculation of rat brain. Acta Neurochir Suppl (Wien) 51:55–57

Del Zoppo GJ, Schmid-Schönbein GW, Mori E, Copeland BR, Chang C-M (1991) Polymorphonuclear leukocytes occlude capillaries following middle cerebral artery occlusion and reperfusion in baboons. Stroke 22:1276–1283

Dirnagl U, Villringer A, Gebhardt R, Haberl R, Schmiedek P, Einhäupl KM (1991) Three-dimensional reconstruction of the rat brain cortical microcirculation in vivo. J Cereb Blood Flow Metab 11:353–360

Engler RL, Schmid-Schönbein GW, Pavelec RS (1983) Leukocyte capillary plugging in myocardial ischemia and reperfusion in the dog. Am J Pathol 111:98–111

Fiebig E, Ley K, Arfors K-E (1991) Rapid leukocyte accumulation by "spontaneous" rolling and adhesion in the exteriorized rabbit mesentery. Int J Microcirc Clin Exp 10:127–144

Garcia JH, Klatzo I, Archer T, Lossinski AS (1981) Arterial air embolism: structural effects on gerbil brain. Stroke 12:414–421

Granger DN, Benoit JN, Suzuki M, Grisham MB (1989) Leukocyte adherence to venular endothelium during ischemia-reperfusion. Am J Physiol 257:G683–G688

Grögaard B, Schürer L, Gerdin B, Arfors KE (1989) Delayed hypoperfusion after incomplete forebrain ischemia in the rat. The role of polymorphonuclear leukocytes. J Cereb Blood Flow Metab 9:500–505

Hallenbeck JM, Dutka AJ, Tanishima T, Kochanek PM, Kumaroo KK, Thompson CB, Obrenovitch TP, Contreras TJ (1986) Polymorphonuclear leukocyte accumulation in brain regions with low blood flow during the early postischemic period. Stroke 17:246–253

Harlan JM (1987) Neutrophil-mediated vascular injury. Acta Med Scand 715 Suppl:123–129

Helps SC, Gorman DF (1991) Air embolism of the brain in rabbits pretreated with mechlorethamine. Stroke 22:351–354

Kawamura S, Schürer L, Goetz A, Kempski O, Schmucker B, Baethmann A (1990) An improved closed cranial window technique for investigation of blood-barrier function and cerebral vasomotor control in the rat. Int J Microcirc Clin Exp 9:369–383

Mori E, del Zoppo GJ, Chambers JD, Copeland BR, Arfors KE (1992) Inhibition of polymorphonuclear leukocyte adherence suppresses no-reflow after focal cerebral ischemia. Stroke 23:712–718

Nordström CH, Siesjö BK (1978) Effects of phenobarbital in cerebral ischemia. I. Cerebral energy metabolism during pronounced incomplete ischemia. Stroke 9:327–335

Pulsinelli WA, Buchan AM (1988) The four-vessel occlusion rat model: method for complete occlusion of vertebral arteries and control of collateral circulation. Stroke 19:913–914

Schürer L, Grögaard B, Gerdin B, Kempski O, Arfors KE (1991) Leukocyte depletion does not affect postischaemic nerve cell damage in the rat. Acta Neurochir (Wien) 111:54–60

Takeshima R, Kirsch JR, Koehler RC, Gomoll AW, Traystman RJ (1992) Monoclonal leukocyte antibody does not decrease the injury of transient focal cerebral ischemia in cats. Stroke 23:247–252

Weiss SJ (1989) Tissue destruction by neutrophils. N Engl J Med 320:365–376

Do Leukocytes Contribute to the Brain Lesion That Begins When an Intracranial Artery Is Occluded?

J.H. Garcia, J. Lian, S. Chen, Y. Yoshida, H. Chen, Y. Li, Z. Zhang, and M. Chopp

Introduction

The arrival of leukocytes in areas of focal ischemia (secondary to an arterial occlusion) may be one of the factors contributing to the progression of the ischemic lesion. Several laboratory studies have concluded that leukocytic depletion may protect skeletal muscle [16], intestine [12], myocardium [18, 23, 25], and spinal cord [5] from the effects of ischemia, in particular during the reperfusion period.

Dutka et al. [7] have measured the beneficial effects that granulocytopenia may have on cerebral blood flow as altered by cerebral air embolism. Previously, Hallenbeck et al. [14] had defined, in the same model of canine cerebral air embolism, the extent of granulocyte accumulation in the brain during the acute phase of the injury. However, to date there have been very few studies that define both the nature and the time course of leukocytic infiltrates in areas of focal cerebral ischemia, secondary to the occlusion of a single intracranial artery. Del Zoppo et al. [6] have demonstrated, in a model of middle cerebral artery (MCA) occlusion (3 h) and reperfusion (3 h), the presence of polymorphonuclear (PMN) leukocytes in the lenticulostriate arteries and arterioles. In addition, the same research group has demonstrated improved microcirculatory alterations by inhibiting the endothelial–leukocyte adherence [17].

Precise estimates of leukocytic infiltrates in human brain infarcts (secondary to the occlusion of a large artery) are difficult to obtain for a variety of reasons. Very few patients die within the first 6 h of having had an intracranial artery occluded, and the brain lesion is extremely difficult to identify before 24–72 h have elapsed [19]. Moreover, ample sampling that would ensure representation of all regions involved by the lesion is logistically difficult to achieve in a human brain. Finally, factors (other than the intracranial arterial occlusion) that may influence the inflammatory response cannot be controlled in a human study of this kind. For these reasons, attempts to define the nature and the timing of leukocytic responses

Departments of Pathology and Neurology, Henry Ford Hospital K6, 2799 West Grand Boulevard, Detroit, MI 48202-2689, USA

Cerebral Ischemia and Basic Mechanisms
Ed. by A. Hartmann, F. Yatsu, and W. Kuschinsky

in areas of focal ischemia are based on estimates derived form personal experience in the evaluation of many human specimens [19]. Preliminary to attempts at making therapeutic modifications that involve leukocytic responses in focal cerebral ischemia, we completed two experimental studies in which we defined the leukocytic responses in two slightly different laboratory models of focal cerebral ischemia. The intervals included in these studies covered the period comprised between 15 min to 30 days after the arterial occlusion [6a, 11].

Preliminary to the studies of leukocytic inflammation we had defined in a group of 58 animals the progression of the neuronal lesion from *ischemic changes*, largely involving neurons and astrocytes in a small region of the ischemic hemisphere, to *brain infarct* (or area of pannecrosis) involving the entire territory of the occluded artery. In that study, we noted the bimodal nature of the neuronal responses (acute at 30 min and delayed at (12 h), the topographic spread of the lesion from a small region (30 min) to the entire territory of the artery (48–72 h), and the growth of the lesion to include <10% (30 min) to >70% (72 h) of the hemispheric surface [12].

One of the methods used to occlude the middle cerebral artery is minimally invasive; because it does not require craniotomy, intracranial pressure is affected only by the ischemia-induced changes, and the methods by which the vessel is occluded are compatible with the easy reopening of the occluded artery.

Materials and Methods

In the first experiment each of 57 male Fisher rats (190–250 g) had a large branch of the MCA electrocauterized by means of a subtemporal craniotomy; this was combined with ligation of the ipsilateral carotid artery according to the method described in detail by Brint et al. [3]. In the second group of experiments we applied, in 58 Wistar male rats (290 ± 5 g), the method described by Zea-Longa et al. [27], which is based on the insertion of an 18.5-mm segment nylon monofilament (caliber 4-0) into the internal carotid artery for the purpose of occluding the origin of the corresponding MCA. About 115 adult rats have been evaluated in these combined studies. The time intervals at which the observations were made span the period of 15 min to 30 days following the arterial occlusion. Appropriate control animals included sham-operated rats in which the filament was left in place less than 1 min, and the animals were killed either 1 or 24 h after the sham operation.

At the completion of the prescribed occlusion time, each animal was placed again under general anesthesia for the final procedure: either intravascular brain fixation or administration of an overdose of anesthetics. All brains were fixed in aldehydes (either by immersion or cardiovascular perfusion), embedded either in paraffin for light microscopy or in araldite for both light and electron microscopy. Quantitation of the abnormalities observed in over 2000 histologic preparations derived from these brains was done with the aid of a CCD video camera attached to an image analysis system (IMAGIST, PGT, Princeton, NJ). Additional details concerning the methods and materials used in these experiments have been included in the companion publications [6a, 12].

Results

We have reported that the lesion created through the occlusion of the MCA grows so that it involves less than 10% of the hemispheric surface (30 min after the arterial occlusion) and eventually spreads to include over 70% of the coronal hemispheric surface (72 h) or the entire territory of the occluded MCA [12]. In addition to this spread of the ischemic lesion, we noted two temporally separate types of neuronal responses: *acute* neuronal changes (scalloping, shrinkage, swelling) were detected within 30 min, whereas *delayed* neuronal changes (eosinophilia, ghost cells) appeared approximately 12 h after the arterial occlusion [6a, 12]. The type of neuronal change that predominates at the delayed interval has been traditionally known as red or eosinophilic change, and the original description of this type of cellular alteration in human specimens is attributed to Spielmeyer [24]. The eosinophilic (red) neuron has also been called "ischemic cell change"; this practice is discouraged because cell necrosis characterized by pyknosis and eosinophilia is not specific to ischemia; viruses, chemicals, and other agents can cause the same type of cell necrosis [10]. The development and the nature of the cellular eosinophilic change have been well defined by Farber [10] in the liver, and Jennings and Reimer [15] in the myocardium, under the designation of *coagulative (or coagulation) necrosis*. In those studies and in ours, it has been noted that two significant abnormalities characterize this irreversible cell change in its initial stages: discontinuities in the plasma membrane and calcium-containing precipitates in the mitochondrial matrix.

Based on the quantitative analysis of over 2000 histologic preparations derived from the experiments described above, we evaluated the time-dependent arrival of inflammatory cells into the area of spreading or growing ischemia. Specifically, we identified PMN leukocytes and occasional monocytes in various regions of the ischemic hemisphere as early as 30 min after the arterial occlusion. Of interest, leukocytes remain within the vascular lumen (capillaries and venules) for the first 12 h, after which time they are found in increasing numbers in the brain parenchyma; outside the vessels, PMN can be seen in close apposition to dying or dead neurons [6a]. A certain specificity seems to exist between the type of inflammatory cell and the type of vessel in which these cells are contained: leukocytes are seen almost exclusively in capillaries and venules (caliber 5.5–15.0 μm), whereas

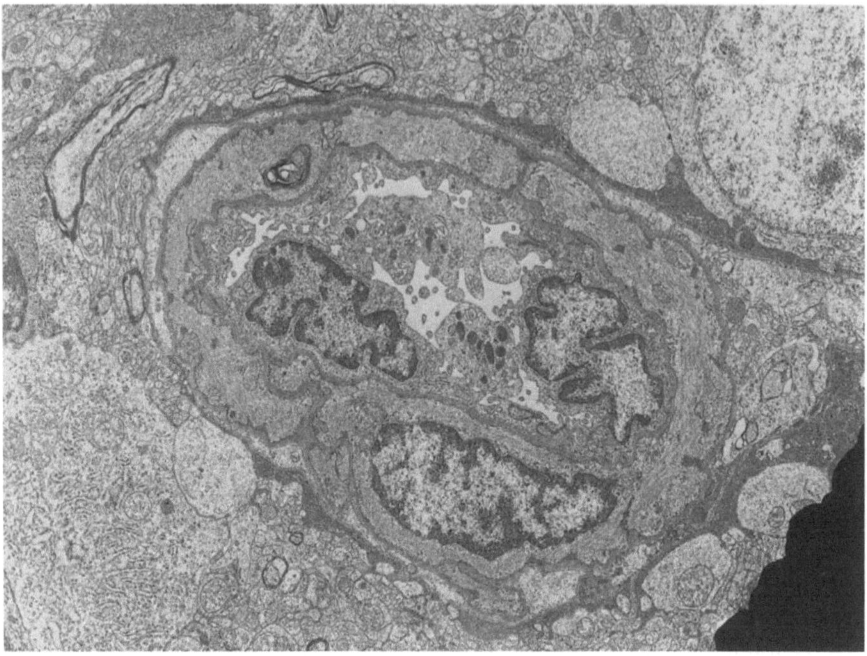

Fig. 1. Cross-section of an arteriole from the area of the striatum 30 min after ipsilateral MCA occlusion in the rat. In addition to possible spasm, there is adhesion of platelets to the endothelial surface. (original magnification ×4500)

platelets are preferentially found attached to the endothelial surface of arterial branches (20–100 μm) (Figs. 1, 2).

The earliest and most numerous leukocytic responses were observed at the same site where the brain parenchymal lesion appeared first, namely the supraoptic area; from this site both the ischemic lesion and the leukocytic infiltrate spread, over a period of 72 h, to the adjacent striatum and cerebral cortex [12]. The two processes (coagulation necrosis of the neurons and marked neutrophilic infiltration) appear to be either simultaneous or to have a close temporal succession, but we have not yet defined their temporal sequence with respect to one another.

Discussion

Dutrochet [8] first reported that circulating leukocytes adhere to vessel walls and migrate as part of the inflammatory response. The inflammatory response, including granulocyte infiltration, has been relatively well documented in experimental myocardial infarcts in dogs. In these animals, leukocytes begin to infiltrate the ischemic myocardium within 60 min of an

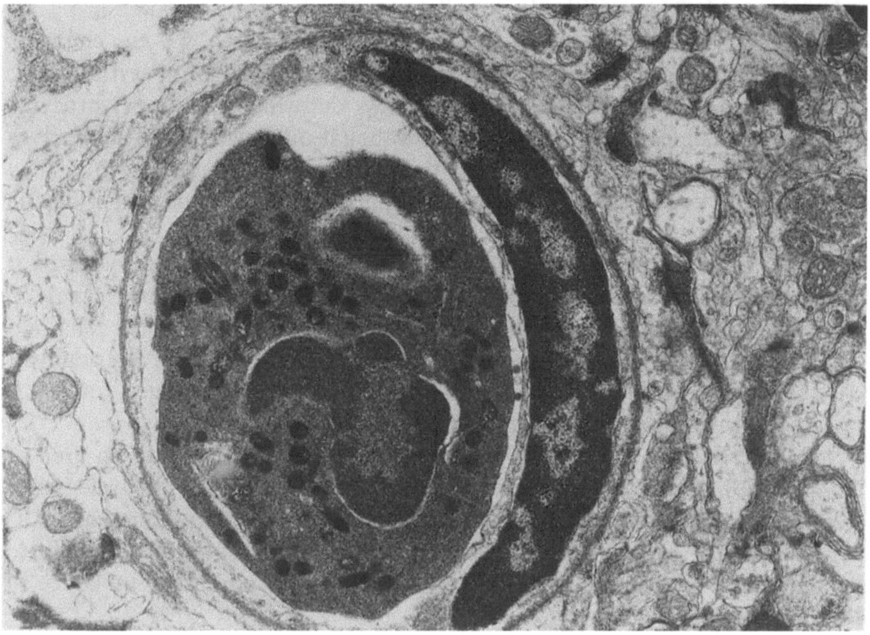

Fig. 2. Cerebral cortex (rat) from the territory of a MCA occluded 1 h before death. A polymorphonuclear leukocyte is adhered to the endothelial surface. (original magnification ×10 000)

experimental coronary occlusion; furthermore, reperfusion significantly increases the number of leukocytes in the ischemic region [26]. This may imply that prior to reperfusing an ischemic territory one may have to inhibit the leukocytic responses.

During the acute ischemic state, leukocytes alter the rheological properties of blood that cause sluggishness in the microcirculation; in addition, leukocytes become increasingly adherent to endothelial surfaces and damage the endothelium, both during the acute and the chronic stages of ischemia [9]. The effect that this may have is to offer an added impediment to the circulation of erythrocytes.

Leukocytes interfere with blood perfusion in areas of compromised circulation by virtue of the mechanical obstruction that their presence creates. This is because these cells are large, stiff, and viscous elements whose capillary transit times are significantly longer than those of erythrocytes. In this manner, the presence of leukocytes in the myocardial microcirculation may contribute to the progression of experimental myocardial infarction [9].

The cerebral microcirculation, by definition, begins at the point where arteries and arterioles penetrate the brain parenchyma; at this point, these vessels measure 30–70 μm in diameter. Precapillary arteriolar diameter is reduced to 14–25 μm with the smallest cerebral capillaries measuring 4–6 μm [1]. Leukocytes are not able to circulate through 5.0-μm channels because their effective viscosity is 1000 times higher than that of erythrocytes; moreover, their nuclei make leukocytes less deformable than erythrocytes [4]. Both the deformability and the endothelial adhesiveness of leukocytes are adversely affected by ischemia, thus the existence of low blood flow enhances the progressive microvascular impairment that occurs after the occlusion of the parent vessel [4].

In addition to the leukocyte-derived procoagulants known to be present in areas of decreased blood flow, endothelial cells could play a crucial role in the induction of local thrombosis and hemorrhage. Endothelial cells and possibly smooth muscle cells probably release cytokines such as interleukin 1 (IL-1), a process that is thought to be initiated by macrophages, and later sustained by circulating neutrophils and monocytes [9].

In the absence of bacterial infection, neutrophils move into an area of inflammation (such an ischemic territory) in response to the release of chemotactic substances such as: leukotriene LTB-4, serum complement C 5a, platelet aggregating factor (PAF), and IL-1. Neutrophils are thought to be stimulated to migrate across the endothelial wall under the influence of IL-1 [2, 17, 21].

Once leukocytes cross the endothelium, they can initiate rapid degradation of extracellular matrix macromolecules by localizing the destructive process to sites of cell–substrate contact.

Plasma and its filtrates contain multiple protease inhibitors, but these inhibitors do not prevent neutrophils from attacking either underlying fibronectin or elastin. However, subjacent substrates could be protected from neutrophils by a leucoprotease inhibitor, a structurally unique serine protease inhibitor whose natural counterpart is normally confined to human mucous secretions [22].

We report evidence of a time-dependent participation by leukocytes and platelets in areas of evolving focal cerebral ischemia in the rat. Leukocytes can be visualized in brain lesions as early as 30 min after an intracranial artery is occluded. The largest numbers of leukocytes, in the vessels of the ischemic hemisphere, appear at approximately 12 h after MCA occlusion; this is the same time interval when red (eosinophilic) neurons appear in large numbers for the first time. Red neurons are the first indisputable microscopic sign of necrosis or infarct. These observations lend support to the hypothesis that the presence of leukocytes in an area of *focal* brain ischemia may contribute to the progression of the ischemic lesion into a bona fide infarct or area of pan-necrosis. However, the completion of this process seems to require a minimum of 6–12 h.

Acknowledgement. Financial support was given by USPHS grants: NS 29463 and NS 10677.

References

1. Bäer T (1978) Morphometric evaluation of capillaries in different laminae of rat cerebral cortex by automatic image analysis: changes during development and aging. Adv Neurol 20:1–9
2. Baggiolini M, Walz A, Kunkel SL (1989) Neutrophil-activating peptide-1/interleukin 8, a novel cytokine that activates neutrophils. J Clin Invest 84:1045–1049
3. Brint S, Jacewicz M, Kiessling M et al. (1988) Focal brain ischemia in the rat: methods for reproducible neocortical infarction using tandem occlusion of the distal middle cerebral and ipsilateral common carotid arteries. J Cereb Blood Flow Metab 8:474–485
4. Chien S (1987) The role of white blood cells in the control of blood rheology. In: Hartmann A, Kuschinsky W (eds) Cerebral ischemia and hemorrheology. Springer, Berlin Heidelberg New York, pp 56–68
5. Clark WM, Madden KP, Rothlein R et al. (1991) Reduction of central nervous system ischemic injury in rabbits using leukocyte adhesion antibody treatment. Stroke 22:877–883
6. Del Zoppo GJ, Schmid-Schönbein GW, Mori E et al. (1991) Polymorphonuclear leukocytes occlude capillaries following middle cerebral artery occlusion and reperfusion in baboons. Stroke 22:1276–1283

6a. Dereski MO, Chopp M, Knight RA, Chen H, Garcia JH (1992) Focal cerebral ischemia in the rat: temporal profile of neutrophil responses. Neurosc Res Comm 11:179–186

7. Dutka AJ, Kochanek PM, Hallenbeck JM (1989) Influence of granulocytopenia on canine cerebral ischemia induced by air embolism. Stroke 20:390–395
8. Dutrochet M (1824) Recherches asnatomiques et physiologiques sur la structure intime des animaux et des végtaux, et sur leur motilité. Bailliere, Paris
9. Engler RL, Dahlgren MD, Morris DD et al. (1986) Role of leukocytes in response to acute myocardial ischemia and reflow in dogs. Am J Physiol 251:H314–H322
10. Farber JL (1982) Biology of disease. Membrance injury and calcium homeostasis in the pathogenesis of coagulative necrosis. Lab Invest 47:114–123
11. Garcia JH, Yoshida Y, Lian J et al. (1992) Brain vasculature in focal cerebral ischemia in the rat (Abstr). 2nd European Stroke Conference
12. Garcia JH, Yoshida Y, Lian J et al. (1993) Progression from ischemic injury to infarct following middle-cerebral-artery occlusion in the rat. Am J Pathol 142:623–645
13. Grisham MB, Hernandez LA, Granger DN (1986) Xanthine oxidase and neutrophil infiltration in intestinal ischemia. Am J Physiol 251:G567–G574
14. Hallenbeck JM, Dutka AJ, Tanishima T et al. (1986) Polymorphonuclear leukocyte accumulation in brain regions with low blood flow during the early postischemic period. Stroke 17:246–253
15. Jennings RB, Reimer KA (1991) The cell biology of acute myocardial ischemia. Annu Rev Med 42:225–246
16. Korthuis RJ, Grisham MB, Granger DN (1988) Leukocyte depletion attenuates vascular injury in post-ischemic skeletal muscle. Am J Physiol 254:H823–H827
17. Lehrer RI, Ganz T, Selsted ME, Babior BM et al. (1988) Neutrophils and host defense. Ann Intern Med 109:127–142
18. Lucchesi BR, Mullane KM (1986) Leukocytes and ischemia-induced myocardial injury. Annu Rev Pharmacol Toxicol 26:201–224
19. Moossy J (1985) Pathology of ischemic cerebrovascular disease. In: Wilkins RA, Rengachary SS (eds) Neurosurgery. McGraw-Hill, New York, pp 1193–1198
20. Mori E, del Zoppo GJ, Chambers JD et al. (1992) Inhibition of polymorphonuclear leukocyte adherence suppresses no-reflow after focal cerebral ischemia in baboons. Stroke 23:712–718

21. Movat HZ, Burrowes CE, Cybulsky MI et al. (1987) Acute inflammation and a Schwatzman-like reaction induced by interleukin-1 and tumor necrosis factor. Synergistic action of the cytokines in the induction of inflammation and microvascular injury. Am J Pathol 129:463–475
22. Rice WG, Weiss SJ (1990) Regulation of proteolysis at the neutrophil-substrate interface by secretary leukoprotease inhibitor. Science 249:249–250
23. Romson JL, Hook BG, Kunkel SL et al. (1983) Reduction of the extent of ischemic myocardial injury by neutrophil depletion in the dog. Circulation 67:1016–1023
24. Scholz W (1953) Selective neuronal necrosis and its topistic patterns in hypoxemia and oligemia. J Neurophathol Exp Neurol 12:249–261
25. Simpson PH, Todd RF, Fantone JC et al. (1988) Reduction of experimental canine myocardial reperfusion injury by a monoclonal antibody (anti-Mol, anti-CD 11b) that inhibits leukocyte adhesion J Clin Invest 81:614–629
26. Vaage J, Semb AG (1989) Myocardial reperfusion injury: an inflammatory reaction. Biomed Biochim Acta 48:S63–S68
27. Zea Longa E, Weinstein PR, Carlson S, Cummins R (1989) Reversible middle cerebral artery occlusion without craniectomy in rats. Stroke 20:84–91

Contribution of Neutrophils to Ischemic Brain Damage

Y. Shiga, H. Onodera, and K. Kogure

Introduction

Neutrophils play a major role in the defense against bacterial infection and modulate inflammation by producing chemical mediators and their immune responses. They also play a critical role in inflammatory edema formation by increasing vascular permeability.

Clinically and experimentally, ischemic brain injury accompanies acute inflammatory responses which are characterized by neutrophil infiltration and edema formation. Neutrophils infiltrate into an ischemically injured area in the brain prior to phagocytes, such as monocytess and macrophages. We can easily assume infiltrating neutrophils can contribute to the ischemic damage (Fig. 1).

Reduction of infarct size by depletion of circulating neutrophils has been reported [1]. We assessed the contribution of neutrophils to ischemic brain damage by means of selective depletion of circulating neutrophils using a monoclonal antibody, RP3. In addition to neutrophils, the contribution of macrophages [2] and complements [3] to ischemic damage has also been reported.

Therefore we next assessed the protective effect of an immunosuppressant, cyclospolin A (CSA) against ischemic brain damage. CSA is a secondary metabolite of fungal origin and is a potent immunosuppressant widely used to reduce rejection after transplantation surgery and to treat autoimmune diseases. Unlike other drugs, it lacks significant myelotoxicity [4].

Materials and Method

We used specific pathogen-free, adult male Wistar rats weighing about 300 g. Ischemic insult was given as follows. Under 2% halothane anesthesia, an 18-mm long 4–0 nylon thread coated with silicone was inserted from

Department of Neurology, Institute of Brain Diseases, Tohoku University School of Medicine, 1-1 Seiryo-machi, Aoba-ku, 980 Sendai, Japan

Cerebral Ischemia and Basic Mechanisms
Ed. by A. Hartmann, F. Yatsu, and W. Kuschinsky

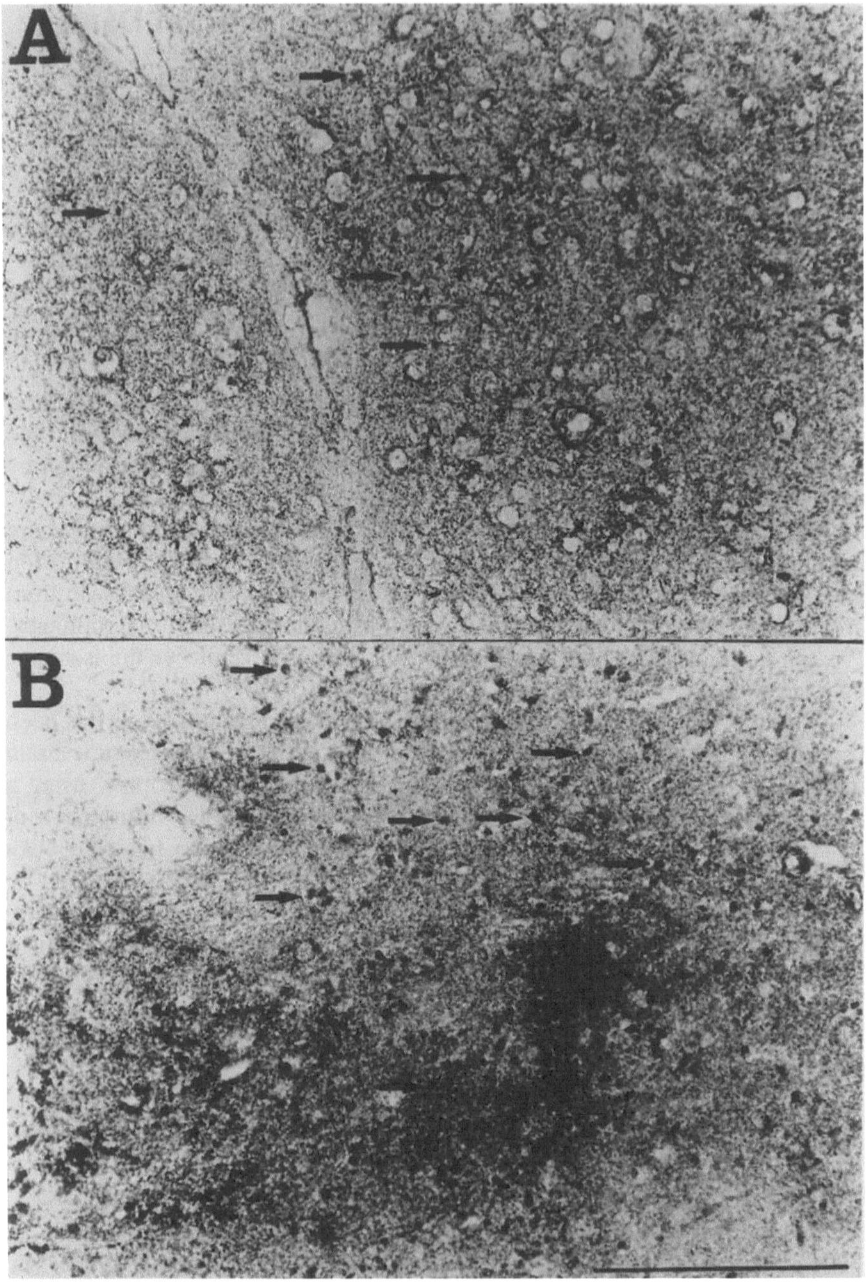

Fig. 1A,B. Infiltrating neutrophils in the DCP 3h after reperfusion (**A**) and 1 day after reperfusion (**B**). Neutrophils were detected immunohistochemically by mouse monoclonal antibody to rat and human granulocytes using the avidin–biotin complex method. Infiltration of neutrophils was much more intense 1 day after reperfusion. *Arrows* indicate indicate infiltrating neutrophils; *bar*, 50 μm the right external carotid artery to the right internal carotid artery to occlude the origin of the right middle cerebral artery (MCA) [5]. After 1 h of the MCA occlusion, the thread was removed to allow reperfusion of the ischemic region via the right common carotid artery. Neurologic deficits characterized by severe left-sided hemiparesis and right Horner's sign were used as criteria for ischemic insult. Convulsive rats and unconscious rats were excluded from further study.

To deplete the rats' circulating neutrophils, a monoclonal antibody, RP3[6], was prepared from ascites fluid.

To assess the influence of depletion of circulating neutrophils on ischemic brain damage, ten rats received intraperitoneal RP3 treatment. Each rat received two 13.3-ml/kg RP3 injections 24 h before and immediately after ischemic insult. The numbers of neutrophils were counted prior to RP3 treatment, 24 h after the first RP3 treatment, and 24 h after the second RP3 treatment. Saline-treated rats served as control.

To assess the effect of the immunosuppressant CSA on ischemic brain damage, rats were orally given 15 mg/kg per day CSA dissolved in olive oil for 5 days preceding the ischemic insult and on the day of ischemia. Postischemic brain edema and infarct size were estimated. Rats given only olive oil served as control.

One day after reperfusion, RP3-treated rats were decapitated and brain samples were dissected into the cerebral cortex perfused by the right MCA, the dorsal area of the caudate putamen (DCP), and the ventral area of the caudate putamen (VCP). To assess brain edema, water contents of these samples were then measured by the wet and dry method [7]. To assess infarct size, rats were perfused with saline and the brains ($n = 4$) were cut into 1-mm wide coronal sections; 1 mm anterior, 1 mm posterior, and 3 mm posterior to the bregma were immersed in 2% triphenyl tetrazoliumchloride (TTC) for 30 min at room temperature. The TTC-unstained area, which reflected the infarct area [8], was measured using a Zeiss IBAS image analyzer system ($n = 6$). Infarct size was expressed as a percentage of the total hemisphere at risk.

In CSA-treated rats, brain edema ($n = 7$) and infarct size ($n = 6$) were assessed in the same manner as stated above.

Results

The numbers of circulating neutrophils prior to RP3 treatment, 24 h after the first treatment, and 24 h after the second treatment were 1429 ± 338 mm^3, 100 ± 70 mm^3, and 201 ± 84 mm^3, respectively (Fig. 2). RP3 treatment effectively suppressed the number of circulating neutrophils throughout the ischemia and reperfusion periods in the present study.

Brain water contents of RP3-treated rats in the MCA, the DCP, and the VCP in the occluded side were 79.93% ± 1.41%, 81.00% ± 1.41%, and

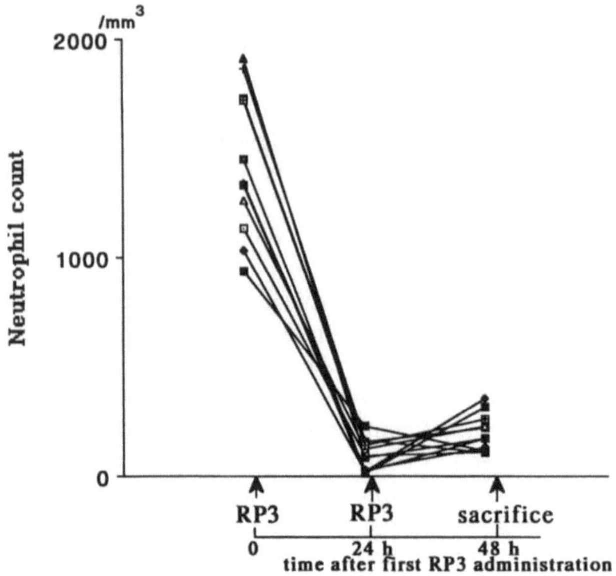

Fig. 2. The numbers of circulating neutrophils. Intraperitoneal injection of RP3 significantly depleted circulating neutrophils during ischemia and reperfusion periods

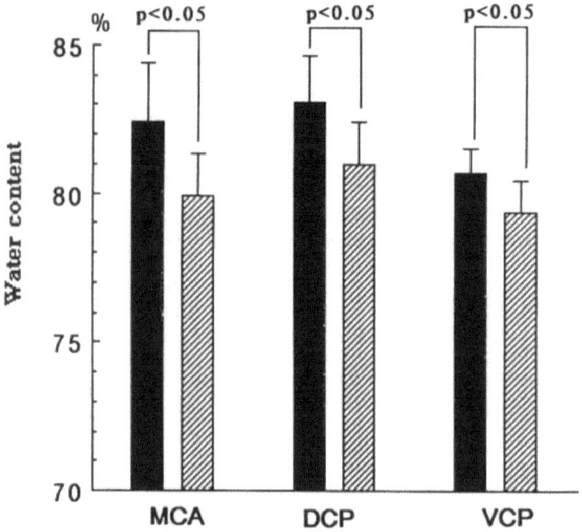

Fig. 3. Brain water contents 1 day after reperfusion in the MCA, the DCP, and the VCP of RP3-treated rats were significantly reduced compared to those of the saline-treated controls. *Solid columns*, saline treatment plus ischemia; *hatched columns*, RP3 treatment plus ischemia

79.38% ± 1.10%, respectively. On the other hand, water content of saline-treated control rats in the MCA, the DCP, and the VCP in the occluded side were 82.64% ± 1.92%, 83.08% ± 1.54%, and 80.70% ± 0.81%, respectively (Fig. 3). This clearly indicates that depletion of neutrophils by RP3 reduced ischemic brain edema formation ($p < 0.05$).

The infarct sizes of RP3-treated rats at the coronal levels 1 mm anterior, 1 mm posterior, and 3 mm posterior to the bregma were 17.48% ± 5.85%, 16.72% ± 4.47%, and 9.50% ± 2.65%, respectively. On the other hand, the infarct sizes of control rats at the coronal levels 1 mm anterior, 1 mm posterior, and 3 mm posterior to the bregma were 35.91% ± 7.51%, 39.98% ± 10.25%, and 28.26% ± 8.42%, respectively (Figs. 4, 5). Depletion of

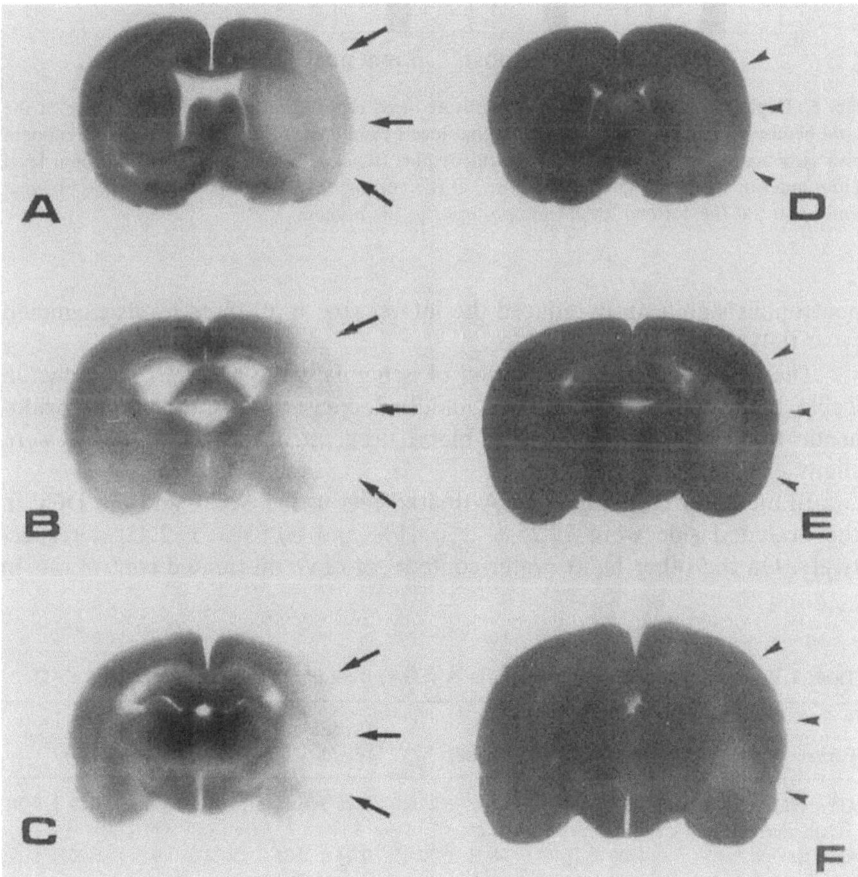

Fig. 4A–F. TTC staining. **A**, **B**, and **C** are saline-treated control rats. **D**, **E**, and **F** are neutrophil-depleted rats. Note that infarct size of neutrophil-depleted rats was smaller than that of control rats. TTC-unstained area reflects infarct area

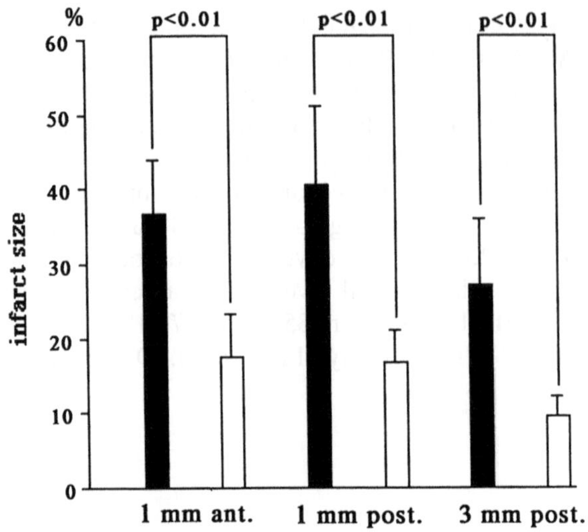

Fig. 5. Depletion of neutrophils significantly reduced that infarct size (as a percentage of the total hemisphere at risk) at all three coronal levels examined. *Solid columns*, saline treatment plus ischemia; *open columns*, RP3 treatment plus ischemia; *1 mm ant.*, at the coronal level 1 mm anterior to the bregma; *1 mm post.*, at the coronal level 1 mm posterior to the bregma; *3 mm post.*, at the coronal level 3 mm posterior to the bregma

neutrophils significantly reduced the infarct size at all three levels examined ($p < 0.01$).

The CSA level at the induction of ischemia was 890 ng/ml. As shown in Table 1, the leukocyte count was mildly decreased. Serum glutamic oxaloacetic transaminase (GOT) and blood urea nitrogen (BUN) levels were sligtly elevated.

Brain water contents of CSA-treated rats in the MCA and the DCP in the occluded side were 79.25% ± 1.41% and 80.53% ± 2.42%, respectively. On the other hand, water contents of olive oil-treated control rats in

Table 1. Laboratory data after cyclosporin A (CSA) treatment (15 mg/kg per day)

Group	*n*	Leukocytes (mm^3)	BUN (mg/dl)	Creatinine (mg/dl)	GOT (IU/l)	GPT (IU/l)
Olive oil-treated ischemia group	5	6525 ± 1043	14.48 ± 2.09	0.46 ± 0.11	10.92 ± 1.90	43.40 ± 8.20
CSA treated ischemia group	5	4600 ± 1046*	23.10 ± 7.64*	0.74 ± 0.25	15.22 ± 0.90*	45.20 ± 15.70

* Significantly different from control group by paired *t* test ($p < 0.05$).
Values are mean ± SD. BUN, blood urea nitrogen; GOT, glutamic oxaloacetic transaminase; GPT, glutamic pyruvic transaminase.

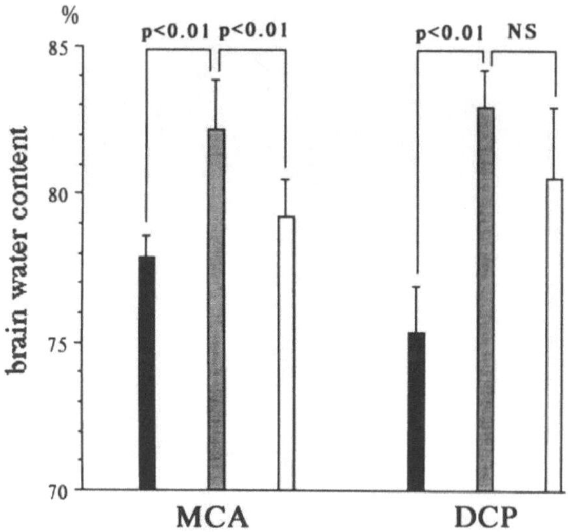

Fig. 6. Brain water contents of CSA-treated rats in the MCA and the DCP. CSA significantly reduced brain edema in the MCA. *Solid columns*, nonischemic normal brain; *shaded columns*, olive oil treatment plus ischemia; *open columns*, CSA treatment plus ischemia

the MCA and the DCP in the occluded side were 82.21% ± 1.62% and 82.93% ± 1.29%, respectively (Fig. 6). CSA significantly reduced brain edema in the MCA ($p < 0.01$)

The infarct sizes of CSA-treated rats at the coronal levels 1 mm anterior, 1 mm posterior, and 3 mm posterior to the bregma were 16.15% ± 4.47%, 12.20% ± 6.40%, and 6.55% ± 1.68%, respectively. On the other hand, the infarct sizes of control rats at the coronal levels 1 mm anterior, 1 mm posterior, and 3 mm posterior to the bregma were 36.82% ± 7.15%, 40.65% ± 10.36%, and 27.17% ± 8.90%, respectively (Figs. 7, 8). CSA significantly reduced the infarct size at all three levels examined ($p < 0.01$). Interestingly, the protective effect of CSA was much more pronounced in the cerebral cortex than in the caudate putamen.

Discussion

Recently, Dukta et al. [9] and Bednar et al. [10] showed that depletion of neutrophils reduced the postischemic brain damage. They explained the reason that depletion of neutrophils might improve the blood flow during the postischemic reperfusion period. del Zoppo et al. [11] showed that postischemic activated neutrophils occluded capillaries and obstructed the postischemic microcirculation. However, we must consider the role of infiltrating neutrophils into the ischemic region.

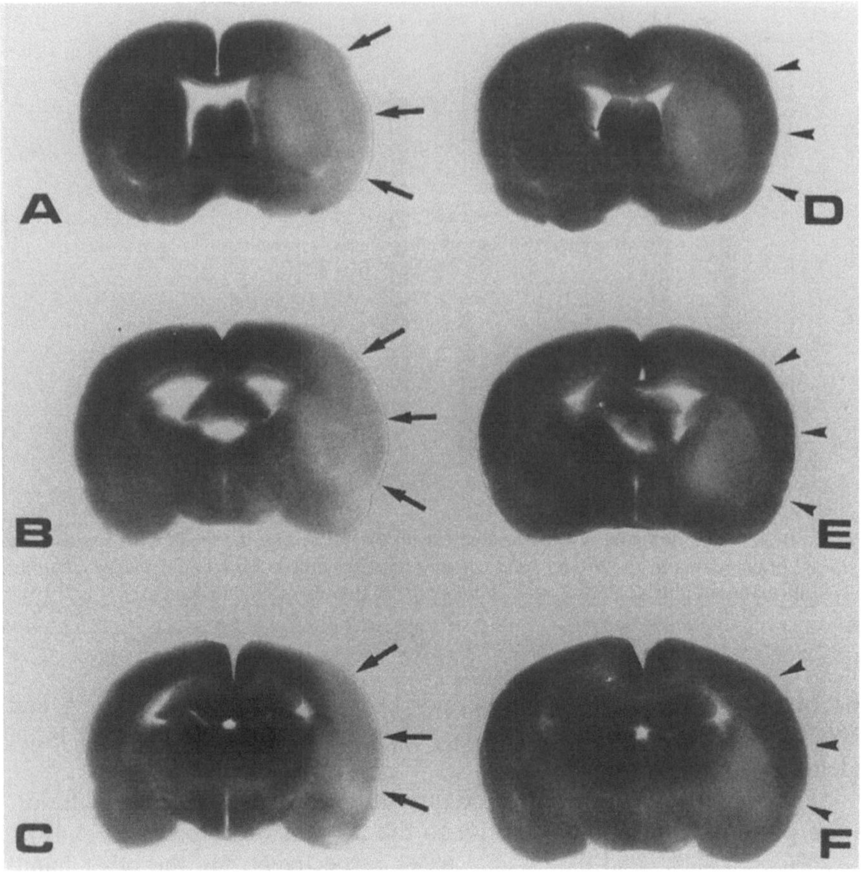

Fig. 7A–F. TTC staining. **A**, **B**, and **C** are olive oil-treated control rats. **D**, **E**, and **F** are CSA-treated rats. Interestingly, the protective effect of CSA was much more pronounced in the cerebral cortex than in the caudate putamen

Activated neutrophils produce chemical mediators of inflammation and immune responses, such as leukotrien B4, platelet-activating factor, interleukin 1, and active oxygen [6].

We believe blood flow disturbance and chemical mediators must be considered for major neutrophil contribution to exacerbating ischemic brain damage.

It was previously reported that CSA at a dose of 14 mg/kg per day was a sufficient immunosuppressive dose in rats. No acute toxicities were recognized at a dose of 15 mg/kg per day [4]. Thus, we believe our dose of CSA treatment had no influence on our experimental results.

The immunosuppressive activities of CSA is believed to be mediated by cyclophilin [12], which is a specific binding protein for CSA. CSA is thought

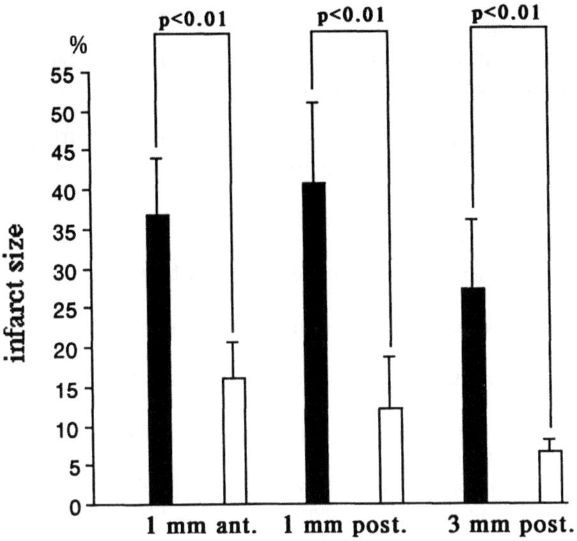

Fig. 8. CSA significantly reduced the infarct size (as a percentage of the total hemisphere at risk) at all three coronal levels examined. *Solid columns*, olive oil treatment plus ischemia; *open columns*, CSA treatment plus ischemia; *1 mm ant.*, at the coronal level 1 mm anterior to the bregma; *1 mm post.*, at the coronal level 1 mm posterior to the bregma; *3 mm post.*, at the coronal level 3 mm posterior to the bregma

to act on the immune systems not only by inhibiting T lymphocyte activation but also by inhibiting the production of interleukin 2, gamma interferon, and other cytokines [13], although its precise mechanism in vivo is not well understood. Since CSA does not influence the blood–brain barrier permeability or brain circulation [14], CSA could reduce postischemic brain damage via the suppression of some immunologic reactions.

There are two possibilities why CSA protected only the cerebral cortex perfused by the MCA but not the caudate putamen. First, the ischemic insult in the MCA is milder than that in the DCP. Second, very much cyclophilin, the specific binding protein of CSA, is concentrated in the cerebral cortex, but is only a trace in the caudate putamen [13].

We showed that both depletion of circulating neutrophils and the immunosuppressant CSA reduced the postischemic brain damage. We believe immunologic reactions have important roles for ischemic brain damage during early reperfusion.

References

1. Romson JL et al. (1983) Reduction of the extent of ischemic myocardial injury by neutrophil depletion in the dog. Circulation 67:1016–1023

2. Gullian D et al. (1990) Inhibition of mononuclear phagocytes reduced ischemic injury in the spinal cord. Ann Neurol 27:33–42
3. Weisman HF et al. (1990) Soluble human component receptor type 1: in vivo inhibitor of complement suppressing post-ischemic myocardial inflammation and necrosis. Science 249:146–151
4. Ryffel B et al. (1983) Toxicological eveluation of cyclospolin A. Arch Toxicol 53:107–141
5. Longa BZ et al. (1989) Reversible middle cerebral artery occlusion without craniectomy in rats. Stroke 20:84–91
6. Sekiya S et al. (1989) Selective depletion of rat neutrophils by in vivo administration of a monoclonal antibody. J Leukoc Biol 46:96–102
7. Hatashita S et al. (1988) Ischemic brain edema and osmotic gradient between blood and brain. J Cereb Blood Flow Metab 8:552–559
8. Liszczak TM et al. (1984) Limitation of tetrazolium salts in delineating infarcted brain. Acta Neuropathol (Berl) 65:150–157
9. Dukta AJ et al. (1989) Influence of granulocytepenia on canine cerebral ischemia induced by air embolism. Stroke 20:390–395
10. Bednar MM et al. (1991) The role of neutrophils and platelets in a rabbit model of thromboembolic stroke. Stroke 22:44–50
11. Del Zoppo GJ et al. (1991) Polymorphonuclear leukocytes occlude capillaries following middle cerebral artery occlusion and reperfusion in baboona. Stroke 22:1276–1283
12. Handshumacher RE et al. (1984) Cyclophilin: a specific cytosolic binding protein for cyclospolin A. Science 226:544–547
13. Rad PP et al. (1991) Molecular cloning and regional distribution of rat brain cyclophilin. Mol Brain Res 9:239–244
14. Owunwanne A et al. (1990) Radionuclide and histophathological assessment of any brain perfusion abnormalities in rats treated with the immunosuppressant drug (CyA). Eur J Nucl Med 16:519

Polymorphonuclear Leukocyte Adherence and Microvascular Reflow After Focal Cerebral Ischemia/Reperfusion

G.J. Del Zoppo[1], E. Mori[1], B.R. Copeland[1], W.S. Thomas[1], G.W. Schmid-Schönbein[2], E.D. Chambers[3], and K.E. Arfors[3]

Interruption of normal cerebral arterial blood flow by atherothromboemboli or thrombi in situ in a territory unprotected by adequate collaterals may lead to neuronal necrosis and tissue disruption (Zülch 1985). Reperfusion of the occluded artery, as may occur following successful use of thrombolytic agents within hours of flow interruption (del Zoppo et al. 1988, 1992; Mori et al. 1988, 1992a,b), introduces significant additional vascular events. Particularly vulnerable is the distally (downstream) dependent microvascular bed which receives the full force of arterial reperfusion. In models of global cerebral ischemia, microvascular perfusion defects ("no-reflow") have been demonstrated following reperfusion (Ames et al. 1968). Similar post-reperfusion obstructions to reflow in the microvascular bed, manifest by loss of patency, have been described in the setting of experimental cardiac and skeletal muscle ischemia (Kloner et al. 1974; Engler et al. 1987; Schmid-Schönbein 1987). It has been postulated that such microvascular perfusion defects may result from perivascular edema or hemorrhage, endothelial cell edema, endothelial cell tags, in situ thrombosis, erythrocyte aggregation, platelet adherence and aggregation, polymorphonuclear (PMN) leukocyte endothelial cell adherence, rheologic properties of PMN leukocytes, and other mechanisms. The appearance of "no-reflow" following single artery occlusion and subsequent reperfusion (O/R), as might occur in focal cerebral ischemia, has been suspected, but only recently described (del Zoppo et al. 1991).

Microvascular Events Following Middle Cerebral Artery O/R

Heretofore, capillary dilatation has been described in anesthetized "open" primate models of middle cerebral artery (MCA) occlusion and reperfusion (Garcia et al. 1983; Little et al. 1975, 1976). We have noted the presence

[1] Department of Molecular and Experimental Medicine, The Scripps Research Institute, 10666 North Torrey Pines Road, La Jolla, CA 92037, USA
[2] AMES Bioengineering, University of California, San Diego, CA, USA
[3] La Jolla Institute of Experimental Medicine, La Jolla, CA, USA

Cerebral Ischemia and Basic Mechanisms
Ed. by A. Hartmann, F. Yatsu, and W. Kuschinsky

of microvascular occlusions in the lenticulostriate artery (LSA) territory following MCA:O/R in an awake "closed" nonhuman primate model (del Zoppo et al. 1986). Here, MCA occlusion lasted 3h, followed by a 1-h period of reperfusion. Deformed erythrocytes, PMN leukocytes, and degranulated platelets constitute the lenticulostriatal microvascular obstructions. Pre-ischemia use of the potent antithrombotic combination heparin/ ticlopidine significantly reduced the deposition of ^{111}In-labelled platelets in the MCA territory, suggesting the presence of platelets in vascular obstructions (Nicolaides and Zukowski 1985). The role of thrombosis has been suggested by the association of tissue factor with cerebral microvessels $\geqslant 10\,\mu$m in diameter and in cerebral gray matter parenchyma (del Zoppo et al. 1991a,b). The relative contributions of platelets and tissue factor-dependent fibrin formation to post-MCA:O/R microvascular perfusion defects is undefined.

The presence of PMN leukocytes in microvascular occlusions has prompted a more consistent study of their location and potential contribution to the development of "no-reflow" in the MCA territory (del Zoppo et al. 1991a,b). For this purpose, two perfusion–fixation schemes were adapted for acute preparation of cerebral tissues under isosmotic conditions to define the presence of microvascular obstructions and their constituents, and of microvascular patency (del Zoppo et al. 1991a,b; Mori et al. 1992a,b). The perfusion protocols consisted of an isosmotic perfusion flush phase, and an isosmotic fixation phase intended for rapid fixation of vascular tissues for light and electron microscopy of microvascular occlusions (scheme 1, colloidal gold), or light microscopy of patent vessels (scheme 2, carbon tracer). Under identical experimental conditions, PMN leukocytes were identified in the territory of microvascular no-reflow. Single leukocytes were documented to (a) completely obstruct microvessels of capillary diameter (4.0–7.5 μm); and (b) partially obstruct microvessels of postcapillary venule diameter. Multiple PMN leukocytes were found in apposition to the endothelium of postcapillary venules (Fig. 1). Among ten microvessels reconstructed along a 110-μm length, 40% displayed a single PMN leukocyte completely obstructing the capillary segment. Reflow occurred in 26.5% and 38.5% of capillaries in the basal ganglia following ischemia in two separate subjects, a significant reduction in reflow. Such reflow abnormalities are striking and similar to those described by Ames et al. (1968).

Indirect evidence of PMN leukocyte involvement in cerebrovascular reflow has been reported (Bednar et al. 1991; Grogaard et al. 1989; Vasthare et al. 1990). An increase in local cerebral blood flow (CBF) following ischemia/reperfusion in a rabbit cerebral embolic/hypotension model accompanied pre-ischemia infusion of an antineutrophil serum and reduction in peripheral PMN leukocyte count (Bednar et al. 1991). In a rat hypotension/ carotid occlusion model of hemispheric ischemia/reperfusion, pre-ischemia depletion of circulating PMN leukocytes following intraperitoneal injection

Fig. 1. PMN leukocytes in postcapillary venule following middle cerebral artery occlusion and reperfusion (see text). Electron-microscopic examination of similar vessels has demonstrated close opposition of leukocytes to endothelial cell membrane

of antineutrophil serum was accompanied by increased local CBF in the cortex, but not the ipsilateral basal ganglia (Grogaard et al. 1989). An improvement in neuronal functional outcome, measured by alterations in EEG and somatosensory evoked potentials, was reported following vinblastine-induced leukopenia in a rat reversible incomplete forebrain ischemia model (Vasthare et al. 1990). Unfortunately, polyclonal antineutrophil sera and antineoplastic agents have complex effects which may also alter platelet number (and function). Antineutrophil strategies in other noncerebral ischemia experimental models have been accompanied by a significant reduction in apparent dysfunction and mortality (Vedder et al. 1989, 1990; Carden et al. 1990; Simpson et al. 1988). Those observations suggest a potential role for the PMN leukocyte in postischemia reperfusion microvascular injury, the no-reflow phenomenon, and perivascular damage during focal cerebral ischemia.

To test the hypothesis that PMN leukocyte adhesion to cerebral microvascular endothelium may contribute to microvascular perfusion defects following MCA occlusion and reperfusion, a study of the effect of the murine anti-CD18 monoclonal antibody (MAb) IB4 on microvascular patency was performed. Adherence of PMN leukocytes to endothelium of the postcapillary venule is mediated by the leukocyte integrin CD11b/CD18

(Mac-1, Mo1) (Harlan et al. 1985). The murine anti-CD18 MAb IB4 (60.3) inhibits aggregation, endothelial adherence, and endothelial injury of activated PMN leukocytes (Hossmann and Olsson 1970; House and Lipowsky 1987; Arfors et al. 1987; Vedder et al. 1990; Suzuki et al. 1989; Hernandez et al. 1987; Ham and Cormack 1979). Well-characterized purified endotoxin-free preparations of IB4 are available, which upon systemic infusion inhibit the "rolling" of PMN leukocytes noted in ex vivo preparations and attenuate reperfusion injury (Vedder et al. 1990).

MCA : O/R – Experimental Paradigm

In the present study, nonhuman primates fitted with a reversible MCA occlusion device were randomly assigned to receive no intervention or IB4 (2–4 mg/kg) as a single intravenous infusion 15 min prior to reperfusion (release of MCA occlusion) following 180 min of LSA territory ischemia. The IB4 dose was derived from in vitro studies demonstrating an 80% inhibition of baboon PMN leukocyte adherence to human umbilical vein endothelium 12 h following a 2-mg/kg in vivo infusion, and 75% inhibition of f-met-leu-phe-stimulated baboon granulocytes by 15 μg/ml IB4.

Preparation of the nonhuman primate model of MCA : O/R and the anesthesia protocol (halothane) are described in detail elsewhere (del Zoppo et al. 1991a,b; Mori et al. 1992a,b). Prior to surgery and study entry, all animals were observed to be disease-free during a mandated quarantine. All procedures were approved by the institutional Animal Research Committee and were performed in accordance with standards promulgated by the National Research Council, the National Institutes of Health in its policy on the Humane Care and Use of Laboratory Animals, and the USDA in the Animal Welfare Act. Following surgical implantation of the MCA occlusion device, all subjects were allowed a 7-day procedure-free interval prior to study entry. Normal neurological status (score = 100) (Spetzler et al. 1980) was required for randomization.

After randomization to no intervention ($n = 6$) or IB4-infusion ($n = 7$), occlusion of the MCA was performed in the awake subject. In the treatment group, IB4 was infused 15 min prior to reperfusion (balloon deflation). Following 1-h reperfusion, each experiment was terminated by pressure perfusion-fixation with an initial perfusion flush to eliminate blood elements under antithrombotic and isosmotic conditions, and a subsequent isosmotic fixation phase with India ink carbon tracer (Pelikan Fount India, Pelikan AG, Hannover, FRG) in 2% paraformaldehyde/0.5% glutaraldehyde (scheme 2) (del Zoppo et al. 1991a,b) at high mean arterial pressure (180–220 torr). Following perfusion-fixation the brain was removed in toto, sectioned in the coronal plane at 1-cm intervals, and further fixed in 87% ethanol, 10% formaldehyde, and 3% glacial acetic acid (volume/volume) (del Zoppo et al. 1991a,b); 1.0-cm × 1.0-cm × 0.2-cm tissue blocks from stereoanatomically identical sites of both basal ganglia and from the left temporal lobe were embedded, sectioned at 10-μm intervals, and stained with basic fuchsin/methylene blue. Sections were analyzed automatically with a video-image acquisition system consisting of a Hamamatsu C2400-07 NTSC video camera staged on a light microscope (Kontron and Carl Zeiss, Munich, FRG). Contiguous fields with a total surface area of 25 mm^2 were processed on each section at 30-μm intervals such that $\geqslant$2000 vessels were counted in the nonischemic basal ganglia. Comparisons of the quantity and size of microvessels from ischemic and nonischemic basal ganglia at stereoanatomically identical sites were made to calculate the "percent reflow." Data are displayed as a continuous function of microvessel size, and in biologically relevant vessel classes: (a) capillaries (4.0–7.5 μm); (b) metarterioles and postcapillary venules (7.5–30 μm); (c) small arterioles and connecting venules (30–50 μm); and (d) muscular arterioles and venules >50-μm diameter.

Serial leukocyte, erythrocyte, and platelet counts and serial determinations of IB4 binding to PMN leukocytes were performed in all animals as described (Mori et al. 1992). Heparinized (100 IU/ml) whole blood specimens at 4°C were immediately analyzed for CD18 receptor availability and receptor occupancy by IB4 by flow cytometry (Bectin Dickinson FAC Scan, San Jose, CA) as described in detail (Mori et al. 1992).

Neurological function was evaluated serially according to a well-described 100-point scale weighted toward motor dysfunction (Spetzler et al. 1980; del Zoppo et al. 1986). Statistical analysis for significance was performed on paired or unpaired data by Student's T test and are reported as p (one-tailed) or $2p$ (two-tailed) values where appropriate. All analyses were performed on discrete data sets, e.g., microvascular diameter.

Prereperfusion Intervention with Anti-CD11b/CD18 Antibody

No significant alteration in peripheral PMN leukocyte count, hematocrit, or platelet count from control ($n = 6$) values was noted following infusion of IB4 ($n = 7$). An abrupt and significant increase in peripheral white cell count and absolute PMN leukocyte count from baseline was seen in both cohorts (Fig. 2).

For analysis of microvascular patency, two subjects in the IB4-treated cohorts were excluded because the gray matter of the parietal or temporal cortex was inhomogeneously stained ($n = 1$), and segmented carbon tracer elements were noted on individual histologic sections suggesting incomplete perfusion ($n = 1$). The resultant ratio of patent microvessels <100-μm diameter in normal basal ganglia to ipsilateral temporal cortex was 0.59 ± 0.27.

Among subjects receiving no treatment, a significant decrease in the quantity of patent incrovessels in the ischemic basal ganglia compared to the

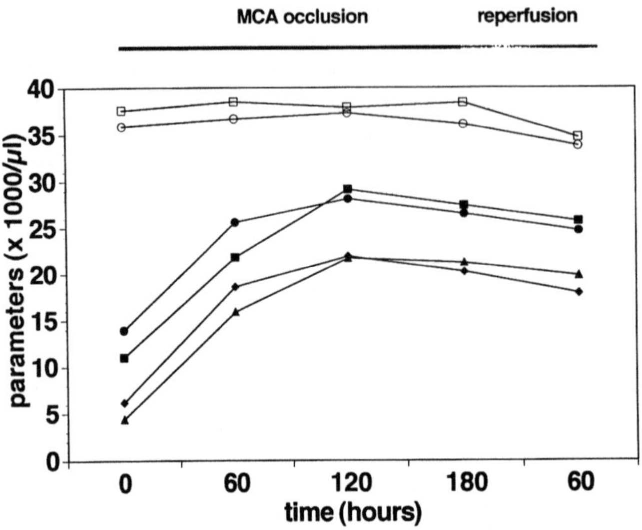

Fig. 2. Serial peripheral total leukocyte count, absolute PMN leukocyte count, and hematocrit. *Solid squares*, total leukocyte count, control; *solid circles*, total leukocyte count, IB4 treated; *solid triangles*, absolute PMN leukocyte count, control; *diamonds*, absolute PMN leukocyte count, IB4 treated; *open squares*, hematocrit, control; *open circles*, hematocrit, IB4 treated

nonischemic basal ganglia following reperfusion was documented supporting previous observations (del Zoppo et al. 1991a,b). This difference was most prominent in both vessel classes 4.0–7.5-μm and 7.5–30-μm diameter. Because of the small number of microvessels in the 30–50-μm and >50-μm diameter classes, percent reflow determinations displayed great variation.

Among subjects receiving IB4, a significant 83% reduction of free PMN leukocyte CD18-IB4-binding sites was determined by flow cytometry. A significant increase in microvascular percent reflow was observed, which was most prominent in microvessels of 7.5–30-μm diameter and 30–50-μm diameter compared with the control cohort (Fig. 3).

No significant increase in the diameter distribution of microvessels in the ischemic territory was observed compared to the nonischemic basal ganglia when the data were normalized for vessel diameter proportion. This is consistent with the view that the improvement in microvascular patency was not an artifact of vasodilatation.

Implications

These data indicate that blockade of the PMN leukocyte adherence complex CD18 may increase microvascular reflow following focal cerebral ischemia and reperfusion. Two observations are relevant: (a) that exposure of circulating PMN leukocytes to IB4, an MAb known to inhibit granulocyte

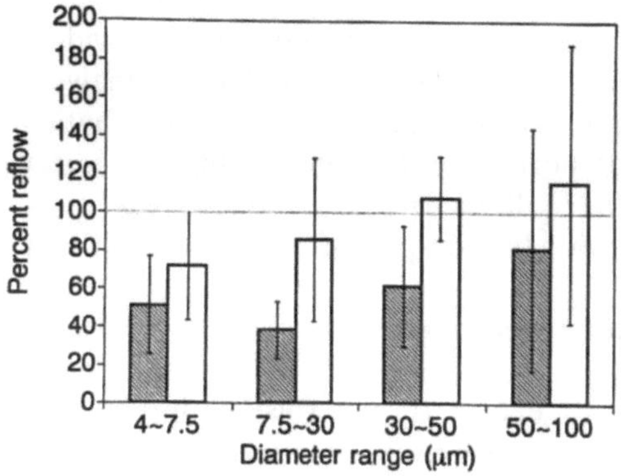

Fig. 3. Improvement in microvascular reflow following infusion of IB4. Significant increase in percent reflow noted in 7.5–30-μm diameter (postcapillary venule/precapillary arteriole, $p = 0.049$) and 30–50-μm diameter (connecting venule/small arterioles, $p = 0.034$). *Hatched columns*, control; *open columns*, IB4 treated

adhesion to postcapillary venules, was associated with a substantial increase in reflow following MCA:O/R; and (b) the effect on microvascular reflow occurred when IB4 was infused prior to reperfusion, but following the onset of ischemia. Indeed, the most significant increase in reflow attended microvessels of the 7.5–30-μm diameter class (postcapillary venules/precapillary arterioles). The use of computer-assisted imaging techniques to evaluate large numbers of microvessels rapidly does not allow distinction between postcapillary venules and precapillary arterioles. However, the findings in these experiments are consistent with the known interaction of activated PMN leukocytes and microvascular endothelium mediated by the CD11/CD18 adherence complex in postcapillary venules in other organ systems (House and Lipowsky 1987; Suzuki et al. 1989; Carden et al. 1990).

The characteristics of this method of microvessel quantitation have been defined in separate manuscripts (del Zoppo et al. 1991a,b; Mori et al. 1992a,b). Coefficients of variance for multiple objects in the size range below 30-μm minimum diameter are at most 5.93% and are independent of size, while interobserver and intraobserver variability are acceptable (del Zoppo et al. 1991a,b). The three-dimensional architecture of the LSA microvascular bed in the LSA territory of humans and nonhuman primates has not been described previously. It is apparent that significant differences in transverse microvascular diameters do not occur unless coronal sections are obtained 30 μm or more from one another in the rostral → caudal direction. Therefore, the minimum interval for section acquisition was chosen at 30 μm. Consistency of data acquisition by automatic methods is dependent upon the morphology of patent microvessels, as well as their distribution. The former requires technically consistent tracer perfusion–fixation, which can be judged by visual inspection of tracer distribution in the cortical gray matter, and the packing of carbon elements in patent vascular lumina. Inhomogeneity of tracer distribution, grossly or microscopically, deemed the preparations technically inappropriate for video-imaging analysis and the subjects were thereby excluded from analysis. Given the above considerations, limits for normal patent vessel distribution have been defined by the ratio of normal basal ganglia/temporal cortex (layers I–VI): capillary = 0.54 ± 0.27, all microvessels = 0.59 ± 0.27 (n = 11). These criteria are consistent with limited published information in a rodent model (Klein et al. 1986).

The implications of this work are several-fold:

1. The no-reflow phenomenon is a feature of reperfusion during focal ischemia. This is significant because microvascular perfusion defects have been described in global ischemia.
2. The microvascular distribution of no-reflow defects may be quantitated. Although the LSA microvascular bed has not been described, it would appear that the defects may be most readily explained by capillary occlusion, although obstruction of postcapillary venules is also involved.
3. Focal no-reflow involves, in part, the PMN leukocyte.
4. Blockade of the PMN leukocyte adherence complex CD11b/CD18 by pharmaceutical grade non-endotoxin-containing monoclonal antibody, known to affect attachment of PMN leukocytes to postcapillary venular endothelium, resulted in a significant increase in perfusion of microvessels of 7.5–50-μm diameter in the ischemia territory. A modest, but insignificant increase in capillary patency was noted. This latter observation is consistent with the presence of PMN leukocytes completely obstructing capillaries early post-reperfusion (del Zoppo et al. 1991a,b) and their rheologic characteristics.
5. Modulation (improvement) in focal no-reflow occurs when IB4 is given just prior to MCA reperfusion. This implies that endothelial adherence leading to microvascular obstruction which is affected by blockade of PMN leukocyte integrins must operate shortly after IB4 infusion, i.e., upon reperfusion. Pre-MCA occlusion exposure to IB4 was necessary.
6. These observations are clinically relevant. Based upon the close analogy of cerebral vascular and parenchymal anatomy (Diaz and Ausman 1980; Garcia and Conger 1987; Garcia et al. 1983), hematology (Hanson and Harker 1987; Taylor et al. 1992), and cross-reaction of antihuman immunoprobes to this nonhuman primate, observations in this model have clinical application.
7. Strategies which block or alter PMN leukocyte adherence, when applied prior to reperfusion *following* cerebral–arterial occlusion, as in atherothrombotic/embolic stroke, may significantly improve microvascular perfusion. This has relevance for patients undergoing major cerebral artery reperfusion with fibrinolytic agents. The above observations have their analogy in experimental focal myocardial (Simpson et al. 1988) and skeletal muscle (Carden et al. 1990) ischemia–reperfusion.

Finally, the long-term effects of such prereperfusion strategies on microvascular integrity, neuronal morphology and function, and clinical neurological outcome are yet to be elucidated.

Acknowledgement. This work was supported in part by grant NS26945 from the National Institutes of Health. This is publication number 7470-mem from the Scripps Clinic and Research Foundation, La Jolla, California, USA.

References

Ames A, Wright LW, Kowade M, Thurston JM, Majors G (1968) Cerebral ischemia. II. The no-reflow phenomenon. Am J Pathol 52:437–453

Arfors K-E, Lundberg C, Lindbom L, Lundberg K, Beatty PG, Harlan JM (1987) A monoclonal antibody to the membrane glycoprotein complex CD18 inhibits polymorphonuclear leukocyte accumulation and plasma leakage in vivo. Blood 69:338–340

Bednar MM, Raymond S, McAuliffe T, Lodge PA, Gross CE (1991) The role of neutrophils and platelets in a rabbit model of thromboembolic stroke. Stroke 22:44–50

Carden DL, Smith JK, Korthuis RJ (1990) Neutrophil-mediated microvascular dysfunction in postischemic canine skeletal muscle. Role of granulocyte adherence. Circ Res 66:1436–1444

Del Zoppo GJ, Copeland BR, Harker LA et al. (1986) Experimental acute thrombotic stroke in baboons. Stroke 17:1254–1265

Del Zoppo GJ, Ferbert A, Otis S et al. (1988) Local intra-arterial fibrinolytic therapy in acute carotid territory stroke: a pilot study. Stroke 19:307–313

Del Zoppo G, Yu J-Q, Morrissey J, Copeland BR (1991a) Tissue factor location in the cerebral gray matter and microvasculature of the non-human primate (Abstr). Thromb Haemost 65:682

Del Zoppo GJ, Schmid-Schönbein GW, Mori E, Copeland BR, Chang C-M (1991b) Polymorphonuclear leukocytes occlude capillaries following middle cerebral artery occlusion and reperfusion in baboons. Stroke 22:1276–1283

Del Zoppo GJ, Poeck K, Pessin MS, 16 co-authors of the rt-PA Acute Stroke Study Group (1992) Recombinant tissue plasminogen activator in acute thrombotic and embolic stroke. Ann Neurol 32:78–86

Diaz FA, Ausman JI (1980) Experimental cerebral ischemia. Neurosurgery 6:436–445

Engler RE, Schmid-Schönbein GW, Pavelec RS (1987) Leukocyte capillary plugging in myocardial ischemia and reperfusion in the dog. Am J Pathol 111:98–111

Garcia JH, Conger KA (1987) Light and electron-microscopic features of brain ischemia. In: Wood JH (ed) Cerebral blood flow. Physiologic and clinical aspects. McGraw-Hill, New York, p 75

Garcia JH, Lowry SL, Briggs L et al. (1983) Brain capillaries expand and repture in areas of ischemia and reperfusion. In: Reivich M, Hurtig HI (eds) Cerebrovascular diseases. Raven, New York, p 169

Grogaard B, Schurer L, Gerdin B, Arfors KE (1989) Delayed hypoperfusion after incomplete forebrain ischemia in the rat. The role of polymorphonuclear leukocytes. J Cereb Blood Flow Metab 9:500–505

Ham AW, Cormack DH (1979) The circulatory system. In: Histology, 8th edn. Lippincott, Philadelphia, p 581

Hanson SR, Harker LA (1987) Baboon models of acute arterial thrombosis. Thromb Haemost 58:801–805

Harlan JM, Killen PD, Senecal FM et al. (1985) The role of neutrophil membrane glycoprotein GMP-140 in neutrophil adherence to endothelium in vitro. Blood 66:167–178

Hernandez LA, Grisham MB, Twohig B, Arfors K-E, Harlan JM, Granger DN (1987) Role of neutrophils in ischemia/reperfusion-induced microvascular injury. Am J Physiol 253: H699–H703

Hossmann KA, Olsson Y (1970) Suppression and recovery of neuronal function in transient cerebral ischemia. Brain Res 22:313–325

House SD, Lipowsky HH (1987) Leukocyte-endothelium adhesion: microhemodynamics in mesentery of the cat. Microvasc Res 34:363–379

Klein B, Kuschinsky W, Schröck H, Vetterlein F (1986) Interdependency of local capillary density, blood flow, and metabolism in rat brains. Am J Physiol 251:H1333–H1340

Kloner RA, Ganote CE, Jennings RB (1974) The "no-reflow" phenomenon after temporary coronary occlusion in the dog. J Clin Invest 54:1496–1508

Little JR, Kerr FWL, Sundt TM Jr (1975) Microcirculatory destruction in focal cerebral ischemia. Mayo Clin Proc 50:264–270

Little JR, Kerr FWL, Sundt TM Jr (1976) Microcirculatory obstruction in focal cerebral ischemia: an electron microscopic investigation in monkeys. Stroke 7:25–30

Mori E, Tabuchi M, Yoshida T, Yamadori A (1988) Intracarotid urokinase with thromboembolic occlusion of the middle cerebral artery. Stroke 19:802–812

Mori E, del Zoppo GJ, Chambers JD, Copeland BR, Arfors KE (1992a) Inhibition of polymorphonuclear leukocyte adherence suppresses no-reflow after focal cerebral ischemia in baboons. Stroke 23:712–718

Mori E, Yoneda Y, Tabuchi M et al. (1992b) Intravenious recombinant tissue plasminogen activator in acute carotid artery territory stroke. Neurology 42:976–982

Nicolaides A, Zukowski A (1985) The place of computerized tomographic brain scanning in the classification of ischemic cerebral disease. In: Courbier R (ed) Basis for a classification of cerebral arterial diseases. Excerpta Medica, Amsterdam, p 59

Schmid-Schönbein GW (1987) Capillary plugging by granulocytes and the no-reflow phenomenon in the microcirculation. Fed Proc 46:2397–2401

Simpson PJ, Todd RF, Fantone JC, Mickelson JK, Griffin JD, Lucchesi BR (1988) Reduction of experimental canine myocardial reperfusion injury by a monoclonal antibody (anti-Mo1, anti-CD11b) that inhibits leukocyte adhesion. J Clin Invest 81:624–629

Spetzler RF, Selman WR, Weinstein P et al. (1980) Chronic reversible cerebral ischemia: evaluation of a new baboon model. J Neurosurg 7:257–261

Suzuki M, Inauen W, Kvietys PR et al. (1989) Superoxide mediates reperfusion-induced leukocyte-endothelial cell interactions. Am J Physiol 257:H1740–H1745

Taylor FB Jr, Hoagendorn H, Chang ACK et al. (1992) Anticoagulant and fibrinolytic activities are promoted, not retarded, in vivo after thrombin generation in the presence of a monoclonal antibody that inhibits activation of protein C. Blood 79:1720–1728

Vasthare US, Heinel LA, Rosenwasser RH, Tuma RF (1990) Leukocyte involvement in cerebral ischemia and reperfusion injury. Surg Neurol 33:261–265

Vedder NB, Fouty BW, Winn RK, Harlan JM, Rice CL (1989) Role of neutrophils in generalized reperfusion injury associated with resuscitation from shock. Surgery 106:509–516

Vedder NB, Winn RK, Rice CL, Chi EY, Arfors K-E, Harlan JM (1990) Inhibition of leukocyte adherence by anti-CD18 monoclonal antibody attenuates reperfusion injury in the rabbit ear. Proc Natl Acad Sci USA 87:2643–2646

Zülch KJ (1985) The cerebral infarct: pathology, pathogenesis, and computed tomography. Springer, Berlin Heidelberg New York

Activation of the Coagulation System and Concomitant Impairment of White Cell Rheology in Patients with Acute Cerebral Infarction

E. Ott and E. Körner

It is now generally accepted that hemorheologic alterations may be present in more than 50% of patients with cerebrovascular disease (Ott et al. 1974; Lechner et al. 1986), and the activation of the coagulation system has been documented in various thrombotic states (Rosc et al. 1990; Uchiyama et al. 1990). Recently attention has been focused on the role of white blood cells (WBC) in ischemic arterial disease (Ernst et al. 1987a). There is evidence from animal experiments that occlusion of capillaries by WBC is a contributing factor to the "non-reflow" phenomenon in the early postischemic period (Hallenbeck et al. 1986; del Zoppo et al. 1991).

The present investigation describes serial measurements of both WBC rheologic parameters and activation markers of blood coagulation in patients with acute cerebral infarction.

Patients and Methods

Twenty female patients (mean age 68 years, range 51–78), admitted within 12 h of onset of acute ischemic infarction (mean 9 h) were compared with 20 control subjects (mean age 62 years, range 36–83) matched for sex and risk factors. The presence of intracranial hemorrhage was excluded by cranial computerized tomography.

Blood samples were drawn without stasis from an antecubital vein immediately after admission and anticoagulated with dry ethylenediaminetetraacetic acid (EDTA) potassium (1.5 mg/ml). Hematologic indexes were determined by an automatic counter (Coulter Electronics Ltd., UK), and leukocyte suspensions were prepared according to methods which were published previously (Mikita et al. 1986). Erythrocyte contamination of the WBC suspension was not detectable, and 3.8×10^6 cells per milliter were used for the filtration measurements. The cell suspensions were filtered with a St. George Filtrometer (Carri-Med Ltd., UK) at 37° Celsius and a driving pressure of −4 cm water using 8-μm polycarbonate membrane filters (Nucleopore, CA., USA). Filter occlusion (filter clogging) was used as

Department of Neurology, University of Graz, Austria

Cerebral Ischemia and Basic Mechanisms
Ed. by A. Hartmann, F. Yatsu, and W. Kuschinsky

an indicator of impaired WBC rheology, whereby the clogging rate (CR) denotes the relative decrease in filtration rate due to filter occlusion at 0.25 ml filtered cell suspension (Matrai and Ernst 1985).

The activation of the blood coagulation system was determined by means of the thrombin–antithrombin III complex (TAT), whereby its increasing concentration is a sensitive marker for the activation of the intravascular coagulation (Seitz et al. 1990).

Serial measurements of WBC-CR and of TAT were performed within 12 h after onset of the acute neurologic symptoms and were repeated 6 days thereafter. The baseline measurements were done before treatment was started. Rheologic modifying agents such as pentoxifylline were then administered intravenously (6.2–9.2 mg/kg bodyweight), but ongoing investigations in our department show that these concentrations are beyond the therapeutic threshold which improves WBC filterability significantly (Ott et al., in preparation). Statistical evaluation was performed by the Student *t* test.

Results

Table 1 displays the data of the patients 12 h and 6 days following onset of acute cerebral infarction and compares them to those obtained in a risk-factored (RF) control group.

It can be seen that the WBC count was significantly higher and the WBC filterability significantly impaired in patients with acute ischemic stroke when compared to the control group. Furthermore there was significant evidence of the activation of the intravascular coagulation system.

However, serial measurements revealed that the quantitative activation of the WBC coincided with the maximal levels of TAT occurring within 12 h while the impairment of the WBC filterability reached its maximal values within 6 days following onset of the neurologic symptoms.

Table 1. Thrombin – antithrombin III complex, WBC count and WBC rheology in patients and in controls

	Acute cerebral infarction (n = 20)		RF controls	
	12 h	6 days	(n = 20)	p
Age (years)	68 ± 13		62 ± 17	NS
TAT (μg/l)	57.0 ± 4.9	20.6 ± 2.9	8.8 ± 2.1	0.01
WBC count (10^{12}/l)	9.44 ± 2.96	7.34 ± 1.80	6.00 ± 1.14	0.05
PMN (%)	72.4 ± 11.7	65.2 ± 9.3	58.6 ± 13.2	0.05
WBC – CR/ml	0.871 ± 0.085	1.659 ± 0.206	0.557 ± 0.043	0.05

WBC, leukocyte; PMN, polymorphonuclear cell; CR, clogging rate; TAT, thrombin – antithrombin III complex.

Discussion

The results presented here confirm previous findings reported by others (Mercuri et al. 1989; Ernst et al. 1987a,b) and indicate that the impaired WBC rheology seems to be connected with a sustained activation of the coagulation system.

There are several mechanisms whereby leukocytes may contribute to microvascular injury (Ernst et al. 1987a). Since already intrinsically normal leukocytes may face difficulties in passing capillaries, it follows that alterations in their rheologic properties such as forming aggregates when stimulated might have a major impact on microvascular flow.

The high number of WBC observed in this study is evidently related to high levels of TAT and might thus in part reflect an acute phase response. However, it might be concluded from the present results that sufficient activation of the leukocytes leading to significant impairment of their filterability was not yet established within 12 h (mean 9 h) after onset of the acute symptoms. Although using another measuring technique Mercuri et al. (1989) showed that significant alteration of WBC filterability can be observed within 24 h of onset of acute ischemic infarction (mean 14.5 h).

Whether acute ischemia is the only underlying process leading to impaired WBC filterability cannot be answered from the present results. However, reports by others (Vermes and Strik 1988) and our own experience showing that rheologic abnormalities of WBC can also be found in chronic stages of cerebrovascular disease (WBC-CR 1.437 ± 0.571/ml) make this assumption unlikely.

Sustained intravascular hypercoagulability or prethrombotic states, which can be detected by TAT measurements (Seitz et al. 1990), might be accompanying rather than causative factors of impaired WBC filterability if one considers that TAT values of healthy subjects are significantly lower than those estimated in the RF control group but values of WBC-CR do not show significant differences (Ott et al., in preparation).

Preliminary studies using rheologic modifying agents ongoing in our department show that a continuous intravenous administration of at least 13.8–18.4 mg/kg bodyweight pentoxifylline is necessary to improve impaired WBC filterability significantly (Ott et al., in preparation).

Conclusion

It can be concluded from the present results that both high TAT levels and a quantitative activation of the WBC count seen within 12 h after onset of the acute ischemic stroke might in part reflect an acute phase response. But a sufficiently qualitative activation of WBC leading to significantly impaired filterability was not yet evident. The significant impairment of WBC filterability occurred within 6 days following onset of the acute symptoms and was connected with a sustained activation of the intravascular coagulation system.

References

del Zoppo GJ, Schmid-Schönbein GW, Mori E et al. (1991) Polymorphnuclear leukocytes occlude capillaries following middle cerebral artery occlusion and reperfusion in baboons. Stroke 22:1276–1283

Ernst E, Hammerschmidt DE, Bagge U et al. (1987a) Leukocytes and the risk of ischemic diseases. JAMA 257:2318–2324

Ernst E, Matrai A, Paulsen F (1987b) Leukocyte rheology in recent stroke. Stroke 18:59–62

Hallenbeck JM, Dutka AJ, Tanishima T et al. (1986) Polymorphnuclear leukocyte accumulation in brain regions with low blood flow during the early postischemic period. Stroke 17:246–253

Lechner H, Ott E, Ossama N et al. (1986) Follow up studies of haemorheological profile in acute ischaemic stroke. Clin Haemorheology 6:3–9

Matrai A, Ernst E (1985) Pentoxifylline improves white cell rheology in claudicants. Clin Haemorheol 5:483–491

Mercuri M, Ciuffetti G, Robinson M et al. (1989) Blood rheology in acute cerebral infarction. Stroke 20:959–962

Mikita J, Nash G, Dormandy J (1986) A simple method of preparing white blood cells for filterability testing. Clin Haemorheol 6:635–639

Ott E, Lechner H, Aranibar A (1974) High blood viscosity syndrome in cerebral infarction. Stroke 5:330–333

Rosc D, Kotschy M, Rewakovicz M et al. (1990) Thrombin/amtithrombin III complex in patients with peripheral occlusive arterial disease. Folia Haematol (Leipz) 117:405–411

Seitz R, Egbring R, Wagner C et al. (1990) Thrombin – Antithrombin III Komplex (TAT): Ein Marker für intravasale Gerinnungsaktivierung. Internist (Berl) 31:69–74

Uchiyama T, Matsumoto M, Kobayashi N (1990) Studies on the pathogenesis of coagulopathy in patients with arterial thromboembolism and malignancy. Thromb Res 59:955–965

Vermes I, Strik F (1988) Altered leukocyte rheology in patients with chronic cerebrovascular disease. Stroke 19:631–633

Microcirculatory Disturbance and Leukocytes: Influence of Leukocyte-Produced Superoxide on Red Blood Cell Aggregation

Y. Shinohara, M. Yamamoto, H. Hamano, and S. Ogawa

Introduction

It is well recognized that cerebral circulation is regulated by perfusion pressure together with cerebrovascular resistance. Cerebrovascular resistance is dependent upon intracranial pressure, organic changes and tonus of cerebral vessels, blood viscosity in the cerebral vessels, and so on. Blood viscosity is determined by factors such as hematocrit, red blood cell (RBC) deformability, RBC aggregability, and plasma viscosity. Among these factors, it is believed that RBC aggregation is one of the major determinants of microcirculation, especially in low-shear flow areas. In other words, the kinetics of RBC aggregation play an important role in the alteration of the microcirculation in regions where the speed of blood flow approaches or falls below the critical level. In our previous observation [1], we conducted a multivariate analysis of the factors influencing cerebral blood flow (CBF) measured by the ^{133}Xe intravenous injection method in patients with cerebral infarction and found that, although the size of cerebral infarction was the most important factor determining CBF value in the acute stage, in the chronic stage (more than 30 days after the onset of cerebral infarction) high Ht was the most important factor and the accelerated value of RBC aggregation time was the second most important. In other words, hemorheological factors for increasing blood viscosity, such as high hematocrit and accelerated RBC aggregation, were more important determinants of CBF value than other factors such as blood pressure, size of lesion, age, and $PaCO_2$ in the chronic stage. Furthermore, the acceleration of RBC aggregation time has been observed in patients with cerebral infarction in acute and chronic stages [1, 2].

Recently, it has been reported [3] that circulating neutrophils contribute to brain injury in animal models of an ischemic event. Several mechanisms have been suggested, for example, impaired leukocyte filtrability or release of deleterious mediators such as leukotrienes from activated neutrophils. In addition, generation of superoxide by neutrophils in the ischemic region is

Department of Neurology, Tokai University School of Medicine, Isehara, Kanagawa, Japan

Cerebral Ischemia and Basic Mechanisms
Ed. by A. Hartmann, F. Yatsu, and W. Kuschinsky

also believed to be important, and it is possible that superoxide released from neutrophils plays a role in causing tissue injury.

The purpose of this study was (a) to investigate whether superoxide generated by activated neutrophils could accelerate RBC aggregation; and (b) to investigate whether the administration of superoxide dismutase (SOD) could inhibit the changes of RBC aggregation induced by activated neutrophils.

Patients and Method

Effect of Superoxide Produced by Activated Neutrophils on RBC Aggregation

There were ten patients with occlusive cerebrovascular disease in the chronic stage (age 16 ± 8 years), and 12 patients with tension headache (age 44 ± 16). As shown in Fig. 1, 10 ml heparinized whole blood from each subject was divided between three test tubes. One of them was centrifuged at 3000 rpm for 10 min to separate plasma and cells. Then, the buffy coat was discarded and the plasma was added back to the cell fraction to make blood without leukocytes (buffy coat-minus group). In order to stimulate the neutrophils to produce superoxide, a solution of 10 μg/ml phorbol-12-myristate-13-acetate (PMA) was added to two test tubes containing buffy coat-minus blood and whole blood. The final concentration of PMA was adjusted to 1 μg/ml. As a control, phosphate-buffered saline (PBS) without PMA was added to whole blood. All materials were adjusted to hematocrit 40%. After incubation for 2 h at 37°C, RBC aggregation time was measured

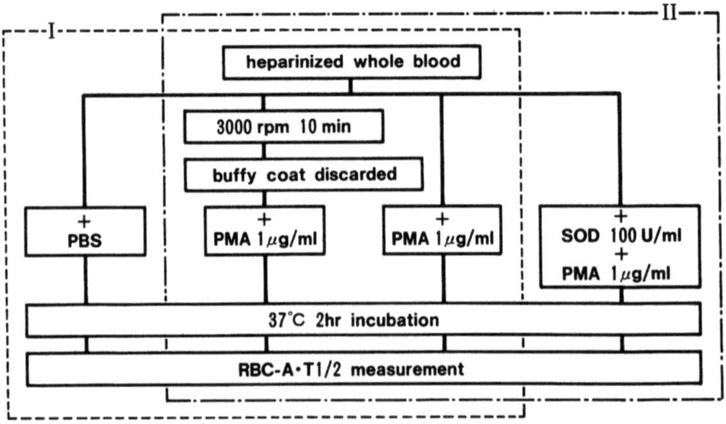

Fig. 1. Method: *I*, effects of neutrophil-derived superoxide on RBC-A; *II*, effects of superoxide dismutase (*SOD*) on the enhanced RBC-A by superoxide. *PMA*, phorbol-12-myristate-13-acetate

by using a photometric aggregometer. The concentration of fibrinogen was measured by using an immunodiffusion method after incubation.

Effect of SOD on Enhanced RBC Aggregation by PMA-Actived Neutrophils

Blood used in the study was obtained from eight healthy volunteers aged 30 ± 2 years. As shown in Fig. 1, 10 ml heparinized whole blood from each subject was divided between three test tubes to give the following test samples: buffy coat-minus blood with 1 μg/ml PMA, whole blood with PMA, and whole blood with PMA plus 100 units per milliliter SOD.

Results

Effect of Superoxide Produced by Activated Neutrophils on RBC Aggregation

Figure 2 shows the effects of PMA on RBC aggregation time in 22 patients with occlusive cerebrovascular disease (CVD) and tension-type headache. After 2 h of incubation at 37°C, the RBC aggregation times of the whole blood with PMA was 6.4 ± 1.9 s which is statistically significantly lower than that of control whole blood (7.1 ± 2.6 s) or that of buffy coat-minus blood with PMA (6.9 ± 2.1 s). Fibrinogen concentration is one of the most im-

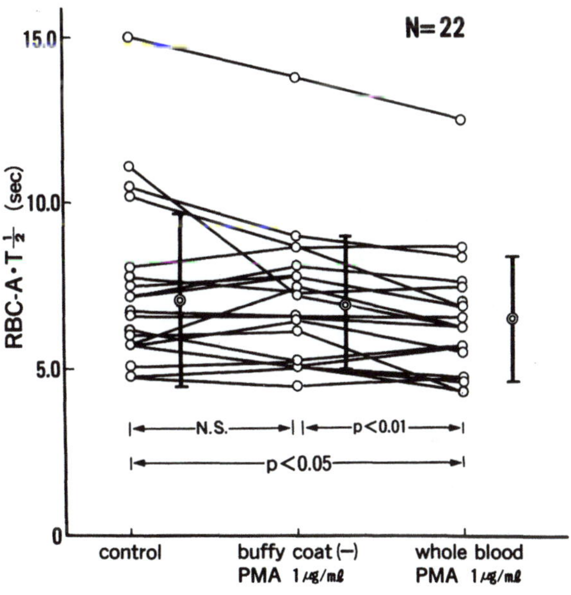

Fig. 2. Effects of PMA on RBC-A in patients with tension-type headache and occlusive CVD

portant factors determining the kinetics of RBC aggregation. However, there was no significant difference in fibrinogen concentration among the three groups. Furthermore, hematocrit was adjusted to 40% in all groups. Thus, we considered that the shortening RBC aggregation time is due to neutrophil-produced superoxide. In order to test this hypothesis, we investigated whether the addition of SOD could inhibit the shortening of RBC aggregation time induced by PMA.

Effect of SOD on Enhanced RBC Aggregation by PMA-Activated Neutrophils

As shown in Fig. 3, the RBC aggregation time of whole blood with PMA (6.5 ± 1.9 s) was statistically significantly accelerated compared with that of buffy coat minus blood with PMA (7.8 ± 2.3 s). However, the RBC aggregation time of whole blood with PMA plus SOD (7.4 ± 2.2 s) was significantly prolonged compared with the value of whole blood with PMA. This result indicates that the addition of SOD inhibits the shortening of RBC aggregation time of whole blood induced by PMA and suggests that this shortening of RBC aggregation time could be accounted for mainly by the release of superoxide from neutrophils stimulated by PMA.

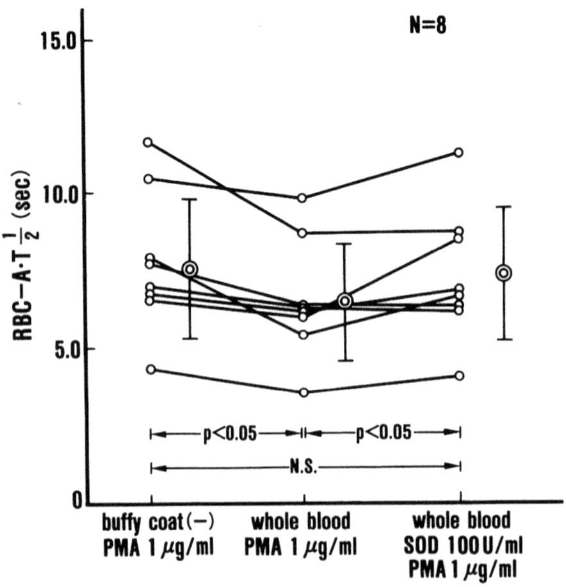

Fig. 3. Effects of PMA and SOD on RBC-A in normal volunteers

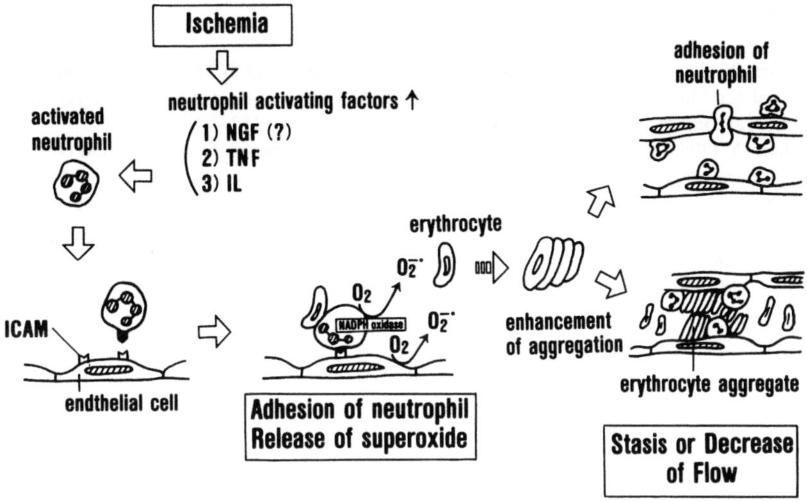

Fig. 4. Hypothetical role of superoxide-enhanced RBC aggregation in ischemia. *NGF*, neuronal growth factor; *TNF*, tumor necrosis factor; *IL*, interleukin; *ICAM*, intracellular cell adhesion molecule

Discussion

If our hypothesis that superoxide generated by activated neutrophils enhances RBC aggregation can be extended to the case of cerebral ischemia, we may speculate that the increase in neutrophil-activating factors, such as interleukin and tumor necrosis factor, in ischemia (Fig. 4) would activate neutrophils. Then adhesion of neutrophils to endothelial cells may occur and release of superoxide may be induced. This superoxide may not only directly injure the brain tissue, but also enhance RBC aggregation. The increase of RBC aggregation may lead to stasis or a decrease of microcirculation, together with capillary plugging by leukocytes themselves, leading to a vicious circle.

Conclusion

The results obtained in this study may be summarized as follows: (a) RBC aggregation was enhanced by adding PMA to whole blood–this effect of PMA could not be observed in the blood after discarding the buffy coat; (b) preadministration of SOD inhibited the enhancement of RBC aggregation induced by PMA. These results suggest that superoxide produced by neutrophils accelerates RBC aggregation and this, together with leukocytes, may result in capillary plugging.

References

1. Shinohara Y, Kobatake K, Yamamoto M, Yoshitoshi M, Saitoh H (1985) Hyperaggregability of red blood cells in patients with occlusive cerebrovascular disease. In: Meyer JS et al. (eds) Cerebral vascular disease 5. Elsevier, Amsterdam, pp 187–190
2. Shinohara Y (1986) Some rheological considerations of the treatment of acute and chronic stages of cerebrovascular disease. In: Gotoh F, Lechner H (eds) Clinical haemorheology: a new approach to cerebrovascular disease. Royal Society of Medicine Services, London, pp 89–95
3. Gran AJ, Berger E, Sung K-L, Schmidt-Schönbein GW (1992) Granulocyte adhesion, deformability, and superoxide formation in acute stroke. Stroke 23:33–39

Small Volume Resuscitation in Hemorrhagic Shock by Hypertonic/Hyperoncontic Saline-Dextran: Effects on the Central Nervous System

S. Berger[1], L. Schürer[2], R. Härtl[1], C. Dautermann[2], R. Murr[3], K, Messmer[1], and A. Baethmann[1]

Introduction

Secondary ischemia of the brain is common in fatal head injury (Graham et al. 1989). Bouma et al. (1991) observed in patients a close correlation between the impairment of cerebral blood flow during the first hours after trauma and final outcome. Based on the United States traumatic coma databank study, Marmarou et al. (1991) reported that periods of elevated intracranial pressure (ICP) (>20 mmHg) or of a decreased blood pressure (<80 mmHg) were highly predictive for poor outcome from severe head injury. Since 10%–20% of head-injured patients are simultaneously affected by severe peripheral injuries causing hemorrhagic hypotension (Miller et al. 1978), primary care must reestablish and maintain an appropriate cerebral as well as systemic circulation.

Fluid resuscitation by a small volume (e.g., 4 ml/kg as rapid bolus) of hypertonic/hyperoncotic solution, such as 7.5% NaCl in 6% dextran 70 has been shown to effectively restore the cardiovascular function of polytraumatized patients in shock (Holcroft et al. 1987). This concept is currently under clinical investigation. Since up to one third of polytraumatized patients are simultaneously affected by head injury (Gennarelli et al. 1989), this form of shock treatment might also be employed in patients with an acute cerebral lesion. While hypertonic saline solutions have frequently been shown to lower ICP both clinically (Worthley et al. 1988) and experimentally (Prough et al. 1991; Zornow et al. 1990), no information is available concerning possible side effects of the mixture of hypertonic/hyperoncotic solutions on the brain. Adverse side effects exerted by hypertonic/hyperoncotic solutions, however, might evolve in the presence of a disrupted blood–brain barrier and/or during impairment of the cerebral autoregulation as in head injury.

[1] Institute for Surgical Research, Klinikum Grosshadern, Ludwig Maximilians University, Marchioninistr. 15, 81377 Munich 70, Germany
[2] Department of Neurosurgery, Klinikum Grosshadern, Ludwig Maximilians University, Marchioninistr. 15, 81377 Munich 70, Germany
[3] Institute of Anesthesiology, Klinikum Grosshadern, Ludwig Maximilians University, Marchioninistr. 15, 81377 Munich 70, Germany

Cerebral Ischemia and Basic Mechanisms
Ed. by A. Hartmann, F. Yatsu, and W. Kuschinsky

In a previous study by our group (Schürer et al. 1992), adverse effects of 7.2% NaCl in 10% dextran 60 (hypertonic/hyperoncotic saline/dextran solution, HHS) were not observed on the cerebral blood flow, cerebral tissue oxygenation, vasomotor control, or the brain water content in shock. In a second experimental series (Härtl et al., in press) we analyzed administration of HHS on the intracranial pressure acutely after a focal cerebral lesion in combination with an intracranial space-occupying process. For the present study, the influence of HHS on intracranial hypertension was studied 24 h after trauma in comparison with mannitol as standard treatment.

Methods

For the first study, six New Zealand rabbits were anesthetized using thiopental, ventilated, and implanted with arterial and venous catheters. Anesthesia was then maintained by α-chloralose. Mean arterial pressure (MAP), central venous pressure (CVP), blood gases, body temperature, plasma electrolytes, osmolality, and colloid osmotic pressure were monitored. After a circular trephination of the skull above the left hemisphere, an epidural balloon was inserted between the skull and intact dura. A burr hole was made over the right hemisphere for manometric measurement of the ICP. A focal cold lesion of the exposed cerebral cortex was then induced by a metal rod cooled with liquid nitrogen. The skull was subsequently closed using dental cement. The ICP was slowly raised (over 30 min) by inflation of the epidural balloon until 15 mmHg was reached. Infusion of 4 ml/kg of HHS into an ear vein followed within 2 min, resulting in a decrease of the ICP. The intracranial pressure increased spontaneously thereafter due to maintenance of inflation of the epidural balloon. As soon as 15 mmHg was reached again, infusion of HHS was repeated.

For the second study, 12 New Zealand rabbits were anesthetized by i.v. ketamine/xylazine under spontaneous respiration. Upon fixation of the skull in a stereotactic frame, a circular trephination was made over the left hemisphere leaving the dura intact. A metal probe (6 mm in diameter), cooled by liquid nitrogen, was attached to the brain surface for 5 min to induce a focal lesion. An epidural balloon was then inserted directly above the lesion. The skull was closed using dental cement. Twenty-one hours later the animals were reanesthetized with thiopental, ventilated, and implanted with catheters including subdural measurement of the ICP. Anesthesia was maintained by chloralose. Physiological variables were monitored as in the first study. The ICP was raised to 17 mmHg by stepwise inflation of the balloon over 60 min. During a control period of 30 min, the ICP remained at 15–20 mmHg without changes in the balloon volume. Then, either 4 ml/kg HHS (six animals) or 9 ml/kg 20% mannitol (six animals) were infused i.v. within 2 min. The osmotic load administered in both groups was equivalent, i.e., 9.6 mosmol/kg bodyweight. The hyperosmolar infusion fluids led to a temporary decrease of the ICP, followed, however, by a subsequent increase. The respective infusions were repeated twice as soon as an ICP level of 15 mmHg was reached again.

Statistics. Differences were tested by the Kruskal-Wallis test in the computer version of Theodorssen-Norheim. The Wilcoxon test for unpaired data was employed to calculate the level of significance; $p < 0.05$ was required to reject the null hypothesis.

Results

In animals with intracranial hypertension (15 mmHg) after a cold lesion and inflation of an epidural balloon, HHS (4 ml/kg) decreased the ICP on average for a period of 90 min. Once ICP exceeded 15 mmHg again, the following injection led to another decrease of the ICP although of shorter duration. The Na^+ concentration in plasma increased by 20–25 m*M*/l after each infusion, but tended to normalize within 30 min thereafter. The hematocrit decreased by 10% after each injection.

Since the above results indicated that HHS is effective in intracranial hypertension and experimental brain injury, the hypertonic/hyperoncotic fluid mixture was subsequently compared with mannitol as a standard treatment. HHS or mannitol were administered at an equimolar dose. In this study MAP or other physiological parameters were not different between the two groups under control conditions. The MAP, however, tended to increase after HHS and decrease after mannitol (Tables 1, 2). The plasma osmolality and Na^+ concentration were more increased by HHS. The hematocrit decreased both after HHS and mannitol, with a tendency towards normalization after mannitol. Hemodilution was seen only in animals receiving HHS. Original registrations of the MAP and ICP of experimental animals with either HHS or mannitol are given in Fig. 1. After the first injection, duration of the lowering of the ICP by HHS was 98 min on average, and 189 min by mannitol. When administration of either regimen was repeated, the ICP was lowered for 142 min by HHS and for 106 min by mannitol. In both groups, the ICP minimum was obtained at 20 min after the start of infusion. A statistical analysis was performed comparing the ICP minima, the time period to reach the minimum, and the duration of the decrease in ICP by either solution. Although significant differences were not found between the two groups, a trend was recognized ($p = 0.054$) indicating that the decrease of ICP by mannitol after the first injection was of longer duration.

Table 1. Administration of hypertonic/hyperoncotic saline/dextran

	MAP (mmHg)	Hct (%)	Na^+ (m*M*/l)	Osm (mosm/l)	COP (mmHg)	pO_2 (mmHg)	pCO_2 (mmHg)	pH ($-\lg [H^+]$)	ICP (mmHg)
Control	88.8 ± 5.7	35.8 ± 1.3	144.4 ± 1.7	298.2 ± 1.7	17.9 ± 1.8	147.2 ± 7.0	32.4 ± 1.7	7.411 ± 0.016	18.1 ± 0.9
10′ after inj. 1	93.0 ± 2.6	29.6 ± 1.1	156.8 ± 2.9	320.0 ± 2.6	18.6 ± 1.6	140.8 ± 10.1	36.1 ± 1.9	7.345 ± 0.016	8.6 ± 2.1
60′ after inj. 1	98.5 ± 3.6	33.0 ± 1.0	147.9 ± 2.3	310.8 ± 1.7	19.7 ± 1.6	151.5 ± 4.5	33.5 ± 0.9	7.383 ± 0.021	12.0 ± 1.6
10′ after inj. 2	98.8 ± 3.9	27.6 ± 8.9	157.5 ± 7.5	328.5 ± 2.2	20.5 ± 0.9	150.9 ± 6.3	34.1 ± 0.9	7.353 ± 0.018	5.9 ± 0.5
60′ after inj. 2	102.7 ± 3.4	29.7 ± 1.0	158.0 ± 3.7	322.0 ± 2.7	19.1 ± 2.0	155.7 ± 4.8	31.3 ± 0.9	7.401 ± 0.018	8.2 ± 1.8
10′ after inj. 3	101.0 ± 3.6	25.9 ± 1.1	169.1 ± 4.9	341.1 ± 4.3	23.1 ± 3.1	150.3 ± 7.6	33.7 ± 0.9	7.365 ± 0.013	4.7 ± 0.3
60′ after inj. 3	100.8 ± 7.5	28.9 ± 1.4	160.1 ± 2.8	333.0 ± 8.1	20.7 ± 2.4	151.8 ± 7.9	32.7 ± 1.0	7.383 ± 0.014	6.4 ± 1.0

Changes of blood pressure, hematocrit; plasma Na^+, osmolality, colloidosmotic pressure; blood gases; and ICP after three subsequent injections of either 4 ml/kg bodyweight HHS (Table 1) or 9 ml/kg bodyweight 20% mannitol (Table 2). MAP: mean arterial pressure; Hct: hematocrit; Na^+: plasma sodium concentration; Osm, plasma osmolality; COP: colloid osmotic pressure of the plasma; pO_2, pCO_2, and pH: arterial blood gas parameters; 10′, 60′ after inj. 1, 2, 3: time in minutes after first, second, and third injection of the respective hypertonic solution.

Table 2. Administration of hypertonic mannitol

	MAP (mmHg)	Hct (%)	Na^+ (m*M*/l)	Osm (mosm/l)	COP (mmHg)	pO_2 (mmHg)	pCO_2 (mmHg)	pH ($-\lg [H^+]$)	ICP (mmHg)
Control	89.2 ± 8.5	39.6 ± 0.7	145.4 ± 4.3	297.2 ± 2.5	17.1 ± 1.3	149.9 ± 5.9	34.2 ± 2.3	7.408 ± 0.010	16.9 ± 0.7
10′ after inj. 1	88.4 ± 8.1	34.5 ± 1.1	135.3 ± 2.8	326.0 ± 4.6	15.9 ± 1.4	115.6 ± 11.8	45.4 ± 4.5	7.287 ± 0.046	5.8 ± 1.3
60′ after inj. 1	91.2 ± 7.7	39.5 ± 1.2	138.1 ± 3.3	309.8 ± 3.7	16.3 ± 2.1	148.1 ± 7.9	33.2 ± 1.3	7.410 ± 0.003	6.6 ± 1.7
10′ after inj. 2	87.8 ± 6.6	34.9 ± 1.8	146.1 ± 5.3	327.0 ± 5.7	17.5 ± 1.7	121.8 ± 9.1	36.5 ± 3.4	7.352 ± 0.028	4.8 ± 1.4
60′ after inj. 2	80.6 ± 7.0	39.2 ± 0.8	146.3 ± 2.8	313.0 ± 3.5	18.1 ± 0.3	134.1 ± 5.2	33.5 ± 2.5	7.414 ± 0.023	9.4 ± 1.6
10′ after inj. 3	83.7 ± 6.6	34.2 ± 1.4	131.9 ± 4.2	327 ± 9.0	16.1 ± 2.0	137.0 ± 1.5	35.7 ± 2.9	7.384 ± 0.016	4.8 ± 2.4
60′ after inj. 3	73.3 ± 7.7	38.3 ± 0.9	150.3 ± 2.9	319 ± 5.6	18.6 ± 1.3	132.9 ± 5.6	30.0 ± 3.2	7.443 ± 0.024	7.9 ± 1.8

Changes of blood pressure, hematocrit; plasma Na^+, osmolality, colloidosmotic pressure; blood gases; and ICP after three subsequent injections of either 4 ml/kg bodyweight HHS (Table 1) or 9 ml/kg bodyweight 20% mannitol (Table 2). MAP: mean arterial pressure; Hct: hematocrit; Na^+: plasma sodium concentration; Osm: plasma osmolality; COP: colloid osmotic pressure of the plasma; pO_2, pCO_2, and pH: arterial blood gas parameters; 10′, 60′ after inj. 1, 2, 3: time in minutes after first, second, and third injection of the respective hypertonic solution.

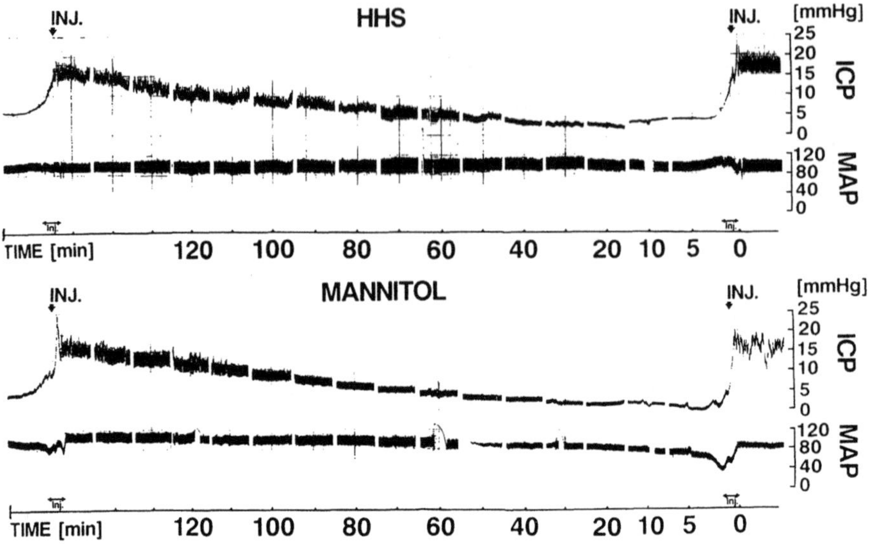

Fig. 1A,B. Original recordings of arterial blood pressure and intracranial pressure in two individual experiments with HHS (**A**) or mannitol (**B**), respectively. The time axis proceeds from right to left. Prior to fluid injection, the ICP was increasing due to a focal brain lesion and inflation of a subdural balloon. The HHS or mannitol bolus led to an immediate decrease of the ICP. When the ICP had increased again to its previous level, administration of HHS or of mannitol was repeated

Discussion

The American neurosurgeon Wilder Penfield (1919) was among the first to report beneficial effects of isotonic saline for resuscitation in hemorrhagic shock. A particular effectiveness of small amounts of hypertonic saline (7.5%) for resuscitation from refractory shock was described by de Felippe et al. (1980). Restoration of cardiovascular function by hypertonic saline was prompt, but nevertheless of limited duration. Smith et al. (1985) compared different saline solutions with or without dextran. They found hypertonic saline/dextran maximally efficient in shock treatment as far as improvement of the cardiovascular function was concerned. Kreimeier and Messmer (1987) observed that combining the hyperoncotic colloid dextran 60 (10%) with hypertonic saline (7.2%) improved cardiac output and oxygen delivery. The intravenous administration of 4 ml/kg 7.2% NaCl and 10% dextran 60 bolus within 2 min was introduced as "small volume resuscitation." During recent years this fluid mixture was studied in an increasing number of clinically relevant conditions. Results of our group (Härtl et al., in press) and clinical observations that a similar fluid mixture increases survival in polytrauma patients with head injury (Vassar et al. 1991) indicate that hypertonic/hyperoncotic saline/dextran is therapeutically useful in patients with head injury and additional peripheral trauma (Murr et al. 1992).

Saline solutions with different hyperosmolarities studied in shock treatment in the presence of a cerebral lesion have been reported to lower cerebral water content and intracranial pressure during the posttraumatic course. Gunnar et al. (1987) found that "in the presence of an intracranial mass lesion, resuscitation by hypertonic (3%) saline is accompanied by lower ICP values and less cerebral edema than by isotonic saline or colloids." Ducey et al. (1989) resuscitated pigs in shock with an epidural mass by either isotonic or hypertonic (6%) saline, hetastarch, or blood. Infusion of hypertonic saline resulted in a lower ICP than infusion of the other fluids. Resuscitation from hemorrhagic shock in rats simultaneously subjected to fluid percussion injury of the brain was investigated by Wisner et al. (1990). The brain water content was more effectively reduced in the uninjured hemisphere – not in the injured hemisphere – by administration of 6.5% saline than by lactated Ringer's solution. In dogs with a subdural mass treated by fluid resuscitation from hemorrhagic shock by either isotonic or hypertonic (7.2%) saline, Prough et al. (1991) found reduction of the ICP and an increase of the rCBF only when employing the hypertonic solution. Again, Battistella and Wisner (1991) succeeded in lowering intracranial pressure more efficiently during the posttraumatic course by 7.5% saline as compared to lactated Ringer's solution. Moreover, the hypertonic fluid led to reduction of the cerebral water content of the uninjured brain hemisphere, whereas the cerebral water content of the injured hemisphere was not differently affected in both groups.

Hypertonic saline is also effective in head injury without peripheral trauma as reported by Worthley et al. (1988). Recently, Fisher et al. (1992) demonstrated in pediatric head injury that 3% NaCl is more effective in lowering the ICP than 0.9% NaCl. An equal efficiency of 3.2% saline and 20% mannitol in lowering posttraumatic ICP was observed in rabbits (Zornow et al. 1990). To the best of our knowledge, HHS has not yet been compared with a standard treatment protocol in intracranial hypertension. The results of our study indicate that this regimen is superior to mannitol in normalizing or maintaining systemic blood pressure. A reliable normalization of the systemic blood pressure is certainly important in hemorrhagic shock. Notwithstanding these favorable properties of HHS, the seemingly more pronounced increase of the plasma Na^+ concentration and osmolality may limit application this treatment. On the other hand, it is unlikely that small volume infusion of hypertonic saline in hemorrhagic shock under acute conditions is associated with deleterious consequences, such as central pontine myelinolysis (McKee et al. 1988) which is characteristic of chronic hypernatremia. Observations of adverse sequelae of systemic hyperosmolality are mostly restricted to chronically ill patients with a secondary increase of the osmo-concentration in plasma over longer periods (Mattar et al. 1973). This category of patients is different from the usually young trauma victim exposed to a short and reversible increase of plasma osmolality by this method of fluid resuscitation (Vassar et al. 1990).

In the present experiments transient changes of blood gases indicating respiratory acidosis were observed in animals receiving mannitol. The systemic response may be attributable to the resulting hemodilution (Winters et al. 1964) together with an increase of cardiac output, leading to a decreased ventilation/perfusion ratio of the lung (Hannon et al. 1990). This explanation, however, does not account for the blood gas changes being more extensive in animals receiving mannitol than in animals receiving HHS, although both solutions were administered at an equimolar dose. It should be mentioned in this context that both solutions are acidic (HHS: pH = 4.15; mannitol: pH = 4.9). Animals in the mannitol group were infused with a larger volume, i.e., 9 ml/kg vs. 4 ml/kg (HHS), which may explain, at least in part, the different pCO_2 response under the current experimental conditions.

Although according to the present findings mannitol lowered intracranial pressure for a somewhat longer period than the hypertonic/hyperoncotic fluid mixture, these actually moderate differences between the two regimens might be of minor importance with regard to restoration of the cardiovascular function in shock. Obviously, patients with severe head injury and peripheral trauma suffering from circulatory failure in the early phase of emergency care may benefit more from a rapid recovery of the systemic perfusion than from a reduction of the intracranial pressure. During the early resuscitation phase of a polytrauma patient with head injury, reduction of intracranial pressure is of less concern than reestablishment of the general circulation. The present experimental findings are promising for further studies to improve and adapt, respectively, administration of hypertonic/hyperoncotic saline/dextran for the treatment of patients with severe head injury and intracranial hypertension.

References

Battistella FD, Wisner DH (1991) Combined hemorrhagic shock and head injury: effects of hypertonic saline (7.5%) resuscitation. J Trauma 31:182–188

Bouma GJ, Muizelaar JP, Choi SC, Newlon PG, Young HF (1991) Cerebral circulation after severe traumatic brain injury: the elusive role of ischemia. J Neurosurg 75:685–693

De Felippe J, Timoner J, Velasco IT, Lopes OU, Rocha E, Silva M (1980) Treatment of refractory shock by 7.5% sodium chloride injections. Lancet 2:1002–1004

Ducey JP, Mozingo DW, Lamiell JM, Okerburg C, Gueller GE (1989) A comparison of the cerebral and cardiovascular effects of complete resuscitation with isotonic and hypertonic saline, hetastarch, and whole blood following hemorrhage. J Trauma 29:1510–1518

Fisher B, Thomas D, Peterson B (1992) Hypertonic saline lowers raised intracranial pressure in children after head trauma. J Neurosurg Anesthesiol 4:4–10

Gennarelli TA, Champion HR, Sacco WJ, Copes WS, Alves WM (1989) Mortality of patients with head injury and extracranial injury treated in trauma centers. J Trauma 29:1193–1202

Graham DI, Ford I, Adams JH, Doyle D, Teasdale GM, Lawrence AE, McLellan DR (1989) Ischemic brain damage is still common in fatal non-missile head injury. J Neurol Neurosurg Psychiatry 52:346–354

Gunnar W, Jonasson O, Merlotti G, Stone J, Barrett J (1987) Head injury and hemorrhagic shock: studies of the blood brain barrier and intracranial pressure after resuscitation with normal saline solution, 3% saline solution, and dextran-40. Surgery 103:398–407

Hannon JP, Wade CE, Bossone CA, Hunt MM, Coppes RI, Loveday JA (1990) Blood gas and acid-base status of conscious pigs subjected to fixed-volume hemorrhage and resuscitated with hypertonic saline dextran. Circ Shock 32:19–29

Härtl R, Schürer L, Dautermann C, Röhrich F, Berger S, Murr R, Messmer K, Baethmann A (1992) Effect of a hypertonic-hyperoncotic solution (HHS) on increased intracranial pressure after a focal brain lesion and inflation of an epidural balloon. Intracranial Pressure vol. 8, Avezaat CJJ (ed.) (in press)

Holcroft JW, Vassar MJ, Turner JE, Derlet RW, Kramer GC (1987) 3% NaCl and 7.5% NaCl/dextran 70 in the resuscitation of severely injured patients. Ann Surg 206:279–288

Kreimeier U, Messmer K (1987) New perspectives in resuscitation and prevention of multiple organ system failure. In: Baethmann A, Messmer K (eds) Surgical research: recent concepts and results. Springer, Berlin Heidelberg New York, pp 39–50

Marmarou A, Anderson RL, Ward JD, Choi SC, Young HF, Eisenberg HM, Foulkes MA, Marshall LF, Jane JA (1991) Impact of ICP instability and hypotension on outcome in patients with severe head trauma. J Neurosurg 75:S59–S66

Mattar JA, Weil MH, Shubin H (1973) A study of the hyperosmolal state in critically ill patients. Crit Care Med 1:293–301

McKee AC, Winkelmann MD, Banker BQ (1988) Central pontine myelinolysis in severely burned patients: relationship to serum osmolality. Neurology 38:1211–1217

Miller JD, Sweet RC, Narayan R, Becker DP (1978) Early insults to the injured brain. JAMA 240:439–442

Murr R, Schürer L, Baethmann A (1992) Acute management of severe head injury. Curr Opin Anesthesiol 5:285–291

Penfield W (1919) The treatment of severe and progressive hemorrhage by intravenous injections. Am J Physiol 48:121–128

Prough DS, Whitley JM, Taylor CL, Deal DD, DeWitt DS (1991) Regional cerebral blood flow following resuscitation from hemorrhagic shock with hypertonic saline. Influence of a subdural mass. Anesthesiology 75:319–327

Schürer L, Dautermann C, Härtl R, Murr R, Berger S, Röhrich F, Messmer K, Baethmann A (1992) Treatment of hemorrhagic hypotension with hypertonic/hyperoncotic solutions: effects on regional cerebral blood flow and brain surface oxygen tension. Eur Surg Res 24:1–12

Smith GJ, Kramer GC, Perron P, Nakayama S, Gunther RA, Holcroft JW (1985) A comparison of several hypertonic solutions for resuscitation of bled sheep. J Surg Res 39: 517–528

Vassar MJ, Perry CA, Holcroft JW (1990) Analysis of potential risks associated with 7.5% sodium chloride resuscitation of traumatic shock. Arch Surg 125:1309–1315

Vassar MJ, Perry CA, Gannaway WL, Holcroft JW (1991) 7.5% sodium chloride/dextran for resuscitation of trauma patients undergoing helicopter transport. Arch Surg 126: 1065–1072

Winters RW, Scaglione PR, Nahas GG, Verosky M (1964) The mechanism of acidosis produced by hyperosmotic infusions. J Clin Invest 43:647–658

Wisner DH, Schuster L, Quinn C (1990) Hypertonic saline resuscitation of head injury: effects on cerebral water content. J Trauma 30:75–78

Worthley LIG, Coopper DJ, Jones N (1988) Treatment of resistant intracranial hypertension with hypertonic saline. J Neurosurg 68:478–481

Zornow MH, Oh YS, Scheller MS (1990) A comparison of the cerebral and haemodynamic effects of mannitol and hypertonic saline in an animal model of brain injury. Acta Neurochir Suppl (Wien) 51:324–325

Effects of Pentoxifylline on Recovery from Cerebral Ischemia and Reperfusion Injury

L.A. Heinel, U.S. Vasthare, R.H. Rosenwasser, and R.F. Tuma

Introduction

White blood cells (WBCs) have been postulated to contribute to post-ischemic damage in a number of tissues including the heart, brain, and skeletal muscle [11]. Results from our laboratory [29] provided evidence that leukocytes are also involved in cerebral ischemia and reperfusion injury. In this investigation a comparison was made of the alteration of cortical electrical activity following transient, incomplete cerebral ischemia between control and leukopenic rats by monitoring somatosensory evoked potentials and EEG activity. Animals that were rendered leukopenic had better preservation of somatosensory evoked potentials both during ischemia and following reperfusion (Figs. 1–3). These results are consistent with the results of similar studies from other laboratories [2, 5].

Pentoxifylline, a hemorheological drug known to alter the deformability of the cellular components of the blood, has also been suggested to modulate the immunological properties of the various WBCs. This drug has been used extensively over the past few years to treat peripheral arterial insufficiency due to its effect on increasing tissue oxygenation by lowering blood viscosity [16, 19–22]. It has also been suggested as a treatment in various cerebrovascular disorders including ischemia and reperfusion injury where microvascular impairment results in compromised blood flow and tissue injury [3, 15, 17, 25]. Classified as a hemorrheologic agent, this drug was primarily thought to exert its effect through increasing the red blood cell deformability, thereby lowering viscosity and improving flow [9, 17]. It was also found to reduce platelet aggregation and fibrinogen levels which are adversely increased following an ischemic insult. Pentoxifylline has also shown its protective influence on WBC rheology and function [1, 4, 7, 8, 23, 28, 29]. These effects occur by either increasing WBC flexibility and/or by reducing their participation in the proinflammatory response which results in the release of injurious inflammatory mediators of tissue damage [8, 23, 28].

Cerebrovascular Laboratory, Departments of Neurosurgery and Physiology, Temple University School of Medicine, Philadelphia, PA 19140, USA

Cerebral Ischemia and Basic Mechanisms
Ed. by A. Hartmann, F. Yatsu, and W. Kuschinsky

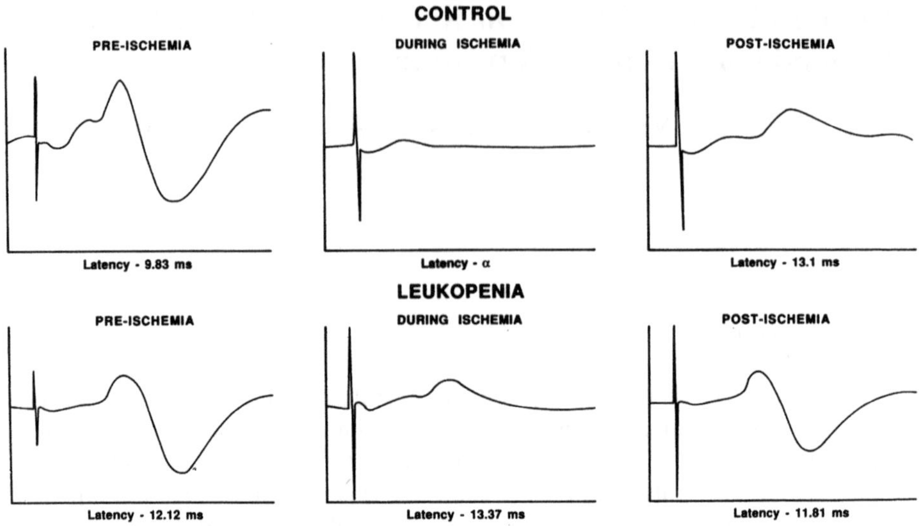

Fig. 1. SSEP recordings of a control and a leukopenic animal before, during, and after ischemia. The cortical peak of the SSEP was maintained during ischemia only in the leukopenic animals [27]

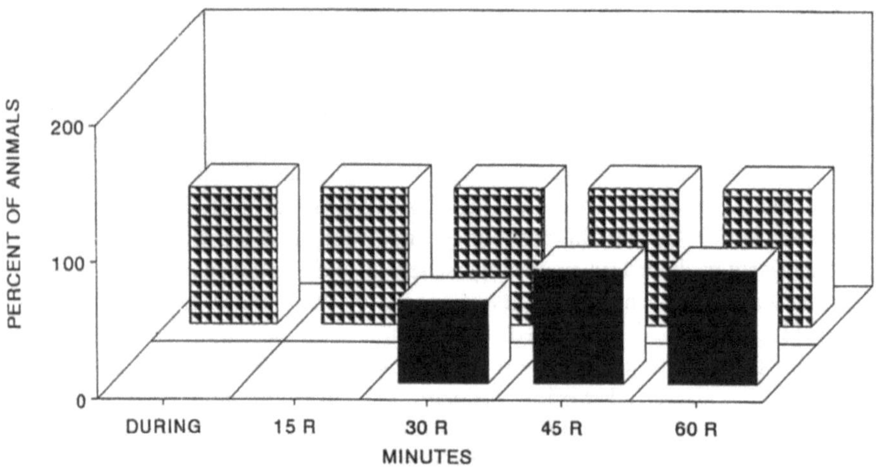

Fig. 2. The percentage of animals with cortical activity during ischemia and 60 min of reperfusion (*R*). *Solid columns*, control animals; *shaded columns*, leukopenic animals

Current experimental evidence from studies of the effect of pentoxifylline on cerebral ischemia and reperfusion injury has been inconsistent. Some studies report improvement in blood flow and various indices of tissue injury while other studies show that pentoxifylline had no effect [9, 10, 15, 17, 18, 26, 27, 29]. The inconsistency of these results must be further

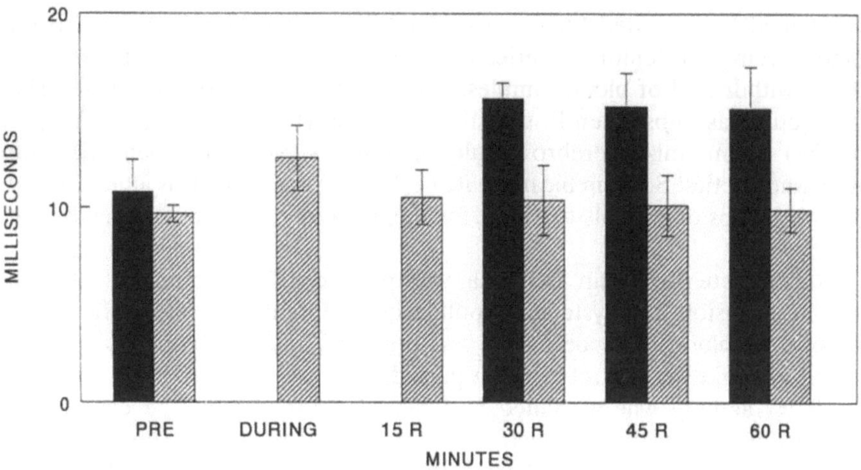

Fig. 3. The latency of the cortical peak of the evoked potential prior to ischemia, during ischemia, and during reperfusion (*R*). *Solid columns*, control animals; *hatched columns*, leukopenic animals

investigated to determine both the mechanisms of pentoxifylline's effect as well as the therapeutic potential of pentoxifylline.

Using a rat model of incomplete global forebrain ischemia with reperfusion, we investigated the influence of pentoxifylline pretreatment on cortical functional activity by monitoring EEG activity and somatosensory evoked potentials.

Methods

Female Barrier-reared Sprague-Dawley rats (225–275 g), maintained on food and water ad libitum, were used in these experiments. Twelve animals were subjected to transient cerebral ischemia for 15 min with subsequent reperfusion. Six animals were treated with either pentoxifylline (25 mg/kg i.v., followed by a continuous i.v. infusion of 8 mg/kg per hour) or saline at similar volumes to control for any effect of hemodilution. Treatment occurred 15 min before the induction of ischemia. The experimental protocol for these experiments included the initial use of an inhalation anesthetic of 1.5% enflurane with a 70%/30% nitrous oxide and oxygen mixture. The animals were placed on a small animal ventilator following tracheostomy, and ventilation was adjusted to maintain arterial blood gases to within physiological range. Atropine sulfate (50 μg/kg, i.m.) was given to reduce bronchial secretions and pancuronium bromide (2 mg/kg, 0.06 mg/kg per minute continuous infusion) was administered to immobilize the animals for electrical recordings and to facilitate breathing. Body temperature was maintained by a heated blanket using a YSI temperature controller. Both femoral veins and femoral arteries were cannulated for the administration of drugs, withdrawal of blood samples, and monitoring of blood pressure. The anesthetic was subsequently switched to fentanyl (50 μg/kg, 1.5 μg/kg per minute) to minimize cerebrovascular alterations commonly detected with other anesthetics. Sodium bicarbonate (1.7 mg/kg per hour) was administered in both groups of animals to adjust for the metabolic acidosis induced by the anesthesia.

Incomplete forebrain ischemia was produced for 15 min by bilateral carotid occlusion and systemic hypotension maintained at 50 mmHg by the removal of blood. The occlusion was removed and the blood volume restored to begin the 1-h reperfusion period.

EEG activity was amplified and recorded throughout the experiment using subdermal platinum electrodes and a Grass recorder. Somatosensory evoked potentials (monitored from the C3, Fz position) were induced by field stimulation of the median nerve in the forepaw. The stimulus needed to generate the evoked potential was adjusted to maximize the amplitude of the evoked response (stimulus of 6–8 V). The frequency of the stimulus was 4 Hz and the duration 0.1 ms. Electrical signals were recorded and processed through a Tracor digital signal analyzer. Cortical peak latency, or the time between the stimulation artifact and the maximum upward deflection of the cortical peak, was used as an index of neuronal function. Cortical peak amplitude was measured from the maximum positive to maximum negative deflection. All recorded data was subjected to blinded analysis by a nonbiased observer. Statistical analysis was completed using an unpaired Student's T test and analysis of variance techniques where applicable. In statistical analysis, to adequately compare both groups of animals, values were assigned where SSEP recordings were not present. These assigned values represented the longest latency measurement able to be recorded for the cortical peak component (25 ms). Otherwise the average latencies would only reflect the values which were able to be recorded and therefore underestimate the latency values. These adjusted values are reflected in the calculation of the mean.

Results

All saline-treated control rats subjected to 15 min of ischemia responded with a cessation of EEG activity within 10–15 s after the initiation of the ischemic insult. The cortical peak component of the somatosensory evoked potentials as well as the EEG activity were abolished in the control animals during the entire 15 min of ischemia, indicating no measurable neuronal activity in the cerebral cortical areas. All pentoxifylline-treated rats, however, maintained EEG activity during the ischemic period. Somatosensory evoked potentials were present during ischemia in two thirds of these pento-

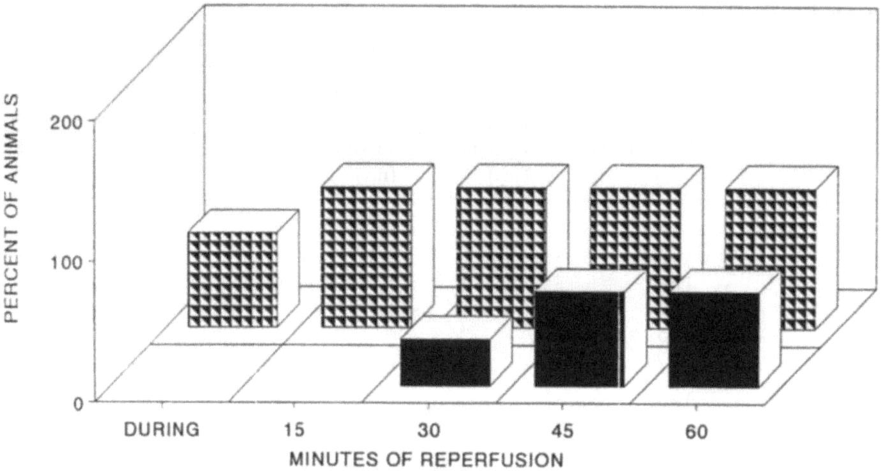

Fig. 4. A comparison between control (*solid columns*) and pentoxifylline-treated (*shaded columns*) animals of the percentage of rats with cortical activity during ischemia and the reperfusion period

xifylline-treated animals (Fig. 4). In animals in whom somatosensory evoked potentials were recorded, the cortical peak component showed an increased latency when compared to the pre-ischemic values. Values are indicated in Table 1.

Following 15 min of reperfusion, the saline-treated rats recovered EEG activity, but no cortical peak component of the SSEP's was present. In contrast, all pentoxifylline-treated rats showed both EEG activity and cortical peaks of the somatosensory evoked potentials. The values recorded for latency indicated an increase when compared to pre-ischemic condition.

By 30 min of reperfusion, EEG activity had returned in all animals of both groups. However, in only 30% of the saline-treated rats were SSEP's present, while they were present in all of the pentoxifylline-treated animals. The latency measurements recorded at 30 min were 21.77 ± 2.15 ms for saline-treated controls and 15.66 ± 1.08 ms for pentoxifylline-treated animals. At 45 and 60 min of reperfusion, in only two thirds of the saline control group were somatosensory evoked potentials present, while they were present in all of the treated animals. The cortical peak latencies of the pentoxifylline-treated rats at these time periods were 14.77 ± 0.62 and 13.75 ± 0.81 ms compared to 19.40 ± 2.08 and 19.28 ± 2.10 ms for saline-treated controls (Fig. 5).

Table 1. Individual latency values of the cortical peak component of the SSEP in saline- and pentoxifylline-treated animals

	Pre-ischemic (ms)	During ischemia (ms)	During reperfusion 15 Min (ms)	30 Min (ms)	45 Min (ms)	60 Min (ms)
Saline control	12.45					
	11.49				14.59	14.98
	11.48			17.90	15.18	15.18
	12.07				21.72	21.41
	10.52					
	9.29			12.74	14.88	14.11
Mean SEM	11.22			21.77	19.40	19.28
	0.47			2.15	2.08	2.10
Pentoxi fylline	11.37		23.91	13.13	13.95	13.44
	11.41	16.09	13.28	12.34	12.03	11.88
	10.60		16.54	15.76	15.37	15.76
	10.80	18.73	15.27	19.36	15.95	13.81
	11.17	18.49	16.15	17.61	15.56	11.38
	11.16	20.13	18.77	15.76	15.76	16.25
Mean SEM	11.09	20.57	17.32	15.66	14.77	13.75
	0.13	1.49	1.51	1.08	0.62	0.81

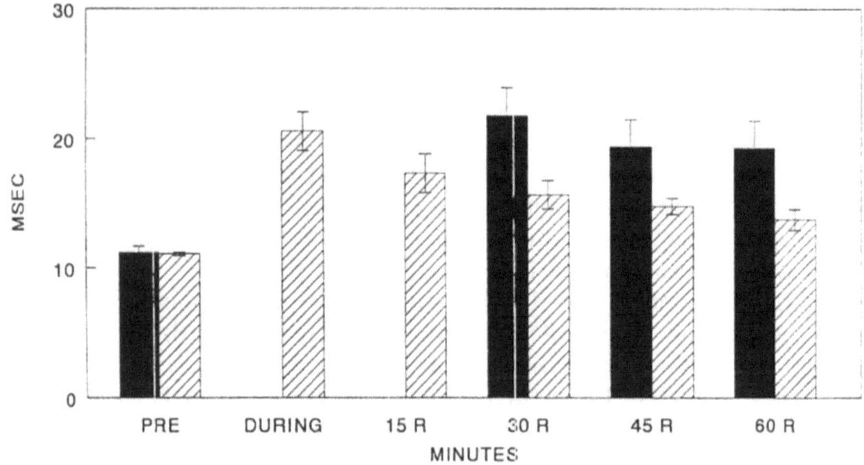

Fig. 5. Mean latency measurement of the cortical peak component of the somatosensory evoked potentials in control (*solid columns*) and pentoxifylline-treated (*hatched columns*) rats. No statistical difference was found between the two groups of animals during the control or preischemic time period. Differences were statistically significant at all other time intervals ($p < 0.05$). *R*, reperfusion

Discussion

Results from this study indicate that treatment with pentoxifylline preserves the cortical functional activity following transient ischemia. We have shown this preservation of neuronal activity by using both EEG activity and somatosensory evoked potentials as indices of cortical activity. Pentoxifylline-treated rats showed a preservation of EEG activity both during ischemia and after 15 min of reperfusion when compared to nontreated control rats. In addition, there was a greater percentage of pentoxifylline-treated rats that maintained the cortical peak component of the SSEP during the same time periods. Latency measurements of the cortical peak components in all pentoxifylline-treated rats were also reduced compared to nontreated animals, indicating a greater preservation of function during both ischemia and reperfusion.

We have shown in previous studies that WBCs affect both cortical functional activity and the generation of infarcts following cerebral ischemia and reperfusion injury [11, 29].

The model of forebrain ischemia utilized in our experiments combines bilateral carotid occlusion with systemic hypotension. Under these conditions of ischemic injury and reduced systemic pressure, inflammatory cytokines such as tumor necrosis factor alpha (TNF-alpha) and interleukin 1 [IL-1] are produced, which may accentuate and stimulate leukocyte activity [7, 28]. These stimulated leukocytes may then participate in ischemia and reperfusion damage by physical and chemical means. Physical plugging of the cerebrovasculature can result from the decreased deformability, the increased adherence to the endothelium, and/or through the reduction of the luminal diameter due to endothelial swelling. The chemical mediators released from leukocytes such as granular release products, reactive oxygen metabolites (i.e., free radicals and associated chemical by products), and products of phospholipids could also exacerbate tissue injury.

It is possible that pentoxifylline exerts its protective effect by modulating leukocyte activity. Pentoxifylline has been reported to alter leukocyte function both in vitro and under various pathophysiological conditions [1, 4, 29]. In addition to increasing leukocyte deformability, there are a number of other effects of pentoxifylline that could influence leukocyte function. It been shown that pentoxifylline decreases the release and effect of TNF-alpha and IL-1 on neutrophils. Pentoxifylline has also been reported to decrease adenosine uptake, which would secondarily inhibit leukocyte activation. In addition, it has been reported that pentoxifylline can increase the production of prostacyclin PGI_2, leading to an augmentation of cerebral blood flow. Other reported actions that could have an effect on ischemic injury are a reduction in platelet aggregation, a decrease in plasma fibrinogen levels, and an increase in cellular cAMP levels. It is not possible to determine which of these effects may have contributed to the results obtained from this study. The results are, however, consistent with the hypothesis that pentoxifylline may have its beneficial effect through an inhibition of leukocyte activity.

There has been conflicting information relating to the efficacy of pentoxifylline on reducing cerebral ischemic and reperfusion injury [9, 15, 17, 18, 26, 27]. Although initial clinical results evaluating the effect of pentoxifylline on various cerebrovascular disorders proved promising [13, 14], further clinical studies suggested an immediate but not long-term benefit of the drug [13]. Other investigators suggest the therapeutic use of pentoxifylline needs further clarification for the treatment of cerebrovascular disorders [20]. The results of our study also support the use pentoxifylline in diminishing cerebral ischemic and reperfusion injury. Additional information from animal studies are also indicated to supplement and clarify the conditions under which this drug may be beneficial, alone or as an adjunct therapy. The conditions to be further clarified include the route of administration, dose, time of the effect, and the type of the experimental model where the effect is beneficial. Once these evaluations are made, the conditions under which it may be useful can be more fully understood and applied.

In conclusion, this study supports the hypothesis that pentoxifylline may be of benefit in reducing the injury caused by cerebral ischemia and reperfusion. Although the mechanisms by which this protection is afforded cannot be distinguished at this time, we suggest the effect may be mediated through alterations in WBCs. Additional studies are being planned to focus on this aspect.

References

1. Barroso-Aranda J, Schmid-Schonbein G (1989) Pentoxifylline pretreatment decreases neutrophil activation during endotoxic shock and improves survival. In: Hakim J, Mandell GL, Novick WJ (eds) Pentoxifylline and analogues: effects on leukocyte function. Karger, Basel, pp 97–104
2. Bednar M, Raymond S, McAuliffe T, Lodge P, Gross C (1991) The role of neutrophils and platelets in a model of thromboembolic stroke 22:44–50
3. Bowton DL, Stump DA, Prough DS, Coker L (1989) Pentoxifylline increases cerebral blood flow in patients with cerebrovascular disease. Stroke 20(12)1662–1666
4. Currie MS, Rao MK, Padmanabhan J, Jones A, Crawford J, Cohen HJ (1990) Stimulus-specific effects of pentoxifylline on neutrophil CR3 expression, degranulation and superoxide production. J Leukoc Biol 47:244–250
5. Dutka A, Kochanek P, Hallenbeck J (1989) Influence of granulocytopenia on canine cerebral ischemia induced by air embolism. Stroke 20:390–395
6. Fossat C, Bonitchi P, David M, Lenoble M, Bory M, Juhan-Vague I (1989) In-vitro effect of pentoxifylline on leukocyte functions in myocardial infarction. In: Hakim J, Mandell GL, Novick WJ (eds) Pentoxifylline and analogues: effects on leukocyte function. Karger, Basel, pp 59–64
7. Hammerschmidt DE, Kotasek D Huh D-W, Freyburger G, Vercellotti GM (1988). Pentoxifylline inhibits granulocyte and platelet function, including granulocyte priming by platelet activating factor. J Lab Clin Med, 112(2)254–263
8. Hartmann A (1983) Effect of pentoxifylline on regional cerebral blood flow in patients with cerebrovascular disease. Eur Neurol 22 Suppl 1:108–115

9. Hartmann A, Tsuda Y (1988) A controlled study on the effect of pentoxifylline and an ergot alkaloid derivative on regional cerebral blood flow in patients with cerebrovascular disease. Angiology 39:449–457
10. Heinel LS, Rubin S, Rosenwasser RH, Vasthare US, Tuma RF (1992) White blood cell involvement in cerebral infarct generation following ischemia and reperfusion injury. Thesis, Temple University School of Medicine
11. Herskovits E, Famulari A, Tamaroff L, Gonzalez AM, Vazquez A, Dominquez R, Fraiman H, Vila J (1985) Preventive treatment of cerebral transient ischemia: comparative randomized trial of pentoxifylline versus conventional antiaggregants. Eur Neurol 24:73–81
12. Hsu CY, Norris JW, Hogan EL, Bladin P, Dinsdale HB, Yatsu FM, Earnest MP, Scheinberg P, Caplan LR, Karp HR, Swanson PD, Feldman RG, Cohen MM, Mayman CI, Cobert B, Savitsky JP (1990) Pentoxifylline in acute nonhemorrhagic stroke. A randomized, placebo-controlled double-blind trial. Stroke 19(6)716–722
13. Janaki S (1980) Pentoxifylline in strokes. A clinical study. J Int Med Res 8:56–62
14. Marcel GA, George C (1983) Pentoxifylline and cerebrovascular diseases. Eur Neurol 22(1)89–97
15. Muller R (1985) On the therapy of disturbances of blood fluidity. Angiology 36:226–234
16. Muller R, Lehrach F (1981) Haemorheology and cerebrovascular disease: multifunctional approach with pentoxifylline. Curr Med Res Opin 7(4)253–263
17. Muller R, Schroer R (1979) Cerebrovascular circulatory disorders: new aspects of pathophysiology and therapy. J Med 10(5)347–364
18. Nierotti M, Longo F, Molaschi M, Macchione C, Pernigotti L (1986) Functional vascular disease: treatment with pentoxifylline. Angiology 37:397
19. Ott E, Chandra B, Hartmann A (1990) Pentoxifylline alters the normal course of acute nonhemorrhagic stroke. (Letter to the Editor). Stroke 21(1)152
20. Perego MA, Sergio G, Artali F, Giunti P, Danese C (1986) Haemorheological improvement by pentoxifylline in patients with peripheral arterial occlusive disease. Curr Med Res Opin 10:135–138
21. Reich T, Cutler BC, Lee BY, Porter JM, Reichle FA, Scogin JT, Strandness DE (1984) Pentoxifylline in the treatment of intermittent claudication of the lower limbs. Angiology 35(7)389–395
22. Schmalzer EA, Chien S (1984) Filterability of subpopulations of leukocytes: effect of pentoxifylline. Blood 64(2)254–263
23. Schneider R, Schmid-Schonbein H, Kiesewetter H (1983) The rheological efficiency of parenteral pentoxifylline (trental) in patients with ischemic brain lesions. Eur Neurol 22(1)98–104
24. Steen PA, Milde JH, Michenfelder JD (1981) Pentoxifylline does not change cerebral blood flow of metabolism in the dog. Acta Anaesthesiol Scand 25:319–322
25. Steen PA, Milde JH, Michenfelder JD (1982) Pentoxifylline in regional cerebral ischemia in cats. Acta Anaesthesiol Scand 26:39–43
26. Sullivan GW, Carper HT, Novick WJ, Mandell GL (1988) Inhibition of the inflammatory action of interleukin-1 and tumor necrosis factor (alpha) on neutrophil function by pentoxifylline. Infect Immun 56(7)1722–1729
27. Vasthare US, Heinel LA, Rossenwasser RH, Tuma RF (1989) Leukocyte involvement in cerebral ischemia and reperfusion injury. Surg Neurol 33:261–265
28. Ward A, Clissold SP (1987) Pentoxifylline. A review of its pharmacodynamic and pharmacokinetic properties, and its therapeutic efficacy. Drugs 34:50–97
29. Waxman K, Clark L, Soliman MH, Parazin S (1989) Pentoxifylline in resuscitation of hemorrhagic shock. In: Hakim J, Mandell GL, Novick WJ (eds) Pentoxifylline and analogues: effects on leukocyte function. Karger, Basel, pp 104–108

Regional Cerebral Blood Flow in Acute Stroke: Comparison of Two Calcium Antagonists

O.B. Paulson, C. Videbæk, S. Vorstrup, and P.S. Sørensen

Introduction

Much effort has been placed in the investigation of the pathophysiological mechanisms that lead to irreversible cell damage following ischemia. Several studies now seem to indicate that disturbances in calcium ion homeostasis may be responsible for the initiation of a series of events leading to irreversible cell damage following cerebral ischemia (for review see Siesjö 1981; Wieloch and Siesjö 1982). It might therefore be expected that treatment with calcium antagonists in the early phase after cerebral ischemia could prevent the intracellular calcium ion accumulation. These drugs might furthermore increase cerebral blood flow (CBF) by dilation of the vessels surrounding the ischemic area. However, conflicting results from experimental studies have been reported.

The present contribution gives a summary of a study undertaken in order to measure the effect of the calcium antagonist PY 108-068 (Vorstrup et al. 1986a) and compare these results to an ongoing study of another dipyrodine calcium antagonist, isradipine (PN 200-110). Both protocols evaluate regional CBF measured by xenon-133 inhalation tomography in patients with acute stroke.

Patients and Methods

Cerebral blood flow measurements were performed in three groups of acute stroke patients before and after the application of the calcium antagonists PY 108-068 or isradipine. Tomography of inhaled xenon-133 was used for the CBF determinations.

In the first group of patients baseline CBF was measured. CBF measurements were repeated 50 min after a dose of 1.5 and 2.5 mg PY 108-069, respectively. The drug was given by intravenous infusion at a rate of 1 mg/10 min.

Department of Neurology, Rigshospitalet, 9 Blegdamsvej 2100 Copenhagen, Denmark

Cerebral Ischemia and Basic Mechanisms
Ed. by A. Hartmann, F. Yatsu, and W. Kuschinsky

In the second group two baseline CBF measurements were performed with a time interval of 20 min. Then a continuous intravenous infusion of PY 108-069 was given at a rate of 1 mg/10 min. CBF measurements were repeated when a total dose of 2.5 and 7.5 mg had been reached.

In the third group a baseline CBF measurements were performed. Thereafter continuous i.v. infusion of isradipine 80 μg/h was started and continued for 72 h followed by oral treatment 1.25 mg given once. CBF was measured after 3, 9, 24, 72 h and 28 days.

Patients

Sixteen patients were studied within 9 days (low-dose PY 108-068, average 5 days, range 1–9 days; high-dose PY 108-068, average 4 days, range 2–7 days; and the isradipine average 6 h, range 3–9 h), after their first acute ischemic stroke. Only patients with symptoms referable to the cerebral hemispheres and hypoperfused areas in accordance were included. Patients with previous cerebrovascular disease, renal insufficiency, severe arterial hypertension, severe cardiac or pulmonary insufficiency, and patients already receiving treatment with calcium antagonists were excluded from the study. Duration of symptoms, age, and computerized tomography (CT) findings are given in Table 1. A fourth to a third of the patients in all the groups studied were hypertensive in the acute phase of their stroke and in some a history of chronic arterial hypertension was obtained. Each patient, or a close relative, gave informed consent to participation in the study,

Table 1. Summary of age, onset of symptoms, and CT scan

		Age		Days from onset of symptoms		
Group	n	Mean	Range	Mean	Range	CT scan
PY 108-068 1.5 + 2.5 mg	6	56	20–81	5	1–9	$n = 6$: Hypodense lesion in agreement with clinical symptoms
PY 108-068 2.5 + 7.5 mg	5	56	43–71	4	2–7	$n = 4$: Hypodense lesion in agreement with clinical symptoms $n = 1$: No focal abnormalities
Isradipine 80 μg/h	5	69	62–76	0.3		$n = 4$: Hypodense lesion in agreement with clinical symptoms $n = 1$: Lacunar infarct in the left hemisphere

which was approved by the Local Ethical Committee and the Danish Health Authorities.

Methods

Cerebral blood flow was measured by xenon-133 inhalation and a rapidly rotating single photon emission computer tomography (SPECT) system (Tomomatic 64, Medimatic, Hellerup, Denmark). This technique, which has been described previously (Stokely et al. 1980), permits repetitive measurements of CBF at 20-min intervals. The instrument consists of four rotating banks of 64 sodium iodide crystals and a sensitive collimator. The patient is studied resting in the supine position and with eyes closed. Each study lasts 4.5 min. During the first 1.5 min the patient inhales xenon-133 from a closed system with an initial xenon concentration of 740 MBq/l and a volume of 4 l. The isotope distribution in the brain is recorded during the xenon inhalation period and during the three following 1-min periods of xenon washout, yielding a sequence of four tomographic pictures. The arterial input to the brain is estimated from the lung curve profile obtained from a single stationary scintillation detector placed above the upper part of the right lung. The algorithm for the calculation of CBF in each pixel is based on a combination of the "sequence of pictures method" and the "early picture method" applying the bolus distribution principle on the sum of the first two pictures (Celsis et al. 1981; Kanno and Lassen 1979). During one single study three horizontal tomographic slices of the brain are obtained simultaneously. Each slice is 2 cm thick and the interslice distance is 2 cm. The resolution in the plane is 1.7–1.9 cm (FWHM) using xenon-133

The patient's head was positioned so that the slices were oriented in parallel to, and 1, 5, and 9 cm above the orbitomeatal (OM) plane. At the end of each CBF measurement procedure, the arterial blood pressure was measured by auscultation. The end-expiratory CO_2 fraction ($FECO_2$) was measured during and at the end of each CBF measurement.

To detect significant changes in a hypoperfused area, the variance in CBF in the low-flow region was calculated in all patients. The mean flow in the visually depicted focal low-flow area as well as the value in the symmetrically placed region in the healthy hemisphere were calculated. The side to side asymmetry ratio (SSAR%) was then calculated as taking the percentage difference between diseased and healthy side, divided by the healthy.

Results

The CBF findings and mean arterial blood pressure (MABP) in the three patient groups are shown in Table 2. The end-expiratory CO_2 fraction was stable in all three groups. In the first PY 108-068 series (low dose), no

Table 2. MABP and CBF before and after calcium antagonist PY 108-068 or isradipine

Group	MABP before (mmHg)	MABP after (mmHg)	CBF (after compared to before)
PY 108-068 1.5 + 2.5 mg	102	101	Mean CBF: Unchanged Focal CBF: $n = 1$: Increase in part of perifocal area. Side to side asymmetry decreased from 31% to 20% $n = 5$: No changes
PY 108-068 2.5 + 7.5 mg	109	96	Mean CBF: Slight increase in nonaffected and affected hemisphere of 5% and 3%, respectively Focal CBF: $n = 3$: Further decrease in the ischemic region with aggravation of the side to side asymmetry, both in infarcted and periinfarcted regions $n = 2$: No changes
Isradipine 80 μg/h	117	109	Mean CBF: Unchanged Focal changes: $n = 5$: No changes

changes in hemispheric CBF, MABP, or clinical symptoms were noted after treatment, but one patient showed an increase of CBF in part of the peri-infarct area. In the second PY 108-068 series (high dose), slight increases in mean hemispheric flow values were seen, but in three of the five patients CBF decreased even further in the ischemic area. MABP decreased by 13%, and the clinical symptoms were unchanged. In the isradipine-treated patients no changes in hemispheric CBF, MABP, or clinical symptoms were noted either after treatment for 3 h or during follow up.

Discussion

Regional CBF in Acute Stroke

Measurements of regional CBF in patients with stroke have shown that in the acute phase hypoperfused areas are present (Paulson 1971) – often involving a larger area of the brain than the hypodense area later evolving on the CT scan (Vorstrup 1988). There is, however, a great inter-individual variability in the flow patterns associated with acute stroke (Vorstrup et al. 1986b; Vorstrup 1988), both due to differences of vascular anatomy and of the actual pathogenesis. Further repetitive regional CBF studies during the early phase of acute stroke have frequently shown transition through a hyperemic phase in all or part of the initially hypoperfused area, probably due to reperfusion following spontaneous thrombolysis (Vorstrup 1988; Skyhøj Olsen et al. 1983; Skyhøj Olsen 1986). The periinfarct zone, "the ischemic penumbra," may at least for a time (hours? days?) have a flow sufficient for cellular viability, but insufficient for cellular function (Astrup et al. 1981). The known abolished/decreased vasoreactivity in acute ischemic stroke (Paulson 1971; Vorstrup et al. 1986b; Vorstrup 1988; Skyhøj Olsen et al. 1983; Skyhøj Olsen 1986) combined with insufficient collateral flow may result in a perfusion pressure in the penumbra below the lower limit of CBF autoregulation yielding a pressure-dependent blood flow within certain limits (Vorstrup 1988; Skyhøj Olsen et al. 1983; Skyhøj Olsen 1986). Measurements with positron emission tomography (PET) have shown an elevated oxygen extraction in the ischemic area for up to some days after the onset of the stroke (Wise et al. 1983), suggesting the presence of potentially viable brain tissue and possibilities for therapeutic intervention.

Calcium Antagonist and CBF and Ischemia

The vasodilating properties of calcium antagonists in nonischemic animal models have been demonstrated in several studies by direct measurements of the vessel diameter (Auer 1981) or by measurement of the resultant flow changes (Brandt et al. 1983; Harper et al. 1981; Haws et al. 1983). However, Harris et al. (1982) found only an insignificant increase of the basal blood flow level following nimodipine treatment in a closed-skull primate model. The study of Harris et al. (1982) also demonstrated that both the CO_2 reactivity, and the autoregulation were severely impaired following nimodipine.

In experimental ischemia conflicting findings have been reported. Promising results were obtained in a dog model of complete cerebral ischemia where *pretreatment* by nimodipine gave an improved neurological outcome in the treated group (Steen et al. 1983). The authors suggested that this effect was caused by the increase in CBF noted during the phase of delayed hypoperfusion. Dilation of constricted cerebral vessels in ischemic territories (Brandt et al. 1983), as well as improved tissue oxygenation, has also been found (Wiernsperger et al. 1984).

In contrast, Hossmann et al. (1983) could not show any positive effect on electrophysiological changes, on metabolic recovery, or on the postischemic accumulation of calcium in the brain tissue during flunarizine treatment in the cat exposed to prolonged complete ischemia. Likewise, Phillis et al. (1986) found no effect upon the basal flow rate following anoxia using either nifedipine or felodipine in a rat model. In a study by Smith et al. (1983), where incomplete ischemia was induced in the rat, it was found that pretreatment with nimodipine increased mean CBF in the postischemic phase, but this increase resulted from very heterogeneous flow values, i.e., zones of hypoperfusion and evident hyperemia within the same structures. No improvement of the electrophysiological parameters was noted in this series. Harris et al. (1982) also found an increase in the residual flow following middle cerebral artery occlusion during nimodipine treatment in an open-skull primate model. However, their results seemed to indicate that the threshold levels of blood flow for the development of cortical edema and for disturbances of ion homeostasis were increased, thereby possibly enhancing the susceptibility of the tissue to ischemic damage.

In the present study, treatment with calcium antagonists did not significantly improve CBF in the ischemic area or its surroundings in any of the three groups. In fact, further CBF decrease was observed in some of the patients from the group given the higher doses of the PY 108-068. On the presumption that the vessels in the ischemic areas are vasoparalytic, this could be explained by a "steal effect": the decrease in perfusion pressure caused by cerebral vasodilation elsewhere along with the reduction in systemic blood pressure noted in these patients could further decrease CBF in the ischemic area and its surroundings, where CBF autoregulation and the vasodilatory effects most likely are abolished. These assumptions are supported by experimental studies on nonischemic rat models, where acute administration of nimodipine nearly suspended cerebral autoregulation (van den Kerchoff and Kazda 1987). The same authors found that nimodipine preserved CBF during moderate hypotension below the lower limit of CBF autoregulation. However, this might be due to simple vasodilation resulting in heterogenous microcirculation with patchy ischemia.

A randomized study on nimodipine in acute ischemic stroke comprising 186 patients showed a decreased mortality and a better neurological outcome during the initial 4-week period, but the long-term effect was not evaluated (Gelmers et al. 1988). The findings in the present study show no effect on CBF of isradipine treatment initiated very soon after primary symptoms in stroke patients. This could suggest that a potentially beneficial effect in stroke of calcium antagonists is not a vascular but rather an effect at the cellular level. It is also of importance to note that calcium antagonists at a high dosage may lead to aggravated ischemia.

Conclusions

The results of the present study did not show any beneficial hemodynamic effect of the examined calcium antagonists either given early or at a high dose. The observed flow reduction in the high-dose group might be a potential risk to the circulation, but there might theoretically still also be a counteracting beneficial effect at cellular level.

References

Astrup J, Siesjö BK, Symon L (1981) Thresholds in cerebral ischemia: the ischemic penumbra (Editorial). Stroke 12:723–725

Auer IM (1981) Pial arterial vasodilation by intravenous nimodipine in cats. Drug Res 31: 1423–1425

Brandt L, Ljunggren B, Andersson KE, Edvinsson L, MacKenzie E, Tamura A, Teasdale G (1983) Effects of topical application of a calcium antagonist (nifedipine) on feline cortical pial microvasculature under normal conditions and in focal ischemia. J Cereb Blood Flow Metab 3:44–50

Celsis P, Goldman T, Henriksen L, Lassen NA (1981) A method for calculating regional cerebral blood flow from emission computerized tomography of inert gas concentrations. J Comput Assist Tomogr 1:349–356

Gelmers HJ, Gorter K, de Weerdt CJ, Wiezer HJA (1988) A controlled trial of nimodipine in acute ischemic stroke. N Engl J Med 318:203–207

Harper AM, Craigen L, Kazda S (1981) Effect of calcium antagonist, nimodipine, on cerebral blood flow and metabolism in the primate. J Cereb Blood Flow Metab 1:349–356

Harris RJ, Branston NM, Symon L, Hayhan M, Watson A (1982) The effects of a calcium antagonist, nimodipine, upon physiological responses of the cerebral vasculature and its possible influence upon cerebral ischaemia. Stroke 13:759–766

Haws CW, Gourley JK, Heistad DD (1983) Effects of nimodipine on cerebral blood flow. J Pharmacol Exp Ther 225:24–28

Hof RP, Vuorela HJ, Neuman P (1982) PY 108-068, a new calcium antagonist, on general hemodynamics and regional blood flow in anesthetized cats: a comparison with nifedipine. J Cardiovasc Pharmacol 4:352–362

Hossmann K-A, Paschen W, Csibal (1983) Relationship between calcium accumulation and recovery of cat brain after prolonged cerebral ischemia. J Cereb Blood Flow Metab 3:346–353

Kanno I, Lassen NA (1979) Two methods for calculating regional cerebral blood flow from emission computed tomography of inert gas concentrations. J Comput Assist Tomogr 3:71–76

Paulson OB (1971) Cerebral apoplexy (stroke): pathogenesis, pathophysiology and therapy as illustrated by regional blood flow measurements in the brain. Stroke 2:327–360

Paulson OB, Jarden JO, Vorstrup S, Holm S, Godtfredsen J (1986) Effect of captopril on the cerebral circulation in chronic heart failure. Eur J Clin Invest 16:124–132

Phillis JW, DeLong RF, Towner JK (1986) The effects of nifedipine and felodipine on cerebral blood flow during anoxic episodes. Stroke 17:229–234

Siesjö BK (1981) Cell damage in the brain: a speculative synthesis. J Cereb Blood Flow Metab 1:155–185

Skyhøj Olsen T (1986) Regional cerebral blood flow after occlusion of the middle cerebral artery. Acta Neurol Scand 73:321–337

Skyhøj Olsen T, Larsen B, Herning M, Bech Skriver E, Lassen NA (1983) Blood flow and vascular reactivity in collaterally perfused brain tissue. Evidence of an ischemic penumbra in patients with acute stroke. Stroke 14:332–341

Smith M, Kågstrøm E, Rosen I, Siesjö BK (1983) Effects of the calcium antagonist nimodipine on the delayed hypoperfusion following incomplete ischemia in the rat. J Cereb Blood Flow Metab 3:543–546

Steen PA, Newberg LA, Milde JH, Michenfelder JD (1983) Nimodipine improves cerebral blood flow and neurological recovery after complete cerebral ischemia in the dog. J Cereb Blood Flow Metab 3:38–43

Stokely EM, Sveinsdottir E, Lassen NA, Rommer P (1980) A single photon dynamic computer assisted tomograph (DCAT) for imaging brain function in multiple cross sections. J Comput Assist Tomogr 4:230–240

Strandgaard S (1978) Autoregulation of cerebral circulation in hypertension. Acta Neurol Scand 57 Suppl 66:S1–S82

van den Kerchoff W, Kazda S (1987) The autoregulation of cerebral blood flow is influenced by calcium antagonists. J Cereb Blood Flow 7 Suppl 1:S169

Vorstrup S (1988) Tomographic cerebral blood flow measurements in patients with ischemic cerebrovascular disease and evaluation of the vasodilatory capacity by the acetazolamide test. Acta Neurol Scand 77:1–48

Vorstrup S, Andersen A, Blegvad N, Paulson OB (1986a) Calcium antagonist (PY 108-068) treatment may further decrease flow in ischemic areas in acute stroke. J Cereb Blood Flow Metab 6:222–229

Vorstrup S, Paulson OB, Lassen NA (1986b) Cerebral blood flow in acute and chronic ischemic stroke using xenon-133 inhalation tomography. Acta Neurol Scand 74:439–451

Wieloch T, Siesjö BK (1982) Ischemic brain injury: the importance of calcium, lipolytic activities, and free fatty acids. Pathol Biol (Paris) 30:269–277

Wiernsperger N, Gygax P, Hofmann A (1984) Calcium antagonist PY 108-068: demonstration of its efficacy in various types of experimental brain ischemia. Stroke 15:679–685

Wise RJS, Bernardi S, Frackowiak RSJ, Legg NJ, Jones T (1983) Serial observations on the pathophysiology of acute stroke. The transition from ischemia to infarction as reflected in regional oxygen extraction. Brain 106:197–222 [illegible] (1988) [illegible] and [illegible] Stroke [illegible]

[illegible] tomographic correlates [illegible]

[illegible]

[illegible]

[illegible]

[illegible]

Subject Index

Springer-Verlag and the Environment

We at Springer-Verlag firmly believe that an international science publisher has a special obligation to the environment, and our corporate policies consistently reflect this conviction.

We also expect our business partners – paper mills, printers, packaging manufacturers, etc. – to commit themselves to using environmentally friendly materials and production processes.

The paper in this book is made from low- or no-chlorine pulp and is acid free, in conformance with international standards for paper permanency.

MIX
Papier aus verantwortungsvollen Quellen
Paper from responsible sources
FSC® C105338

If you have any concerns about our products,
you can contact us on
ProductSafety@springernature.com

In case Publisher is established outside the EU,
the EU authorized representative is:
Springer Nature Customer Service Center GmbH
Europaplatz 3, 69115 Heidelberg, Germany

Printed by Libri Plureos GmbH
in Hamburg, Germany